Potential Theory

Potential Theory

Proceedings of the International Conference on Potential Theory, Nagoya (Japan), August 30 - September 4, 1990

Editor

Masanori Kishi

Walter de Gruyter · Berlin · New York 1992

Editor

Masanori Kishi
Department of Mathematics
College of General Education
Nagoya University
Chikusa-ku, Nagoya, 464-01, Japan

1991 Mathematics Subject Classification: 31-06; 22-06, 30-06, 35-06, 43-06, 44-06, 45-06, 46-06, 53-06, 58-06, 60-06, 73-06.

⊚ Printed on acid free paper which falls within the guidelines of the ANSI to ensure permanence and durability.

Library of Congress Cataloging-in-Publication Data

International Conference on Potential Theory (1990 : Nagoya--shi, Japan)
Potential theory : proceedings of the International Conference on Potential Theory, Nagoya (Japan) August 30–September 4, 1990 / editor, Masanori Kishi.
p. cm.
Includes bibliographical references.
ISBN 3-11-012812-8 (acid free)
1. Potential theory (Mathematics)--Congresses. I. Kishi, Masanori, 1932- . II.Title.
QA404.7.I57 1990 91-39776
515'.9--dc20 CIP

Die Deutsche Bibliothek – Cataloging-in-Publication Data

Potential theory : proceedings of the International Conference on Potential Theory, Nagoya (Japan), August 30 – September 4, 1990 / ed. Masanori Kishi. – Berlin ; New York : de Gruyter, 1992
ISBN 3-11-012812-8
NE: Kishi, Masanori [Hrsg.]; International Conference on Potential Theory <1990, Nagoya>

Printed in Germany.
Printing: Gerike GmbH, Berlin. Binding: Lüderitz & Bauer GmbH, Berlin.
Cover design: Thomas Bonnie, Hamburg.

Preface

This volume collects, in written form, twelve invited lectures and twenty-six selected contributions from short communications delivered at the International Conference on Potential Theory 90.

The Conference was held in the tradition of meetings devoted to potential theory in Nagoya from 30 August to 4 September, 1990, and thus followed the 21st International Congress of Mathematicians in Kyoto. The Conference was co-organized by the Mathematical Society of Japan.

There were 190 participants from twenty-nine countries including fourteen invited speakers and ninety-seven speakers who gave short communications which were arranged in three sessions. The presentations ranged over almost all branches of potential theory, and they discussed various topics from classical to nonlinear potential theory as well as applications of potential theoretic techniques to partial differential equations and other branches of mathematics.

The Organizing Committee acknowledges with appreciation the generous financial support from the Commemorative Association for the Japan World Exposition, IBM Japan Ltd., Nagoya University, the Prefectural Government of Aichi, and the Municipal Office of Nagoya. The Committee expresses gratitude to the enterprises which extended financial support for the Conference, for without these donations, the Conference could not have been held.

The present volume owes much to the selfless service of many referees who carefully read the submitted manuscripts. The careful preparation of camera-ready manuscripts composed using TEX also aided in the timely preparation of this collection. The editor is grateful to each member of the Editorial Board for their constant and indispensable efforts in the course of editing.

Nagoya, October 1991 *Masanori Kishi*

Organizing Committee and Consultative Committee

Organizing Committee

Chairman	Makoto OHTSUKA
Secretary General	Masanori KISHI
Treasurer	Masayuki ITÔ
Committee Members	Teruo IKEGAMI
	Fumi-Yuki MAEDA
	Takafumi MURAI
	Mitsuru NAKAI
	Makoto SAKAI

Consultative Committee

Heinz BAUER
Bent FUGLEDE
Walter Kurt HAYMAN
Gabriel MOKOBODZKI

Contents

Preface V

Organizing Committee and Consultative Committee VI

Invited Lectures

David R. ADAMS
L^p potential theory techniques and nonlinear PDE 1

Jürgen BLIEDTNER
Applications of Choquet theory to potential theory 17

Claude DELLACHERIE
Une version non linéaire du théorème de Hunt 25

Walter K. HAYMAN
Strict isoperimetric inequalities and asymmetry 33

Lars Inge HEDBERG
Nonlinear potential theory 43

Olli MARTIO
Potential theory and quasiconformal mappings 55

Gabriel MOKOBODZKI
Maximal inequalities and potential theory 65

Ivan NETUKA
The boundary behaviour of solutions of the Dirichlet problem 75

Anthony G. O'FARRELL
Capacities in function theory 93

Michael RÖCKNER
Potential theory on non-locally compact space via Dirichlet forms 107

John C. TAYLOR
The Martin compactification of a symmetric space of non-compact type at the bottom of the positive spectrum : An introduction 127

Jang-Mei WU
Level sets and the Green function 141

Contributed Papers

Marco BIROLI
Local properties of solutions to equations involving square Hörmander's operators 147

Andrej V. BITSADZE
On the generalized Neumann problem 155

Abderrahman BOUKRICHA
Principe de Picard pour les mesures invariantes par rotation et applications 161

Vladimir Ya. EIDERMAN
Measure and capacity of exceptional sets arising in estimations of δ-subharmonic functions 171

Sirkka-Liisa A. ERIKSSON-BIQUE
Balayage spaces, standard H-cones and hyperharmonic cones 179

Jerzy A. GAWINECKI
On the thermoelastic potential in three-dimensional hyperbolic thermoelasticity theory 193

Nikolai V. GRACHEV
Representations and estimates for inverse operators of the potential theory integral equations in a polyhedron 201

Ashot HAKOBYAN
Use of single layer potential in mixed problems of elasticity 207

Vadim A. KAIMANOVICH
Discretization of bounded harmonic functions on Riemannian manifolds and entropy 213

Victor I. KRUGLIKOV
Capacities and mappings quasiconformal in the mean 225

Takahide KUROKAWA
Beppo Levi spaces and Riesz potential spaces 229

Yoshihiro MIZUTA
Boundary limits of harmonic functions in Sobolev-Orlicz classes 235

Minoru MURATA
Positive harmonic functions on rotationary symmetric Riemannian manifolds 251

Massimo PICARDELLO and Wolfgang WOESS
Examples of stable Martin boundaries of Markov chains 261

Sundararaja RAMASWAMY
Comparison of liminf and fine liminf of positive superharmonic functions 271

Valeri S. SEROV
Some problems of the potential theory for Schrödinger operator with singular potential 275

Yuri V. SHESTOPALOV
Potential theory for Helmholtz operator and nonlinear eigenvalue problems 281

Keiichi SHIBATA
On stationary points of the energy integral 291

Abdullah SHIDFAR and M. A. HOSSEINI
Application of potential theory in biharmonic analysis 299

Wayne SMITH and David A. STEGENGA
Sobolev imbeddings and integrability of harmonic functions on Hölder domains 303

Noriaki SUZUKI
An estimate of harmonic measure with an application to subharmonic functions 315

Hiroshi TANAKA
Kuramochi boundaries of Riemannian manifolds 321

Frans van GOOL
On axiomatic non-linear potential theory 331

Hisako WATANABE
The Neumann problem and Hausdorff measures 345

WU Jiong Qi and GAO Qi-Ren
On potential extension and capacity on harmonic spaces 361

Stepan S. ZARGARYAN
Singularities of the solution of integral equations of potential theory arising in problems of elasticity for non-homogeneous media 367

Appendices

List of participants 383

List of non-participating contributors 396

List of lectures 397

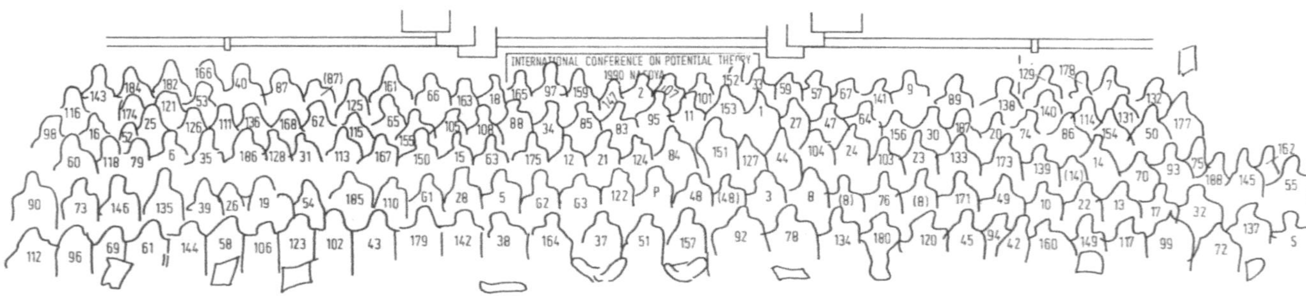
INTERNATIONAL CONFERENCE ON POTENTIAL THEORY
1990 NAGOYA

1 Adams, D.R.
2 Aikawa, H.
3 Akin, Ō
4 Baba, T.
5 Bauer, H.
6 Bendikov, A.D.
7 Berg, C.
8 Bertin, E.M.J.
9 Biroli, M.
10 Bitsadze, A.V.
11 Björck, G.
12 Bliedtner, J.
13 Boboc, N.
14 Boivin, A.
15 Boukricha, A.
16 Brodsky, M.
17 Bucur, G.
18 Bui, H.-Q.
19 Bănulescu, M.
20 Chen, H.
21 Chwang, A.T.
22 Cornea, A.
23 de la Pradelle, A.
24 Dellacherie, C.
25 Eiderman, V.Ya.
26 Eriksson-Bique, S.-L.A.
27 Essén, M.
28 Fuglede, B.
29 Fuji'i'e, T.
30 Fukushima, M.
31 Gawinecki, J.A.
32 Goldstein, M.
33 Gotoh, Y.
34 GowriSankaran, K.
35 Grachov, N.
36 Grigor'yan, A.A.
37 Gulisashvili, A.
38 Hakobyan, A.S.
39 Haliste, A.K.
40 Hansen, W.
41 Hara, H.
42 Hara, M.
43 Hashimoto, Y.
44 Hasumi, M.
45 Hatano, K.
46 Hawkes, J.
47 Hayashi, K.
48 Hayman, W.K.
49 Hedberg, L.I.
50 Heinonen, J.
51 Higuchi, I.
52 Hong, I.
53 Horiuchi, T.
54 Huber, A.O.
55 Hwang, J.S.
56 Ichihara, K.
57 Ikegami, T.
58 Imai, H.
59 Ishizaki, K.
60 Itô, M.
61 Ito, Y.
62 Jacob, N.
63 Janssen, K.
64 Jin, N.
65 Kaimanovich, V.
66 Kamiya, S.
67 Kanda, M.
68 Kaneko, H.
69 Kawamura, M.
70 Kenig, C.E.
71 Kerman, R.A.
72 Kilpeläinen, T.
73 Kishi, M.
74 Kokilashvili, V.
75 Korenblum, B.
76 Korevaar, J.
77 Kori, T.
78 Kruglikov, V.I.
79 Korukawa, Takahide
80 Kurokawa, Toshiko
81 Kusunoki, Y.
82 Kuwae, K.
83 Laine, I.E.
84 Lawrynowicz, J.
85 Leha, G.
86 Lindqvist, L.P.
87 Love, E.R.
88 Lukeš, J.
89 Ma, Z.M.
90 Maeda, F.-Y.
91 Maitani, F.
92 Makarov, N.G.
93 Manfredi, J.J.
94 Marshall, D.E.
95 Martio, O.T.
96 Masaoka, H.
97 Masumoto, M.
98 Matsumoto, S.
99 Mattila, P.
100 Miyazaki, Y.
101 Mizumoto, H.
102 Mizuta, Y.
103 Mokobodzki, G.
104 Moore, C.N.
105 Morinaka, H.
106 Murai, T.
107 Murata, M.
108 Murazawa, T.
109 Nagai, H.
110 Naïm, L.
111 Nakai, M.
112 Narita, J.
113 Netuka, I.
114 Nguyen, D.T.
115 Ninomiya, N.
116 Nishikawa, N.
117 Nishio, M.
118 Nozawa, R.
119 Näkki, R.
120 O'Farrell, A.G.
121 Ohtake, H.
122 Ohtsuka, M.
123 Okada, M.
124 Osada, M.
125 Oshima, Y.
126 Ow, W.H.
127 Pathak, R.S.
128 Pavlov, I.
129 Petrushko, I.M.
130 Picardello, A.M.
131 Prilepko, A.I.
132 Qiu, S.
133 Ramaswamy, S.
134 Rashkovskiĭ, A.Y.
135 Rippon, P.J.
136 Röckner, M.
137 Sakai, M.
138 Sakan, K.
139 Saltykov, E.G.
140 Samko, S.G.
141 Sato, K.
142 Sato, S.
143 Sawai, H.
144 Segawa, S.
145 Serov, V.S.
146 Shestopalov, Y.V.
147 Shiba, M.
148 Shidfar, A.
149 Shimomura, K.
150 Sjögren, P.
151 Smith, W.S.
152 Song, S.H.
153 Stegenga, D.A.
154 Stephenson, K.R.
155 Sturm, K.T.
156 Sugawa, T.
157 Suita, N.
159 Suzuki, N.
160 Tada, T.
161 Takeda, M.
162 Tamrazov, P.M.
163 Tanaka, H.
164 Tanaka, H.
165 Taniguchi, M.
166 Taniguchi, S.
167 Taylor, J.C.
168 Toda, N.
169 Tu, H.
170 Urabe, H.
171 van Gool, F.
172 Veselý, J.
173 Vodop'janov, S.
174 Watanabe, H.
175 Wildenhain, G.
176 Wu, J.
177 Wu, J.-M.
178 Xuan-Loc, N.
179 Yamaguchi, H.
180 Yamamoto, H.
181 Yamano, T.
182 Yamasaki, M.
183 Yanigihara, N.
184 Yoshida, H.
185 Zalcman, L.
186 Zargaryan, S.
187 Zhang, S.
188 Zoriĭ, N.V.
P President S. Hayakawa
G1 Mrs. K. Noshiro
G2 Mr. S. Kawana
G3 Mr. T. Hino
S Miss Y. Sekiya

L^p potential theory techniques and nonlinear PDE

David R. Adams *

A. Introduction

What follows is a revised and expanded version of my talk with the same title at the ICPT '90.

In classical potential theory, there is a duality between the concept of superharmonic function on the one hand, and Newtonian/Green potential of a measure on the other. This principle is usually refered to as the Riesz Decomposition Theorem. It permeates much of the modern theory of linear elliptic and parabolic PDE - solutions can often be given a potential representation. For nonlinear equations, there is no such principle. However, for some nonlinear equations a strong analogy still exists with many of the results of the classical linear theory, results of the linear theory that are usually established with the help of this duality. The methods in the nonlinear case are, of course, much different, but the results are strikingly similar. This is especially true for the p-Laplacian Δ_p and other similar nonlinear operators. Here we are using the notation

$$\Delta_p u = \operatorname{div}\{|\nabla u|^{p-2} \cdot \nabla u\},$$

where ∇ is the usual gradient operator in Euclidean n-space $\boldsymbol{R}^n$, $n \geq 2$, and $1 \leq p < \infty$. Of course, $\Delta_2 = \Delta$, the usual Laplace operator. Δ_p is of interest mainly as a prototype for the study of general quasilinear elliptic PDEs. But, the operator Δ_n also occurs prominantly in the theory of quasiregular mappings.

Thus we have an area of interest and a recent body of knowledge that we shall refer to here as "a potential theory associated to the nonlinear operator", i.e. a study of the regularity properties of supersolutions u, $-\Delta_p u = \mu$ where μ is a positive Borel measure. And dual, in some sense, to this, we have a nonlinear analogue of the Newtonian-Riesz potentials of positive measures, a body of results that has been dubbed "Nonlinear Potential Theory" or NLPT for short. This latter theory is based on the potentials

$$U^{\mu}_{\alpha,p}(x) = I_\alpha * (I_\alpha * \mu)^{1/(p-1)}(x),$$

where $I_\alpha(x) = |x|^{\alpha-n}$, $0 < \alpha < n$, $x \in \boldsymbol{R}^n$ - the Riesz kernel of order α. Newtonian potentials correspond to the case $\alpha = 1$, $p = 2$. So we could also study the regularity properties of these potentials. But there seems to be something missing. There is, so

*partially supported by NSF grant DMS - 8702755 and by the Commonwealth of Kentucky through the Kentucky EPSCoR program.

Potential Theory, ed. by M. Kishi

far, no direct connection between nonlinear supersolutions u and nonlinear potentials $U^{\mu}_{1,p}$. Thus one may argue as to whether or not we have the right potentials, or, as it turns out, seek some more indirect and subtle connection between these two ideas. Such a connection can be made through the concept capacity, the L^p-capacity as a set function. Analogues of classical results occur in each of these two nonlinear theories, and they closely parallel each other. However, one can no longer zip back and forth between the two to prove a common result as is possible in the linear theory.

Consider, for example, the question of the boundary regularity of weak solutions of the Dirichlet problem for $\Delta_p u = 0$ in a bounded domain $\Omega \subset \mathbf{R}^n$, $1 < p \leq n$. When $p = 2$, there are three basic concepts that are equivalent.

(i) $x_0 \in \partial\Omega$ is a regular point for solutions to $\Delta_2 u = 0$ in Ω, $u = \phi$ on $\partial\Omega$, where ϕ is an arbitrarily given continuous function on $\partial\Omega$, i.e.

$$\lim_{\substack{x \to x_0 \\ x \in \Omega}} u(x) = \phi(x_0); \tag{1}$$

(ii) the complement of Ω is not thin at x_0, i.e. the following does not hold at x_0: either x_0 does not belong to the closure of Ω^c or there exists a positive Borel measure μ such that

$$I_2 * \mu(x_0) < \liminf_{\substack{x \to x_0 \\ x \notin \Omega^C}} I_2 * \mu(x);$$

(iii) The Wiener Test integral diverges at x_0 i.e.

$$\int_0 r^{2-n} C_{1,2}(\Omega^c \cap B(x_0, r)) \frac{dr}{r} = +\infty. \tag{2}$$

Here $C_{m,p}(\cdot)$ is the capacity first defined on compact sets $K \subset \mathbf{R}^n$ by

$$C_{m,p}(K) = \inf \int |D^m v|^p dx$$

where the infimum is over all $v \in C_0^\infty(\mathbf{R}^n)$ such that $v \geq 1$ on some neighborhood of K. For general $E \subset \mathbf{R}^n$, we set $C_{m,p}(E) = \sup_{K \subset E} C_{m,p}(K)$. Here $D^m v$ denotes the vector of all m-th order derivatives of v. Or we could have used the capacity

$$R_{m,p}(K) = \inf \int f(x)^p dx$$

where the infimum is over all nonnegative $f \in L^p(\mathbf{R}^n)$ such that $I_m * f(x) \geq 1$ on K. It is a simple consequence of the Fourier Transform technique that $C_{m,p} \sim R_{m,p}$, i.e. the ratio of these two set functions is bounded above and below by finite positive constants depending only on m, p and n, $1 < p < \infty$. It is the equivalence of these two notions of capacity (and other versions as well) that provides the link between the two nonlinear potential theories.

So when $p \neq 2$, (i) is revised so as to speak about solutions to $\Delta_p u = 0$ in Ω; in (ii) the potentials $I_2 * \mu$ are replaced by $U^{\mu}_{1,p}$ or even by

$$W^{\mu}_{1,p}(x) = \int_0^{\infty} \left[r^{p-n}\mu(B(x,r))\right]^{1/(p-1)} \frac{dr}{r};$$

and the Wiener Test integral in (iii) is replaced by

$$\int_0 [r^{p-n} C_{1,p}(\Omega^c \cap B(x_0, r))]^{1/(p-1)} \frac{dr}{r} = +\infty. \tag{3}$$

Now if the revised versions of (i)-(iii) where all known to be equivalent for $1 < p \leq n$, then we would have little to talk about here. But, in fact, they are known to be equivalent for some values of p, not for other values of p, and with a lot of open questions remaining for still other values of p. Thinness on the Nonlinear Potential Theory side will be discussed at this conference by L.I. Hedberg, so we will concentrate here on the differential equation side of things. However, as we shall see, techniques from NLPT will creep into our discussions as well, mostly because of the need to compute the capacity of a set, but nevertheless in a fundamental way. Papers on thinness in NLPT include [AM], [H], [AH], [HW], [A2].

This talk is mainly concerned with the relationships between the statement "x_0 is a boundary regular point for a class of solutions to $Lu = 0$ in Ω" and various versions of the Wiener Test. Here we are mainly interested in nonlinear second order operators L since there is already a large and often times, a quite definitive literature for the linear case; cf. [LSW], [St], [FJK], [B], [EJ], to mention only a few. In the nonlinear case, we mention [Ma1], [GZ], and [Sk] for sufficiency of a Wiener Test, and [LM] and [Sk] with regard to necessity. For the potential theory associated with quasilinear equations, the reader is refered to [KM] and the references contained therein.

The author wishes to thank Professors John Lewis and Ann Heard for many stimulating discussions on the topics of this lecture.

B. Wiener Test for nonlinear PDE

For the p-Laplacian, we have from [Ma1] and [LM]

Theorem 1. *Let $\Phi \in W^{1,p}(\Omega) \cap C^0(\partial\Omega)$, then a sufficient condition that $x_0 \in \partial\Omega$ be a regular point for a weak solution $u \in W^{1,p}(\Omega)$ to*

$$\begin{cases} \Delta_p u = 0, & \text{in } \Omega \\ u - \Phi \in W_0^{1,p}(\Omega) \end{cases} \tag{4}$$

is that (3) holds. This condition is also necessary when $p = 2$ or $n - 1 < p \leq n$.

In this lecture, we unfortunately have no new results to report on the necessity of (3) for boundary regularity for solutions to the p-Laplace equation. And this is by now a well known open problem - for $1 < p < 2$ or $2 < p \leq n-1$. We will instead pass on to other nonlinear results. in particular, Theorem 1 can be extended to p-Laplace type operators, i.e. $Lu = \text{div}\, A(x, \nabla u)$ where $|A(x,w)| \approx |w|^{p-1}$; see e.g. [KM]. But rather than our going into that, we prefer to study some other nonlinear equations. The following result extends Theorem 1, at least as far as sufficiency, to a very general class of quasilinear elliptic PDE; see [GZ]. We shall confine our attention to a special case of their results.

Consider the quasilinear PDE

$$\text{div}\, A(x,u,\nabla u) = B(x,u,\nabla u) \tag{5}$$

satisfying the growth and ellipticity conditions:

$$\begin{cases} |A(x,u,w)| \leq a_0|w|^{p-1} + a_1|u|^{q_0} + a_2 \\ |B(x,u,w)| \leq b_0|w|^p + b_1|u|^{q_1} + b_2 \\ w \cdot A(x,u,w) \geq C_0|w|^p - C_1|u|^{q_2} - C_2, \end{cases} \tag{6}$$

where, for simplicity, we will assume that all the coefficients a_k, b_k, c_k above are positive constants, $1 < p \leq n$, and $q_k > 0$, $k = 0,1,2$.

Theorem 2. *Let $\Phi \in W^{1,p}(\Omega) \cap C^0(\partial\Omega)$, then a sufficient condition that $x_0 \in \partial\Omega$ be a regular point for a weak solution $u \in W^{1,p}(\Omega) \cap L^\infty(\Omega)$ to* (5) *with* (6) *and $u - \Phi \in W_0^{1,p}(\Omega)$ is that* (3) *holds.*

The proof of Theorem 2, found in [GZ], relies on the following fundamental gradient estimate at the boundary.

Lemma 1. *Let $\zeta \in C_0^\infty(B(x_0, r/2))$ with $|\nabla\zeta| \leq C_1/r$. Then there exists a constant C_2 depending only on n, p, C_1, the upper bound for u and the constants of* (6) *such that*

$$\frac{1}{r^{n-p}} \int |\nabla(\zeta v)|^p dx \leq C_2[\mu(r) + r](\mu(r) - \mu(r/2) + r)^{p-1},$$

where $v(x) = \mu(r) - u_k(x) + r$, with $\mu(r) = \sup_x \{u(x) : \ x \in B(x_0,r)\}$, $u_k(x) = (u(x) - k)^+$, $k > \Phi(x_0)$, and u a weak $W^{1,p} \cap L^\infty$ solution to (4).

We should also mention the work of Skrypnik [Sk], with regard to necessity. There, the only result that this author was able to verify as being both necessary and sufficient is the so called "linear growth" case. This corresponds to $p = 2$ in (6), although Skrypnik's lower order terms - especially with regard to gradient growth in the B term - are not as general as those of [GZ]. We do not need his results for what follows and hence put them aside for now. However, there are many interesting ideas there that need to be explored further.

Now before we go any further, a word should be said about what has been proved and what is being sought. We say that $x_0 \in \partial\Omega$ is a regular point for the equation $Lu = 0$ in Ω if (1) holds for that particular x_0 and Ω. We note also that the Wiener Test integral has a local character: if it diverges for Ω^c it also must diverge for Ω_1^c, where Ω_1 is any subdomain of Ω for which $x_0 \in \partial\Omega \cap \partial\Omega_1$. Thus if equation $Lu = 0$ has a Wiener Test for boundary regularity that is both necessary and sufficient, boundary regularity at x_0 must be a local property, i.e. it is true for any subdomain. It is basicly a consequence of the maximum (minimum) principle. However, it might fail for certain nonlinear operators. Thus we need to formulate our results in terms of local regularity.

Definition. $x_0 \in \partial\Omega$ is a *local regular point* for Ω and the operator L if for all subdomains $\Omega_1 \subset \Omega$, with $x_0 \in \partial\Omega_1 \cap \partial\Omega$, x_0 is a regular point for Ω_1.

Thus with regard to the Wiener Test, the sufficiency is: the Wiener Test integral, for Ω^c at x_0, diverges implies that x_0 is a local regular point for Ω; the necessity is: the Wiener Test integral, for Ω^c at x_0, converges implies that there exists a subdomain $\Omega_1 \subset \Omega, x_0 \in \partial\Omega_1 \cap \partial\Omega$, such that x_0 is irregular for Ω_1.

We remark that for the p-Laplacian there is a maximum or comparision principle and in particular all regular points are local regular points.

We now turn to the Dirichlet problem

$$\begin{cases} \Delta u = f(x, u, \nabla u), & \text{in } \Omega \\ u - \Phi \in W_0^{1,2}(\Omega), \end{cases} \tag{7}$$

where $|f(x, u, w)| \leq a|w|^p + b|u|^q + c$, for some $p < 2$ and any $q > 0$. We will say that $Lu = \Delta u - f(x, u, \nabla u) = 0$ is *locally solvable* in Ω if for any subdomain $\Omega_1 \subset \Omega$, there exists a $W^{1,2} \cap L^\infty$ solution in Ω_1 corresponding to arbitrarily chosen boundary data $\Phi \in W^{1,2}(\Omega_1) \cap C^0(\partial\Omega_1)$, the L^∞ bound being independent of the measure of Ω_1.

Theorem 3. *If $Lu = \Delta u - f(x, u, \nabla u) = 0$ is locally solvable in Ω, then a necessary and sufficient condition for $x_0 \in \partial\Omega$ to be a local regular point for bounded $W^{1,2}$-solutions to* (7) *in Ω is that* (2) *holds.*

Proof. The sufficiency follows from Theorem 2, so we concentrate on the necessity. Our argument is a modification of that given in [EJ]. For $0 < r < s < R \leq R_0$,

$$\begin{aligned} \frac{d}{ds}\left(\fint_{B(x_0,s)} u(x)dx\right) &= s^{-1}\fint_{B(x_0,s)} \nabla u(x) \cdot (x - x_0)dx \\ &= -\frac{1}{2s}\fint_{B(x_0,s)} \nabla u(x) \cdot \nabla(s^2 - |x - x_0|^2) \cdot (1 - \eta(x))dx \\ &+\frac{1}{s}\fint_{B(x_0,s)} \nabla u(x) \cdot (x - x_0)\eta(x)dx, \end{aligned} \tag{8}$$

where the barred integral denotes "integral average", and η is the $C_{1,2}$-capacitary extremal for $\Omega^c \cap B(x_0, s)$. Integrating the first integral on the right side of (8) by parts, we have that its absolute value does not exceed

$$\frac{C}{s^{n-1}}\left(\int_{B(x_0,s)}|\nabla u|^2dx\right)^{1/2}\left(\int|\nabla\eta|^2dx\right)^{1/2}$$

$$+\frac{C}{s^{n-1}}\int_{B(x_0,s)}|f|dx$$

$$\leq \epsilon\cdot s^{1-n}\int_{B(x_0,s)}|\nabla u|^2dx + C_\epsilon s + C_\epsilon\frac{\gamma(s)}{s},$$

by Young's inequality for some $\epsilon > 0$ to be specified. Here

$$\gamma(s) = s^{2-n}C_{1,2}(\Omega^c \cap B(x_0, s)).$$

The second integral in (8), in absolute value, never exceeds

$$\frac{C}{s^n}\left(\int_{B(x_0,s)}|\nabla u|^2dx\right)^{1/2}\left(\int\eta^{2n/(n-2)}dx\right)^{1/2-1/n}\cdot s$$

$$\leq \epsilon\cdot s^{1-n}\int_{B(x_0,s)}|\nabla u|^2dx + C_\epsilon\frac{\gamma(s)}{s},$$

by Young's inequality and also using Sobolev's inequality. Thus

$$\frac{d}{ds}\left(⨍_{B(x_0,s)}udx\right) \geq -\frac{2\epsilon}{s^{n-1}}\int_{B(x_0,s)}|\nabla u|^2dx - C_\epsilon s - C_\epsilon\frac{\gamma(s)}{s}. \tag{9}$$

Integration of (9) over the interval $[r, R]$ yields

$$\begin{aligned}⨍_{B(x_0,r)}udx \leq ⨍_{B(x_0,R)}udx + 2\epsilon\int_r^R s^{2-n}\int_{B(x_0,s)}|\nabla u|^2dx\frac{ds}{s}\\ +C_\epsilon R^2 + C_\epsilon\int_r^R\frac{\gamma(s)}{s}ds.\end{aligned} \tag{10}$$

Now we need

Lemma 2. *Suppose u is a $W^{1,2} \cap L^\infty$ solution to $\Delta u = f(x, u, \nabla u)$ in $\Omega_1 \subset \Omega$ with boundary values*

$$\Phi(x) = \begin{cases} 1, & x \in \partial\Omega_1 \cap B(x_0, R_0) \\ 0, & x \in \partial\Omega_1 \cap B(x_0, 2R_0)^c \end{cases}$$

and $\Phi \in W^{1,2}(\Omega_1) \cap C^0(\partial\Omega_1)$, $x_0 \in \partial\Omega_1$. Extend u to $\Omega_1^c \cap B(x_0, R_0)$ by setting $u \equiv 1$ there. Then there is a constant C_3 independent of R_0 such that

$$\int_0^{R_0} r^{2-n} \int_{B(x_0,r)} |\nabla u|^2 dx \frac{dr}{r} \leq C_3.$$

Proof of lemma. From Lemma 1, we get

$$\frac{1}{s^{n-2}} \int_{B(x_0,s/4)\cap[u>1]} |\nabla u|^2 dx \leq C[\mu(s) - \mu(s/2) + s]$$

by choosing $\zeta = 1$ on $B(x_0, s/4)$. Thus

$$\int_r^R s^{2-n} \int_{B(x_0,s/4)\cap[u>1]} |\nabla u|^2 dx \frac{ds}{s} \leq C \int_r^R [\mu(s) - \mu(s/2) + s] \frac{ds}{s}$$
$$\leq C.$$

Now we replace u by $-u$ in the above argument. This gives

$$\int_r^R s^{2-n} \int_{B(x_0,s/4)\cap[u<1]} |\nabla u|^2 dx \leq C,$$

since $|u| \leq M$ on Ω for some M independent of R_0. Thus since $\nabla u = 0$ a.e. on $[u = 1] \cap B(x_0, s/4)$, we are done.

Now continuing with the proof of Theorem 3, we see that the right side of (10) can be made less than 1 for some $\epsilon = \epsilon_0$ and $R = R_0$ all chosen sufficiently small provided we can show that

$$\left| ⨍_{B(x_0,R_0)} u dx \right| < 1/100, \quad \text{for small enough } R_0. \tag{11}$$

With (11) the conclusion of the theorem follows by fixing such a value of R_0 and then using the arbitrariness of $r < R_0$ to derive

$$\liminf_{\substack{x \to x_0 \\ x \in \Omega_1}} u(x) < 1 = \Phi(x_0).$$

Thus x_0 is irregular for Ω_1.

To get (11), we will require that Ω_1 be small enough. In particular, we take $\Omega_1 = \Omega \cap B(x_0, 3R_0)$ and $\eta =$ the $C_{1,2}$-capacitary extremal for $B(x_0, 3R_0) \cap \Omega^c$. Then $u(1-\eta) = 0$ on $\partial B(x_0, 3R_0)$, hence the Poincaré inequality gives

$$⨍_{B(x_0,3R_0)} u^2(1-\eta)^2 dx \leq cR_0^{2-n} \int_{B(x_0,3R_0)} |\nabla u|^2 (1-\eta)^2 dx$$
$$+ cR_0^{2-n} \int_{B(x_0,3R_0)} u^2 |\nabla \eta|^2 dx \tag{12}$$

For the first integral on the right of (12), we integrate by parts getting

$$\int_{B(x_0,3R_0)} |\nabla u|^2(1-\eta)^2 dx \le c\int_{B(x_0,3R_0)} |f| dx$$

$$+c\left(\int_{B(x_0,3R_0)} |\nabla u|^2(1-\eta)^2 dx\right)^{1/2} \left(\int_{B(x_0,3R_0)} u^2|\nabla\eta|^2 dx\right)^{1/2}$$

$$\le \epsilon_1 \int_{B(x_0,3R_0)} |\nabla u|^2 dx + \epsilon_2 \int_{B(x_0,3R_0)} |\nabla u|^2(1-\eta)^2 dx$$

$$+c_{\epsilon_1} R_0^n + C_{\epsilon_2}\gamma(3R_0)R_0^{n-2}$$

for small numbers ϵ_1 and ϵ_2 by Young's inequality. Thus the left side of (12) does not exceed

$$\epsilon_1 R_0^{2-n} \int_{B(x_0,3R_0)} |\nabla u|^2 dx + C_{\epsilon_1} R_0^2 + C\gamma(3R_0),$$

which again can be made small by proper choice of ϵ_1 and R_0. Also via Sobolev's inequality, we have

$$\fint_{B(x_0,R_0)} u^2\eta^2 dx \le C R_0^{2-n}\left(\int \eta^{2n/(n-2)} dx\right)^{1-2/n} \le c_1\gamma(3R_0). \square$$

C. Removable singularities

Next we turn to the question of removable singularities for certain nonlinear PDE and their connection with the Wiener Test integral and the Kellogg property in NLPT. We say that a compact subset K of the domain $\Omega \subset \boldsymbol{R}^n$ is removable for the operator L and the class of solutions $\boldsymbol{P}$ if each $u \in \boldsymbol{P}$ solving $Lu = 0$ in $\Omega \setminus K$ also solves $Lu = 0$ in Ω. For the p-Laplacian, we have from [Ma2] or [Se],

Theorem 4. *A sufficient condition that K be removable for $W^{1,p}$-solutions for the p-Laplace equation that also belong to $L^\infty_{loc}(\Omega)$ is that $C_{1,p}(K) = 0$.*

For the converse of Theorem 4, we use the following version of the Kellogg property from NLPT. The classical Kellogg property corresponds to the case $p = 2$. See [HW], [H].

Theorem 5. *Let*

$$e(K) = \{x \in \Omega : \int_0 [C_{1,p}(K \cap B(x,r))r^{p-n}]^{1/(p-1)} \frac{dr}{r} < \infty\},$$

then $C_{1,p}(e(K) \cap K) = 0$.

Now let us suppose, as we may, that $K^0 = \emptyset$, and that K is removable for $W^{1,p}$-solutions to $\Delta_p u = 0$ in $\Omega \setminus K$, $u \in L^\infty_{\text{loc}}(\Omega)$. Then each $x \in K$ must be an irregular point. But by Theorem 2, the Wiener Test integral must converge. But then by Theorem 5, $C_{1,p}(K) = 0$.

When $p = 2$, the above reduces to the well known fact that $C_{1,2}(K) = 0$ is both necessary and sufficient for a set K to be removable for bounded harmonic functions on $\Omega \setminus K$. In this case, we usually don't use a result quite as strong as the Kellogg property to prove necessity. It generally suffices to solve the problem

$$\begin{cases} -\Delta u = \mu, & \text{in } \Omega \\ u = 0, & \text{on } \partial\Omega \end{cases} \tag{13}$$

where μ is a positive Borel measure supported on K, $C_{1,2}(K) > 0$. In fact, by letting μ be the $R_{1,2}$-capacitary extremal measure, it follows that $I_2 * \mu$ is bounded and since $u = G\mu$, $G =$ Green function for Ω, it follows that the solution to (13) will also be bounded. Hence K is not removable. Here by the $R_{1,2}$-capacitary extremal measure we mean the unique positive measure μ that solves

$$\min \{\mu(K) : \ \|I_1 * \mu\|_{L^2} \le 1\}. \tag{14}$$

The value obtained in (14) is an equivalent definition of $C_{1,2}(K)^{1/2}$. See [Me].

Next, we contrast this with the removability question for the semilinear problem

$$\begin{aligned} &-\Delta u = u^{p'}, \quad u > 0, \quad \text{in } \Omega \\ &u = 0, \quad \text{on } \partial\Omega \end{aligned} \tag{15}$$

By the techniques discussed above including the classical version of the Kellogg property, it is easy to show that a necessary and sufficient condition for a compact set $K \subset \Omega$ to be removable for distribution solutions to (15) in $\Omega \setminus K$ that are also bounded is that $C_{1,2}(K) = 0$. This is a standard varient of our earlier methods and is unremarkable. However, it is also clear that boundedness is clearly not essential in formulating this problem. What would happen if we replaced boundedness by merely $L^{p'}$ integrability, $1 < p < \infty$, $p' = p/(p-1)$? In a series of papers by P. Baras, M. Pierre and later including the present author, ([BP1], [BP2], [AP]), this question was addressed.

Theorem 6. *Any solution of*

$$\begin{cases} -\Delta u = u^{p'} & \textit{in } \mathcal{D}'(\Omega \setminus K) \\ u \in L^{p'}_{loc}(\Omega \setminus K) \cap W^{1,1}_{loc}(\Omega \setminus K), & u \ge 0 \end{cases} \tag{16}$$

is a solution of

$$\begin{cases} -\Delta u = u^{p'} & \textit{in } \mathcal{D}'(\Omega) \\ u \in L^{p'}_{loc}(\Omega) \cap W^{1,1}_{loc}(\Omega), & u \ge 0 \end{cases} \tag{17}$$

if and only if $C_{2,p}(K) = 0$.

An interesting feature of this result is that it respects the structure conditions of the equation, i.e. the $L^{p'}_{\text{loc}}$ integrability is needed to formulate the problem and it is exactly this structure that is used in the metric condition on K. In an analogous result for harmonic functions, $C_{2,p}(K) = 0$ iff K is removable for harmonic functions in $L^{p'}_{\text{loc}}(\Omega)$, the p does not appear in what it means to be a solution, and local $L^{p'}$ integrability is usually assumed on all compact subsets of Ω, not just those of $\Omega \setminus K$.

The proof of the sufficiency of $C_{2,p}(K) = 0$ in Theorem 7 occurs in [BP1] for a much more general class of semilinear equations: $-\text{div}\,(A\nabla u) + g(u) = 0$. What is interesting for us here is the proof of the necessity. It not only seems to depend on the precise structure of the equation, but involves the more direct approach taken for the Laplacian. Of course, looking for nonnegative solutions adds extra interest. So here we study the existance of solutions to the problem

$$\begin{cases} -\Delta u = u^{p'} + \mu, \quad u \geq 0, \quad \text{on } \Omega \\ \\ u = 0, \quad \text{on } \partial\Omega \end{cases} \tag{18}$$

where μ is a positive Borel measure, like, for example, the $R_{2,p}$ capacitary extremal measure for K when $C_{2,p}(K) > 0$. Showing that solutions of (18) exist for small μ depends on deep results for singular integral operators and the Muckenhoupt theory of A_p-weights. This construction shows that $C_{2,p}(K) = 0$ is in fact necessary for removablility in Theorem 6. We refer the interested reader to [AP] for details.

Thus we are lead to ask if it isn't possible to give a direct proof of the necessity of $C_{1,p}(K) = 0$ in the converse of Theorem 4 by finding a bounded solution to

$$\begin{cases} -\Delta_p u = \mu, \quad \text{in } \Omega \\ u = 0, \quad \text{on } \partial\Omega \end{cases} \tag{19}$$

where μ is a Borel measure concentrated on K compactly contained in Ω, i.e. especially when μ is the $R_{1,p}$-capacitary extremal measure for K. This is like seeking a PDE approach to the converse of Theorem 4 as opposed to the NLPT approach that was discussed above. We can almost do this with existing PDE methods: the Capacitary Strong-type Inequality (CSI), Theorem 7 below, and the Moser iteration method. They fail but the methods are very instructive.

Let μ be the measure that solves: $\max_\nu\{\nu(K) : \ ||I_1 * \nu||_{p'} \leq 1\}$. Then from [Me] we have that $U^\mu_{1,p}$ is bounded. Now if we could write the solution of (14) interm of $U^\mu_{1,p}$, as we can do when $p = 2$, then we would be done. However, this is the missing key ingredient. Instead, we attempt to approach the boundedness of solutions to (14) thru PDE methods.

Indeed, $U^\mu_{1,p}$ bounded easily implies

$$\mu(e) \leq C \cdot R_{1,p}(e) \tag{20}$$

for all Borel sets $e \subset \mathbb{R}^n$. But because of the relationship $R_{1,p} \sim C_{1,p}$, (22) below holds. Thus by CSI, the $W_0^{1,p}(\Omega)$ solution to (19) satisfies

$$q \int_\Omega |\nabla u|^p u^{q-1} dx = \int_\Omega (u^{q/p})^p d\mu \leq C \int_\Omega |\nabla u^{q/p}|^p dx,$$

for any $q > 1$. But then

$$\int_\Omega |\nabla u^r|^p dx \leq C \frac{r^p}{q} \int |\nabla u^{q/p}|^p dx,$$

where $r = q/p - 1/p + 1$. Iterating this inequality and using the Sobolev inequality yields

$$\left(\int_\Omega u^{k\sigma} dx \right)^{1/\sigma} \leq C^{N-1} Q_N \cdot \int_\Omega |\nabla u|^p dx \tag{21}$$

where $\sigma = n/(n-p)$, $1 < p < n$, $k = N(p-1) + 1$, $N = 2, 3, 4, \ldots$, and

$$Q_N = \prod_{j=2}^{N} \frac{[pj - (j-1)]^p}{p^p[(j-1)p - (j-2)]}.$$

Now since Q_N behaves like $N^{N(p-1)}$ as $N \to \infty$ we can sum an appropriate power series in u and show that there are positive constants C_0 and β_0 independent of u such that

$$\int_\Omega (e^{\beta_0 u} - 1) u^\sigma dx \leq C_0 \left(\int |\nabla u|^p dx \right)^\sigma .$$

Had Q_N been bounded independent of N, then a passage to the limit in (21) would bound u.

By CSI, we mean

Theorem 7. (CSI) *Let $u \in W_0^{1,p}(\Omega)$, then the following are equivalent:*

(i) *there is a constant C_1 such that*

$$\mu(e) \leq C_1 \cdot C_{1,p}(e) \tag{22}$$

for all Borel sets $e \subset \mathbb{R}^n$,

(ii) *there is a constant C_2 such that*

$$\int_\Omega u^p d\mu \leq C_2 \cdot \int_\Omega |\nabla u|^p dx \tag{23}$$

for all $u \in W_0^{1,p}(\Omega)$.

Here μ is a positive Borel measure supported on K, a compact subset of Ω.

Thus the solution to (19) is probably at least of class BMO in Ω, perhaps even bmo.

Notice that (20) can not imply the boundedness of u, since it doesn't, even when $p = 2$, i.e. there are mearsures μ for which (20) holds but for which $I_2 * \mu$ is not bounded.

T. Kilpelainen has informed the author that he can show that if u of (19) is bounded by a multiple of the bound for $W_{1,p}^{\mu}$, then he can prove the necessity of the Wiener test in Theorem 1 for that p. In particular, he has such an estimate when $p > n-1$. Of course this is just the range for necessity given already in Theorem 1.

Finally, we close this section with a question. All of the above considerations seem to suggest that there might be a version of the Wiener Test for problem (15) that strongly reflects the $L^{p'}$ growth in the equation. With the sufficiency of such a test, we would expect to recover the necessity in Theorem 7. In fact, since one can consider removability results as a limiting case of boundary regularity - in some very formal sense the boundary pokes inside the domain - one might even expect a whole scale of boundary regularity results paralleling the situation with removable singularity results. Is this the case? In the next section, we refer to one such situation ... perhaps.

D. The harmonic obstacle problem

Let ψ be defined on the bounded domain $\Omega \subset \mathbb{R}^n$, with $\max_\Omega \psi > 0$ and $\psi \leq 0$ near $\partial\Omega$. By the harmonic obstacle problem we mean: find the least super-harmonic function u on Ω that lies above ψ and vanishes at $\partial\Omega$. This can be accomplished by minimizing the Dirichlet integral $\int_\Omega |\nabla v|^2 dx$ over all $v \in W_0^{1,2}(\Omega)$ such that $v \geq \psi$ a.e. on Ω. We call this the "harmonic" obstacle problem because the solution u is harmonic whenever $u > \psi$ on Ω. The set where $u = \psi$, a set that must have positive $C_{1,2}$ capacity, is called the *contact set* and will be denoted by Γ. The boundary of Γ is the free boundary. A good readable account of the obstacle problem can be found in [C].

Now, it is well known that the solution can only inherit a limited amount of smoothness from the data, ψ. In fact, even if ψ is C^∞, the corresponding solution u will be at best $W^{2,\infty}$, i.e. its second derivatives are all bounded. Under these circumstances, both u and ∇u are continuous at the free boundary. However, if ψ is less smooth, then it may be that u is no longer continuous at $\partial\Gamma$. Such can easily be the case when ψ is the characteristic function of a compact $K \subset \Omega$. Then the solution to the harmonic obstacle problem would be precisely the $C_{1,2}$-capacitary extremal. In this special case, the free boundary is essentially the boundary of K since $u = 1$ $C_{1,2}$ - a.e. on K and u is continuous at $x_0 \in \partial K$ precisely when K is not thin at x_0.

This last result is a special case of an extensive study of the behavior of solutions to linear elliptic obstacle type problems at the contact set which has culminated recently in the definitive paper [DMV]. In particular, it follows that each local solution to the harmonic obstacle problem is continuous at exactly those points where the set $\{x \in \Omega : \psi(x) > \bar{\psi}(x_0) - \epsilon\}$ is not thin for every $\epsilon > 0$. Here $\bar{\psi}(x_0)$ is the essential limsup at x_0 of ψ, "essential" in that sets of capacity zero are ignored. For an extension to p-Laplacian type operators, see [HK].

One thing that follows from all of this is that $C_{1,2}$ - a.e. point of the free boundary is a regular point; i.e. the solution u tends to ψ at the free boundary $C_{1,2}$-a.e. All that is needed to guarantee this is the existance of a solution in $W_0^{1,2}(\Omega)$ for the given obstacle. For this it is sufficient that

$$\int_0^\infty C_{1,2}(\{x \in \Omega : \ \psi(x) > \lambda\})\lambda d\lambda < \infty;$$

see [A1]. In particular, ψ bounded suffices. For smoother obstacles, one might expect a little better behavior at the free boundary. For example, if $\psi \in W^{2,2}(\Omega)$, then it is known, see [C], that $u \in W^{2,2}_{\text{loc}}(\Omega)$. What can be said for when and how $u \to \psi$ at $\partial\Gamma$ is this case? We want to discuss the following

Conjecture. $u(x) \to \psi(x_0)$ as $x \to x_0$, $x \in \Omega \setminus \Gamma$, for $C_{2,2}$ - a.e. $x_0 \in \partial\Gamma$ when $\psi \in W^{2,2}(\Omega)$.

We give some evidence for this idea. In fact, we plan to argue that a condition like (2) at the free boundary is all that is needed - see (24) below. We are going to restrict attention to obstacles $\psi_0 = I_4 * \mu - m/2$, where μ is a positive Borel measure with compact support in Ω and $\sup_\Omega I_4 * \mu = m < \infty$. Notice that since $I_4 * \mu$ is bounded, $I_2 * \mu \in L^2$ so that $\psi_0 \in W^{2,2}(\Omega)$ follows, although ψ_0 need not be continuous at points on supp μ.

Theorem 8. *The conjecture is true for* $\psi = \psi_0$ *and* $n = 4, 5, 6$.

The proof we have in mind here is a modification of an argument due to Maz'ja [Ma4] for solutions to the biharmonic equation and only holds for $n \le 6$; dimensions $n < 4$ are not interesting. It is expected that Theorem 8 holds for all $n \ge 4$ using the arguments of [L]; this work is presently in progress.

Thus, if we let $w = u - \psi$, then w satisfies

$$-\Delta w = \Delta\psi \quad \text{in } \Omega \setminus \Gamma$$

$$\zeta \cdot w \in W_0^{2,2}(B(x_0, \rho) \cap (\Omega \setminus \Gamma)),$$

where $\zeta \in C_0^\infty(B(x_0, \rho))$, $\zeta = 1$ on $B(x_0, \rho/2)$ and $x_0 \in \partial\Gamma$. We now apply the methods of [Ma4] (which the author can only verify for $n \le 6$ but which are claimed for $n \le 7$) to $\zeta \cdot u$. Because $-\Delta^2\psi \le 0$, the Maz'ja argument goes thru with no changes. Thus if

$$\int_0 r^{4-n} C_{2,2}(\Gamma \cap B(x_0, r))\frac{dr}{r} = +\infty, \tag{24}$$

then $u(x) \to \psi(x_0)$ as $x \to x_0$, $x \in \Omega \setminus \Gamma$, i.e. $C_{2,2}$ - a.e. on $\partial\Gamma$ by an extension of the Kellogg property; cf. [HW].

So in particular, we might ask: is it in general possible to upgrade boundary regularity results from assumption of boundary values $C_{1,2}$ - a.e. to something like $C_{2,2}$ - a.e. for second order elliptic Dirichlet problems in arbitrary domains with smooth data on the boundary?

References

[A1] Adams, D.R., *Capacity and the obstacle problem,* Appl. Math. Optim **8** (1981), 39-57.

[A2] Adams, D.R., *Weighted nonlinear potential theory,* Trans. Amer. Math. Soc. **297** (1986), 73-94.

[AH] Adams, D.R., Hedberg, L.I., *Inclusion relations among fine topologies in nonlinear potential theory,* Ind. U. Math. J. **33** (1984), 117-126.

[AM] Adams, D.R., Meyers, N. *Thinness and Weiner Criteria for nonlinear potentials,* Ind. U. Math. J. **22** (1972), 169-197.

[AP] Adams, D.R., Pierre, M., *Capacitary strong type estimates in semilinear problems,* submitted to Ann. Inst. Fourier.

[BP1] Baras, P., *Singularities eliminables pour de equations semilineares,* Ann. Inst. Fourier (Grenoble) **34** (1984), 185-206.

[BP2] Baras, P., *Critere d'existence de solutions positives pour des equations semilineaires non monotones,* Ann. I.H.P. **2** (1985), 185-212.

[B] Bauman, P., *A Weiner test for nondivergence structure second-order elliptic equations,* Ind. U. Math. J. **34** (1985), 825-844.

[C] Chipot, M., *Variational inequalities and flow in porous media,* Applied Math. Sc. **52,** Springer-Verlag 1984.

[DMV] Dal Maso, G., Mosco, U., Vivaldi, M.A., *A pointwise regularity theory for the two-obstacle problem,* Acta Math. **163** (1989), 57-107.

[EJ] Evans, L.C., Jensen, R., *A boundary gradient estimate for harmonic functions and applications,* Research Notes in Math., *Nonlinear partial differential equations and their appl.,* College de France, Vol. 1, 1982, ed. H. Brezis, J.L. Lions, Pitman Press.

[FJK] Fabes, E., Jerison, D., Kenig, C., *The Weiner test for degenerate elliptic equations,* Ann. Inst. Fourier (Grenoble) **32** (1982), 151-182.

[GZ] Gariepy, R., Ziemer, W.P., *A regularity condition at the boundary for solutions of quasilinear elliptic equations,* Arch. Rat. Mech. Anal. **67** (1977), 25-39.

[H] Hedberg, L.I., *Nonlinear potentials and approximations in the mean by analytic functions,* Math. Z. **129** (1972), 299-319.

[HW] Hedberg, L.I., Wolff, T., *Thin sets in nonlinear potential theory,* Ann. Inst. Fourier (Grenoble) **33** (1983), 161-187.

[HK] Heinonen, J., Kipelainen, T., *On the Weiner Criterion and quasilinear obstacle problems,* to appear in Trans. Amer. Math. Soc.

[KM] Kilpelainen, T., Maly, J., *Generalized Dirichlet problem in nonlinear potential theory,* Manuscripta Math. **66** (1989), 25-44.

[LM] Lindqvist, P., Martio, O., *Two theorems of N. Weiner for solutions of quasilinear elliptic equations,* Acta. Math. **155** (1985), 153-171.

[LSW] Littman, W., Stampacchia, G., Wienberger, H., *Regular points for elliptic equations with discontinuous coefficients,* Ann. Sco. Norm. Sup. Pisa. **17** (1963), 45-79.

[L] Lewis, J., *Uniformly fat sets,* Trans. Amer. Math. **308** (1988), 177-196.

[Ma1] Maz'ja, V.G., *On the continuity at the boundary of the solutions of quasilinear elliptic equations,* Vestnik Leningradskogo Universiteta **25** (1970), 42-55.

[Ma2] Maz'ja, V.G., *Removable singularities of bounded solutions of quasilinear elliptic equations of any order*, Zap. Nauch. Semin. Leningr. Otd. Mat. Inst. Steklova **26** (1972), 116-130 (J. Soviet Math. **3** (1975), 480-492).

[Ma3] Maz'ja, V.G., *Sobolev Spaces*, Springer-Verlag, 1985.

[Ma4] Maz'ja, V.G., *Behavior of solutions to the Dirichlet problem for the biharmonic operator at the boundary point*, Wissenschaftliche Inform. **10**, Tech. Hoch. Karl-Marx-Stadt, 1979.

[Me] Meyers, N.G., *A theory of capacities for potential of functions in Lebesgue classes,*, Math. Scand. **26** (1970), 255-292.

[Se] Serrin, J. *Local behavior of solutions of quasilinear equations*, Acta Math. III (1964), 247-302.

[Sk] Skrypnik, I.V., *Nonlinear elliptic boundary value problems*, Teubner-Texte zur Math. **91**, Teubner Verlag, Leipzig, 1986.

[St] Stampacchia, G., *Le probleme de Dirichlet pour les equationes elliptiques du second ordre a coefficients discontinus*, Ann. Inst. Fourier (Grenoble) **15** (1965), 189-258.

Applications of Choquet theory to potential theory

Jürgen Bliedtner

Introduction

The notion of a (Choquet) simplex grew out of potential theory in connection with integral representations of (super) harmonic functions. In this paper we shall present certain aspects of potential theory which are closely related to the Choquet theory of function cones, the solvability of the weak Dirichlet problem, and its consequences to approximation theory.

To make these considerations more precise, fix an open relatively compact subset U of $\mathbb{R}^n$, $n \geq 3$. Given a continuous function f on the boundary U^* of U, the *classical Dirichlet problem* asks for a continuous extension of f to a function $h \in \mathcal{C}(\overline{U})$ which is harmonic on U. Therefore, one is interested in the function space

$$H(U) := \{h \in \mathcal{C}(\overline{U}) : h \quad \text{harmonic on} \quad U\} .$$

This space is called *simplicial* if its state space, i.e. the set of normed positive linear functionals on $H(U)$, is a Choquet simplex. This notion is so important to potential theory by a theorem of D.A. Edwards ([1]) since $H(U)$ is simplicial if and only if the *weak Dirichlet problem* is solvable, i.e. for any compact subset K of the Choquet boundary and any continuous function f on K there exists a function $h \in H(U)$ such that

$$h_{|K} = f \quad \text{and} \quad \|h\|_{\overline{U}} = \|f\|_K .$$

The most surprising consequence of the simpliciality of $H(U)$ is for me the following approximation property:

Theorem. *Let $\mathcal{P}$ denote the convex cone of real continuous potentials on $\mathbb{R}^n$. Then there exists a $p \in \mathcal{P}$ such that $H(U)$ is the closed linear subspace of $\mathcal{C}(\overline{U})$ generated by the set $\{q_{|\overline{U}} : q \in \mathcal{P},\ p - q \in \mathcal{P}\}$.*

The proof of this theorem can be found in [4], [5] where purely potential-theoretic methods are used to obtain a very general theorem of this type. In this paper certain results in this direction will be announced. The treatments make consequent use of the Choquet theory of function cones on locally compact spaces, developed e.g. in [3], [7], [8]. This theory is an extension of the geometric situation of compact convex subsets of locally convex spaces (see [1], [2]). It is interesting to note that this procedure

Potential Theory, ed. by M. Kishi

allows applications to the more general situation of potential cones introduced by G. Mokobodzki and D. Sibony [7], [8] and gives more insight into the background of approximation theory. Proofs and more results in this direction will appear in a subsequent paper.

1. Function cones

Let X be a locally compact (Hausdorff) space with a countable base and let $\mathcal{P} \subset \mathcal{C}^+(X)$ be a convex cone of real continuous functions on X such that

(P1) $\mathcal{P}$ is *linearly separating*, i.e. for every pair of points $x \neq y$ of X and every $\lambda \geq 0$ there exists $p \in \mathcal{P}$ such that

$$p(x) \neq \lambda p(y) .$$

(P2) Every $p \in \mathcal{P}$ is *dominated* by a function $q \in \mathcal{P}$, i.e. for any $\varepsilon > 0$ there exists a compact set $K \subset X$ such that

$$p(x) \leq \varepsilon q(x) , \quad x \in X \setminus K .$$

The vector space

$$\mathcal{C}_{\mathcal{P}}(X) := \{f \in \mathcal{C}(X) : |f| \leq p \quad \text{for some} \quad p \in \mathcal{P}\}$$

of all $\mathcal{P}$*-bounded* continuous functions will be endowed with the *order topology*, e.g. a function $g \in \mathcal{C}_{\mathcal{P}}(X)$ belongs to the closure $\overline{\mathcal{F}}$ of $\mathcal{F} \subset \mathcal{C}_{\mathcal{P}}(X)$ if there exists $p \in \mathcal{P}$ such that for any $\varepsilon > 0$ there is a function $f \in \mathcal{F}$ such that

$$|g - f| \leq \varepsilon p .$$

Introducing

$$\hat{\mathcal{P}} := \{\sup p_n : (p_n) \subset \mathcal{P} \text{ increasing}\} ,$$

we define the *reduced function* of a positive numerical function $f : X \longrightarrow \overline{\mathbb{R}}_+$ and an arbitrary subset $A \subset X$ by

$$R_f^A := \inf\{u \in \hat{\mathcal{P}} : f \leq u \quad \text{on} \quad A\} .$$

and $R_f := R_f^X$.

In the sequel we shall assume that $\mathcal{P}$ satisfies in addition the following condition

(P3) (*Continuity of the reduced function*) : $R_f \in \mathcal{P}$ for every $f \in \mathcal{C}_{\mathcal{P}}{}^+(X)$.

Then

$$\overline{\mathcal{P} - \mathcal{P}} = \mathcal{C}_{\mathcal{P}}(X)$$

and

$$\mathcal{C}_{\mathcal{P}}(X)' = \mathcal{M}(\mathcal{P}) - \mathcal{M}(\mathcal{P})$$

where

$$\mathcal{M}(\mathcal{P}) := \{\mu \in \mathcal{M}_+(X) : \mathcal{P} \subset L^1(X,\mu)\}$$

is the set of positive (Radon) measures on X for which the functions of $\mathcal{P}$ are integrable.

Any convex cone S such that

$$\mathcal{P} \subset S \subset \mathcal{C}_{\mathcal{P}}(X)$$

is called a *function cone.* Every function cone S generates an *inf-stable* function cone $W(S)$ consisting of all functions $\inf(s_1, \ldots, s_n)$ where $s_1, \ldots, s_n \in S$, $n \in \mathbb{N}$.

Examples 1.1.

(1) For any $p \in \mathcal{P}$ the convex cone

$$S(p) := \mathcal{P} - \mathbb{R}_+ p$$

is an inf-stable function cone.

(2) A kernel T on X is a *$\widehat{\mathcal{P}}$-dilation* if for every $p \in \mathcal{P}$

$$p \geq Tp \in \widehat{\mathcal{P}} .$$

For any family $\mathcal{T}$ of $\widehat{\mathcal{P}}$-dilations, the set

$$S(\mathcal{T}) := \{s \in \mathcal{C}_{\mathcal{P}}(X) : Ts \leq s \quad \text{for all} \quad T \in \mathcal{T}\}$$

is an inf-stable function cone.

(3) If $\mathcal{P}$ is the convex cone of real continuous potentials of a "nice" potential theory (e.g. elliptic-parabolic differential equations, harmonic spaces, balayage spaces) then for every open relatively compact subset $V \subset X$ the kernel

$$T^V = (\varepsilon_x^{X \setminus V})_{x \in X}$$

of balayage measures on $X \setminus V$ defines a $\widehat{\mathcal{P}}$-dilation.

Hence for every open subset U the set

$$\mathcal{T}^U = \{T^V : V \text{ open, relatively compact, } \overline{V} \subset U\}$$

defines the function cone

$$S(U) := \{s \in \mathcal{C}_{\mathcal{P}}(U) : s \text{ superharmonic on } U\}$$

such that

$$S(U) = S(\mathcal{T}^U) .$$

Let S be a function cone. For any $x \in X$, the elements of the set

$$\mathcal{M}_x(S) := \{\mu \in \mathcal{M}(\mathcal{P}) : \int s\, d\mu \leq s(x) \text{ for all } s \in S\}$$

are called *representing measures* for x (with respect to S).

The set

$$\delta(S) = \{x \in X : \mathcal{M}_x(S) = \{\varepsilon_x\}\}$$

is called the *Choquet boundary* of S . Then $\delta(S)$ is a G_δ–set such that the following *minimum principle* holds:

$$s \in S,\ s \geq 0 \quad \text{on} \quad \delta(S) \Longrightarrow s \geq 0 \quad \text{on} \quad X\ .$$

Furthermore, for any $x \in X$ there exists a *boundary measure* $\mu \in \mathcal{M}_x(S)$, i.e.

$$\mu(X \setminus \delta(S)) = 0\ .$$

The set

$$H(S) := \overline{W(S)} \cap (\overline{-W(S)})$$

is the linear space of all S–*affine continuous functions* on X, i.e.

$$H(S) = \{h \in \mathcal{C}_{\mathcal{P}}(X) : \int h\, d\mu = h(x) \quad \text{for all} \quad x \in X,\ \mu \in \mathcal{M}_x(S)\}\ .$$

Examples 1.2.

(1) For any $p \in \mathcal{P}$ the Choquet boundary of $S(p)$ is also called *fine support* of p, i.e.

$$\delta(p) := \delta(S(p))\ .$$

Furthermore, let $H(p) := H(S(p))$.

(2) For any set of $\widehat{\mathcal{P}}$–dilations we have

$$H(\mathcal{T}) := H(S(\mathcal{T})) = \{h \in \mathcal{C}_{\mathcal{P}}(X) : Th = h \text{ for all } T \in \mathcal{T}\}\ .$$

2. Simplicial cones

Let $\mathcal{P} \subset \mathcal{C}^+(X)$ again satisfy (P1) – (P3). A function cone S is called a *simplicial cone* if for every $x \in X$ there exists a unique measure $m_x^S \in \mathcal{M}_x(S)$ such that $m_x^S(X \setminus \delta(S)) = 0$. This notion is very important to potential theory by a theorem of D.A. Edwards which states the equivalence of simpliciality by the solution of the *weak Dirichlet problem:* For every closed set $F \subset \delta(S)$ and any $f \in \mathcal{C}_{\mathcal{P}}(X)$ with $|f| \leq s$ for some $s \in S^+$ there exists $h \in H(S)$ such that

(1) $h_{|F} = f$

(2) $|h| \leq s$.

The family $D^S := (m_x^S)_{x\in X}$ of boundary measures of a simplicial cone S defines a kernel on X such that

$$D^S s \leq s \quad \text{for any} \quad s \in S \ .$$

Moreover, S–affine continuous functions are characterized by D^S since

$$H(S) = \{h \in \mathcal{C}_{\mathcal{P}}(X) : D^S h = h\} \ .$$

Example 2.1. The function cones $S(\mathcal{T})$ of (1.1) are simplicial cones. This follows from the fundamental lemma 3.3 of [3]. The idea of the proof leads to the following notion.

A simplicial cone S is called a $\mathcal{P}$*–simplicial* cone if D^S is a $\widehat{\mathcal{P}}$–dilation, i.e. $D^S(\mathcal{P}) \subset \widehat{\mathcal{P}}$. The following theorem characterizes these cones by properties of the reduced function.

Theorem 2.2. *For any function cone* S *, the following statements are equivalent:*

(1) S *is a* $\mathcal{P}$*–simplicial cone.*

(2) $R_t \in H(S)$ *for every* $t \in -W(S)$.

Example 2.3. The function cones $S(\mathcal{T})$ of example 1.1 are $\mathcal{P}$–simplicial cones.

3. Potential cones

The treatment of the function cones $S(p)$ of example 1.1 leads to the notion of a potential cone introduced by Mokobodzki and Sibony [7].

A convex cone $\mathcal{P} \subset \mathcal{C}^+(X)$ is called a *potential cone* if in addition to (P1) – (P3) the following property holds:

(P4) $p - R_{p-q} \in \mathcal{P}$ for every $p, q \in \mathcal{P}$.

Example 3.1. Let $\mathcal{P}$ be a potential cone and $p \in \mathcal{P}$. Then $S(p)$ is a $\mathcal{P}$–simplicial cone such that $H(p)$ is the closed linear subspace of $\mathcal{C}_{\mathcal{P}}(X)$ generated by $\{q \in \mathcal{P} : p - q \in \mathcal{P}\}$.

The property (P4) is equivalent to the *additivity of the reduced function:*

For any compact set $K \subset X$ and any $p, q \in \mathcal{P}$, the following property holds:

$$R_{p+q}^K = R_p^K + R_q^K \ .$$

This property allows the introduction of *reduced measures,* i.e. for any compact set $K \subset X$ and any $x \in X$ there exists a unique measure $\varepsilon_x^K \in \mathcal{M}(\mathcal{P})$ such that

$$\int p\, d\varepsilon_x^K = R_p^K(x) \,, \quad p \in \mathcal{P} \,.$$

Furthermore, ε_x^K is supported by K. This notion will be extended to Borel sets $B \subset X$ by the relation

$$\int p\, d\varepsilon_x^B := \sup\{\int p\, d\varepsilon_x^K : K \subset B \text{ compact}\} \,, \quad p \in \mathcal{P} \,.$$

Note that in general only $\int p\, d\varepsilon_x^B \leq R_p^B(x)$ holds. The connection to simplicial cones is stated in the following proposition.

Proposition 3.2. *Let S be a $\mathcal{P}$-simplicial cone. Then the following properties hold:*

(1) $m_x^S = \varepsilon_x^{\delta(S)}$ *for any* $x \in X$.

(2) $D^S p = R_p^{\delta(S)}$ *for any* $p \in \mathcal{P}$.

For any $x \in X$, the set $\mathcal{M}_x(\mathcal{P})$ is a convex set which is compact with respect to the weak topology $\sigma(\mathcal{M}(\mathcal{P}) - \mathcal{M}(\mathcal{P}),\, \mathcal{C}_{\mathcal{P}}(X))$. G. Mokobodzki has determined in [6] the set $[\mathcal{M}_x(\mathcal{P})]_e$ of its extreme points:

$$[\mathcal{M}_x(\mathcal{P})]_e = \{\varepsilon_x^B : B \subset X \text{ Borel}\}.$$

This result allows to describe the set $\mathcal{M}_x(S)$ of extreme points for certain function cones.

Example 3.3. Let $p \in \mathcal{P}$. Then for every $x \in X$,

$$[\mathcal{M}_x(S(p))]_e = \{\varepsilon_x^B : \delta(p) \subset B \subset X \text{ Borel}\}.$$

4. Approximation theorems

In this section let $\mathcal{P} \subset \mathcal{C}^+(X)$ be a potential cone on X and S be a simplicial cone. We shall consider the following problem: Find $p \in \mathcal{P}$ such that

(1) $H(p) = H(S)$,

(2) $\overline{S(p)} = \overline{W(S)}$.

A necessary condition for these properties is certainly that $\delta(p) = \delta(S)$ holds. By the results of (2.3) this condition is satisfied by the function cones of type $S(T)$ as the following proposition shows.

Proposition 4.1. *The following statements are equivalent:*

(1) S *is a* $\mathcal{P}$*–simplicial cone.*

(2) *There exists* $p \in \mathcal{P}$ *such that* $\delta(p) = \delta(S)$.

(3) *There exists* $p \in \mathcal{P}$ *such that* $H(p) = H(S)$.

An example in [5] shows that the conditions of proposition 4.1 for a simplicial cone are not sufficient for the approximation of $W(S)$. However, suggested by (3.2) the following proposition solves this problem. Note that by results of [4], these conditions are satisfied by the function cones $S(U)$ of example 1.1.

Proposition 4.2. *Let* $p \in \mathcal{P}$ *such that* $\delta(p) = \delta(S)$. *Then the following statements are equivalent:*

(1) $\overline{S(p)} = \overline{W(S)}$

(2) $\mathcal{M}_x(S) = \mathcal{M}_x(\mathcal{P}) \cap \mathcal{M}_x(H(S))$ *for every* $x \in X$.

(3) $\mathcal{M}_x(S) = \mathcal{M}_x(\mathcal{P}) \cap \mathcal{M}_x(H(p))$ *for every* $x \in X$.

(4) $\varepsilon_x^B \in \mathcal{M}_x(S)$ *for every* $x \in X$, $\delta(S) \subset B \subset X$ *Borel.*

(5) *The following* local minimum principle *holds:*
For any Borel set B *such that* $\delta(S) \subset B \subset X$ *and any compact set* $K \subset B$, $u \in \widehat{W(S)}$, $q \in \mathcal{P}$:

$$u \geq R_q^K \text{ on } A \Longrightarrow u \geq R_q^K \text{ on } X.$$

The last two propositions lead to the following problems:

(1) In connection with a characterization of ellipticity in potential theory it would be interesting to characterize those functions $p \in \mathcal{P}$ such that $H(p)$ is either a lattice or an anti–lattice.

(2) The local minimum principle of (4.2) suggests the validity of a *local* theory of the function cones $S(p)$, $p \in \mathcal{P}$, in the sense that the collection $(S(p))_{p \in \mathcal{P}}$ has a property similar to an appropriate sheaf property.

References

[1] E.M. ALFSEN: *Compact Convex Sets and Boundary Integrals.* Ergebn. d. Math. **57**, Springer–Verlag (1971)

[2] L. ASIMOW, A.J. ELLIS: *Convexity Theory and its Applications in Functional Analysis.* Academic Press (1980)

[3] J. BLIEDTNER, W. HANSEN: *Simplicial Cones in Potential Theory.* Inventiones math. **29** (1975), 83–110

[4] J. BLIEDTNER, W. HANSEN: *Potential Theory. An Analytic and Probabilistic Approach to Balayage.* Universitext, Springer–Verlag (1986)

[5] J. BLIEDTNER, W. HANSEN: *Approximation by Continuous Potentials.* In: Proceedings of a Conference on Potential Theory 1987, Plenum Press (1988), 53–58

[6] G. MOKOBODZKI: *Eléments extrémaux pour le balayage.* Sem. Brelot–Choquet–Deny "Théorie du potentiel" **13** (1970), 5.01–5.14

[7] G. MOKOBODZKI, D. SIBONY: *Cônes adaptés de fonctions continues et théorie du potentiel.* In: Sem. Choquet "Initiation à l'analyse" **6** (1968), 5.01–5.35

[8] D. SIBONY: *Cônes de fonctions et potentiels.* Lecture Notes McGill University Montreal (1968)

Une version non linéaire du théorème de Hunt

Claude Dellacherie

Abstract

In the classical topological context (l.c.c.b. space, etc.) we prove a nonlinear version of Hunt's theorem characterizing the proper kernels verifying the complete maximum principle as those closing a submarkovian resolvant and also, as a main step in the proof, a non linear version of Meyer's theorem characterizing the proper kernels verifying the reinforced complete maximum principle as those being the elementary potential kernel of a submarkovian kernel. See [2] or [3] for an exposition of both theorems in the linear case.

1980 Mathematics Subject Classification (1985 Revision) : 31C99.

1. Énoncé des théorèmes (H) et (M)

Soit E un espace localement compact à base dénombrable. Nous désignons resp. par $\mathcal{C}_K$, $\mathcal{C}_c$, $\mathcal{C}_o$ l'espace des fonctions (réelles) continues sur E resp. nulles hors du compact K, à support compact, tendant vers 0 à l'∞ ; ces espaces sont munis de la norme uniforme. On appelle *opérateur* toute application V de $\mathcal{C}_K$, $\mathcal{C}_c$ ou $\mathcal{C}_o$ dans $\mathcal{C}_o$; l'opérateur V est dit *borné* s'il est lipschitzien, et sa meilleure constante de Lipschitz est alors appelée sa *norme* $\|V\|$ (quoique cela ne soit pas tout à fait une norme).

Un opérateur V de $\mathcal{C}_c$ dans $\mathcal{C}_o$ est un *opérateur propre* si pour tout compact K de E la restriction de V à $\mathcal{C}_K$ est un opérateur borné ; un opérateur linéaire, croissant et propre est exactement ce qu'on appelle un noyau propre en théorie linéaire du potentiel sur $\mathcal{C}_o$. Par ailleurs on dira qu'un opérateur est *sousmarkovien* s'il est croissant et de norme ≤ 1, autrement dit si c'est une contraction croissante. Enfin, une *résolvante sousmarkovienne* sera une famille $(V_p)_{p>0}$ d'opérateurs sur $\mathcal{C}_o$ telle que pV_p soit sousmarkovien pour tout $p > 0$ et que l'on ait *l'équation résolvante* : pour tout $p, q > 0$

$$[I - (p-q)V_p][I + (p-q)V_q] = I$$

où I est l'identité, soit encore

$$V_q = V_p[I + (p-q)V_q].$$

En non linéaire, on ne peut pas développer plus ces expressions ; en particulier V_p et V_q ne commutent pas en général.

Ces petits ajustements de vocabulaire nous permettent de formuler notre version non linéaire du théorème de Hunt à peu près dans les mêmes termes que la version linéaire classique.

Potential Theory, ed. by M. Kishi

Théorème (H). *Soit V un opérateur propre vérifiant le* principe complet du maximum *(en abrégé, le* PCM*), i.e. vérifiant :* $\forall f, g \in \mathcal{C}_c \ \forall \lambda \geq 0$

$$Vf \leq Vg + \lambda \text{ sur } \{f > g\} \Rightarrow Vf \leq Vg + \lambda \text{ partout.}$$

Alors

1) Il existe une unique résolvante sousmarkovienne $(V_p)_{p>0}$ telle qu'on ait pour chaque $p > 0$, sur $\mathcal{C}_c$,

$$(I - pV_p)(I + pV) = I \textit{ soit encore } V = V_p(I + pV)$$

De plus on a $V_p = V(I - pV_p)$ sur $(I - pV_p)^{-1}(\mathcal{C}_c)$.

2) Pour tout $g \in \mathcal{C}_o$ coincé entre deux éléments de $V(\mathcal{C}_c)$, on a

$$Vf = \lim_{p \downarrow 0} V_p(f + pg)$$

pour tout $f \in \mathcal{C}_c$. En particulier $Vf = \lim_{p\downarrow 0} V_p f$ pour tout $f \in \mathcal{C}_c$ si $V0 = 0$.

Remarques. a) L'énoncé du PCM est en tout point conforme à l'un des énoncés classiques en linéaire. Bien entendu, il n'est pas question ici d'introduire la différence $f - g$ pour n'avoir affaire qu'à une fonction. Noter par ailleurs que V est nécessairement croissant (prendre $f \leq g$ et $\lambda = 0$).

b) Les deux points du théorème assurent en quelque sorte que V ferme la résolvante algébriquement et analytiquement.

c) Pour retrouver pleinement l'énoncé linéaire, il faudrait ajouter un troisième point : *si $V(\mathcal{C}_c)$ est dense dans $\mathcal{C}_o$, alors V est le potentiel d'un unique semigroupe $(P_t)_{t \geq 0}$ fellerien et sousmarkovien.* Cela peut se faire grâce au théorème de Crandall-Liggett, analogue non linéaire du théorème de Hille-Yosida. Encore faut-il s'entendre sur la signification de "potentiel" ; la formule exponentielle de Crandall-Liggett (analogue de la formule de Hille) permet de passer de la résolvante au semigroupe mais je ne connais pas de formule permettant de passer du semigroupe à la résolvante comme le permet la transformation de Laplace en linéaire.

Nous passons à l'énoncé de notre version non linéaire du théorème de Meyer (en situation topologique). Nous notons $\mathcal{B}$ l'espace des fonctions bornées (sans autre régularité) sur E, et $\mathcal{B}_o$ le sous-espace des fonctions bornées tendant vers 0 à l'∞, tous deux munis de la norme uniforme.

Théorème (M). *Soit V un opérateur propre vérifiant le* principe complet du maximum affaibli *(en abrégé, le* PCMA*), i.e. vérifiant :* $\forall f, g \in \mathcal{C}_c \ \forall \lambda \geq 0$

$$f + Vf \leq g + Vg + \lambda \text{ sur } \{f > g\} \Rightarrow Vf \leq Vg + \lambda \text{ partout.}$$

Alors

1) Il existe un unique opérateur sousmarkovien N sur $\mathcal{C}_o$ tel qu'on ait

$$(I - N)(I + V) = I \textit{ soit encore } V = N(I + V)$$

2) Pour toute extension de N en une contraction croissante $\tilde{N}$ sur $\mathcal{B}$ et pour tout $f \in \mathcal{C}_c$, l'équation de Poisson élémentaire en $v \in \mathcal{B}_o$

$$(I - \tilde{N})v = f$$

a pour seule solution $v = f + Vf$.

Remarques. a) Ni le nom ni la formulation du PCMA ne sont classiques en linéaire. Cependant V vérifie le PCMA ssi $G = I + V$ vérifie : $\forall f, g \in \mathcal{C}_c\ \forall \lambda \geq 0$

$$Gf \leq Gg + \lambda \text{ sur } \{f > g\} \Rightarrow Gf - f \leq Gg - g + \lambda \text{ partout,}$$

qui est l'une des formulations possibles du *principe complet du maximum renforcé* (en abrégé, le *PCMR*). On voit sans peine qu'un opérateur vérifie le PCMA (resp PCM) dès qu'il vérifie le PCM (resp PCMR), ce qui justifie les noms donnés. Il y a d'autres relations concernant ces différents principes, mais nous nous contenterons ici du minimum indispensable pour la compréhension des démonstrations des deux théorèmes énoncés. Noter par ailleurs que, comme plus haut pour le PCM, un opérateur vérifiant le PCMA est nécessairement croissant.

b) Le deuxième point donne un sens simple et précis au fait que $I + V$ est l'opérateur potentiel (élémentaire) de N. Il implique que pour tout $f, \varphi \in \mathcal{C}_c$ avec $f \geq \varphi$ la fonction $f + Vf$ est le plus petit élément u de $\mathcal{B}$ vérifiant $u \geq \varphi + V\varphi$ et $(I - \tilde{N})u \geq f$, d'où l'on peut déduire que $f + Vf$ est la limite de la suite croissante (u_n) où $u_0 = \varphi + V\varphi$ et $u_{n+1} = f + Nu_n$.

c) Une application croissante M d'une partie $\mathcal{D}_M$ de $\mathcal{B}$ dans $\mathcal{B}$ est une contraction ssi on a pour tout $f, g \in \mathcal{D}_M$ et tout $\lambda \geq 0$

$$f \leq g + \lambda \Rightarrow Mf \leq Mg + \lambda$$

On en déduit sans peine que toute contraction croissante définie sur une partie de $\mathcal{B}$ admet une extension en une contraction croissante définie sur tout $\mathcal{B}$: le point 2) a toujours un contenu non trivial.

2. Déduction de (H) à partir de (M)

Si V vérifie le PCM, alors pV vérifie le PCM et donc le PCMA pour tout $p > 0$, d'où l'existence d'après le point 1) de (M), pour chaque $p > 0$, d'un unique opérateur sousmarkovien N_p sur $\mathcal{C}_o$ tel que $pV = N_p(I + pV)$ sur $\mathcal{C}_c$. Posons, pour tout $p > 0$,

$$V_p = \frac{1}{p} N_p \text{ d'où } V = V_p(I + pV);$$

on a alors sur $\mathcal{C}_c$, pour tout $p, q > 0$,

$$V_p (I + (p-q)V_q)(I + qV) = V_p (I + qV + (p-q)V) = V$$

et l'unicité de N_q entraine alors que $V_q = V_p (I + (p-q)V_q)$. Ainsi (V_p) est une résolvante sousmarkovienne. Reste à prouver du 1) de (H) le "De plus" : il s'agit d'une propriété d'injectivité, conséquence immédiate du 2) de (M) appliqué à pV.

Passons au 2) de (H). Supposons d'abord $V0 = 0$, $g = 0$ et $f \geq 0$. Alors, d'après l'équation résolvante, $V_p f$ est ≥ 0 et croît quand p décroît tout en restant majorée par Vf. Nous appelons $V_0 f$ la fonction semicontinue inférieurement (en abrégé, s.c.i.) limite simple des $V_p f$ quand p tend vers 0 ; comme elle est comprise entre 0 et Vf, elle tend vers 0 à l'∞. Prolongeons d'autre part l'opérateur sousmarkovien V_1 aux fonctions s.c.i. tendant vers 0 à l'∞ par

$$\overline{V}_1 h = \sup \{V_1\varphi, \varphi \leq h, \varphi \in \mathcal{C}_o\};$$

un argument classique fondé sur le lemme de Dini-Cartan assure que $\overline{V}_1$ est *montant*, i.e. que, si (h_n) est une suite croissante de limite simple h, alors $(\overline{V}_1 h_n)$ tend simplement en croissant vers $\overline{V}_1 h$. En faisant tendre p vers 0 dans l'instance de l'équation résolvante

$$V_p f = V_1(f + (1-p)V_p f)$$

on obtient alors grâce à la montée de $\overline{V}_1$

$$V_0 f = \overline{V}_1(f + V_0 f).$$

Mais on voit sans peine que $\overline{V}_1$ est une contraction croissante, prolongeant V_1 ; il résulte alors de cette dernière égalité et du point 2) de (M) appliqué à $N = V_1$ et $v = f + V_0 f$ qu'on a $V_0 f = Vf$. Du coup, d'après le lemme de Dini, la convergence de $V_p f$ vers $V_0 f = Vf$ est uniforme. Cela clôt le cas particulier $V0 = 0$, $g = 0$ et $f \geq 0$ et, de manière analogue, le cas $V0 = 0$, $g = 0$ et $f \leq 0$. Passons au cas général. Fixons $\varphi \in \mathcal{C}_c$ et, pour $h \in \mathcal{C}_c$, posons

$$\hat{V}h = V(h + \varphi) - V\varphi.$$

L'opérateur $\hat{V}$ ainsi défini est évidemment propre, et vérifie le PCM, comme V, et le point 1) de (H) lui associe la résolvante $(\hat{V}_p)$ avec, pour $h \in \mathcal{C}_o$,

$$\hat{V}_p h = V_p(h + \varphi + pV\varphi) - V\varphi$$

(d'après l'unicité, il suffit de vérifier que $\hat{V}h = \hat{V}_p(h + \hat{V}h)$). Comme $\hat{V}0 = 0$, on déduit de ce qui précède appliqué à $(\hat{V}_p)$ que, pour tout $f \in \mathcal{C}_c$ tel que $f \geq \varphi$ ou $f \leq \varphi$, $V_p(f + pV\varphi)$ tend (uniformément) vers Vf quand p tend vers 0. Fixons $f \in \mathcal{C}_c$ et, si f n'est pas comparable à φ, choisissons φ_0 et φ_1 dans $\mathcal{C}_c$ de sorte qu'on ait $\varphi_0 \leq f \wedge \varphi$ et $f \vee \varphi \leq \varphi_1$; on a

$$V_p(f + pV\varphi_0) \leq V_p(f + pV\varphi) \leq V_p(f + pV\varphi_1)$$

et comme, d'après ce qui précède, les membres extrêmes de ces inégalités tendent vers Vf quand p tend vers 0, il en est de même pour le membre médian. Ainsi $V_p(f + pg)$ tend vers Vf quand p tend vers 0 pour tout $f \in \mathcal{C}_c$ et tout $g \in V(\mathcal{C}_c)$ ou plus généralement tout $g \in \mathcal{C}_o$ coincé entre deux éléments ψ_0 et ψ_1 de $V(\mathcal{C}_c)$ (appliquer le raisonnement précédent à $V_p(f + pV\psi_0) \leq V_p(f + pg) \leq V_p(f + pV\psi_1)$). Il n'est sans doute pas possible en général de prendre pour g n'importe quel élément de $\mathcal{C}_o$ (en particulier 0), mais je n'ai pas cherché de contre-exemple.

3. Preuve du point 2) du théorème (M)

On a au départ un opérateur sousmarkovien N sur $\mathcal{C}_o$ et un opérateur propre V vérifiant le PCMA tels qu'on ait

$$(I-N)(I+V)=I \text{ sur } \mathcal{C}_c.$$

On se donne une extension $\tilde{N}$ de N en une contraction croissante sur $\mathcal{B}$, un élément f de $\mathcal{C}_c$ et une fonction bornée v vérifiant $(I-\tilde{N})v=f$ et tendant vers 0 à l'∞. Il s'agit de prouver que $v=f+Vf$.

Fixons un $\varepsilon>0$ et choisissons un compact K de E assez grand pour qu'on ait $f=0$, $|Vf|<\varepsilon$ et $|v|<\varepsilon$ hors de K, puis $\eta\in\mathcal{C}_c^+$ valant 1 sur K. Posons pour $\lambda>0$

$$u=(f-\lambda\eta)+V(f-\lambda\eta)\ ,\ w=(f+\lambda\eta)+V(f+\lambda\eta)$$

et choisissons enfin λ suffisammment petit pour que $|u|$ et $|w|$ soient $<2\varepsilon$ hors de K (on utilise ici la continuité de la restriction de V au support de η). Soit $a=u-v, b=v-w$; des égalités $u=(\text{f}-\lambda\eta)+\tilde{N}u$, $v=f+\tilde{N}v$, $w=(\text{f}+\lambda\eta)+\tilde{N}w$ on tire par différence

$$a=-\lambda\eta+\tilde{N}(v+a)-\tilde{N}(v)\ ,\ b=-\lambda\eta+\tilde{N}(w+b)-\tilde{N}w$$

et donc, en tenant compte du fait que $\tilde{N}$ est une contraction croissante,

$$a\leq-\lambda\eta+\|a\vee 0\|\ ,\ b\leq-\lambda\eta+\|b\vee 0\|.$$

Comme $\lambda\eta$ vaut $\lambda>0$ sur K, a^+ et b^+ atteignent leur maximum hors de K, et comme $|u|$,$|v|$ et$|w|$ sont $<2\varepsilon$ hors de K, on en déduit

$$f-\lambda\eta+V(f-\lambda\eta)\leq v+4\varepsilon\ ,\ v\leq f+\lambda\eta+V(f+\lambda\eta)+4\varepsilon$$

et il ne reste plus qu'à faire tendre λ puis ε vers 0 pour conclure.

4. Preuve du point 1) du théorème (M) dans le cas borné

Plus précisément, nous allons donner les grandes lignes de la démonstration du résultat suivant que nous avons publié dans [1] : *si V est un opérateur borné sur $\mathcal{C}_o$ vérifiant le PCMA, il existe un unique opérateur sousmarkovien N sur $\mathcal{C}_o$ tel que $(I-N)(I+V)=I$ et on a alors aussi $(I+V)(I-N)=I$.* Nous restons dans notre cadre topologique, mais en fait l'énoncé est valable dans un cadre abstrait très large.

Disons que deux opérateurs bornés M et W sur $\mathcal{C}_o$ sont *conjugués* si on a

$$(I-M)(I+W)=I=(I+W)(I-M),$$

avatar de l'équation résolvante qui s'écrit encore

$$M=W(I-M)\ ,\ W=M(I+W);$$

la relation n'étant pas symétrique, on dira que W est le *conjugué* de M tandis que M est l'*anticonjugué* de W. On établit d'abord que, si M et W sont des opérateurs bornés conjugués, alors M est sousmarkovien ssi W vérifie le PCMA. Ceci fait, comme notre opérateur V est borné, d'après le théorème de point fixe de Banach tV admet un anticonjugué N_t pour $0 < t < \|V\|^{-1}$, et comme V vérifie le PCMA, tV le vérifie aussi pour $t \leq 1$; par conséquent, N_t est sousmarkovien pour $t < \tau = 1 \wedge \|V\|^{-1}$. Cela permet de définir $W_t = \frac{1}{t}N_t$ pour $0 < t < \tau$ et donc un début $(W_t)_{0<t<\tau}$ de résolvante sousmarkovienne. Soit alors ρ le sup des $s > 0$ tels que l'anticonjugué N_t de tV existe et soit sousmarkovien : on montre par l'absurde qu'on ne peut avoir $\rho < 1$ (c'est la même idée que dans la démonstration du théorème de Hunt classique dans le cas borné ; la seule différence est que le PCMA nous bloque à 1 alors que le PCM permet d'aller jusqu'à l'∞). On obtient ainsi une résolvante sousmarkovienne "courte" $(W_t)_{0<t<1}$ qu'il n'est pas difficile de prolonger en 0 par $W_0 = V$ et en 1 par un opérateur sousmarkovien W_1 ; et on vérifie que $N = W_1$ convient.

5. Preuve du point 1) du théorème (M) dans le cas propre

On a au départ un opérateur propre V vérifiant le PCMA et il s'agit de prouver qu'il existe un unique opérateur sousmarkovien N tel qu'on ait $V = N(I + V)$ sur $\mathcal{C}_c$.

D'abord quelques notations. On pose $\Phi = \{\varphi \in \mathcal{C}_c : 0 \leq \varphi \leq 1\}$ et φ,ψ désigneront désormais des éléments génériques de Φ. Pour $f \in \mathcal{C}_c$ la notation $f|\varphi$ signifie que $\varphi \in \Phi$ vaut 1 sur le support de f. Sur Φ la relation $\varphi|\psi$ est une relation d'ordre strict et on symbolisera par lim_φ le fait qu'on regarde une limite (simple si c'est dit, uniforme sinon) le long du filtre engendré par les intervalles d'ordre $[\varphi, 1[$ pour cette relation ; on pourrait tout aussi bien considérer des suites croissantes pour cette relation, convergeant simplement vers 1.

Comme V est propre, on sait, d'après le cas borné, qu'il existe pour tout φ un unique opérateur sousmarkovien N_φ tel qu'on ait sur $\mathcal{C}_o$

$$VI_\varphi = N_\varphi(I + VI_\varphi) \quad , \quad N_\varphi = VI_\varphi(I - N_\varphi)$$

Pour $\varphi|\psi$ on a

$$N_\varphi = VI_\varphi(I - N_\varphi) = N_\psi(I + VI_\psi)I_\varphi(I - N_\varphi) = N_\psi[I_\varphi(I - N_\varphi) + VI_\varphi(I - N_\varphi)]$$

si bien que, pour $\varphi|\psi$, on a l'avatar d'équation résolvante

$$N_\varphi = N_\psi(I_\varphi + I_{1-\varphi}N_\varphi).$$

Nous allons bientôt montrer, par petites étapes, que $lim_\varphi N_\varphi f = Nf$ existe pour tout $f \in \mathcal{C}_o$. Mais avant cela, nous allons vérifier que cet opérateur N, supposé ainsi défini, a les propriétés requises.

D'abord, on a $V = N(I + V)$ sur $\mathcal{C}_c$. En effet, on a $VI_\varphi = N_\varphi(I + VI_\varphi)$ et, pour $f \in \mathcal{C}_c$, on a $VI_\varphi f = Vf$ si $f|\varphi$ alors que $lim_\varphi N_\varphi(f + Vf) = N(f + Vf)$. Puis, N est le seul opérateur borné (même continu) ayant cette propriété. En effet, si M est un opérateur sur $\mathcal{C}_o$ tel que $V = M(I + V)$ sur $\mathcal{C}_c$, on a à peu près comme ci-dessus

$$N_\varphi = VI_\varphi(I - N_\varphi) = M(I + V)I_\varphi(I - N_\varphi) = M[I_\varphi(I - N_\varphi) + VI_\varphi(I - N_\varphi)]$$

si bien qu'on a pour tout φ

$$N_\varphi = M(I_\varphi + I_{1-\varphi}N_\varphi).$$

Si M est continu, il n'y a plus qu'à passer à la lim_φ pour conclure.

Revenons à la définition de Nf comme $lim_\varphi N_\varphi f$. Nous commençons par supposer que $V0 = 0$, ce qui implique $N_\varphi 0 = 0$ pour tout φ, et que f est ≥ 0. On va successivement établir que $lim_\varphi N_\varphi f$ existe au sens de la convergence simple pour $f \in \mathcal{C}_c^+$, puis pour $f \in \mathcal{C}_o^+$, puis au sens de la convergence uniforme pour $f \in \mathcal{C}_c^+$ et enfin pour $f \in \mathcal{C}_o^+$. Soit donc $f \in \mathcal{C}_c^+$; pour $\varphi|\psi$ on a d'après ce qu'on a vu plus haut, pour $f|\varphi$,

$$N_\varphi f = N_\psi(f + I_{1-\varphi}N_\varphi f)$$

et donc $N_\varphi f$ décroît quand φ croît pour l'ordre "|" considéré : d'où une limite simple $lim_\varphi N_\varphi f = Nf \geq 0$. Prenons maintenant $f \in \mathcal{C}_o^+$, et, pour $\varepsilon > 0$ donné, soit $f^\varepsilon \in \mathcal{C}_c^+$ tel que $\|f - f^\varepsilon\| < \varepsilon$: comme N_φ est une contraction, on peut écrire $N_\varphi f = \eta_\varphi^\varepsilon + N_\varphi f^\varepsilon$ avec $\|\eta_\varphi^\varepsilon\| < \varepsilon$ et, comme Nf^ε existe, on a au sens de la convergence simple

$$\|lim\,sup_\varphi N_\varphi f - lim\,inf_\varphi N_\varphi f\| < 2\varepsilon$$

d'où l'existence de la limite simple voulue $lim_\varphi N_\varphi f = Nf \geq 0$ et la définition ainsi d'une contraction croissante N sur $\mathcal{C}_o^+$. Supposons de nouveau $f \in \mathcal{C}_c^+$ et récrivons $N_\varphi f = N_\psi(f + I_{1-\varphi}N_\varphi f)$ avec $f|\varphi|\psi$; comme la fonction $(f + I_{1-\varphi}N_\varphi f)$ appartient à $\mathcal{C}_o^+$, on peut passer à la lim_ψ simple (en ψ), d'où $N_\varphi f = N(f + I_{1-\varphi}N_\varphi f)$ et donc $\|Nf - N_\varphi f\| \leq \|I_{1-\varphi}N_\varphi f\|$. Comme $N_\varphi f$ est ≥ 0 et décroît quand φ croît, on a bien convergence uniforme de $N_\varphi f$ vers Nf. Enfin, revenons à $f \in \mathcal{C}_o^+$: pour un $\varepsilon > 0$ on approche comme ci-dessus f par $f^\varepsilon \in \mathcal{C}_c^+$, d'où $N_\varphi f = \eta_\varphi^\varepsilon + N_\varphi f^\varepsilon$ avec $\|\eta_\varphi^\varepsilon\| < \varepsilon$; et comme $N_\varphi f^\varepsilon$ tend uniformément vers Nf^ε tandis que $N_\varphi f$ tend simplement vers Nf, on en déduit que $N_\varphi f$ tend bien uniformément vers Nf.

Passons pour finir au cas général : comme pour la démonstration du 2) de (H), on va utiliser un argument de translation. On commence par supposer $f \in \mathcal{C}_c$ et on va définir $N(f + V0)$. Définissons $g \in \mathcal{C}_o^+$ par

$$g = f - (f \wedge 0) + V0 - V(f \wedge 0)$$

et, pour tout $h \in \mathcal{C}_c$, posons

$$\hat{V}h = V[h + (f \wedge 0)] - V(f \wedge 0)$$

ce qui définit un opérateur $\hat{V}$ vérifiant le PCMA tel que $\hat{V}0 = 0$ auquel on associe comme plus haut des opérateurs $\hat{V}I_\varphi$ et $\hat{N}_\varphi$ pour $\varphi \in \Phi$. Lorsque l'on a $f|\varphi$ (le seul cas qui nous intéresse) cela donne pour $h \in \mathcal{C}_o$

$$\hat{V}I_\varphi h = VI_\varphi[h + (f \wedge 0)] - VI_\varphi(f \wedge 0)$$

d'où l'on déduit (par exemple, en vérifiant que ça marche)

$$\hat{N}_\varphi h = N_\varphi[h + (f \wedge 0) + V(f \wedge 0)] - V(f \wedge 0)$$

Appliquons cette dernière égalité à la fonction g : on a

$$\hat{N}_\varphi g = N_\varphi(f + V0) - V(f \wedge 0)$$

et, appliquant la première partie de la démonstration à $\hat{V}$ et g, on trouve alors que $lim_\varphi N_\varphi(f + V0) = N(f + V0)$ existe. Enfin, pour $f \in \mathcal{C}_o$, on procède comme par deux fois ci-dessus en approchant f par des éléments de $\mathcal{C}_c$.

6. En guise de conclusion

Le théorème classique de Hunt est justement célèbre pour avoir jeté un pont entre la théorie du potentiel et la théorie des processus de Markov. Actuellement il est toujours utilisé sous une forme plus ou moins raffinée pour passer d'une théorie axiomatique du potentiel à une résolvante après utilisation d'un avatar du théorème de Mme Hervé pour obtenir un "noyau potentiel" vérifiant le principe complet du maximum.

En non linéaire, on connait actuellement plusieurs opérateurs "naturels" vérifiant le principe complet du maximum, mais ils sont construits à partir d'opérateurs d'équations aux dérivées partielles (par exemple, l'équation de Hamilton-Jacobi) et il n'y a sans doute pas de difficulté (je sors là de mon domaine de compétence !) à construire toute la résolvante plutôt que le seul opérateur potentiel. J'espère néanmoins que le théorème présenté ici se révèlera utile, ne serait-ce qu'en signe qu'il peut y avoir une théorie du potentiel non linéaire substantielle dans un cadre assez large.

Références

[1] Dellacherie C., Théorie élémentaire du potentiel non linéaire. In : Séminaire Initiation à l'Analyse 25e année, 1985/86 (ed. by G.Choquet, M.Rogalski, J.Saint-Raymond), 32 p. Publication n°82 de l'Université Paris VI, Paris 1987

[2] Dellacherie C., Sur la caractérisation des noyaux potentiels. In : Séminaire de Théorie du Potentiel Paris, n°9 (ed. by N.Bouleau, D.Feyel, F.Hirsch, G.Mokobodzki), 78-95. Lecture Notes in Mathematics n°1393, Springer-Verlag, New York-Berlin-Heidelberg 1989

[3] Revuz D., Markov Chains. North Holland, Amsterdam 1975

Strict isoperimetric inequalities and asymmetry

Walter K. Hayman

1. Introduction

This talk describes some joint work with R. R. Hall and A. Weitsman [3].

Consider the following statements. Of all plane bodies Ω with given area A, the disk has

a) The smallest circumference $L = (4\pi A)^{1/2}$;

b) The smallest capacity $\operatorname{Cap}\Omega = (A/\pi)^{1/2}$.

L. E. Fraenkel has asked whether the above and similar statements can be sharpened if we know that our body is not a disk. In order to make this question more precise we introduce some notation.

Let Ω be a measurable set in $\mathbf{R}^n$ having finite n-dimensional measure $V_n(\Omega)$. We denote by $\omega_n = \pi^{n/2}/\Gamma(1+\frac{1}{2}n)$ the corresponding measure of the unit ball. We write

$$B_\rho^n(x) = \{\xi : |\xi - x| < \rho\} \tag{1.1}$$

for the open ball of centre x and radius ρ. Thus $V_n\{B_\rho^n(x)\} = \omega_n\rho^n$. We choose $\rho = \{V_n(\Omega)/\omega_n\}^{1/n}$ so that $V_n(\Omega) = \omega_n\rho^n$. Then following an (unpublished) suggestion of Fraenkel we define the asymmetry $\alpha(\Omega)$ of Ω by

$$\alpha = \alpha(\Omega) = \inf_x \frac{V_n\{\Omega \backslash B_\rho^n(x)\}}{\omega_n\rho^n}. \tag{1.2}$$

Thus $1-\alpha$ is the maximum proportion of the volume of Ω which can be contained in any ball of radius ρ. The quantity $\alpha(\Omega)$ is zero if and only if Ω differs from a ball by sets of zero volume.

We denote by $\operatorname{Cap}\Omega$ the logarithmic capacity if $n = 2$, and Newtonian capacity if $n > 2$, normalised so that $\operatorname{Cap}\{B_\rho^n(x)\} = \rho$. We also denote the frontier of Ω by $\partial\Omega$ and $(n-1)$-dimensional Hausdorff measure by S_n. If $n = 2$, we write A, L instead of V_2, S_2. We denote by $c_1, c_2, \ldots$ constants depending only on n.

We conjecture that

$$S_n(\partial\Omega) \geq n\omega_n\rho^{n-1}(1 + c_1\alpha^2); \tag{1.3}$$

$$\operatorname{Cap}\Omega \geq \rho(1 + c_2\alpha^2). \tag{1.4}$$

Potential Theory, ed. by M. Kishi

Unfortunately we are far from being able to obtain results of such precision and generality, except for (1.3), when $n = 2$. It is an unpublished results of G. T. Keady that, if Ω is connected, then (1.4) also holds for $n = 2$. Fraenkel has given some fairly strong restrictions, under which (1.4) holds for $n = 3$. We can however prove

Theorem 1. *If $n = 2$, we have*

$$L^2 \geq 4\pi A(1 + \frac{\alpha^2}{6}); \tag{1.5}$$

$$\operatorname{Cap}\Omega \geq (A/\pi)^{1/2}(1 + \frac{\alpha^3}{331}). \tag{1.6}$$

Theorem 2. *If $n \geq 3$ and Ω is convex, we have*

$$S_n(\partial\Omega) \geq n\omega_n\rho^{n-1}(1 + c_3\alpha^n); \tag{1.7}$$

$$\operatorname{Cap}\Omega \geq \rho(1 + c_4\alpha^{n+1}). \tag{1.8}$$

In the next section I sketch a proof of Theorem 1. The proof of Theorem 2 is similar. In section 3 I describe our results for α close to 1, which are more complete. In the last section a few related theorems and conjectures are described.

2. Isoperimetric inequalities in the plane

It can be shown by approximation that we may confine ourselves to smooth sets, for instance sets Ω bounded by a finite number of mutually disjoint analytic hypersurfaces. General sets Ω can be approximated by smooth sets Ω_m in such a way that $V_n(\Omega_m) \to V(\Omega)$, $\alpha(\Omega_m) \to \alpha(\Omega)$ and $S_n(\partial\Omega_m) \to S_n(\partial\Omega)$. If $S_n(\partial\Omega) = \infty$, (1.3) is trivially true, so to prove (1.3) we may assume that $S_n(\partial\Omega) < \infty$. To test the accuracy of (1.3) and (1.4) we consider the following

Example. Suppose that $\varepsilon > 0$ and that Ω is the ellipse $x = (1+\varepsilon)\cos t$, $y = \sin t$, $0 \leq t \leq 2\pi$. Then $A(\Omega) = (1+\varepsilon)\pi$, $\operatorname{Cap}(\Omega) = 1 + \frac{1}{2}\varepsilon$, and $L = L(\Omega)$, $\alpha = \alpha(\Omega)$ satisfy

$$L = 2\pi + \pi\varepsilon + \frac{\pi\varepsilon^2}{8} + O(\varepsilon^3), \quad \alpha(\Omega) = \frac{\varepsilon}{\pi} + O(\varepsilon^2).$$

This yields

$$L^2 - 4\pi A = \frac{3\pi^2\varepsilon^2}{2} + O(\varepsilon^3) \quad \text{as} \quad \varepsilon \to 0,$$

i.e.

$$L^2 = 4\pi A(1 + \frac{3\varepsilon^2}{8} + \cdots), \quad L = (4\pi A)^{1/2}(1 + \frac{3\pi^2\alpha^2}{16} + O(\alpha^3)),$$

$$\operatorname{Cap}(\Omega) = (\frac{A}{\pi})^{1/2} + \frac{\varepsilon^2}{8} + \cdots = (\frac{A}{\pi})^{1/2}(1 + \frac{\pi^2\alpha^2}{8} + \cdots).$$

Thus the orders of magnitude in (1.3) and (1.4) are correct if $n = 2$ and $c_1 \leq 3\pi^2/16$, $c_2 \leq \pi^2/8$ in this case.

We proceed to sketch a proof of (1.5). We suppose first that Ω is simply connected and has inradius r, i.e. the largest disk contained in Ω has radius r. Then an inequality of Osserman [5, p.3] states that

$$L^2 - 4\pi A \geq \frac{(A - \pi r^2)^2}{r^2}.$$

Since $\alpha A \leq A - \pi r^2$, it follows that

$$L^2 - 4\pi A \geq \frac{\pi\alpha^2 A}{1-\alpha},$$

$$L^2 \geq 4\pi A(1 + \frac{\alpha^2}{4(1-\alpha)}).$$

This proves (1.5) when Ω is simply connected with 1/4 instead of 1/6. By a series of not very complicated manipulations we can extend the inequality first to the case when Ω is multiply connected and then to the general case at the expense of reducing the constant to 1/6.

Next I should like to indicate how (1.5) leads to (1.6). Let $G(z) = G(z, \infty)$ be the Green's function for $\widehat{\mathbf{C}}\backslash\Omega$ with pole at ∞ and write

$$\Omega_\lambda = \{z \in \widehat{\mathbf{C}} : G(z) > \lambda\}.$$

Let C_λ be the condenser with inner set $\overline{\Omega}$ and outer set $\overline{\Omega}_\lambda$, that is $C_\lambda = (\overline{\Omega}, \overline{\Omega}_\lambda)$. Then the capacity of the condenser is given by

$$\operatorname{Cap} C = \int\int_{\mathbf{C}} |\nabla u|^2 dx dy$$

where u is continuous in $\mathbf{C}$, $u = 0$ on $\overline{\Omega}$, $u = 1$ on $\overline{\Omega}_\lambda$, and u is harmonic elsewhere; thus $u = G(z)/\lambda$ elsewhere.

For $0 < t < 1$, we write $F(t) = \{z : u(z) \leq t\}$ and $A_t = A\{F(t)\}$. We also write $\alpha_0 = \alpha(\Omega)$ and proceed to show that $\alpha\{F(t)\} \geq \alpha_0/3$ for all t such that $A(\Omega) \leq A\{F(t)\} \leq (1 + \frac{1}{2}\alpha_0)A(\Omega)$.

In fact for any such t let B' be a disk of area $A\{F(t)\}$ and let B be the disk concentric with B' and having area $A(\Omega)$. Then

$$\begin{aligned} A\{F(t)\backslash B'\} &\geq A\{F(t)\backslash B\} - A(B'\backslash B) \\ &\geq A(\Omega\backslash B) - (A\{F(t)\} - A(\Omega)) \\ &\geq \frac{\alpha_0}{2}A(\Omega) \geq \frac{\alpha}{2+\alpha_0}A\{F(t)\} \geq \frac{\alpha_0}{3}A\{F(t)\}. \end{aligned}$$

We now define τ by

$$\tau = \inf\{t : A_t > (1 + \frac{1}{2}\alpha_0)A(\Omega)\}, \tag{2.1}$$

so that $A_\tau = (1 + \frac{1}{2}\alpha_0)A(\Omega)$. Then by the above argument

$$\alpha(F(t)) \geq \frac{\alpha_0}{3}, \quad 0 < t < \tau. \tag{2.2}$$

Following an argument of Gehring [2], we now complete our proof by the use of the Coarea formula [1, p.248]. Suppose that $0 = t_0 < t_1 < \cdots < t_n = \tau$. Then this formula shows that

$$\int_{t_{j-1}}^{t_j} L(\{u = t\})dt = \iint_{F(t_j)\setminus F(t_{j-1})} |\nabla(u)|dxdy.$$

By Schwarz's inequality the right hand side is at most

$$(\iint_{F(t_j)\setminus F(t_{j-1})} |\nabla(u)|^2 dxdy)^{1/2}(A_{t_j} - A_{t_{j-1}})^{1/2},$$

while by (1.5) the left hand side is at least

$$(t_j - t_{j-1})(1 + \frac{1}{6}\frac{\alpha_0^2}{9})^{1/2}(4\pi A_{t_{j-1}})^{1/2} = (t_j - t_{j-1})\{4\pi A_{t_{j-1}}(1 + \frac{\alpha_0^2}{54})\}^{1/2}.$$

Thus

$$\begin{aligned}\tau &= \sum_{j=1}^{n}(t_j - t_{j-1}) \\ &\leq \sum_{j=1}^{n}(\iint_{F(t_j)\setminus F(t_{j-1})} |\nabla(u)|^2 dxdy)^{1/2}(\frac{A_{t_j} - A_{t_{j-1}}}{4\pi A_{t_{j-1}}})^{1/2}\frac{1}{(1 + \alpha_0^2/54)^{1/2}}.\end{aligned}$$

Using Schwarz's inequality, we obtain

$$\tau \leq \frac{1}{(1 + \alpha_0^2/54)^{1/2}}(\iint_{F(\tau)} |\nabla(u)|^2 dxdy)^{1/2}(\sum_{j=1}^{n}\frac{A_{t_j} - A_{t_{j-1}}}{4\pi A_{t_{j-1}}})^{1/2}.$$

Letting $\max(t_j - t_{j-1})$ tend to zero, we deduce

$$\tau \leq \frac{1}{(1 + \alpha_0^2/54)^{1/2}}(\iint_{F(\tau)} |\nabla(u)|^2 dxdy)^{1/2}(\frac{1}{4\pi}\log\frac{A_\tau}{A(\Omega)})^{1/2}.$$

Using the ordinary isoperimetric inequality, we obtain similarly

$$1-\tau \le \Big(\iint_{F(1)\backslash F(\tau)} |\nabla(u)|^2 dxdy\Big)^{1/2}\Big(\frac{1}{4\pi}\log\frac{A(\mathbf{C}\backslash\Omega_\lambda)}{A_\tau}\Big)^{1/2}.$$

Adding these and using Schwarz's inequality again yield

$$4\pi \le \Big(\iint_{\mathbf{C}} |\nabla(u)|^2 dxdy\Big)\log\Big\{\Big(\frac{A_\tau}{A(\Omega)}\Big)^{1/(1+\alpha_0^2/54)}\frac{A(\mathbf{C}\backslash\Omega_\lambda)}{A_\tau}\Big\}.$$

We now recall (2.1), i.e. $A_\tau = (1+\frac{1}{2}\alpha_0)A(\Omega)$. Thus

$$\Big(\frac{A_\tau}{A(\Omega)}\Big)^{1/(1+\alpha_0^2/54)}\frac{A(\mathbf{C}\backslash\Omega_\lambda)}{A_\tau} = \frac{A(\mathbf{C}\backslash\Omega)}{A(\Omega)}\Big(1+\frac{1}{2}\alpha_0\Big)^{-\alpha_0^2/(54+\alpha_0^2)} \le \frac{A(\mathbf{C}\backslash\Omega_\lambda)}{A(\Omega)(1+c_5\alpha_0^3)},$$

where $c_5 = 1/165$. In fact if $0 < a \le \frac{1}{2}$, $b > 0$, it follows from the mean value theorem that there exists a_1, where $0 < a_1 < a \le \frac{1}{2}$, such that

$$(1+a)^b = 1 + ab(1+a_1)^{b-1} > 1 + \frac{ab}{1+a_1} > 1 + \frac{2}{3}ab.$$

Writing $a = \frac{1}{2}\alpha_0, b = \alpha_0^2/(54+\alpha_0^2) > \alpha_0^2/55$, we obtain

$$\Big(1+\frac{1}{2}\alpha_0\Big)^{\alpha_0^2/(54+\alpha_0^2)} > 1 + \frac{\alpha_0^3}{165}.$$

Thus

$$4\pi \le \Big(\iint_{\mathbf{C}} |\nabla(u)|^2 dxdy\Big)\Big(\log\frac{A(\mathbf{C}\backslash\Omega_\lambda)}{A(\Omega)(1+c_5\alpha_0^3)}\Big).$$

We now recall that $u(z) = G(z)/\lambda$ in $\mathbf{C}\backslash(\Omega\cup\Omega_\lambda)$, and that u is constant in Ω and in Ω_λ. Thus Green's Theorem yields

$$\iint_{\mathbf{C}} |\nabla(u)|^2 dxdy = \int_{\partial\Omega_\lambda}\frac{\partial u}{\partial n}ds = \frac{1}{\lambda}\int_{\partial\Omega_\lambda}\frac{\partial G}{\partial n}ds.$$

Also, since $G(z) - \log|z|$ is harmonic near ∞,

$$G(z) = \log|z| - \log C + \frac{O(1)}{|z|},$$

$$\frac{\partial G}{\partial n} = \frac{1}{|z|} + \frac{O(1)}{|z|^2} \quad \text{as } z \to \infty,$$

where $C = \operatorname{Cap}\Omega$. Thus, if λ is large, then $r = |z|$ satisfies $r = Ce^{\lambda} + O(1)$ on Ω_λ. Hence

$$\int_{\partial\Omega_\lambda} \frac{\partial G}{\partial n} ds = \frac{2\pi r}{r} + \frac{O(1)}{r} = 2\pi + O(e^{-\lambda}), \quad \lambda \to \infty.$$

Also $(\mathbf{C}\backslash\Omega_\lambda) = \pi C^2 e^{2\lambda}(1 + O(e^{-\lambda}))$ as $\lambda \to \infty$. We deduce that

$$\frac{2\pi}{\lambda}(1 + O(e^{-\lambda})) \geq \frac{4\pi}{\log\{\frac{\pi C^2 e^{2\lambda}(1 + O(e^{-\lambda}))}{A(\Omega)(1 + c_5\alpha_0^3)}\}}$$

or

$$2\lambda + \log\{\frac{\pi C^2}{A(\Omega)(1 + c_5\alpha_0^3)}\} \geq 2\lambda + O(1), \quad \text{with } c_5 = \frac{1}{165}.$$

Letting λ tend to ∞ we obtain

$$C \geq \{\frac{A(\Omega)}{\pi}\}^{1/2}(1 + \frac{\alpha_0^3}{165})^{1/2} \geq \{\frac{A(\Omega)}{\pi}\}^{1/2}(1 + \frac{\alpha_0^3}{331}), \quad \text{since } \alpha_0 \leq 1,$$

and this proves (1.6).

A corresponding argument goes through in $\mathbf{R}^n$, provided that we can obtain a suitable version of (1.7). If Ω is convex, another inequality of Ossermann [5, (114)] enables us to do this, but in general we have been unable to obtain any result of the form

$$S_n(\partial\Omega) \geq n\omega_n\rho^{n-1}(1 + \varepsilon(\alpha)),$$

where $\varepsilon(\alpha)$ is a positive function of α, where $\alpha > 0$.

It would be enough to obtain such an inequality when Ω is a solid of revolution. Let $O(x_i)$ be coordinate axes in $\mathbf{R}^n$, for $i = 1$ to n, and let $\widehat{\Omega}_i$ be the body obtained from Ω by symmetrising about the axis $O(x_i)$. This means that $\widehat{\Omega}_i$ meets each hyperplane $H(c) : x_i = c$, in an $(n-1)$-dimensional ball, whose $(n-1)$-dimensional volume is equal to that of $H(c) \cap \Omega$ and whose centre is on $O(x_i)$. Then we have

Theorem 3.

$$\max_i \alpha(\widehat{\Omega}_i) \geq c_6\alpha(\Omega)^2.$$

The order of magnitude is sharp in all dimensions.

Thus if for instance we could show that, for some positive increasing f,

$$S_n(\partial\widehat{\Omega}) \geq n\omega_n\rho^{n-1}(1 + f(\hat{\alpha})),$$

where $\hat{\alpha}$ is the asymmetry of an axisymmetric body $\widehat{\Omega}$, we could symmetrise a general body Ω of asymmetry α, to obtain an axisymmetric body $\widehat{\Omega}$ of asymmetry $\hat{\alpha}$ and deduce

$$S_n(\partial\Omega) \geq S_n(\partial\widehat{\Omega}) \geq n\omega_n\rho^{n-1}(1 + f(\hat{\alpha})) \geq n\omega_n\rho^{n-1}\{1 + f[c_6\alpha(\Omega)^2]\},$$

for a general body Ω. We could then obtain a corresponding inequality for $\operatorname{Cap}\Omega$, using the technique outlined above.

3. Inequalities for α close to 1

Suppose now that α is close to 1, so that $\alpha = 1 - \delta$, where δ is small. Thus a ball of radius ρ intersects Ω in a set of measure at most $\delta\omega_n\rho^n$. In this section we normalise so that

$$V_n(\Omega) = \omega_n,$$

and, for $0 < \rho < \infty$, we define

$$\hat{\delta} = \hat{\delta}(\rho) = \sup_x \frac{V_n(\Omega \cap B_\rho^n(x))}{\omega_n}. \tag{3.1}$$

Evidently $0 < \hat{\delta} \le 1$ and $\hat{\delta}^{1/n} \le \rho < \infty$. Subject to these conditions we seek to find

$$C(\delta, \rho) = \inf \operatorname{Cap}(\Omega), \tag{3.2}$$

where the infimum is taken over general bodies Ω, which have volume ω_n and are such that for a given fixed ρ, where $0 < \rho < \infty$, we have $\hat{\delta}(\rho) = \delta$.

We note that the asymmetry of Ω is given by $\alpha(\Omega) = 1 - \hat{\delta}(1)$. So the lower bound of capacities of bodies with asymmetry α and volume ω_n is $C(1-\alpha, 1)$. For general bodies Ω of finite volume $V_n(\Omega)$ and asymmetry α, the lower bound for the capacity is given by

$$\operatorname{Cap}\Omega \ge \{\frac{V_n(\Omega)}{\omega_n}\}^{1/n} C(1-\alpha, 1). \tag{3.3}$$

For we may replace Ω by $\Omega' = t^{-1}\Omega$, where $V_n(\Omega) = \omega_n t^n$. Then Ω' has volume ω_n,asymmetry $\alpha(\Omega') = \alpha(\Omega)$ and capacity $\operatorname{Cap}(\Omega') = t^{-1}\operatorname{Cap}(\Omega)$.

Theorem 4. *If $n \ge 3$, we have*

$$c_7 \inf\{\rho\delta^{-1/n}, \delta^{-2/[n(n-2)]}\} \le C(\delta, \rho) \le c_8 \inf\{\rho\delta^{-1/n}, \delta^{-2/[n(n-2)]}\}. \tag{3.4}$$

If $n = 2$, we have

$$C(\delta, \rho) \ge c_9 \frac{\rho^{1-\delta}}{\delta^{1/2}}, \tag{3.5}$$

while if M is the positive integer such that $M - 1 < 1/\delta \le M$, we have

$$C(\delta, \rho) \le c_{10} \frac{\rho^{1-1/M}}{\delta^{1/2}}. \tag{3.6}$$

Corollary. *If α, ρ are defined as in section 1, we have*

$$\text{Cap}(\Omega) \geq c_{11}\rho(1-\alpha)^{-1/n}, \quad \textit{if} \ \ n = 2,3,4. \tag{3.7}$$

$$\text{Cap}(\Omega) \geq c_{12}\rho(1-\alpha)^{-2/[n(n-2)]}, \quad \textit{if} \ \ n \geq 5. \tag{3.8}$$

The orders of magnitude implied in (3.7) and (3.8) are sharp.

I should like to sketch some of the ideas on which the proof of Theorem 4 depends. The hypotheses of that theorem are that Ω has volume ω_n and that a ball of radius ρ intersects Ω in a set whose maximum volume is $\omega_n\delta$. Then $C(\delta, \rho)$ is the lower bound of Cap(Ω) under these conditions. Intuitively we expect the capacity to be minimal if Ω is packed as close as possible subject to the above hypotheses, and such a packing does turn out to give the right order of magnitude for $C(\delta, \rho)$ if $n \geq 3$, or if $n = 2$ and $\delta = 1/M$, where M is a positive integer. If $\delta = 1/M$, we achieve such a packing by choosing M balls B_j, each of volume $\omega_n\delta$, i.e. radius $\delta^{1/n}$ and whose centres lie as close as possible to each other on a rectangular lattice, generated by vectors which have magnitude 4ρ, and are mutually orthogonal. Since two distinct lattice points have a mutual distance of at least 4ρ, and since $\delta^{1/n} \leq \rho$, this means that no ball of radius ρ can meet more than one of the B_j, so that the hypotheses are satisfied.

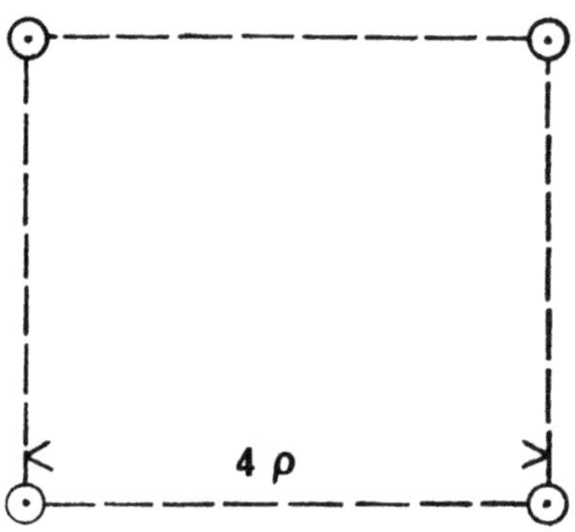

$\delta = \frac{1}{4}$, M = 4

If $1/M < \delta < 1/(M-1)$, we choose the first $M-1$ balls as above but make the last ball a bit smaller, so that the total volume is still ω_n. This explains the discrepancy between (3.5) and (3.6) when $n = 2$ and ρ is large. For instance the upper estimate (3.6) jumps from $O(\rho^{2/3})$ to $O(\rho^{3/4})$ as δ decreases through the value $\frac{1}{3}$. In higher dimensions such a small change in δ cannot affect the order of magnitude of $C(\delta, \rho)$, as (3.4) shows.

We can deduce (3.6) from the subadditivity of logarithmic capacity [4, Theorem 7.17, p.437]. In this case Ω consists of M disks in the above configuration, where

$M - 1 < 1/\delta \le M$. We can achieve this by a set Ω, which is contained in a disk of radius R, where $R < c_{13}\rho M^{1/2}$. Each disk has capacity at most $\delta^{1/2}$. Thus

$$\{\log[\frac{2R}{\mathrm{Cap}(\Omega)}]\}^{-1} \le \sum_1^M \{\log[\frac{2R}{\delta^{1/2}}]\}^{-1}$$

i.e.

$$\mathrm{Cap}\,\Omega \le (2R)^{1-1/M}\delta^{1/(2M)} < c_{10}\rho^{1-1/M}\delta^{-1/2}.$$

(We use the fact that $M^{1/M} \to 1$ as $M \to \infty$.) This proves (3.6).

If $n > 2$, subadditivity yields too weak a result in general. However we can deduce (3.4) and (3.5) from the following

Lemma. *Suppose that ν is a unit mass distribution on $\overline{\Omega}$ and that*

$$v(x) = \int_{\overline{\Omega}} K(x-y)d\nu(e_y)$$

is the corresponding potential, where

$$K(x) = \log|x|, \text{ if } n = 2, \quad K(x) = -|x|^{2-n}, \text{ if } n > 2,$$

Then

$$\inf_{x\in\partial\Omega} v(x) \le V_0 \le \sup_{x\in\partial\Omega} v(x),$$

where

$$V_0 = \log \mathrm{Cap}(\Omega) \text{ if } n = 2; \quad V_0 = -(\mathrm{Cap}\,\Omega)^{2-n}, \text{ if } n > 2.$$

In our case it is convenient to take $\nu(e) = V_n(e \cap \Omega)/\omega_n$. We write

$$\widehat{K}(t) = \log t, \text{ if } n = 2; \quad \widehat{K}(t) = -t^{2-n}, \text{ if } n > 2$$

Suppose now that Ω is bounded and that R is chosen so large that Ω lies in $B^n_R(x)$, where $x \in \partial\Omega$. Then if $\hat{\nu}(t) = \nu(B^n_t(x))$, we have

$$v(x) = \int_0^R \widehat{K}(t)d\hat{\nu}(t) = \hat{\nu}(R)\widehat{K}(R) - d_n\int_0^R t^{1-n}\hat{\nu}(t)dt,$$

where $d_n = \max(1, n-2)$, and $\hat{\nu}(R) = 1$. This shows that in order to obtain lower bounds for $v(x)$, we need upper bounds for $\hat{\nu}(t)$. Such bounds are readily obtained from our basic hypothesis (3.1). Feeding these lower bounds for $v(x)$ into the Lemma yields lower bounds for V_0 and so for $\mathrm{Cap}(\Omega)$ which in turn give the left hand inequality in (3.4) and also (3.5). We obtain the right hand inequality in (3.4) by obtaining lower bounds for $\hat{\nu}(t)$ in the example for Ω constructed above. The corresponding upper bounds for V_0 and so for $\mathrm{Cap}(\Omega)$ complete the proof of

(3.4) and so of Theorem 4. The Corollary is an easy consequence of the case $\rho = 1$ of Theorem 4, with $c_{11} = c_9$, if $n = 2$, $c_{11} = c_{12} = c_7$, if $n > 2$.

4. Some other problems

Clearly conjectures similar to (1.3) and (1.4) can be made whenever, in some class of sets with given volume in $\mathbf{R}^n$, the balls are extremal. For instance if Ω is a convex body of volume $V_n(\Omega) = \omega_n R^n$ and diameter $D(\Omega) = (2 + \varepsilon(\Omega))R$, we can ask for inequalities between $\varepsilon(\Omega)$ and the asymmetry $\alpha(\Omega)$. It turns out that

$$\alpha \geq C_n \varepsilon^{(n+1)/2} + O(\varepsilon^{(n+3)/2}), \quad \text{where } C_n = \frac{2\omega_{n-1}}{(n+1)^2\omega_n}(\frac{n-1}{n+1})^{(n-1)/2}.$$

Also the above relation fails if C_n is increased by a factor greater than 2.

In the opposite direction we can prove for $n = 2$

$$\alpha^2 \leq \Lambda_0 \varepsilon + O(\varepsilon^{3/2}). \tag{4.1}$$

The constant Λ_0 is defined as follows. Let S be the class of functions $s(\theta)$ in $[-\pi, \pi]$, such that $s(\theta) = \pm 1$ almost everywhere and let

$$s(\theta) \sim \frac{1}{2}c_0 + \sum_{n=1}^{\infty}(c_n \cos n\theta + d_n \sin n\theta)$$

be the Fourier series of $s(\theta)$. Then

$$\Lambda_0 = \sup_S \sum_{n=1}^{\infty} \frac{(c_n^2 + d_n^2)}{2(n^2 - 1)}.$$

We conjecture that $\Lambda_0 = (4 - \pi)/\pi$, with equality when $s(\theta) = \text{sgn}(\cos 2\pi\theta)$. In this case the constant in (4.1) would also be sharp, with equality when the boundary of Ω consists of certain set of 4 circular arcs.

If $n = 3$, (4.1) still holds, provided that Λ_0 is replaced by 3000. I conjecture that (4.1) holds for all n, and for general sets, when Λ_0 is replaced by a suitable constant.

References

1. H. Federer, *Geometric measure theory*, Springer-Berlin 1969.
2. F. W. Gehring, Inequalities for condensers, hyperbolic capacity and extremal length, Michigan Math. J. 18 (1971), 1-20.
3. R. R. Hall, W. K. Hayman, and A. Weitsman, On asymmetry and capacity, submitted to J. d'Analyse.
4. W. K. Hayman, *Subharmonic functions*, volume 2, Academic Press, London, 1989.
5. F. Ossermann, Bonnesen-style isoperimetric inequalities, Amer. Math. Monthly, 86 (1979), 1-29.

Nonlinear potential theory

Lars Inge Hedberg

Abstract
A theory of nonlinear potentials was created around 1970 by a number of people. It took a long time before a reasonably satisfactory corresponding theory of thin sets and fine topologies was developed, and it is only recently that this material has reached a state of beauty and completeness. We give a brief survey of these developments.

1980 Mathematics Subject Classification (1985 Revision): 31B15, 46E35.

1. Introduction

About 20 years ago a theory of nonlinear potentials was created by B. Fuglede, N. G. Meyers, Yu. G. Reshetnyak, V. P. Khavin, and V. G. Maz'ya. We formulate the central notions in terms of Bessel potentials on $\mathbf{R}^N$. These are defined by means of the Bessel kernels, G_α, which are defined most easily by means of their Fourier transforms

$$\hat{G}_\alpha(\xi) = (1 + |\xi|^2)^{-\alpha/2}, \quad \alpha \in \mathbf{R}.$$

For $0 < \alpha < N$ the Bessel kernels behave like the M. Riesz kernels, $I_\alpha(x) = C_\alpha |x|^{\alpha - N}$, near the origin, but they have the advantage that they decay exponentially at infinity and thus belong to $L^1(\mathbf{R}^N)$. See e. g. A. P. Calderón [6] or E. M. Stein [23].

The set of positive Radon measures with support in a set $E \subset \mathbf{R}^N$ are denoted by $\mathcal{M}^+(E)$. For $\alpha > 0$ the Bessel potentials of measures $\mu \in \mathcal{M}^+(\mathbf{R}^N) = \mathcal{M}^+$ are defined by

$$(G_\alpha * \mu)(x) = \int G_\alpha(x - y)\, d\mu(y).$$

If μ is absolutely continuous with density f we write, with the usual abuse of notation,

$$(G_\alpha * f)(x) = \int G_\alpha(x - y) f(y)\, dy.$$

We now define a two parameter family of capacities $C_{\alpha,p}$ on $\mathbf{R}^N$ for $1 < p < \infty$, $0 < \alpha \le N/p$.

Definition 1. Let $E \subset \mathbf{R}^N$ be an arbitrary set.

$$C_{\alpha,p}(E) = \inf\left\{\int f^p\, dx;\ f \ge 0,\ (G_\alpha * f)(x) \ge 1 \text{ on } E\right\}.$$

Potential Theory, ed. by M. Kishi

If the set of such f is empty we set $C_{\alpha,p}(E) = \infty$.

A statement is said to hold (α, p)-quasieverywhere ((α, p)-q.e.) if it is true except on a set E with $C_{\alpha,p}(E) = 0$.

The reason for not allowing $\alpha p > N$ is that in that case the only set with zero capacity is the empty set, and we are at present not interested in this situation.

It is a cornerstone in the theory that there is a dual characterization of these capacities. We let $p^{-1} + (p')^{-1} = 1$.

Theorem 1. *Let $K \subset \mathbf{R}^N$ be compact. Then*

$$C_{\alpha,p}(K)^{1/p} = \sup\left\{\frac{\mu(K)}{\|G_\alpha * \mu\|_{p'}};\ \mu \in \mathcal{M}^+(K)\right\}.$$

Moreover, there exist extremals μ^K and f^K such that

$$\begin{aligned}
&\mu^K(K) = C_{\alpha,p}(K);\\
&\|f^K\|_p^p = C_{\alpha,p}(K);\\
&f^K = (G_\alpha * \mu^K)^{p'-1};\\
&G_\alpha * f^K(x) \geq 1 \quad (\alpha, p)\text{-}q.e.\ on\ K.
\end{aligned}$$

For any $\mu \in \mathcal{M}^+$ the function

$$V^\mu_{\alpha,p} = G_\alpha * (G_\alpha * \mu)^{p'-1}$$

is called a *nonlinear potential* of μ. Note that in the case $p = 2$ the nonlinear potential is linear, and because of the group property of $\{G_\alpha\}$,

$$V^\mu_{\alpha,2} = G_\alpha * (G_\alpha * \mu) = G_{2\alpha} * \mu.$$

There are two different proofs of these results. One is variational, and is found in the papers Reshetnyak [22] and Maz'ya-Khavin [17]. The other one depends on the minimax theorem, and is found in Fuglede [7] and Meyers [18].

The notion of a *thin set* is central in classical potential theory. It is a remarkable fact that there is a quite satisfactory extension of this theory to the nonlinear situation.

It is not obvious what it should mean that a set E is (α, p)-thin at a point a. The following two statements are natural candidates for a definition.

(A) E is such that there exists a $\mu \in \mathcal{M}^+$ with

$$V^\mu_{\alpha,p}(a) < \liminf_{x \to a,\, x \in E \setminus \{a\}} V^\mu_{\alpha,p}(x).$$

(B) E is such that

$$\int_0^1 \left(\frac{C_{\alpha,p}(E \cap B(a,r))}{r^{N-\alpha p}}\right)^{p'-1} \frac{dr}{r} < \infty,$$

or equivalently

$$\sum_0^\infty \left(2^{n(N-\alpha p)} C_{\alpha,p}(E \cup B_n(a))\right)^{p'-1} < \infty.$$

Here $B(a,r)$ denotes the (open) ball with center at a and radius r, and $B_n(a) = B_{2^{-n}}(a)$. It should be noticed that for $p = 2$ both statements reduce to well-known characterizations of thin sets. The integral in (B) first appears (for $\alpha = 1$) in the work of Maz'ya [16].

For $p \neq 2$ (A) and (B) are no longer equivalent. In fact, (A) implies (B), but the converse is not true in general. See D. R. Adams and N. G. Meyers [5] and L. I. Hedberg [10].

The statement (B) was first chosen as the basis of a definition of a fine topology by Meyers [19]. It turned out later that this topology has all the properties that one can reasonably desire. This was clarified in [12], where the crucial role was played by an estimate proved by Th. H. Wolff. Thanks mainly to the insights of D. R. Adams (see especially [1] and [2]) it is now possible to present this material in a much more satisfactory way than in [12]. The purpose of this paper is to give a brief such presentation. More details will be given in [3].

In order to explain this we need to make an excursion into a more general situation considered by Fuglede [7] and Meyers [18].

2. General nonlinear potentials

Let Y be a measure space, and consider a *kernel* g on $\mathbf{R}^N \times Y$, i. e. a nonnegative function $g(x,y)$ which is lower semicontinuous on $\mathbf{R}^N$ for each $y \in Y$, and measurable on Y for each $x \in \mathbf{R}^N$. Fix a positive measure ν on Y, let $\mu \in \mathcal{M}^+(\mathbf{R}^N)$, and let f be a ν-measurable function. Slightly changing the notation of Fuglede [7], we define potentials $\mathcal{G}f$ and $\check{\mathcal{G}}\mu$ by

$$\mathcal{G}f(x) = \int_Y g(x,y)\, f(y)\, d\nu(y);$$

$$\check{\mathcal{G}}\mu(y) = \int_{\mathbf{R}^N} g(x,y)\, d\mu(x).$$

Notice that by Fubini's theorem

$$\int_{\mathbf{R}^N} \mathcal{G}f\, d\mu = \int_Y \check{\mathcal{G}}\mu\, f\, d\nu.$$

Definition 1 and Theorem 1 then extend in the following way (Fuglede [7], Meyers [18]).

Definition 2. Let $E \subset \mathbf{R}^N$ be an arbitrary set.

$$C_{g,p}(E) = \inf\left\{\int_Y f^p \, d\nu;\ f \geq 0,\ \mathcal{G}f(x) \geq 1 \text{ on } E\right\}.$$

Theorem 2. *Let $K \subset \mathbf{R}^N$ be compact. Then*

$$C_{g,p}(K)^{1/p} = \sup\left\{\frac{\mu(K)}{\|\check{\mathcal{G}}\mu\|_{L^{p'}(\nu)}};\ \mu \in \mathcal{M}^+(K)\right\}.$$

Moreover, there exist extremals μ^K and f^K such that

$$\begin{aligned}
&\mu^K(K) = C_{g,p}(K);\\
&\|f^K\|^p_{L^p(\nu)} = C_{g,p}(K);\\
&f^K = (\check{\mathcal{G}}\mu^K)^{p'-1};\\
&\mathcal{G}f^K(x) \geq 1 \quad (g,p)\text{-q.e. on } K.
\end{aligned}$$

The function $\mathcal{V}^\mu_{g,p} = \mathcal{G}(\check{\mathcal{G}}\mu)^{p'-1}$ is again called a nonlinear potential of the measure μ, and by Fubini's theorem

$$\int \mathcal{V}^\mu_{g,p} \, d\mu = \int \mathcal{G}(\check{\mathcal{G}}\mu)^{p'-1} \, d\mu = \|\check{\mathcal{G}}\mu\|^{p'}_{L^{p'}(\nu)}.$$

We are interested in a special case. Let $Y = \mathbf{R}^N \times \mathbf{Z}^+$. Define the measure ν on $\mathbf{R}^N \times \mathbf{Z}^+$ by

$$\nu = m \otimes \sum_0^\infty \delta_n,$$

where m is Lebesgue measure and δ_n denotes unit mass at n. Thus $f \in L^p(\nu)$ means $f = \{f_n\}_0^\infty$, with

$$\|f\|^p_{L^p(\nu)} = \sum_0^\infty \|f_n\|_p^p < \infty.$$

In other words, $f = \{f_n\}_0^\infty \in L^p(l^p) = l^p(L^p)$, and $\|f\|_{L^p(\nu)} = \|f\|_{l^p(L^p)}$.

Define a kernel g_α on $\mathbf{R}^N \times Y = \mathbf{R}^N \times \mathbf{R}^N \times \mathbf{Z}^+$ by

$$g_\alpha(x, y, n) = 2^{-n\alpha}\chi_n(x-y),$$

where $\chi_0 = \chi$ is a nonnegative, integrable, lower semicontinuous function, and $\chi_n(x) = 2^{nN}\chi(2^n x)$. We now denote

$$\mathcal{G}_\alpha f = \sum_0^\infty 2^{-n\alpha}\chi_n * f_n,$$

and, if $\mu \in \mathcal{M}^+(\mathbf{R}^N)$,

$$\check{\mathcal{G}}_\alpha \mu = \{2^{-n\alpha}\chi_n * \mu\}_0^\infty.$$

As in Definition 2 we define

$$\mathcal{C}_{\alpha,p}(E) = \inf\left\{\|f\|^p_{l^p(L^p)};\ f \geq 0,\ \mathcal{G}_\alpha f(x) \geq 1 \text{ on } E\right\},$$

and the dual characterization is

$$\mathcal{C}_{\alpha,p}(K)^{1/p} = \sup\left\{\frac{\mu(K)}{\|\check{\mathcal{G}}_\alpha \mu\|_{l^{p'}(L^{p'})}};\ \mu \in \mathcal{M}^+(K)\right\}.$$

It is quite remarkable that, at least under some restrictions on the function χ, the capacity we have just defined is equivalent both to the Bessel capacity $C_{\alpha,p}$, and to a capacity associated to the Besov space $B_\alpha^{p,p}$.

3. Maximal functions and potentials

The equivalence of $C_{\alpha,p}$ and $\mathcal{C}_{\alpha,p}$ is a consequence of a theorem of B. Muckenhoupt and R. Wheeden [20], estimating a fractional maximal function. The latter is defined for $\mu \in \mathcal{M}^+(\mathbf{R}^N)$ by

$$M_\alpha \mu(x) = \sup_{0<r\leq 1} \frac{\mu(B(x,r))}{r^{N-\alpha}}.$$

Then it is easy to see that for a suitable constant A

$$M_\alpha \mu(x) \leq A\, G_\alpha * \mu(x).$$

The estimate of Muckenhoupt and Wheeden goes in the other direction, and says that for any p, $1 < p < \infty$, there is a constant A such that

$$\|G_\alpha * \mu\|_p \leq A\|M_\alpha \mu\|_p.$$

This is, in fact, a slight extension of their theorem, since they considered Riesz, and not Bessel, potentials, and because of this allowed all $r > 0$ in the definition of M_α. On the other hand they proved their theorem in weighted L^p-spaces, but this case does not concern us here. (See however D. R. Adams [1].)

Now assume that $\chi(x)$ is a decreasing function of $|x|$, and that $\operatorname{supp}\chi \subset B(0,1)$. Then it is easy to see that

$$M_\alpha\mu(x) \approx \sup_{n\geq 0} 2^{-n\alpha}\chi_n * \mu(x) = \|\check{\mathcal{G}}_\alpha\mu(x)\|_{l^\infty},$$

so by the theorem of Muckenhoupt and Wheeden

$$\|M_\alpha\mu\|_p \approx \|\check{\mathcal{G}}_\alpha\mu\|_{L^p(l^\infty)}.$$

It is also easy to see that

$$G_\alpha * \mu(x) \geq A\sum_0^\infty 2^{-n\alpha}\chi_n * \mu(x) = \|\check{\mathcal{G}}_\alpha\mu(x)\|_{l^1},$$

and thus, by combining this with the elementary inequality

$$\|\cdot\|_{l^\infty} \leq \|\cdot\|_{l^q} \leq \|\cdot\|_{l^1}, \quad 1 < q < \infty,$$

we find that there are positive constants A_1, A_2, A_3 such that for $1 \leq q \leq \infty$

$$\|M_\alpha\mu\|_p \leq A_1\|\check{\mathcal{G}}_\alpha\mu\|_{L^p(l^q)} \leq A_2\|G_\alpha * \mu\|_p \leq A_3\|M_\alpha\mu\|_p.$$

The following theorem is now an easy consequence.

Theorem 3. *For any* $\alpha > 0$, $1 < p \leq N/\alpha$, *there is a constant* A *such that*

$$A^{-1}C_{\alpha,p}(E) \leq \mathcal{C}_{\alpha,p}(E) \leq A\,C_{\alpha,p}(E)$$

for all $E \in \mathbf{R}^N$.

Elementary estimates show that, except for a few terms in the beginning, $\mathcal{V}^\mu_{\alpha,p}(x)$ is comparable to another nonlinear function of μ,

$$W^\mu_{\alpha,p} = \sum_{n=0}^\infty \left(2^{n(N-\alpha p)}\mu(B_n(x))\right)^{p'-1}.$$

Theorem 4. (Wolff's inequality) *There are positive constants such that for any* $\mu \in \mathcal{M}^+(\mathbf{R}^N)$

$$\int W^\mu_{\alpha,p}\,d\mu \leq A_1\int \mathcal{V}^\mu_{\alpha,p}\,d\mu = A_1\|\check{\mathcal{G}}_\alpha\mu\|^{p'}_{L^{p'}(l^{p'})} \leq A_2\|G_\alpha * \mu\|^{p'}_{p'} \leq A_3\int W^\mu_{\alpha,p}\,d\mu.$$

This inequality is the reason why the statement previously denoted (B) gives a good definition for thinness. We state the definition formally.

Definition 3. A set $E \subset \mathbf{R}^N$ is (α,p)-thin at a, or $a \in e_{\alpha,p}(E)$, if

$$\sum_{n=0}^{\infty}\left(2^{n(N-\alpha p)}C_{\alpha,p}(E\cap(B_n(x))\right)^{p'-1}<\infty.$$

Before presenting the theory of thin sets we shall put the above in the context of Besov spaces.

4. Besov spaces and potentials

Let Φ belong to the Schwartz class $\mathcal{S}(\mathbf{R}^N)$, denote its Fourier transform by $\hat{\Phi}$, and assume that $\operatorname{supp}\hat{\Phi}\subset B(0,1)=B_0$ and $\hat{\Phi}(\xi)=1$ on $B(0,\frac{1}{2})=B_1$. Define $\{\varphi_n\}_0^\infty$ by setting $\varphi_0=\Phi$ and

$$\hat{\varphi}_n(\xi)=\hat{\Phi}(2^{-n}\xi)-\hat{\Phi}(2^{-n+1}\xi),\quad n=1,2,\ldots.$$

It follows that

$$\sum_{n=0}^{\infty}\hat{\varphi}_n(\xi)\equiv 1,$$

$$\operatorname{supp}\hat{\varphi}_n\subset B(0,2^n)\setminus B(0,2^{n-2}),\quad n=1,2,\ldots,$$

and

$$\varphi_n(x)=2^{(n-1)N}\varphi_1(2^{n-1}x),\quad n=2,3,\ldots.$$

Definition 4. Let $\alpha\in\mathbf{R}$, $1\le p\le\infty$, and $1\le q\le\infty$. The Besov space corresponding to these indices is

$$B_\alpha^{p,q}=B_\alpha^{p,q}(\mathbf{R}^N)=\left\{u\in\mathcal{S}';\ \textstyle\sum_0^\infty\|2^{n\alpha}\varphi_n*u\|_p^q<\infty\right\},\quad 1\le q<\infty,$$

$$B_\alpha^{p,\infty}=B_\alpha^{p,\infty}(\mathbf{R}^N)=\{u\in\mathcal{S}';\ \sup_n\|2^{n\alpha}\varphi_n*u\|_p<\infty\},$$

and

$$\|u\|_{B_\alpha^{p,q}}=\|\{2^{n\alpha}\varphi_n*u\}\|_{l^q(L^p)}.$$

It is well-known that the dual space $(B_\alpha^{p,q})^*$ is $B_{-\alpha}^{p',q'}$ for $1\le p<\infty$, $1\le q<\infty$.

The functions $\{\varphi_n\}_0^\infty$ in the definition can be replaced by much more general functions. The following important theorem is a special case of a theorem found in J. Peetre's book [21], Ch. 8, Theorems 1 and 2, and Remark 1.

Theorem 5. *Let $\chi \in \mathcal{S}(\mathbf{R}^N)$ and set $\chi_n(x) = 2^{nN}\chi(2^n x)$ for $n = 0, 1, 2, \ldots$. Let $\alpha > 0$, $1 \le p \le \infty$, and $1 \le q \le \infty$. Then there is a constant A such that for any $u \in B_{-\alpha}^{p,q}$*

$$\|\{2^{-n\alpha}\chi_n * u\}\|_{l^q(L^p)} = \left(\textstyle\sum_0^\infty \|2^{-n\alpha}\chi_n * u\|_p^q\right)^{1/q} \le A\,\|\chi\|_{B_\alpha^{1,1}}\|u\|_{B_{-\alpha}^{p,q}}, \quad 1 \le q < \infty,$$

and

$$\|\{2^{-n\alpha}\chi_n * u\}\|_{l^\infty(L^p)} = \sup_{n\ge 0}\|2^{-n\alpha}\chi_n * u\|_p \le A\,\|\chi\|_{B_\alpha^{1,1}}\|u\|_{B_{-\alpha}^{p,\infty}}.$$

If, in addition, $\hat{\chi}(0) \ne 0$, then

$$\|\{2^{-n\alpha}\chi_n * u\}\|_{l^q(L^p)}$$

is an equivalent norm on $B_{-\alpha}^{p,q}$.

Peetre gives the detailed proof only for the case $q = \infty$ (and for the "homogeneous Besov spaces"), and it seems difficult to find the details written down anywhere in the published literature. But it is not difficult to complete Peetre's proof along lines given elsewhere in his book, e. g. on p. 73. The case $p = q = 2$ is found already in L. Hörmander's book [14], Corollary 2.4.1. See also [3].

It was apparently first observed by D. R. Adams [2] that the above theorem implies that the functions in Besov spaces have a potential representation. The following theorem, which shows that these potentials are the $\mathcal{G}_\alpha f$ defined in section 2, is a slight modification of his result.

Theorem 6. *Let $\{\chi_n\}_0^\infty$ be as in Theorem 5, and assume that $\hat{\chi}(0) \ne 0$. Let $\alpha > 0$, $1 < p \le \infty$, and $1 < q \le \infty$. Then a function (or tempered distribution) F belongs to $B_\alpha^{p,q}$ if and only if there is a function sequence $f = \{f_n\}_0^\infty \in l^q(L^p)$ such that*

$$F = \mathcal{G}_\alpha f = \sum_{n=0}^{\infty} 2^{-n\alpha}\chi_n * f_n.$$

Moreover, there is a constant A such that

$$A^{-1}\|F\|_{B_\alpha^{p,q}} \le \|f\|_{l^q(L^p)} \le A\,\|F\|_{B_\alpha^{p,q}}.$$

Proof. First assume that $F = \sum_{n=0}^\infty 2^{-n\alpha}\chi_n * f_n$, $f = \{f_n\}_0^\infty \in l^q(L^p)$. Let $u \in B_{-\alpha}^{p',q'}$. Then by the first part of Theorem 5

$$\begin{aligned}|\langle u, F\rangle| &\le \sum |\langle u, 2^{-n\alpha}\chi_n * f_n\rangle| = \sum |\langle 2^{-n\alpha}\chi_n * u, f_n\rangle| \le \sum \|2^{-n\alpha}\chi_n * u\|_{p'}\|f_n\|_p \\ &\le \|\{2^{-n\alpha}\chi_n * u\}\|_{l^{q'}(L^{p'})}\|\{f_n\}\|_{l^q(L^p)} \le A\|\chi\|_{B_\alpha^{1,1}}\|u\|_{B_{-\alpha}^{p',q'}}\|\{f_n\}\|_{l^q(L^p)}.\end{aligned}$$

It follows that $f \in (B_{-\alpha}^{p',q'})^* = B_\alpha^{p,q}$.

To prove the converse we let $F \in B_\alpha^{p,q}$. We have

$$F = \sum_0^\infty \varphi_n * F.$$

It is no loss of generality to assume that $\hat{\chi}(\xi) \neq 0$ on $\operatorname{supp} \hat{\Phi}$. It follows that for a constant $A > 0$

$$A^{-1} \leq \hat{\chi}_n(\xi) \leq A \text{ on } \operatorname{supp} \hat{\varphi}_n.$$

By Wiener's theorem there are functions $\psi_n \in L^1$, $n = 0, 1, 2, \ldots$, such that $\hat{\psi}_n \hat{\chi}_n = 1$ on $\operatorname{supp} \hat{\varphi}_n$, and $\|\psi_n\|_1 = \|\psi_1\|_1 \leq A$, $n = 2, 3, \ldots$. (See e. g. Y. Katznelson [15], VIII.6.3, p. 228. Cf. Peetre [21], p. 158, and p. 175, where the use of Wiener's theorem in this context is attributed to H. S. Shapiro.)

Define f_n by

$$\hat{f}_n = 2^{n\alpha} \hat{\psi}_n \hat{\varphi}_n \hat{F}.$$

Then

$$\hat{F} = \sum_0^\infty \hat{\varphi}_n \hat{F} = \sum_0^\infty \hat{\chi}_n \hat{\psi}_n \hat{\varphi}_n \hat{F} = \sum_0^\infty 2^{-n\alpha} \hat{\chi}_n \hat{f}_n,$$

and thus

$$F = \sum_0^\infty 2^{-n\alpha} \chi_n * f_n.$$

Moreover

$$\|f_n\|_p \leq 2^{n\alpha} \|\psi_n\|_1 \|\varphi_n * F\|_p \leq A\, 2^{n\alpha} \|\varphi_n * F\|_p,$$

which proves that $f = \{f_n\}_0^\infty \in l^q(L^p)$ and

$$\|f\|_{l^q(L^p)} \leq A\, \|F\|_{B_\alpha^{p,q}}.$$

5. The (α, p)-fine topology

We shall now justify the claim made in the introduction by reviewing a few more or less recent results. The first important point is that the potentials $\mathcal{V}_{\alpha,p}^\mu$ give us a very good substitute for the statement (A). The following theorem can be proved much as in [12], using easy estimates for $\mathcal{V}_{\alpha,p}^\mu$. See also [3].

Theorem 7. *Let $E \subset \mathbf{R}^N$. Then $a \in e_{\alpha,p}(E)$ if and only if there exists $\mu \in \mathcal{M}^+$ such that*

$$\mathcal{V}_{\alpha,p}^\mu(a) < \liminf_{x \to a,\, x \in E \setminus \{a\}} \mathcal{V}_{\alpha,p}^\mu(x).$$

Then let $\{B_n\}_1^\infty$ be an enumeration of the balls with rational centers and radii. Set U_n be the subset of $\bar{E} \cap B_n$ where the capacitary potential $\mathcal{V}^{\mu}_{\alpha,p}$ for $E \cap B_n$ is < 1. The following theorem and its corollaries extend classical results.

Theorem 8. [12] *Let $E \subset \mathbf{R}^N$. Then $e_{\alpha,p}(E) \cap \bar{E} = \bigcup_1^\infty U_n$.*

Corollary 1. (Kellogg property, [12]) *Let $E \subset \mathbf{R}^N$. Then $C_{\alpha,p}(e_{\alpha,p}(E) \cap E) = 0$.*

Corollary 2. (Choquet property, [12]) *Let $E \subset \mathbf{R}^N$. For any $\varepsilon > 0$ there is an open G such that $e_{\alpha,p}(E) \subset G$ and $C_{\alpha,p}(G \cap E) < \varepsilon$.*

Definition 5. A set U is an *(α, p)-fine neighborhood* of x if $x \in U$ and $x \in e_{\alpha,p}(U^c)$. The associated topology is called the *(α, p)-fine topology* , and denoted $\tau_{\alpha,p}$.

Definition 6. A function f, defined (α, p)-quasieverywhere, is said to be *(α, p)-quasicontinuous* if for every $\varepsilon > 0$ there is an open set G such that $C_{\alpha,p}(G) < \varepsilon$, and the restriction of f to the complement G^c is continuous on G^c in the induced topology.

There is also a corresponding "(α, p)-quasitopology", but we omit its definition. It is a consequence of the above results that the assumptions in the abstract theory of Fuglede [8], [9] are satisfied. The following theorem is a consequence.

Theorem 9. ([12]) *A function f, defined (α, p)-quasieverywhere, is (α, p)-finely continuous (α, p)-quasieverywhere if and only if it is (α, p)-quasicontinuous.*

The classical fine and quasi topologies have a number of connectedness properties, which also extend to the general situation, sometimes with quite difficult proofs.

Theorem 10. (D. R. Adams and J. L. Lewis [4]) *Let $G \subset \mathbf{R}^N$ be open and connected. Then G is (α, p)-quasiconnected and (α, p)-finely connected if $\alpha p > 1$.*

Theorem 11. (D. R. Adams and J. L. Lewis [4]) *If $G \subset \mathbf{R}^N$ is (α, p)-finely open and (α, p)-finely connected, then G is arcwise connected, if $\alpha p > 1$.*

Theorem 12. (J. Heinonen, T. Kilpeläinen and J. Malý [13]) *If $G \subset \mathbf{R}^N$ is open and connected, then G is locally (α, p)-finely connected if $\alpha p \geq 1$.*

These connectedness properties are false for $\alpha p \leq 1$, since in this case a hyperplane has zero capacity.

6. Problems

We finally indicate a few problems, closely related to the material presented here.

Problem 1. Let f belong to a Sobolev space $W^{m,p}(\mathbf{R}^N)$, where m is a positive integer, and $1 < p < \infty$. This is equivalent to saying that f can be represented as a Bessel potential, $f = G_m * g$ with $g \in L^p$. We can assume that f and its derivatives of order k, $k \leq m-1$, are defined $(m-k,p)$-q.e., and in this sense we define a trace operator for any closed set E by

$$\mathrm{Tr}|_E f = \{D^\beta f|_E\}_{|\beta|\leq m-1}.$$

It was proved in [11] and [12], using nonlinear potential theory, that the kernel of this operator is the closure of the test functions with support away from E, i. e.

$$\mathrm{Ker}\ \mathrm{Tr}|_E = W_0^{m,p}(E^c).$$

However, the proof is very long, and it would be desirable to have a shorter one.

Problem 2. In the proof of the above result f is approximated by functions ωf, where $0 \leq \omega \leq 1$, and $\mathrm{supp}\,\omega \cap E = \emptyset$, but in the general case ω depends on the function f itself, and not only on E. The problem is to find a sequence of linear operators $\{P_n\}_1^\infty$

$$P_n : \mathrm{Ker}\ \mathrm{Tr}|_E \longrightarrow C_0^\infty(E^c)$$

such that $P_n f$ approximate f.

Problem 3. To construct an axiomatic nonlinear potential theory that contains the theory presented here. The existing axiomatic theories seem to be modeled on the case $\alpha = 1$, which corresponds to second order partial differential equations.

References

[1] Adams, D. R., Weighted nonlinear potential theory. Trans. Amer. Math. Soc. 297 (1986), 73–94

[2] —, The classification problem for the capacities associated with the Besov and Triebel-Lizorkin spaces. In: Approximation and Function Spaces, Banach Center Publications 22, 9–24. PWN Polish Scientific Publishers, Warsaw 1989

[3] Adams, D. R., Hedberg, L. I., Book. In preparation

[4] Adams, D. R., Lewis, J. L., Fine and quasi connectedness in nonlinear potential theory. Ann. Inst. Fourier (Grenoble) 35:1 (1985), 57–73

[5] Adams, D. R., Meyers, N. G., Thinness and Wiener criteria for non-linear potentials. Indiana Univ. Math. J. 22 (1972), 169–197

[6] Calderón, A. P., Lebesgue spaces of differentiable functions and distributions. In: Proc. Sympos. Pure Math. 4, 33–49. Amer. Math. Soc., Providence, R.I. 1961

[7] Fuglede, B., Applications du théorème minimax à l'étude de diverses capacités. C. R. Acad. Sci. Paris Sér. A-B 266 (1968), 921–923

[8] —, The quasi topology associated with a countably subadditive set function. Ann. Inst. Fourier (Grenoble) 21:1 (1971), 123–169

[9] —, Connexion en topologie fine et balayage des mesures, Ann. Inst. Fourier (Grenoble) 21:3 (1971), 227–244
[10] Hedberg, L. I., Non-linear potentials and approximation in the mean by analytic functions. Math. Z. 129 (1972), 299–319
[11] Hedberg, L. I., Spectral synthesis in Sobolev spaces, and uniqueness of solutions of the Dirichlet problem. Acta Math. 147 (1981), 237–264
[12] Hedberg, L. I., Wolff, Th. H., Thin sets in nonlinear potential theory. Ann. Inst. Fourier (Grenoble) 33:4 (1983), 161–187
[13] Heinonen, J., Kilpeläinen, T., Malý, J., Connectedness in fine topologies. Ann. Acad. Sci. Fenn. Ser. A. I. Math. 15 (1990), 107–123
[14] Hörmander, L., Linear Partial Differential Operators. Springer-Verlag, Berlin-Heidelberg 1963
[15] Katznelson, Y., An Introduction to Harmonic Analysis. Wiley, New York 1968
[16] Maz'ya, V. G., On the continuity at a boundary point of solutions of quasilinear equations (Russian). Vestnik Leningrad. Univ. 25:13 (1970), 42–55
[17] Maz'ya, V. G., Khavin V. P., Nonlinear potential theory (Russian). Uspekhi Mat. Nauk 27:6 (1972), 67–138
[18] Meyers, N. G., A theory of capacities for potentials of functions in Lebesgue classes. Math. Scand. 26 (1970), 255–292
[19] —, Continuity properties of potentials. Duke Math. J. 42 (1975), 157–166
[20] Muckenhoupt, B., Wheeden, R. L., Weighted norm inequalities for fractional integrals. Trans. Amer. Math. Soc. 192 (1974), 261–274
[21] Peetre, J., New Thoughts on Besov Spaces. Mathematics Department, Duke University, Durham, North Carolina 1976
[22] Reshetnyak, Yu. G., On the concept of capacity in the theory of functions with generalized derivatives (Russian). Sibirsk. Mat. Zh. 10 (1969), 1109–1138
[23] Stein, E. M., Singular Integrals and Differentiability of Functions. Princeton University Press, Princeton, N.J. 1970

Potential theory and quasiconformal mappings

Olli Martio

Abstract

$\mathcal{A}$-harmonic functions and quasiregular mappings are generalizations of harmonic and analytic functions, respectively, to higher dimensional euclidean spaces. Also the basic tools of the Potential Theory, harmonic measure, capacity and fine topology, have their counterparts in this non-linear situation. It is shown how these tools are used to prove results in the theory of quasiregular mappings.

1980 Mathematics Subject Classification(1985 Revision): 31C99, 30C60, 31A15.

1. Introduction

B. Riemann based the theory of analytic functions on the Potential Theory and since then the subject has grown enormously; M. Tsuji's substantial book [T] is a well-known proof. Quasiconformal mappings were introduced by C. B. Morrey in 1938 to study second order partial differential equations in the plane. This connection remained unnoticed almost twenty years. However, the planar case is now rather well understood [BJS], [V]. The study of non-homeomorphic quasiconformal mappings, called quasiregular mappings, in higher dimensional euclidean spaces began in the late 1960's. The classical relation between harmonic and analytic functions was extended by Yu. G. Resetnyak [R] to quasiregular mappings and solutions of certain partial differential equations. In general, these equations are non-linear. It was then realized, cf. [GLM1], [HKM2], that these equations support a potential theory similar to that of harmonic functions. The purpose of this survey is to consider recent proceedings in this area.

2. Equations

We consider mappings $\mathcal{A}\colon \mathbb{R}^n \times \mathbb{R}^n \to \mathbb{R}^n$ which satisfy the following assumptions for some $p > 1$ and $0 < \alpha \leq \beta$:

$$\begin{aligned} &x \mapsto \mathcal{A}(x,h) \quad \text{is measurable for all } h \in \mathbb{R}^n \quad \text{and} \\ &h \mapsto \mathcal{A}(x,h) \quad \text{is continuous for a.e. } x, \end{aligned} \tag{a}$$

Potential Theory, ed. by M. Kishi

for all $h \in \mathbb{R}^n$ and a.e. $x \in \mathbb{R}^n$

$$\mathcal{A}(x,h)\cdot h \geq \alpha\,|h|^p, \tag{b}$$
$$|\mathcal{A}(x,h)| \leq \beta\,|h|^{p-1}, \tag{c}$$
$$(\mathcal{A}(x,h_1)-\mathcal{A}(x,h_2))\cdot(h_1-h_2) > 0, \quad h_1 \neq h_2, \tag{d}$$

and

$$\mathcal{A}(x,\lambda\,h) \equiv |\lambda|^{p-2}\,\lambda\,\mathcal{A}(x,h); \quad \lambda \in \mathbb{R}. \tag{e}$$

Let Ω be an open set in $\mathbb{R}^n$. A function $u \in C(\Omega)\cap \operatorname{loc} W_p^1(\Omega)$ is a solution of the equation

$$\nabla\cdot\mathcal{A}(x,\nabla u(x)) = 0 \tag{2.1}$$

induced by the operator $\mathcal{A}$ if

$$\int_\Omega \mathcal{A}(x,\nabla u(x))\cdot\nabla\varphi(x)\,dm(x) = 0 \tag{2.2}$$

for all $\varphi \in C_0^\infty(\Omega)$. Here $\operatorname{loc} W_p^1(\Omega)$ is the Sobolev space whose functions are locally p-integrable and have first locally p-integrable partial distributional derivatives; $\nabla u(x) = (\partial_1 u(x),\dots,\partial_n u(x))$ is the gradient of u. Solutions u of (2.1) are called *$\mathcal{A}$-harmonic.*

Conditions (a)–(e) mean that the equation (2.1) is a second order quasi-linear elliptic partial differential equation which is, in general, degenerate and homogeneous in ∇u.

2.3. Example. The mapping $\mathcal{A}(x,h) = |h|^{p-2}\,h$ generates the p-harmonic equation

$$\nabla\cdot(|\nabla u|^{p-2}\,\nabla u) = 0;$$

$\mathcal{A}$ satisfies the assumptions (a)–(e). For $p = 2$ we obtain the classical Laplace equation $\Delta u = 0$.

A lower semicontinuous function $v\colon \Omega \to \mathbb{R}\cup\{\infty\}$ is *$\mathcal{A}$-superharmonic* if it satisfies the comparison principle: For each domain $G \Subset \Omega$ and each $u \in C(\overline{G})$, $\mathcal{A}$-harmonic in G, the condition $u \leq v$ in ∂G implies $u \leq v$ in G.

$\mathcal{A}$-harmonic and -superharmonic functions form a potential theory similar to the classical potential theory of ordinary harmonic functions, see [HKM2]. The features of (2.1) which make the theory work are:

(i) The unique solvability of the Dirichlet problem in smooth domains D, and
(ii) if $u_1,\ u_2 \in C(\overline{D})$ are $\mathcal{A}$-harmonic in D and if $u_1 \leq u_2$ in ∂D, then $u_1 \leq u_2$ in D.

In general, the theory is non-linear. In this article we are mainly interested in the case $p = n$; this is the conformally invariant situation. Note that for $n \geq 3$ all the equations $\nabla \cdot \mathcal{A}(x, \nabla u) = 0$ for $p = n$ are non-linear and even for $p = n = 2$ there are non-linear equations where $\mathcal{A}$ satisfies (a)–(e).

3. Quasiregular mappings

Let G be a domain in $\mathbb{R}^n$. A continuous mapping $f\colon G \to \mathbb{R}^n$ is said to be K-*quasiregular*, abbreviated QR, if the coordinate functions of f belong to $\operatorname{loc} W_n^1(G)$ and if

$$|f'(x)|^n \le K\, J(x, f) \tag{3.1}$$

a.e. in G. Here $|f'(x)|$ denotes the supremum norm of the (formal) derivative of f, i.e.

$$|f'(x)| = \sup_{|h|=1} |f'(x)\, h|,$$

and $J(x, f)$ is the jacobian determinant of f at x. A homeomorphic QR mapping is called *quasiconformal* (QC).

3.2. Remarks. (a) For $n = 2$ a mapping f is $1 - QR$ if and only if f is analytic. A $1 - QC$ mapping is conformal. A planar QR mapping f can always be represented in the form $f = g \circ h$ where g is analytic and h is QC. Such a representation is not possible in $\mathbb{R}^n$, $n \ge 3$.

(b) For $n \ge 3$ a mapping $f\colon G \to \mathbb{R}^n$ is $1 - QR$ if and only if f is either constant or the restriction of a Möbius transformation of $\mathbb{R}^n \cup \{\infty\}$ to G. For this deep result see [R].

(c) Although the class of $1 - QR$ mappings is rather narrow in $\mathbb{R}^n$, $n \ge 3$, many interesting examples of QR mappings are now known. The simplest non-injective QR mapping $f\colon \mathbb{R}^3 \to \mathbb{R}^3$ is $(r, \varphi, z) \mapsto (r, 2\varphi, z)$ in cylinder coordinates of $\mathbb{R}^3$; f is $4 - QR$.

(d) For the theory of QR mappings see [R], [MRV1], and [Ri2].

4. Equations and quasiregular mappings

Suppose that $\mathcal{A}\colon \mathbb{R}^n \times \mathbb{R}^n \to \mathbb{R}^n$ is a mapping satisfying (a)–(e) for $p = n$. Let $f\colon G \to \mathbb{R}^n$ be QR. Define the *pull back* $f^{\#}\mathcal{A}$ of $\mathcal{A}$ as follows

$$f^{\#}\mathcal{A}(x, h) = J(x, f)\, f'(x)^{-1}\, \mathcal{A}(f(x), f'(x)^{-1^*}\, h)$$

if $J(x, f) \ne 0$ and $f^{\#}\mathcal{A}(x, h) = \mathcal{A}(x, h)$ if $J(x, f) = 0$ or $f'(x)$ does not exist or $x \notin D$. Here $f'(x)^{-1}$ denotes the inverse of the linear mapping $f'(x)$ and $f'(x)^{-1^*}$ is the transpose of $f'(x)^{-1}$. Then $f^{\#}\mathcal{A}\colon \mathbb{R}^n \times \mathbb{R}^n \to \mathbb{R}^n$ and the proof of the next lemma is not difficult.

4.1. Lemma. [MVa], [HKM2] *The mapping $f^{\#}\mathcal{A}$ satisfies the assumptions (a)–(e) for $p = n$.*

4.2. Remark. If $f\colon G \to \mathbb{R}^2$ is analytic or if $f\colon G \to \mathbb{R}^n$ is a Möbius transformation, then $f^{\#}\mathcal{A} = \mathcal{A}$ for the n-harmonic operator

$$\mathcal{A}(x,h) = |h|^{n-2}\,h.$$

The next theorem states that QR mappings are harmonic morphisms in this framework.

4.3. Theorem. *Let $\mathcal{A}$, f, and $f^{\#}\mathcal{A}$ be as above. If u is $\mathcal{A}$-(super)harmonic in Ω, then $u \circ f$ is $f^{\#}\mathcal{A}$-(super)harmonic in $f^{-1}(\Omega)$.*

Theorem 4.3 is due to Resetnyak [R] in the $\mathcal{A}$-harmonic case under slightly more restrictive conditions for $\mathcal{A}$ than (a)–(e). For the superharmonic case see [GLM1] and [HKM2]. At present there is no simple method to prove the invariance of harmonic functions from the definition of QR mappings.

4.4. Remarks. (a) Consider the n-harmonic operator $\mathcal{A}(x,h) = |h|^{n-2}\,h$. It is easy to see that the coordinate functions $u(x) = x_i$, $i = 1,\dots,n$, are $\mathcal{A}$-harmonic. Thus Theorem 4.3 says that the coordinate functions of f are $f^{\#}\mathcal{A}$-harmonic. This is a generalization of the well-known fact that the coordinate functions of an analytic function are harmonic.

(b) For $n \geq 3$ it is not possible, in general, for a given $\mathcal{A}_1$-harmonic function u_1 to construct $\mathcal{A}_i$-harmonic functions u_i, $i = 2,\dots,n$, such that $f = (u_1,\dots,u_n)$ is QR. To construct such an operator $\mathcal{A}_1$ and an $\mathcal{A}_1$-harmonic function u_1 note that there is an operator $\mathcal{A}_1$ without the unique continuation property, see [M2]. More precisely, there is a non-constant $\mathcal{A}_1$-harmonic function $u_1\colon \mathbb{R}^n \to \mathbb{R}$, $n \geq 3$, such that $u_1 \equiv 0$ in an open non-empty set. Then u_1 cannot be a coordinate function of a QR mapping since a non-constant QR mapping is an open mapping. For $n = 2$ the situation is different because the Beltrami system always has a QC solution.

(c) For the converse of Theorem 4.3 see [HKM3].

5. $\mathcal{A}$-harmonic measures

Suppose that Ω is an open set, $E \subset \partial\Omega$, and $\mathcal{A}\colon \mathbb{R}^n \times \mathbb{R}^n \to \mathbb{R}^n$ satisfies (a)–(d) for $p > 1$. The $\mathcal{A}$-harmonic measure of E with respect to Ω at $x \in \Omega$ is defined as

$$\omega(x) = \omega(E,\Omega;\mathcal{A})(x) = \inf\{v(x)\} \tag{5.1}$$

where the infimum is taken over all non-negative $\mathcal{A}$-superharmonic functions v in Ω such that

$$\liminf_{y\to z} v(y) \geq 1$$

for all $z \in E$. Such a function v is called admissible for ω.

5.2. Remarks. (a) If $\mathcal{A}(x,h) = h$ is the ordinary Laplace operator, then ω is the classical (outer) harmonic measure of E with respect to Ω. Definition (5.1) is completely analogous to the classical case.

(b) It can be shown that ω is $\mathcal{A}$-harmonic in Ω and that $0 \le \omega \le 1$, see [GLM2], [HKM2]. In general, for a fixed $x \in G$, $E \mapsto \omega(E,G;\mathcal{A})(x)$ does not define a measure on the Borel subsets E of $\partial\Omega$; this is due to the non-linearity of the equation $\nabla \cdot \mathcal{A} = 0$.

With a proper interpretation the $\mathcal{A}$-harmonic measure is increased under QR mappings. For this fix a mapping $\mathcal{A}$ with $p = n$ and let $f\colon G \to \mathbb{R}^n$ be QR and non-constant. For $E \subset \partial G$ let $C(E,f)$ denote the cluster set of E under f, i.e. $y \in C(E,f)$ if there is a sequence of points $x_i \in G$ such that $x_i \to E$ and $f(x_i) \to y$. The cluster set $C(E,f)$ may contain ∞; the definition of $\omega(E,\Omega;\mathcal{A})$ can easily be extended to the case where E contains ∞.

The next theorem is a very general form of the principle of the harmonic measure for QR mappings. Note that since f is non-constant, $f(G)$ is open and that $C(E,f)$ is always closed.

5.3. Theorem. *Under the above assumptions*

$$\omega(E,G;f^{\#}\mathcal{A})(x) \le \omega(C(E,f), f(G) \setminus C(E,f);\mathcal{A})(f(x)) \tag{5.4}$$

whenever $x \in G$ and $f(x) \notin C(E,f)$.

For the proof note that $C(E,f)$ need not be a subset of the boundary of $\Omega = f(G) \setminus C(E,f)$ – the right hand side is actually $\omega^* = \omega(E',\Omega;\mathcal{A})$ where $E' = C(E,f) \cap \partial\Omega$. We will use this obvious abbreviation. Write $\omega = \omega(E,G;f^{\#}\mathcal{A})$. To show that $\omega(x) \le \omega^*(f(x))$ let u be an $\mathcal{A}$-superharmonic function admissible for ω^*. By Theorem 4.3, $u \circ f$ is admissible for $\omega_1 = \omega(E^*, f^{-1}(\Omega); f^{\#}\mathcal{A})$ where $E^* = E \cup f^{-1}(C(E,f))$. Now for each $\varepsilon > 0$, $v_\varepsilon = \min(1, u \circ f + \varepsilon)$ is $f^{\#}\mathcal{A}$-superharmonic in G and admissible for ω. Thus $\omega(x) \le v_\varepsilon(x) \le u(f(x)) + \varepsilon$ for each $x \in G$ with $f(x) \in \Omega$ and since u and ε were arbitrary, we obtain $\omega(x) \le \omega^*(f(x))$ as required.

Suppose that f is QC. Then $f^{-1\#} f^{\#}\mathcal{A} = \mathcal{A}$ and hence Theorem 5.3 yields

5.5. Corollary. *Suppose that $f\colon \overline{G} \to \overline{G}'$ is a homeomorphism and that $f|G$ is QC. Then*

$$\omega(E,G;f^{\#}\mathcal{A})(x) = \omega(f(E), f(G);\mathcal{A})(f(x)) \tag{5.6}$$

for every $E \subset \partial G$ and for every $x \in G$.

For $n = 2$ and $\mathcal{A}(x,h) = h$ it is a famous result due to Beurling and Ahlfors that there is no relation between the left and the right hand sides of (5.6) if $f^{\#}\mathcal{A}$ is replaced by $\mathcal{A}$, see also [CFK]. The corresponding question is open in $\mathbb{R}^n$, $n \ge 3$.

Since an $\mathcal{A}$-harmonic measure is not a measure in general, $\mathcal{A}$-harmonic measures ω have been mostly used in the situations where ω can be estimated in geometric terms of E. Applications to QR mappings include

(a) two constant theorem [Ri1], [GLM2],
(b) Phragmen–Lindelöf type results [GLM3],
(c) Lindelöf's theorem on asymptotic limits [GLM3], [Ri1],
(d) Milloux's problem [M1], and
(e) Øksendal's theorem for QC mappings [HM].

An interesting problem, also connected to the aforementioned applications, is to determine sets of $\mathcal{A}$-harmonic measure zero. In general, the class of sets $E \subset \partial\Omega$ such that $\omega(E,\Omega;\mathcal{A}) = 0$ depends on $\mathcal{A}$. However, there is a stable subclass: A set $E \subset \partial\Omega$ is said to be of *total harmonic measure zero* if $\omega(E,\Omega;\mathcal{A}) = 0$ for all $\mathcal{A}$. Note that we have not restricted the value of $p > 1$. There is abundance sets of total harmonic measure zero. For simplicity consider the unit ball B in $\mathbb{R}^n$. Let $E \subset \partial B$ and $r > 0$. Let E_r denote the r-inflation of E, i.e.

$$E_r = \bigcup_{x \in E} B(x,r),$$

and write $\delta(r) = \sup_{y\in E} \operatorname{dist}(y, \partial B \setminus E_r)$.

5.7. Theorem. [M5] *If* $\liminf_{r\to 0} \delta(r)/r < \infty$, *then E is of total harmonic measure zero.*

Theorem 5.7 implies, for example, that a smooth curve on ∂B in $\mathbb{R}^n$, $n \geq 3$, is of total harmonic measure zero. For $n = 2$ it has always positive $\mathcal{A}$-harmonic measure. For this reason the boundary behavior of QR mappings in B is quite different for $n = 2$ and for $n \geq 3$. A typical example is the application (c) above. Let γ be a curve in the unit ball B of $\mathbb{R}^n$ with one endpoint b in ∂B. For $n = 2$ the classical theorem of Lindelöf says that if $f\colon B \to \mathbb{R}^2$ is a bounded analytic function and if $f(x)$ tends to a limit as x tends to b along γ, then f has the same limit in every Stolz angle at b. This holds for plane QR mappings as well. Now for $n \geq 3$ this is not true, see [Ri1]. However, there is a version of this result in $\mathbb{R}^n$, $n \geq 2$. Let L be the x_1-axis in $\mathbb{R}^n$ and let γ be a curve in $B \setminus L$ ending at 0. If $f\colon B \setminus L \to \mathbb{R}^n$ is a bounded QR mapping and if $f(x)$ tends to a limit as x tends to 0 along γ, then $f(x)$ tends to the same limit in each Stolz "cone" at 0 in $B \setminus L$; for $n = 2$, $B \setminus L$ is disconnected and the Stolz angle must lie in the same halfplane as γ. The reason for this result is that L and γ are of the same size from the $\mathcal{A}$-harmonic measure theoretic point of view, see [GLM3].

Note also that sets of total harmonic measure zero cannot be characterized in terms of the Hausdorff dimension. For $n = 2$ there are sets E on the boundary of the unit disk B whose Hausdorff dimension is arbitrary near 0 but E is not of total harmonic measure zero, see [Tu], [M4]. On the other hand it follows from Theorem 5.7 that $\frac{1}{3}$-Cantor set constructed on ∂B is of total harmonic measure zero.

6. Fine limits

Let $\mathcal{A}$ be an operator as before with $p > 1$. The coarsest topology making all $\mathcal{A}$-superharmonic functions in $\mathbb{R}^n$ continuous is the $\mathcal{A}$-fine topology $\tau_{\mathcal{A}}$.

6.1. Theorem. [HKM1] *$\tau_{\mathcal{A}}$ depends only on p not on $\mathcal{A}$.*

Thus we can write $\tau_p = \tau_{\mathcal{A}}$ and it turns out, see [HKM1], that τ_p is the well-known $(1,p)$-fine topology studied by N. G. Meyers, D. R. Adams, L.-I. Hedberg, and others. Theorem 6.1 shows that this topology also has an interpretation similar to the classical fine topology τ_2 introduced by H. Cartan in 1940. The case $p = n$ is of interest for QR mappings.

The classical fine topology has been used to study limits of analytic functions. The following result is due to B. Fuglede [F]: Suppose that $f: G \to \mathbb{R}^2$ is analytic and omits a set of positive 2-capacity. If $x_0 \in \partial G$ and if $G \cup \{x_o\}$ is a τ_2-fine neighborhood of x_0, then f has a τ_2-fine limit at x_0. This result has a QR counterpart:

6.2. Theorem. [HKM1] *Suppose that $f: G \to \mathbb{R}^n$ is QR and omits a set of positive n-capacity. If $G \cup \{x_0\}$ is a τ_n-fine neighborhood of x_0, then f has a τ_n-fine limit at x_0.*

The proof for Theorem 6.2 employs the basic capacity inequality $\operatorname{cap}_n fE \leq K^{n-1} \operatorname{cap}_n E$ for $K - QR$ mappings f.

Theorem 6.2 has a version in the usual topology: There exists an open set $\mathcal{U} \subset G$ such that $\mathcal{U} \cup \{x_0\}$ is a τ_n-fine neighborhood of x_0 and $f(x)$ tends to a limit as $x \to x_0$ in $\mathcal{U}$. This is a typical situation in the fine topology. Note that in Theorem 6.2, f need not have a limit as x tends to x_0 in G.

If f is conformal or QC, then much more can be said about the behavior of f at x_0 whenever $G \cup \{x_0\}$ is a τ_n-fine neighborhood of x_0, see [MS]. In particular, f always has a limit y_0 at x_0 and $f(G) \cup \{y_0\}$ is a τ_n-fine neighborhood of y_0.

7. Removability results

Suppose that C is a relatively closed subset of a plane domain G. The removability theorem of Painleve and Besicovitch says that if $\mathcal{H}^1(C) = 0$ and if $f: G \setminus C \to \mathbb{R}^2$ is a bounded analytic function, then f extends to an analytic function of G. For QR mappings $f: G \setminus C \to \mathbb{R}^n$, $n \geq 2$, removability questions seem to be more problematic. Since QC mappings do not preserve the Hausdorff dimension of a set, removability results can be divided into two categories:

(a) The size of C is measured in terms of K, for example using the Hausdorff dimension.

(b) The set C is removable for all QR mappings regardless of K.

Clearly, letting $K \to \infty$ we should derive results in (b) from results in (a). However, there are a number of results which are independent of K.

For a long time the following result was the only known; it belongs to (b).

7.1. Theorem. [MRV2] *Suppose that* $f: G \backslash C \to \mathbb{R}^n$ *is a bounded* $K-QR$ *mapping and that* $\mathrm{cap}_n(C) = 0$. *Then* f *has a* $K - QR$ *extension to* G.

The proof for this result is potential theoretic indeed: Consider the n-harmonic equation $\nabla \cdot \mathcal{A} = 0$ where $\mathcal{A}(x, h) = |h|^{n-2} h$. Since the coordinate functions $u(x) = x_i$ are $\mathcal{A}$-harmonic, the coordinate functions f_i, $i = 1, \ldots, n$, of f are $f^{\#}\mathcal{A}$-harmonic, see Theorem 4.3. Now $f_i: G \setminus C \to \mathbb{R}$ is a bounded $f^{\#}\mathcal{A}$-harmonic function and $\mathrm{cap}_n(C) = 0$. Hence f_i can be extended to G as a $f^{\#}\mathcal{A}$-harmonic function; this is a basic removability theorem in the non-linear potential theory, see [HKM2]. Since $\mathrm{cap}_n(C) = 0$, the set C is of measure zero and hence the inequality $|f'(x)|^n \leq K\, J(x, f)$ holds a.e. in G as well. Thus the extended f is $K - QR$ in G.

The boundedness of f in Theorem 7.1 can be replaced by the condition that f omits the set of positive n-capacity – in this case the extended mapping may take the value ∞ and quasimeromorphic mappings must be considered. These mappings generalize the concept of a QR mapping in the same way as meromorphic mappings generalize analytic functions, see [MRV2]. This result is usually proved through function theoretic means – capacity or modulus inequalities for QR play an important role.

In 1990 several new results were obtained on the removability in higher dimensions. These seem to provide a major step forward. In *even* dimensional spaces T. Iwaniec and G. Martin [IM] show that sets C whose Hausdorff dimension is less than a positive number (which depends on K and n) are removable for bounded $K - QR$ mappings. The proof uses Donaldson's ideas and Caccioppoli's estimates (or standard estimates) for a lower exponent p than the natural exponent $p = n$. Thus the proof seems to open a new potential theoretic approach to removability results in category (a).

T. Järvi and M. Vuorinen [JV] show that certain Cantor sets C with $\dim_{\mathcal{H}} C \leq q(n, K) > 0$ are removable for $K - QR$ mappings which need not be bounded. The proof employs Schottky-type estimates due to S. Rickman [Ri]. In [KM] P. Koskela and Martio provide an analytic condition for a set C such that C is removable for all QR mappings which omit a set of positive n-capacity. The set C is quite thin, e.g. $\dim_{\mathcal{H}} C = 0$, but could still be of positive n-capacity. The result belongs to category (b). The proof is function theoretic.

Note that in the planar case it is possible to provide quite accurate removability results for QR mappings f due to the representation $f = g \circ h$ where g is analytic and h is a QC mapping. For details see [KM].

Removability results for locally Hölder continuous mappings $f: G \to \mathbb{R}^n$ which are QR in $G \backslash C$ are considered in [KM]; f is locally Hölder continuous with exponent $\alpha \in (0, 1]$ if each $x_0 \in G$ has a neighborhood $\mathcal{U}$ and a constant $M < \infty$ such that

$$|f(x) - f(y)| \leq M\, |x - y|^{\alpha}$$

for all $x, y \in \mathcal{U}$. If f is analytic in $G \backslash C$ and locally Hölder continuous with exponent α in G, then the removability result of L. Carleson says that C is removable provided that $\mathcal{H}^{1+\alpha}(C) = 0$. In [KM] it is shown that if $\dim_{\mathcal{M}}(C) \leq \lambda < n - 1$ and if

$f: G \to \mathbb{R}^n$ is locally Hölder continuous with exponent $\alpha > \lambda/n$ and $K - QR$ in $G \setminus C$, then f is $K - QR$ in G. Here $\dim_{\mathcal{M}}(C)$ refers to the Minkowski dimension of C, i.e. $\dim_{\mathcal{M}}(C)$ is similar to $\dim_{\mathcal{H}}(C)$ but coverings of C by balls of equal radii are only used. To be more precise, $\dim_{\mathcal{M}}(C) \leq \lambda$ if for all compact subsets E of C, $\mathcal{M}^t(E) = 0$ for all $t > \lambda$;

$$\mathcal{M}^t(E) = \inf\Big\{k\, r^t\colon \bigcup_{i=1}^{k} B(x_i, r) \supset E\Big\}.$$

In general, $\dim_{\mathcal{H}}(C) \leq \dim_{\mathcal{M}}(C)$ but for self similar and other regular sets we have $\dim_{\mathcal{H}}(C) = \dim_{\mathcal{M}}(C)$, see [MV]. This result belongs to category (b) since the condition for α does not involve K. The proof is potential theoretic. It uses the standard estimate in balls outside C and the fact that the number of cubes in Whitney decomposition of $\mathbb{R}^n \setminus C$ and $\dim_{\mathcal{M}}(C)$ are related. There are also results, see [KM], which only employ $\dim_{\mathcal{H}}(C)$ but these are weaker than those involving $\dim_{\mathcal{M}}(C)$.

References

[BJS] Bers, L., F. John, and M. Schechter, *Partial Differential Equations*, Interscience Publishers, New York, 1964.

[BI] Bojarski, B., and T. Iwaniec, *Analytic foundations of the theory of quasiconformal mappings in* $\mathbb{R}^n$, Ann. Acad. Sci. Fenn. Ser. A I Math. **8** (1983), 234–257.

[CFK] Caffarelli, L., E. Fabes, and C. Kenig, *Completely singular elliptic harmonic measures*, Indiana Univ. Math. J. **30** (1981), 917–924.

[F] Fuglede, B., *Value distribution of harmonic and finely harmonic morphisms and applications in complex analysis*, Ann. Acad. Sci. Fenn. Ser. A I Math. **11** (1986), 111–135.

[GLM1] Granlund, S., P. Lindqvist, and O. Martio, *Conformally invariant variational integrals*, Trans. Amer. Math. Soc. **277** (1983), 43–47.

[GLM2] Granlund, S., P. Lindqvist, and O. Martio, *F-harmonic measure in space*, Ann. Acad. Sci. Fenn. Ser. A I Math. **7** (1982), 233–247.

[GLM3] Granlund, S., P. Lindqvist, and O. Martio, *Phragmén-Lindelöf's and Lindelöf's theorems*, Ark. Mat. **23** (1985), 103–128.

[HM] Heinonen, J., and O. Martio, *Estimates for F-harmonic measures and Oksendal's theorem for quasiconformal mappings*, Indiana Univ. Math. J. **38** (1987), 659–683.

[HKM1] Heinonen, J., T. Kilpeläinen, and O. Martio, *Fine topology and quasilinear elliptic equations*, Ann. Inst. Fourier (Grenoble) **39, 2** (1989), 293–318.

[HKM2] Heinonen, J., T. Kilpeläinen, and O. Martio, *Non-linear Potential Theory*, (to appear).

[HKM3] Heinonen, J., T. Kilpeläinen, and O. Martio, *Harmonic morphisms for quasilinear elliptic partial differential equations*, (to appear).

[IM] Iwaniec, T., and G. J. Martin, *Quasiregular mappings in even dimensions*, (to appear).

[JV] Järvi, P., and M. Vuorinen, *Self-similar Cantor sets and quasiregular mappings*, (to appear).

[KM] Koskela, P., and O. Martio, *Removability theorems for quasiregular mappings*, Ann. Acad. Sci. Fenn. Ser. A I Math., to appear.

[M1] Martio, O., *F-harmonic measures, quasihyperbolic distance and Milloux's problem*, Ann. Acad. Sci. Fenn. Ser. A I Math. **12** (1987), 151–162.

[M2] Martio, O., *Counterexamples for unique continuation*, Manuscripta Math. **60** (1988), 21-47.

[M3] Martio, O., *Sets of zero elliptic harmonic measures*, Ann. Acad. Sci. Fenn. Ser. A I Math. **14** (1989), 47–55.

[M4] Martio, O., *Potential theoretic aspects of non-linear elliptic partial differential equations*, Dept. of Math., Univ. of Jyväskylä, Report **44** (1989), 1–23.

[M5] Martio, O., *Harmonic measures for second order non-linear partial differential equations*, Function spaces, differential operators and non-linear analysis, L. Päivärinta ed., Pitman Research Notes in Mathematics 211, 1989, 271–279.

[MRV1] Martio, O., S. Rickman, and J. Väisälä, *Definitions for quasiregular mappings*, Ann. Acad. Sci. Fenn. Ser. A I Math. **448** (1969), 1–40.

[MRV2] Martio, O., S. Rickman, and J. Väisälä, *Distortion and singularities of quasiregular mappings*, Ann. Acad. Sci. Fenn. Ser. A I Math. **465** (1970), 1–13.

[MS] Martio, O., and J. Sarvas, *Density conditions in the n-capacity*, Indiana Univ. Math. J. **26** (1977), 761–776.

[MVa] Martio, O., and J. Väisälä, *Elliptic equations and maps of bounded length distortion*, Math. Ann. **282** (1988), 423–443.

[MV] Martio, O., and M. Vuorinen, *Whitney cubes, p-capacity, and Minkowski content*, Exposition. Math. **5** (1987), 17–40.

[R] Reshetnyak, Yu. G., *Space mappings with bounded distortion*, AMS Trans. of Math. Monographs vol. 73, 1980.

[Ri1] Rickman, S., *Asymptotic values and angular limits of quasiregular mappings of a ball*, Ann. Acad. Sci. Fenn. Ser. A I Math. **5** (1980), 185–196.

[Ri2] Rickman, S., *Quasiregular mappings*, (to appear).

[T] Tsuji, M., *Potential Theory in Modern Function Theory*, Maruzen Co., Tokyo, 1959.

[Tu] Tukia, P., *Hausdorff dimension and quasisymmetric mappings*, Math. Scand. **65** (1989), 152–160.

[V] Vekua, N., *Generalized Analytic Functions*, Pure and Applied Math. 25, Pergamon Press, 1962.

Maximal inequalities and potential theory

Gabriel Mokobodzki

Introduction

This paper is devoted to the study of the relationship between maximal inequalities in ergodic theory and the classical principles of potential theory.

We start with semi-groups $(P_t)_{t>0}$ of positive linear contractions on a space $L^1(X,\sigma)$. Of course, the same theory may be developed for discrete semi-groups.

We shall say that a family $(Q_t)_{t>0}$ of operators on $L^1(X,\sigma)$ is subordinated to $(P_t)_{t>0}$ if

a) Q_t is positive linear, and $Q_t \leq P_t$ for every $t > 0$,

b) $Q_t P_s \geq Q_{t+s}$ for all $t, s > 0$.

This definition extends the classical concept of subordinated semi-groups, but families $(Q_t)_{t>0}$ need not be semi-groups. Nethertheless, they yield the same Hopf-type maximal inequality :

$$\int_{(f>0)\cup(S_t f>0)} f d\sigma \geq 0 \quad \text{with } f \in L^1(X,\sigma), \quad S_t f = \int_0^t Q_s f ds$$

or

$$\int_{(f>0)\cup(M_t f>0)} f d\sigma \geq 0 \quad \text{with } M_t f = \sup_{r\leq t} S_r f.$$

The key to obtain such inequality lies in associating to every $f \in L^1(X,\sigma)$ a set $A(f)$, depending only on f and $(P_t)_{t>0}$, such that $(S_t f > 0) \cap A(-f) = \emptyset$ for every $t > 0$ and such that the a priori maximal inequality holds :

$$\int_{(f>0)\cup A(-f)^c} d\sigma \geq 0.$$

The construction and the properties of the set $A(f)$ may be related to the so called filling scheme.

For interpreting these facts in terms of potential theory, it would be better to work with semi-groups and operators defined as true kernels acting on a compact space or a measurable space.

This lecture was delivered with the title "Potential theory and ergodic pointwise convergence theorems". The new title is more closely related to the central part of this paper.

Potential Theory, ed. by M. Kishi

This is no longer restrictive since it is always possible to represent pseudo-kernels as kernels, using standard techniques and this happens to be sufficient in most questions involving almost pointwise convergence.

Moreover, if we assume that the semi-group $(P_t)_{t>0}$ is submarkovian, then we can also prove in this general framework that, for $f, g \in L^1_+(X, \sigma)$

$\lim_{t\to\infty} \frac{S_t f}{S_t g}$ exists σ-a.e. on the set $(g > 0)$ and we identify the limit.

Plan

A. Subordinated families.

B. The set $A(f)$ and its properties.

C. Convergence theorems.

D. Discussion : Do the subordinated families give the optimal framework for obtaining maximal inequalities ?

A. Subordinated families

Let $(X, \mathcal{B})$ be a measurable space, $(P_t)_{t>0}$ a semi-group of positive bounded kernels on $(X, \mathcal{B})$ such that for any $h \geq 0$ measurable on $X, (x,t) \mapsto P_t(h)(x)$ is measurable on $X \times \mathbb{R}^+_*$. We assume that the corresponding resolvent family $V_\lambda = \int_0^\infty e^{-\lambda t} P_t dt$ is well defined and bounded for each $\lambda > 0$.

Let $(Q_t)_{t>0}$ be a measurable family (see above) of positive kernels on $(X, \mathcal{B})$.

Definition 1. The family $(Q_t)_{t>0}$ is *subordinated* to (P_t) if

a) $Q_t \leq P_t$ for every $t > 0$ and

b) $Q_t P_s \geq Q_{t+s}$ for all $t, s > 0$.

We denote by $\Gamma((P_t))$ or simply Γ the set of subordinated families to $(P_t)_{t>0}$, by Γ_λ the set corresponding to $(e^{-\lambda t} P_t)_{t>0}$, for $\lambda > 0$. Notice that in condition b) the factor P_s appears on the right.

First properties. Let $F = (Q_t)_{t>0} \in \Gamma$.

1) $F_\lambda = (e^{-\lambda t} Q_t)_{t>0} \in \Gamma_\lambda$.

2) Let $(\varphi_t)_{t>0}$ be a non-increasing family of positive measurable functions with $0 \leq \varphi_t \leq \varphi_{t'} \leq 1$ for $0 < t' < t$. Then $F' = (\varphi_t Q_t)_{t>0} \in \Gamma$.

3) Let $D = (0 < t_1 < t_2 < \cdots < t_n < \cdots)$ be a sequence of real numbers tending to ∞. Define

$$Q_t^D = \begin{cases} P_t \text{ for } t < t_1 \\ Q_{t_n} \circ P_{t-t_n} \text{ for } t_n \leq t < t_{n+1}. \end{cases}$$

Then $F^D = (Q_t^D)_{t>0} \in \Gamma$; for another sequence D' with $D \subset D'$, $Q_t \leq Q_t^{D'} \leq Q_t^D$ for all $t > 0$.

4) For an excessive function $u, t \mapsto Q_t u$ is non-increasing.

5) Let $(D_n)_{n\in\mathbb{N}}$ be a family of discrete sequences contained in $\mathbb{R}^+$, with $D_n \subset D_{n+1}$ and $\bigcup_n D_n$ dense in $\mathbb{R}^+$. Assume that there exists an excessive function $u \geq 0$ for which $V_0 u = \int_0^\infty P_t u dt$ is finite. Then for all $f, 0 \leq f \leq u, \int_0^\infty Q_t f dt = \inf_n \int_0^\infty Q_t^{D_n} f dt$.

6) Let $(F_k)_{k\leq n}, F_k = (Q_t^k)_{t>0} \in \Gamma$ and let $(\varphi_k)_{k\leq n}$ be a finite measurable partition of unity on $(X, \mathcal{B})$. For each t, define $Q_t = \sum_{k=1}^{n} \varphi_k Q_t^k$. Then $F = (Q_t)_{t>0} \in \Gamma$.

Proposition 2. *Suppose that $V_0 1 = \int_0^\infty P_t 1 dt$ is finite. Let $F = (Q_t)_{t>0} \in \Gamma$ and $D = (t_n)_{n\in\mathbb{N}}$ an increasing sequence in $\mathbb{R}_*^+$, tending to ∞. Then*

$$W = \int_0^\infty Q_t^D dt = V - P^D V,$$

with $P^D = (P_{t_1} - Q_{t_1}) + \sum_{n\geq 1} (Q_{t_n} P_{t_{n+1}-t_n} - Q_{t_{n+1}})$.

Remark. Replacing (P_t) and (Q_t) by $(e^{-\lambda t} P_t)$ and $(e^{-\lambda t} Q_t)$, the corresponding P_λ^D is given by

$$P_\lambda^D = e^{-\lambda t_1}(P_{t_1} - Q_{t_1}) + \sum_{n\geq 1} e^{-\lambda t_{n+1}} (Q_{t_n} P_{t_{n+1}-t_n} - Q_{t_{n+1}}).$$

Since $\lambda \mapsto P_\lambda^D$ is decreasing, even with $V_0 1$ not finite we may write $V_0 = W + P^D V_0$.

Corollary 3. *Suppose that X is compact and $(V_\lambda)_{\lambda>0}$ is a Ray resolvent, $V_\lambda = \int_0^\infty e^{-\lambda t} P_t dt$.*

For each $W_\lambda = \int_0^\infty e^{-\lambda t} Q_t dt$ there exists $P_\lambda \geq 0$ such that $W_\lambda = V_\lambda + P_\lambda V_\lambda$ and

$$\lambda \mapsto P_\lambda \quad \textit{non-increasing}.$$

Returning now to the general situation, let

$$S_t = \int_0^t Q_s ds = \int_0^\infty Q_s' ds$$

with

$$Q_s' = \begin{cases} Q_s \text{ for } s \leq t \text{ and } (Q_t')_{t>0} \in \Gamma \\ 0 \text{ for } s > t. \end{cases}$$

Since for f bounded, we have $S_t f = \lim_\lambda \int_0^t e^{-\lambda s} Q_s f ds$, we have the following

Theorem 4. *For any fixed $t > 0$, there exists a family (P^n_λ) of positive kernels on $(X, \mathcal{B})$ such that $S_t f = \lim_{\substack{\lambda\to 0\\ n\to\infty}} (V_\lambda - P^n_\lambda V_\lambda) f$ for all bounded f and $n \in \mathbb{N}, \lambda > 0$ $V_\lambda - P^n_\lambda V_\lambda \geq 0$ for all $n \in \mathbb{N}$, $\lambda > 0$.*

B. Definition and properties of $A(f)$

We remain with the hypotheses of Section A.

For every $\lambda > 0$, let $\mathcal{S}_\lambda$ be the convex cone of λ-excessive functions and ${}^\lambda R$ the corresponding reduite operator defined by ${}^\lambda Ru = \inf\{w | w \in \mathcal{S}_\lambda; w \geq u\}$.

A bounded measure σ will be called *supermedian* if $\sigma \geq 0$ and $\sigma\lambda V_\lambda \leq \sigma$ for all $\lambda > 0$.

Definition 5. For f bounded measurable on X, or more generally, for f such that $V_\lambda(f^+)$ is finite for all $\lambda > 0$, we say that $A_\lambda(f) = \{{}^\lambda R(V_\lambda f) = V_\lambda f\}$ is the *contact set for* f (relative to $\mathcal{S}_\lambda$).

Main Lemma 6. *Let f be measurable and bounded on $(X, \mathcal{B})$. For every $\lambda > 0$, and every measure $\nu \geq 0$ on X, there exists $h_\lambda \geq 0$ such that*

a) $(h_\lambda > 0) \subset A_\lambda(f)$, $h_\lambda \leq f^+$, $Vh_\lambda \leq {}^\lambda RV_\lambda f$ and $\int({}^\lambda R(V_\lambda f) - Vh_\lambda)d\nu = 0$,

b) for every supermedian measure $\sigma \geq 0$, and $\lambda > 0$

$$\int_{(f>0)\cup A_\lambda(-f)^c} f d\sigma \geq 0.$$

Part a) was first proved by the author for a submarkovian resolvent, but the same method applies here.

Part b) of the lemma is a simple consequence of the inequality ${}^\lambda R(V_\lambda f) \geq V_\lambda f$.

<u>Another characterization of</u> $A_\lambda(f)$.

Let us consider the extended resolvent equation of Neveu $V_k = V_h + V_h(k-h)V_k$, where h, k may be taken as bounded measurable functions on $(X, \mathcal{B})$.

Theorem 7. *1) For f bounded measurable on $(X, \mathcal{B})$*

$${}^\lambda R(V_\lambda f) = \sup_{k\geq 0} V_{\lambda+k}(k.V_\lambda f) = \sup_{n\in\mathbb{N}} nV_{\lambda+1_{A_\lambda}}(1_{A_\lambda}.V_\lambda f)$$

where A_λ means $A_\lambda(f)$.

2) $A_\lambda(f) = \bigcap_{k\geq 0}(V_{\lambda+k} f \geq 0)$ and $A_\lambda(f) = \bigcap_{\lambda'>\lambda} A_{\lambda'}(f)$.

(*Sketch of proof*: by the domination principle, ${}^{\lambda}RV_\lambda f$ is carried by $A_\lambda(f)$; that means it is equal to its reduite on $A_\lambda(f)$, and by the extended resolvent equation, $(V_\lambda)_k = V_{\lambda+k}$).

Definition 8. The set $A(f)$ is by definition

$$A(f) = \bigcap_{\lambda>0} A_\lambda(f) = \bigcap_{h>0} \{V_h \geq 0\}.$$

If $V_0 = \sup_{\lambda>0} V_\lambda$ is bounded, then $A(f) = A_0(f)$ is the contact set for f relative to the cône of excessive functions $\mathcal{S} = \{u \geq 0 | u = \sup \lambda V_\lambda u\}$.

Hence we have

Theorem 9. *Let $\sigma \geq 0$ be a bounded supermedian measure on $(X, \mathcal{B})$ and let f be bounded measurable (resp. $V_\lambda(f^-)$ finite for all $\lambda > 0$). Then*

$$\int_{(f>0)\cup A(-f)^c} f d\sigma \geq 0.$$

Let $(Q_t)_{t>0}$ be a subordinated family to $(P_t)_{t>0}$. Let $S_t = \int_0^t Q_s ds$. By Theorem 4, Section A, we obtain that $S_t f(x) \leq 0$ for all $x \in A(-f)$. Then we obtain

Theorem 10. *For any set $D \subset \mathbb{R}^+$, every bounded f and every supermedian measure $\sigma \geq 0$,*

$$\int_{(f>0)\cup \bigcup_{t\in D}(S_t f>0)} f d\sigma \geq 0.$$

Notice that for $f_1 \leq f_2$, we have $A(f_1) \subset A(f_2)$, and thus it is sufficient to consider the maximal inequality for bounded functions and then proceed by approximation.

C. Convergence theorems

In this section, we suppose that X is compact and $(V_\lambda)_{\lambda>0}$ is a submarkovian Feller resolvent family which is associated to a measurable submarkovian semi-group $(P_t)_{t>0}$ of positive kernels.

Let $(Q_t)_{t>0}$ be a measurable family of kernels subordinated to $(P_t)_{t>0}$ and put $S_t = \int_0^t Q_s ds, W = \sup_t S_t$.

For a bounded supermedian measure $\sigma \geq 0$ on X, the Riesz decomposition gives two supermedian measures σ_1, σ_2 such that $\sigma = \sigma_1 + \sigma_2$, $\sigma_1 = \sigma_1 \lambda V_\lambda$ for all $\lambda > 0$ and $\inf_{\lambda>0} \sigma_2 \lambda V_\lambda = 0$. Then for an invariant measure $\sigma \geq 0$ there is an integral representation

$$\sigma = \int \sigma_\alpha d\mu(\alpha)$$

on the set of invariant extremal measures on X, and here we use the assumption that X is compact and $(V_\lambda)_{\lambda>0}$ is a Feller resolvent family.

We fix now an invariant extremal bounded measure $\sigma \geq 0$, that is to say an ergodic measure for $(P_t)_{t>0}$.

Lemma 11. *Let $f, g \in \mathcal{L}^1_+(X,\sigma), g \geq 0$, and for $k \in \mathbb{R}^+_*$ let*

$$M_k(f) = \bigcup_{t>0} (S_t f < k S_t g);$$

then $\int_{M_k(f)} g d\sigma \leq k^{-1} \int |f| d\sigma$ and $\sup_{t>0} \frac{S_t f}{S_t g}$ is finite on the set $(g > 0)$.

Theorem 12. *There exists a partition of X in $X = D + C$ (dissipative and conservative part for $(Q_t)_{t>0}$), with $C = \{\int_0^\infty Q_t 1 dt = +\infty\}$ such that for all $f \in \mathcal{L}^1_+, Wf = \int_0^\infty Q_t f dt$ is finite almost everywhere on D and infinite almost everywhere on $C \cap (Wf > 0)$.*

These lemma and theorem are the classical consequences of the Hopf-type inequality. We need another classical lemma.

Lemma 13. *Let ν be an invariant positive bounded measure which is absolutely continuous relative to σ. Then, since σ is extremal, ν is of the form $\nu = k.\sigma$ for a positive number k.*

We now give steps to identify $\lim_{t\to\infty} \frac{S_t f}{S_t g}$ on the conservative part C. For sake of simplicity we assume that $g \geq 1$ and let $H = \{f | \exists k > 0, |f| \leq kg\}$, let $h \geq 0$ be bounded such that $(h > 0) \subset C$ and $\int h d\sigma > 0$.

We define K on a H by $K(f) = \int h. \limsup_{t\to\infty} \frac{S_t f}{S_t g} d\sigma$. Clearly, K is a bounded sublinear functional on H . The following lemma shows that K is linear.

Lemma 14. *Let ν be a linear form on H with $\nu \leq K$. Then*

a) *ν is positive ; $\nu(1) \leq \nu(g) = \int h d\sigma$;*

b) *for a decreasing family $(f_n) \subset H^+$ with $\inf_n \int f_n d\sigma = 0$, $\inf_n K(f_n) = 0$ (by the maximal inequality) hence ν is a measure dominated by σ ;*

c) *for $f \in H^+, \nu(f - P_t f) \leq 0$; hence $\nu \lambda V_\lambda \geq \nu$ for all $\lambda > 0$. Since $\lambda V_\lambda \leq 1$, $\nu = \nu \lambda V_\lambda$ and ν is invariant.*

d) *The preceding lemma gives $\nu = k\sigma$, so*

$$\nu(f) = \frac{\int h d\sigma}{\int g d\sigma} \cdot \int f d\sigma.$$

By the Hahn-Banach Theorem, this lemma implies that K is linear and equal to ν on H.

Now, if we make h vary and fix f, g we obtain the following

Theorem 15. *For every $f \in \mathcal{L}^1_+(X, \sigma)$*

$$\limsup_{t\to\infty} \frac{S_t f}{S_t g} = \liminf_{t\to\infty} \frac{S_t f}{S_t g} = \frac{\int f d\sigma}{\int g d\sigma}$$

almost everywhere on the conservative part C.

This theorem is first proved for $f \in H$ and, applying the maximal inequality again, is also valid for $f \in \mathcal{L}^1_+(X, \sigma)$.

For a general invariant and non extremal measure $\sigma \geq 0$, let (V^*_λ) be the dual resolvent acting on $L^\infty(\sigma)$.

Let $\mathcal{F}$ be the sub-σ-field generated on C by the invariant bounded functions $h \in L^\infty$, $\lambda V^*_\lambda h = h$, $\forall \lambda > 0$. We then have the following

Theorem 16. *For $f, g \in \mathcal{L}^1_+(X, \sigma)$*

$$\lim_{t\to\infty} \frac{S_t f}{S_t g} = \frac{E(f/\mathcal{F})}{E(g/\mathcal{F})} \quad \sigma\text{-almost everywhere}$$

on the set $C \cap (\bigcup_{t>0} S_t g > 0)$.

<u>Nota</u> : If $\sigma \geq 0$ is supermedian and $\inf_\lambda \sigma \lambda V_\lambda = 0$, then $\sigma(C) = 0$.

D. Discussion

What kind of a framework for the existence of maximal inequalities?

Let X be a compact space, $(V_\lambda)_{\lambda \geq 0}$ a Feller resolvent family of positive kernels on X. Assume that $V = V_0$ is bounded, $V1 > 0$ on X, and that $V(\mathcal{C}(X))$ linearly separates points of X. This last condition enables us to make use of the theory of Choquet's and Silov's boundaries.

Let $(P_t)_{t>0}$ be the semi-group associated to $(V_\lambda)_{\lambda>0}$ which is well defined on the set of non-branching points of X.

Definition 17. We say that a positive kernel W on X is *compatible* with $V_0 = V$ if for every supermedian measure $\sigma \geq 0$ and every $f \in \mathcal{C}(X)$

$$\int_{(f>0)\cup(Vf>0)\cup(Wf>0)} f\,d\sigma \geq 0.$$

We have shown that for a family $(Q_t)_{t>0}$ subordinated to $(P_t)_{t>0}$, $W = \int_0^\infty Q_t dt$ is compatible with V.

Lemma 18. *Let W be compatible with V and suppose that W is invertible as a bounded linear operator. Then $K = VW^{-1}$ satisfies the domination principle.*

This particular result suggested the following

Proposition 19. *Let W be a Feller kernel on X. The following conditions are equivalent :*

a) *W is compatible with V ;*

b) *for every $f \in \mathcal{C}(X)$, $(Wf > 0) \cap A(-f) = \emptyset$;*

c) *there exist sequences (W_n) and (P_n) of positive kernels, a sequence (k_n) of positive functions such that $W_n = k_n(V - P_nV)$ and that for every $f \in \mathcal{C}(X)$, $Wf = \lim_{n\to\infty} k_n(V - P_nV)(f)$.*

Proposition 20. *Let W be a Feller positive kernel on X. The following conditions are equivalent :*

a) *$W = V - PV$ with P being a positive measurable kernel;*

b) *there exists a measurable family $(Q_t)_{t>0}$ subordinated to $(P_t)_{t>0}$ such that $W = \int_0^\infty Q_t dt$.*

Among the many examples of the above studied structures, let us just mention a queer one. For $0 < \alpha < 1$, let V_λ^α be the fractional power of V_λ. Then for any supermedian measure $\sigma \geq 0$

$$\int_{(f>0)\cup\bigcup_{\alpha,\lambda}(V_\lambda^\alpha f>0)} f d\sigma \geq 0.$$

Bibliography

[A-M] Azema J. et Meyer P.A., *Une nouvelle représentation du type de Skorohod.* Sem. Strasbourg L.N. in Maths. n381, Springer.

[H] Heath D., *Skorohod stopping times via potential theory.* L.N. in Maths. n° 381. Springer.

[H] Heath D., *Skorohod stopping in discrete time.* Séminaire Strasbourg , L.N. in Maths. n° 465, Springer.

[N] Neveu J., *Potentiel Markovien recurrent des chaines de Harris.* Ann. Inst. Fourier 1972.

[M] Meyer P.A., *Travaux de Rost en théorie du balayage.* Séminaire de probabilités Strasbourg. L.N. in Maths. n° 191. Springer.

[R] Rost H., *Markov-Ketten bei sich füllenden Löchern im Zustrandsraum.* Ann. Institut Fourier vol. XXI, 1971.

[K] Krengel U., *Ergodic theorems.* W. d. G.

[F] Feyel D., *Théorèmes de convergence presque-sre. Existence de semi-groupes.* Advances in Math. vol. 34 n° 2, 1979.

[M] Mokobodzki G., *Densité relative de deux potentiels comparables.* Sem. proba. Strasbourg IV. L.N. in Maths. n° 124, Springer.

The boundary behaviour of solutions of the Dirichlet problem

Ivan Netuka

Abstract

This survey concentrates on the boundary behaviour of solutions of the Dirichlet problem in harmonic spaces. In this connection, the vague convergence properties of harmonic measures with respect to the original topology as well as to the fine topology are investigated and various important subsets of boundary points are studied.

1980 Mathematics Subject Classification(1985 Revision): 31B25, 31D05, 35J25, 35K20.

1. Introduction

In what follows, X will denote a *$\mathcal{P}$-harmonic space* with countable base in the sense of [20] and $U \subset X$ will be an open relatively compact set with the boundary U^*. For a *resolutive* function f on U^*, $H_U f$ denotes the solution of the *generalized Dirichlet problem in the sense of Perron–Wiener–Brelot*. Recall that, for $x \in U$, the *harmonic measure* $f \mapsto H_U f(x)$, $f \in C(U^*)$, coincides with $\varepsilon_x^{\complement U}$, the *balayage* of the Dirac measure at x on $\complement U$, the complement of U.

2. The Dirichlet problem and cluster sets of harmonic measures

Recall that an open relatively compact set $U \subset X$ is said to be *regular*, if

$$\lim_{x \to z,\, x \in U} H_U f(x) = f(z),$$

whenever $z \in U^*$ and $f \in C(U^*)$.

To say that U is regular obviously means that a solution of the *classical Dirichlet problem* exists for all continuous boundary data. There is another useful formulation in terms of harmonic measure: *U is regular, if and only if*

$$\lim_{x \to z,\, x \in U} \varepsilon_x^{\complement U} = \varepsilon_z$$

for every $z \in U^*$.

Potential Theory, ed. by M. Kishi

The existence of non-regular sets was discovered around 1910 and was considered to be a rather unexpected fact. We quote from Sec. 2 of [35]: "*Limitations on the Dirichlet Problem.* Thus far, writers on the subject had regarded the Dirichlet problem as always possible, and their endeavors had been to find methods of proof powerful enough to establish its possibility. A new chapter begins with recognition of the fact that domains exist which are not normal." A similar description of the situation can be found in [51]: "On a cru pendant longtemps que ce problème était toujours possible, ce qui veut dire, qu'il admettait toujours une solution, ainsi que l'affirmait le princip de Dirichlet. On ne parvenait pas à la résoudre dans le cas d'un domaine quelconque, mais on attribuait cela à l'extrême variété que présentait la notion de domaine."

Another type of domain with good properties from the boundary behaviour point of view was introduced in [1]. An open relatively compact set $U \subset X$ is said to be *semiregular* if, for every $f \in C(U^*)$, $H_U f$ admits a continuous extension to $\overline{U}$.

A punctured disc in $\mathbb{R}^2$ (classical potential theory) or an open interval in $\mathbb{R} \times \mathbb{R}$ (parabolic potential theory) are both semiregular but not regular sets. Of course, *$U \subset X$ is semiregular, if and only if*

$$\lim_{x \to z,\, x \in U} \varepsilon_x^{\complement U}$$

exists for every $z \in U^*$. It will be seen in a moment how this limit measure can be identified.

Let $U \subset X$ be again an open relatively compact set with $z \in U^*$. Then z is said to be *regular*, if

$$\lim_{x \to z,\, x \in U} \varepsilon_x^{\complement U} = \varepsilon_z .$$

The set of all regular points will be denoted by U^*_{reg} and we put $U^*_{\text{irr}} = U^* \setminus U^*_{\text{reg}}$. The points of U^*_{irr} are called *irregular*. Recall that U^*_{reg} is a G_δ-set and U^*_{irr} is semipolar; see [20, p. 169]. Obviously, *U is regular, if and only if $U^* = U^*_{\text{reg}}$.*

A point $z \in U^*$ is known to be regular, if and only if $\varepsilon_z^{\complement U} = \varepsilon_z$; the last condition expresses the fact that $\complement U$ is not thin at z; see [20, pp. 155, 164]. For the related notion of a *barrier*, the reader is referred to [20, p. 53]; cf. also [2, p. 136].

A point $z \in U^*$ will be termed *semiregular*, if $z \in U^*_{\text{irr}}$ and the limit

$$\lim_{x \to z,\, x \in U} \varepsilon_x^{\complement U}$$

exists. The set of all semiregular points will be denoted U^*_{sem}. Thus *U is semiregular, if and only if every boundary point is either regular or semiregular.* For a better understanding of semiregular points, the following result is important.

Proposition 1. *Suppose that $z \in U^*$. Then there is a sequence (x_n) in U such that $x_n \to z$ and*

$$\varepsilon_{x_n}^{\complement U} \to \varepsilon_z^{\complement U} .$$

Corollary 1. *A point $z \in U^*$ is semiregular, if and only if*

$$\lim_{x \to z,\, x \in U} \varepsilon_x^{\complement U} = \varepsilon_z^{\complement U}$$

and $\varepsilon_z^{\complement U} \neq \varepsilon_z$.

We shall give a sketch of the *proof* of Proposition 1. Denote by $\mathcal{P}$ the set of continuous potentials on X. Fix a strict potential $q \in \mathcal{P}$. Since $R_q^{\complement U} = q$ on $\complement U$ and $q(x) \to q(z) \geqq \hat{R}_q^{\complement U}(z)$ as $x \to z$, we have

$$\hat{R}_q^{\complement U}(z) = \liminf_{x \to z} R_q^{\complement U}(x) = \liminf_{x \to z,\, x \in U} \hat{R}_q^{\complement U}(x).$$

Now take a sequence (x_n) in U such that $x_n \to z$ and $\hat{R}_q^{\complement U}(x_n) \to \hat{R}_q^{\complement U}(z)$. If λ is an accumulation point of $(\varepsilon_{x_n}^{\complement U})$, then $\lambda(q) = \varepsilon_z^{\complement U}(q)$. For $p \in \mathcal{P}$ we get $\lambda(p) \geqq \varepsilon_z^{\complement U}(p)$ and so, since q is strict, $\lambda = \varepsilon_z^{\complement U}$; cf. [8, p. 328].

Proposition 1 is sometimes referred as the *Köhn–Sieveking lemma*, see [36]. Another proof based on a fine topology argument is given in [23]; see also Sec. 3 of this survey. There is a close connection with the notion of *suite maximale*; see Sec. 3 and references in [47].

Before giving a characterization of semiregular points and semiregular sets, let us introduce an important subset of U^*: Denote by T the smallest closed set containing the support of every $\varepsilon_x^{\complement U}$, $x \in U$. Obviously, $U_{\text{reg}}^* \subset T$.

The connection between T and the semiregular points of U^* is contained in the following result.

Theorem 1. *For $z \in U^*$, the following statements are equivalent:*

(i) *z is not semiregular;*
(ii) *$z \in T$;*
(iii) *for every $f \in C(U^*)$,*

$$\liminf_{x \to z,\, x \in U} H_U f(x) \leqq f(z) \leqq \limsup_{x \to z,\, x \in U} H_U f(x);$$

(iv) *there exists a sequence (x_n) in U such that $x_n \to z$ and $\varepsilon_{x_n}^{\complement U} \to \varepsilon_z$.*

For the proof, see [43], where a barrier test for semiregular points is also mentioned; cf. [42], [7], [8, p. 357]. The implication (ii)$\Rightarrow$(iii) was proved in [30]; for an analogous result in the context of a resolutive compactification, see [31], [32]. Semiregular points in *elliptic* harmonic spaces are investigated in [54].

Theorem 2. *The following statements are equivalent:*

(i) *U is semiregular;*

(ii) $U^*_{\text{reg}} = T$;
(iii) U^*_{reg} *is closed and* U^*_{irr} *is negligible, in the sense that* $\varepsilon_x^{\complement U}(U^*_{\text{irr}}) = 0$ *for every* $x \in U$;
(iv) *the mapping* $x \mapsto \varepsilon_x^{\complement U}$ *is vaguely continuous on* $\overline{U}$.

The proof can be found in [7], see also [8, p. 356]. The implication (i)$\Rightarrow$(iii) was proved in [1], the equivalence (i)$\Longleftrightarrow$(iv) in [36]. In [37], the following ***problem*** was posed. Prove or disprove: *In Brelot spaces,* U^*_{reg} *closed implies that* U *is semiregular.* An affirmative answer was given in [45] under the additional assumption that U^*_{irr} is polar. This assumption cannot be omitted in general; for an example, see [46]. The implication (iii)$\Longrightarrow$(i) was, under additional hypotheses, deduced in [41] using deep results on simpliciality from [6]. For a simplicial characterization of semiregular sets, see [7].

Let us mention briefly an important subset of U^*_{reg}. Following [1], cf. also [20, p. 59], a point $z \in U^*$ is said to be *extremal*, if ε_z is the only measure μ on $\overline{U}$ satisfying

$$\mu(s) \leqq s(z)$$

for every $s \in C(\overline{U})$ which is hyperharmonic on U. Let U^*_{ext} be the set of extremal points. By [1], $U^*_{\text{ext}} \subset U^*_{\text{reg}}$ and the equality does not hold in general, as shown in [36]. For semiregular sets, however, we do have equality, see [1]. More generally, $U^*_{\text{ext}} = U^*_{\text{reg}}$, if U^*_{irr} is negligible; see [6] and [8, p. 365] where deep relations to the Choquet theory are studied. We only mention that every point of U^*_{ext} possesses a "***high quality***" barrier, provided that the set $H(U)$ of functions continuous on $\overline{U}$ and harmonic on U is linearly separating, as follows: *If* $z \in U^*_{\text{ext}}$, *then there exists* $h \in H(U)$ *such that* $h(z) = 0$ *and* $h > 0$ *on* $\overline{U} \setminus \{z\}$; see [6] and [8, p. 360]. We remark that all ***stable points*** of $\overline{U}$ are contained in U^*_{ext}; see [6], [23], [54].

So far we know that, by definitions, $H_U f$ shows a nice boundary behaviour at every $z \in U^*_{\text{reg}} \cup U^*_{\text{sem}}$. If $z \notin U^*_{\text{reg}} \cup U^*_{\text{sem}}$, however, then $\varepsilon_z^{\complement U} \neq \varepsilon_z$ and there are sequences (x_n), (y_n) in U such that $x_n \to z$, $y_n \to z$ and

$$\varepsilon_{x_n}^{\complement U} \to \varepsilon_z^{\complement U}, \quad \varepsilon_{y_n}^{\complement U} \to \varepsilon_z .$$

The result on *accumulation points of balayage measures* given below provides more complete information on the boundary behaviour of $H_U f$.

Definition 1. For $A, B \subset X, z \in X$, denote by $\Lambda_B^A(z)$ the set of all measures λ on X such that $(\varepsilon_{x_n}^A)$ converges to λ for some sequence (x_n) in B converging to z.

Theorem 3. *For* $A \subset X, z \in X$, *we have either*

$$\Lambda_X^A(z) \subset \{\varepsilon_z, \varepsilon_z^A\} \quad or \quad \Lambda_X^A = \{\alpha\varepsilon_z + (1-\alpha)\varepsilon_z^A; \alpha \in [0,1]\}.$$

Also, we have either

$$\Lambda_{\complement A}^A \subset \{\varepsilon_z, \varepsilon_z^A\} \quad or \quad \Lambda_{\complement A}^A = \{\alpha\varepsilon_z + (1-\alpha)\varepsilon_z^A; \alpha \in [0,1]\}.$$

The proof can be found in [26], see also [8, pp. 333, 335]. It depends on certain *fine properties of balayage*. The second statement, for $A = \complement U$, was proved in [43] and the proof uses properties of PWB-solutions. A self-contained presentation of Theorem 3 for $A = \complement U$ is compiled in [48]. The following assertion is proved in [9]; see also [20, p. 174]:

$$\Lambda_X^A(z) \subset \{\alpha\varepsilon_z + (1-\alpha)\varepsilon_z^{A\setminus\{z\}};\ \alpha \in [0,1]\}.$$

In classical potential theory, for $A = \complement U$, the second part of Theorem 3 was proved in [24] for U connected and the first part independently in [39]; cf. [38], [40]. (Following this talk at ICPT 90, Professor N. Boboc informed me of his joint work with Gh. Bucur which proves the above inclusion in the setting of H-cones; see [53].)

It should be mentioned, however, that a deeper investigation of the cluster set of $H_U f$ at an irregular point z seems to go back to [33] (a preliminary communication) and [34, p. 259]. The cluster set of $H_U f$ at z is shown there to be included between $f(z)$ and the *quasi-limit* of $H_U f$ at z; cf. [10] for further references. (Note that the notion of a fine limit appeared later.) A short proof of this assertion (with the fine limit, of course) is to be found in [14, p. 106].

From the historical point of view, it seems to be interesting (also with regard to the material of Sec. 4) to include the following quotation from [34, p. 258]: "The Wiener theorem provides a complete solution of the question about conditions which have to be fulfilled for a domain in order to be normal, i.e. the Dirichlet problem is solvable for arbitrary continuous boundary data. If the boundary of a domain contains irregular points, a question arises about conditions which have to be fulfilled by boundary values in order that the Dirichlet problem would be solvable. From this point of view, a more detailed investigation of the Dirichlet problem in a neighbourhood of an irregular point is of interest."

A careful analysis (done for $A = \complement U$ in [43] and for a general set A in [26]) leads to the following *classification of boundary points* (here we write $\Lambda(z)$ instead of $\Lambda_U^{\complement U}(z)$):

Theorem 4. *For $z \in U^*$, exactly one of the following possibilities occurs:*

(i) $\Lambda(z) = \{\varepsilon_z\}$;
(ii) $\Lambda(z) = \{\varepsilon_z^{\complement U}\}$ *and* $\varepsilon_z^{\complement U} \neq \varepsilon_z$;
(iii) $\Lambda(z) = \{\varepsilon_z, \varepsilon_z^{\complement U}\}$ *and* $\varepsilon_z^{\complement U} \neq \varepsilon_z$;
(iv) $\Lambda(z) = \{\alpha\varepsilon_z + (1-\alpha)\varepsilon_z^{\complement U}; 0 \leqq \alpha \leqq 1\}$ *and* $\varepsilon_z^{\complement U} \neq \varepsilon_z$.

Of course, (i) means that $z \in U^*_{\mathrm{reg}}$, (ii) corresponds to $z \in U^*_{\mathrm{sem}}$.

For the harmonic space associated to the heat equation in $\mathbb{R}\times\mathbb{R}$, (iii) is illustrated by the following example:

$$U =]-2,2[\times]-1,1[\setminus([-1,1]\times\{0\}),\ z = (0,0).$$

Then $\varepsilon_x^{\complement U}$ tends to $\varepsilon_z^{\complement U}$ as x converges to z from "*below*", whereas $\varepsilon_x^{\complement U}$ tends to ε_z as x approaches z from "*above*". This example is, in a way, typical. As shown in [43], (iii) is equivalent with the following statement: *There exist an open neighbourhood V of z and disjoint non-empty open sets R and S such that $V \cap U = R \cup S$, $z \in R^*_{\text{reg}}$ and $z \in S^*_{\text{sem}}$.*

A point $z \in U^*$ is said to be *weakly irregular*, if (iii) holds and *strongly irregular* if (iv) occurs; see [43], [8, p. 355]. There are no weakly irregular points in *elliptic* harmonic spaces, see [43] or [26]. An analogous classification of boundary points in the context of a resolutive compactification of a harmonic space can be found in [32].

In [26] or [8, p. 328], the following result is proved: *Let $A, A' \subset X, z \in X, V$ be a neighbourhood of z such that $A \cap V = A' \cap V, (x_n)$ be a sequence in X converging to z and $\alpha \in [0,1]$. Then $\varepsilon_{x_n}^A \to \alpha\varepsilon_z + (1-\alpha)\varepsilon_z^A$ if and only if $\varepsilon_{x_n}^{A'} \to \alpha\varepsilon_z + (1-\alpha)\varepsilon_z^{A'}$.* It follows that regularity, semiregularity, weak and strong irregularity are *local properties* of U and z. To be an extremal point is a local property, too; see [8, p. 358].

We conclude this section by a consequence of Theorem 4 which will be important in Sec. 4.

Corollary 2. *For $z \in U^*$ and $f \in C(U^*)$, the following statements are equivalent:*

(i) $\lim\limits_{x \to z,\, x \in U} H_U f(x) = f(z)$;

(ii) $f(z) = \varepsilon_z^{\complement U}(f)$.

3. Fine boundary behaviour of the PWB-solution

In this section we shall suppose that $U \subset X$ is an open relatively compact set and $z \in U^*_{\text{irr}}$ is such that $\{z\}$ is *polar*.

It follows from Sec. 2 that the boundary behaviour of $H_U f$ is, in general, very complicated near z. We shall see that the situation is completely different, if the fine topology is considered instead of the original topology . Then $H_U g$ behaves well near z even for a large class of (possibly discontinuous) resolutive functions g on U^*.

Recall that the *fine topology* on X is the coarsest topology which is finer than the initial topology of X and in which every hyperharmonic function on every open set of X is continuous.

We shall start with a simple fact.

Observation 1. Let p be a continuous potential on X, let $g = p\,|_{U^*}$, the restriction of p to U^*. Then the equality $H_U g = \hat{R}_p^{\complement U}$ holds on U. Since z belongs to the fine closure of U, $\hat{R}_p^{\complement U}$ is finely continuous on X and $\hat{R}_p^{\complement U}(z) = \int p \, d\varepsilon_z^{\complement U}$, we have the following expression for the fine limit:

$$(*) \qquad f-\lim_{y\to z,\, y\in U} H_U g(y) = \int g\, d\varepsilon_z^{\complement U}.$$

A usual approximation argument leads to the following simple result.

Proposition 2. *If $g \in C(U^*)$, then $(*)$ holds.*

We are interested in the extent to which the formula $(*)$ holds for a general resolutive function. There are three partial problems included in $(*)$: *the existence of the fine limit, integrability with respect to $\varepsilon_z^{\complement U}$ and validity of the equality.*

Let us look back to the history of the problem. In classical potential theory, Proposition 2 was known (though unpublished) to H. Cartan in the 1940s; see [10, the footnote on p. 22]. Later $(*)$ was proved for a bounded resolutive function g continuous at z in [10] and the continuity assumption was removed in [11]. In [22] the fine limit of $H_U g$ at z is proved (also for parabolic potential theory) to determine a Radon measure which is not, however, identified with $\varepsilon_z^{\complement U}$.

For Brelot harmonic spaces satisfying additional hypotheses, results of [13] were generalized in [49], [50]. Let us remark here that, for a class of Brelot spaces, a related non-tangential behaviour of $H_U g$ was studied in [52].

In [4], the equality $(*)$ is proved for arbitrary *bounded* resolutive functions in harmonic spaces with Doob's convergence axiom and in [5] in harmonic spaces with Bauer's convergence axiom. Results in the same spirit but concerning the fine boundary behaviour of *finely* harmonic functions are obtained in [27].

As mentioned in [5], " ... it is not clear what can be said about unbounded resolutive functions (with or without Doob's convergence axiom). Even for classical potential theory this seems to be an open question."

Results reproduced below (unless otherwise stated) are from [47] and give information in the direction of the above mentioned problem.

First observation: Some restriction on a resolutive g have to be imposed. Indeed, for the harmonic space associated with the heat equation, a resolutive function g can be constructed such that

$$f-\liminf_{y\to z,\, y\in U} H_U g(y) = -\infty, \quad f-\limsup_{y\to z,\, y\in U} H_U g(y) = \infty.$$

Second observation: For classical potential theory, the following result holds: *If $U \subset \mathbb{R}^m$ is a bounded domain, $U^* = (\overline{U})^*$, then there exists a resolutive function $g \geqq 0$ such that $\int g\, d\varepsilon_z^{\complement U} = \infty$.*

Thus resolutivity does not imply $\varepsilon_z^{\complement U}$-integrability. (Note that *for Brelot spaces,* however, *with U connected, every $\varepsilon_z^{\complement U}$-integrable function is resolutive*. For classical potential theory this was established in [10].)

Unbounded resolutive functions are discussed in the theorem below and its corollary.

Theorem 5. *Let g be a lower bounded resolutive function. Then g is $\varepsilon_z^{\complement U}$-measurable and $(*)$ holds.*

Corollary 3. *Suppose that g as well as $|g|$ are resolutive. If g is $\varepsilon_z^{\complement U}$-integrable, then $(*)$ holds.*

Before giving some comments concerning the proof of Theorem 5, we propose the following problem.

Problem 1. Prove or disprove: *There is a resolutive function g on U^* such that*

$$f-\lim_{y\to z,\, y\in U} H_U g(y)$$

exists and is finite, but $g \notin L^1(\varepsilon_z^{\complement U})$.

The proof of Theorem 5 is based on the following result for fine limits. For classical potential theory, it was mentioned without proof in a footnote on p. 207 of [12] and then proved for Brelot harmonic spaces and bounded sequences in [50, p. 172] and for Bauer harmonic spaces and sequences of nonnegative hyperharmonic functions in [17, p. 450]. A proof is included in [47].

Proposition 3. *Let $(u_k)_1^\infty$ be a lower bounded increasing sequence of hyperharmonic functions on U and $u_0 = \lim_{k\to\infty} u_k$ Then the fine limits*

$$a_k = f-\lim_{y\to z} u_k(y), \quad k \in \mathbb{N}\cup\{0\},$$

exist and $\lim_{k\to\infty} a_k = a_0$.

This *monotone convergence property of fine limits* makes it possible to represent the fine limit by means of the theorem of Daniell–Stone as the integral with respect to a measure, which turns out to be $\varepsilon_z^{\complement U}$.

The fine behaviour of $H_U g$ is closely connected with the notion of a maximal sequence.

Definition 2. A sequence (x_n) in U is said to be *maximal* (with respect to z), if $x_n \to z$ and $\varepsilon_{x_n}^{\complement U} \to \varepsilon_z^{\complement U}$.

As a matter of fact, we have already met maximal sequences in Sec. 2 (cf. the lemma of Köhn–Sieveking). The original definition of a maximal sequence in [13] used the Green function. In abstract potential theory, this notion was investigated in [49], [50], [16] and later in [4], [5], [27] in connection with fine limits of bounded harmonic functions.

Under various generality, the following result was proved in the above quoted papers.

Theorem 6. *Let* (x_n) *be a maximal sequence with respect to* z. *Then*

$$\lim_{n\to\infty} H_U g(x_n) = f - \lim_{y\to z,\, y\in U} H_U g(y),$$

whenever g *is a bounded resolutive function.*

In other words, the vague convergence $\varepsilon_{x_n}^{\complement U} \to \varepsilon_z^{\complement U}$ is actually *much stronger* here, namely $\int f \, d\varepsilon_{x_n}^{\complement U} \to \int f \, d\varepsilon_z^{\complement U}$ for every bounded Borel measurable function f on U^*. An attempt to explain this remarkable fact using a result of [21] on convergence of measures can be found in [47].

A closer look shows, however, that the convergence is *actually much stronger*. We conclude this section by a result which does not seem to have been mentioned explicitly before; cf. [44] for a similar investigation.

Theorem 7. *Let* (x_n) *be a maximal sequence with respect to* z. *Then* $\|\varepsilon_{x_n}^{\complement U} - \varepsilon_z^{\complement U}\| \to 0$ *as* $n \to \infty$.

Proof. (cf. [27, p. 644]). We shall denote by $\mathcal{K}$ the set of all continuous functions with compact support defined on X. Fix $\delta > 0$, take a strict continuous potential q and put $L = \{x_n;\, n \in \mathbb{N}\} \cup \{z\}$. It is sufficient to show that there exists a neighbourhood W of z such that

$$\left|\varepsilon_x^{\complement U}(f) - \varepsilon_z^{\complement U}(f)\right| < \delta,$$

whenever $f \in \mathcal{K}$, $|f| \leqq q$, and $x \in L \cap W$. We shall write $\mathcal{K}_1 = \{f \in \mathcal{K};\, |f| \leqq q\}$.

Choose a decreasing sequence (V_k) of open relatively compact neighbourhoods of z such that $V_k \searrow \{z\}$. Construct a decreasing sequence (f_k) of functions from $\mathcal{K}_1$ such that $f_k = q$ on V_k and $f_k = 0$ on $\complement V_{k-1}$, for every $k \in \mathbb{N}$, $k > 1$.

Since (x_n) is a maximal sequence, the function

$$g_k : x \mapsto \varepsilon_x^{\complement U}(f_k), \quad x \in L,$$

is continuous for every $k \in \mathbb{N}$ and the sequence (g_k) is decreasing. The set $\{z\}$ is polar by hypothesis, so that $\varepsilon_x^{\complement U}(\{z\}) = 0$ for every $x \in L$. Since, for every $x \in L$, $f_k \searrow 0$ $\varepsilon_x^{\complement U}$-almost everywhere, we have $g_k \searrow 0$ on L. By Dini's theorem, therefore $g_k \to 0$ uniformly on L.

Consequently, there is an open relatively compact neighbourhood V of z and $g \in \mathcal{K}_1$ such that $g = q$ on V and

$$\varepsilon_x^{\complement U}(g) < \tfrac{1}{8}\delta, \quad x \in L.$$

For $x \in L$, define

$$\mu_x = \varepsilon_x^{\complement U} |_V .$$

Then, for every $x \in L$,

$$\mu_x(q) = \int_V q \, d\varepsilon_x^{\complement U} \leqq \int g \, d\varepsilon_x^{\complement U} < \tfrac{1}{8}\delta$$

and

$$\mu_x^{\complement U \setminus V}(q) = \int \hat{R}_q^{\complement U \setminus V} \, d\mu_x \leqq \mu_x(q) < \tfrac{1}{8}\delta .$$

Define, for every $f \in \mathcal{K}$,

$$h_f : y \mapsto \varepsilon_y^{\complement U \setminus V}(f), \quad y \in V.$$

Then $\{h_f;\ f \in \mathcal{K}_1\}$ is a uniformly bounded family of harmonic functions on V and so is equicontinuous at z; see [20, p. 273]. Consequently, there is an open neighbourhood W of z such that $\overline{W} \subset V$ and

$$\left|\varepsilon_x^{\complement U \setminus V}(f) - \varepsilon_z^{\complement U \setminus V}(f)\right| < \tfrac{1}{2}\delta ,$$

whenever $f \in \mathcal{K}_1$ and $x \in W$.

For every $x \in V$,by [8, p. 317],

$$\varepsilon_x^{\complement U \setminus V} = \varepsilon_x^{\complement U} - \mu_x + \mu_x^{\complement U \setminus V} .$$

It follows that

$$\left|\varepsilon_x^{\complement U}(f) - \varepsilon_x^{\complement U \setminus V}(f)\right| \leqq \mu_x(q) + \mu_x^{\complement U \setminus V}(q) < \tfrac{1}{4}\delta ,$$

whenever $x \in L$ and $f \in \mathcal{K}_1$.

Thus for $f \in \mathcal{K}_1$ and $x \in L \cap W$,

$$\begin{aligned} &|\varepsilon_x^{\complement U}(f) - \varepsilon_z^{\complement U}(f)| \leqq |\varepsilon_x^{\complement U}(f) - \varepsilon_x^{\complement U \setminus V}(f)| + \\ &+ |\varepsilon_x^{\complement U \setminus V}(f) - \varepsilon_z^{\complement U \setminus V}(f)| + |\varepsilon_z^{\complement U \setminus V}(f) - \varepsilon_z^{\complement U}(f)| < \\ &< \tfrac{1}{4}\delta + \tfrac{1}{2}\delta + \tfrac{1}{4}\delta = \delta . \end{aligned}$$

□

4. Regularizing sets of irregular points

Problem 2. Suppose that $f \in C(U^*)$. How can we tell whether f admits a solution of the classical Dirichlet problem? An immediate but obvious answer reads as follows: *f admits a solution of the classical Dirichlet problem, if and only if*

$$(*) \qquad \lim_{x \to z,\, x \in U} H_U f(x) = f(z),$$

for every irregular point $z \in U^*$. But do we really have to verify $(*)$ for *all* irregular points? To answer this question, we should like to find subsets A of U^*_{irr}, as small as possible, possessing the following property: *If $f \in C(U^*)$ and $(*)$ holds for every $z \in A$, then $(*)$ actually holds for every $z \in U^*_{\text{irr}}$.*

Are there always proper subsets of U^*_{irr} having this property? How can these sets be described?

The aim of this section is to summarize results obtained in this area; to a large extent, the exposition is based on [28].

Recall first that $(*)$ holds if and only if $\varepsilon_z^{\complement U}(f) = f(z)$; see the end of Sec. 2.

In view of this fact, the following definition is natural: For every $f \in C(U^*)$, let us *extend* $H_U f$ to $\overline{U}$ by

$$H_U f(z) = \varepsilon_z^{\complement U}(f), \quad z \in U^*.$$

Thus $f \in C(U^*)$ admits a solution of the classical Dirichlet problem if and only if $H_U f = f$ on U^*.

Definition 3. A set $A \subset U^*_{\text{irr}}$ is said to be *regularizing* if $H_U f = f$ on U^*_{irr}, for every $f \in C(U^*)$ satisfying $H_U f = f$ on A.

Proper regularizing sets do exist, at least in the case that the set of irregular points is uncountable.

Theorem 8. *There exists a countable regularizing set.*

To prove this, consider the set $\{\varepsilon_z - \varepsilon_z^{\complement U};\, z \in U^*_{\text{irr}}\}$. This is a subset of a compact, metrizable subspace of the dual of $C(U^*)$ endowed with the weak* topology. Choosing a countable set $A \subset U^*_{\text{irr}}$ such that $\{\varepsilon_z - \varepsilon_z^{\complement U};\, z \in A\}$ is a dense subset of $\{\varepsilon_z - \varepsilon_z^{\complement U};\, z \in U^*_{\text{irr}}\}$, we obtain a regularizing set.

The existence of a countable regularizing set was, in classical potential theory, already stated in [33] and proved in [34]. The proof sketched above is from [14, p. 106]. From [19] we take the following quotation: "La démonstration de Keldych n'est pas simple. Depuis, Brelot en a donné une démonstration fort courte, basée sur un théorème de Banach; toutefois sa démonstration, purement existentielle, ne montre pas comment il faut choisir les points de A." In that paper, three proofs of Theorem 8 are given, the first and the second in an abstract framework. As mentioned, " ... mais la seconde met en évidence, ... , le fait que A doit être assez riche pour représenter tous les types possible d'irregularité."

This second proof is based on the following notion: For a topological space E and a real function φ on E, a subset $P \subset E$ is said to be *a piquetage* of φ, if P is dense in E and, for every $x \in E$, $\varphi(x)$ is an accumulation point of $\varphi|_P$ at x. Obviously, P is a piquetage of φ, if and only if there exists in $E \times \mathbb{R}$ a dense subset of the graph of φ which projects onto P. Thus, *if E is second countable, then there exists a countable piquetage of φ.*

For an application of this notion, let us consider the following situation: Y is a Hausdorff topological space, B is a nowhere dense compact subspace of Y, $V = Y \setminus B$ and $T : C(B) \to C(V)$ is a positive linear mapping such that $\sup T\mathbf{1}(V) < \infty$. (Of course, in our potential theoretic context, $Y = \overline{U}$, $B = U^*$ and $T = H_U$.)

A function $f \in C(B)$ is said to be *regular* at a point $b \in B$, if

$$\lim_{v \to b} Tf(v) = f(b).$$

Now the assertion from [19] reads as follows: *If B is metrizable, then there is a countable set $A \subset B$ such that every $f \in C(B)$, which is regular at every point of A, is actually regular at every point of B.*

To see this, for $f \in C(B)$ and $b \in B$, denote by $\omega_f(b)$ the oscillation of Tf at b. As mentioned in [19], " ... on peut dire, de façon imagée, que $\omega_f(b)$ est la mesure, par la fonction f, de l'irregularité de b."

A countable set with the required property can now be constructed very quickly: Take a countable dense subset F in $C(B)$ and, for every $f \in F$, fix a countable piquetage P_f of the function ω_f. Then $A = \bigcup\{P_f;\ f \in F\}$ is a suitable set.

Let us return to potential theory. Are there points that *have to* appear in an arbitrary regularizing set? A natural guess would be that e.g. isolated points of U^*_{irr} are of this kind. This follows from the following result.

Proposition 4. *Every regularizing set is dense in U^*_{irr}.*

For a proof using "*smearing*" of balayage on a suitably chosen family of closed sets, see [28].

The more careful analysis which follows shows that the denseness should not be expected to be a sufficient condition. On the other hand, given a regularizing set A, it seems plausible that some points of A are redundant, i.e. after having omitted them, the set obtained is still regularizing. The following observation is useful for making this idea more precise.

Observation 2. Suppose that $A \subset U^*_{\text{irr}}$, $z \in \overline{A} \cap U^*_{\text{irr}}$, $f \in C(U^*)$ and $\varepsilon_y^{\complement U}(f) = f(y)$, whenever $y \in A$. We investigate whether $\varepsilon_z^{\complement U}(f) = f(z)$. To that end, take $z_n \in A$ such that $z_n \to z$ and $\varepsilon_{z_n}^{\complement U} \to \mu$. From Sec. 2 we know that there exists $\alpha \in [0,1]$

such that $\mu = \alpha\varepsilon_z + (1-\alpha)\varepsilon_z^{\complement U}$. On the one hand, $\varepsilon_{z_n}^{\complement U}(f) \to \alpha f(z) + (1-\alpha)\varepsilon_z^{\complement U}(f)$, but on the other hand, $\varepsilon_{z_n}^{\complement U}(f) = f(z_n) \to f(z)$. Consequently, $(1-\alpha)f(z) = (1-\alpha)\varepsilon_z^{\complement U}(f)$. We arrive at the following conclusion: *if* $\alpha \neq 1$, *then* $f(z) = \varepsilon_z^{\complement U}(f)$, but *no conclusion can be drawn if* $\alpha = 1$. So we are led to the following definition.

Definition 4. For every subset A of U_{irr}^*, let $\lambda_0(A)$ be the set of all $z \in \overline{A} \cap U_{\mathrm{irr}}^*$ such that $\varepsilon_x^{\complement U}$ *does not* tend to ε_z as $x \in A$ tends to z.

Proposition 5. *Suppose that* $A \subset U_{\mathrm{irr}}^*, B \subset U_{\mathrm{irr}}^*$. *If* $f \in C(U^*)$ *satisfies* $H_U f = f$ *on* A, *then* $H_U f = f$ *on* $\lambda_0(A)$. *Further,* $A \subset \lambda_0(A)$ *and* $\lambda_0(A \cup B) = \lambda_0(A) \cup \lambda_0(B)$.

This follows easily from the above observation and the definition of λ_0. It is clear now that a special role is played by points which satisfy the following definition: A point $z \in U_{\mathrm{irr}}^*$ is said to be *quasi-isolated*, if

$$\lim_{x \to z,\, x \in U^* \setminus \{z\}} \varepsilon_x^{\complement U} = \varepsilon_z.$$

Obviously, every isolated point of U_{irr}^* is quasi-isolated. To say that $z \in U_{\mathrm{irr}}^*$ is quasi-isolated means that $z \notin \lambda_0(U_{\mathrm{irr}}^* \setminus \{z\})$. In classical potential theory, the notion of a quasi-isolated point was introduced in [38], [39]; cf. also [40]. In [39] (where another proof of the existence of a countable regularizing set is given) the following result is proved: Given a countable set $D \subset U_{\mathrm{irr}}^*$ and a quasi-isolated point $z \in U_{\mathrm{irr}}^* \setminus D$, there is a bounded function g on U^* which is continuous except for a polar set, continuous at the points of $D \cup \{z\}$ and which satisfies $g(y) = \varepsilon_y^{\complement U}(g)$ for all $y \in D$, whereas $g(z) \neq \varepsilon_z^{\complement U}(g)$. This indicates that, from our point of view, the point z is, in some sense, far from D.

The following type of subset of U_{irr}^* is important for our investigations: A set $A \subset U_{\mathrm{irr}}^*$ is said to be *a piquetage faible*, if $\lambda_0(A) = U_{\mathrm{irr}}^*$. It follows from Proposition 5 that every piquetage faible is regularizing. In [19], the following problem was proposed:

Problem 3. *Is every regularizing set a piquetage faible?*

The notion of a piquetage faible in classical potential theory was introduced in [19]. As a matter of fact, the original definition is expressed in terms of the Green function and only irregular points having the property that each of their neighbourhood meets U^* in a non-polar set are considered. The reason for the terminology lies in the fact that a piquetage faible is a "special" piquetage for the function defined at every irregular point as a fine limit of the Green function. A *negative answer* to Problem 3 is contained in the following proposition.

Proposition 6. *For classical potential theory in* $\mathbb{R}^3$, *there is a bounded open set* U, *a countable set* $A \subset U_{\mathrm{irr}}^*$ *and a point* $z_0 \in U_{\mathrm{irr}}^*$ *such that* $\lambda_0(A) = U_{\mathrm{irr}}^* \setminus \{z_0\}$ *and* $\lambda_0(\lambda_0(A)) = U_{\mathrm{irr}}^*$. *Thus* A *is a regularizing set which is not a piquetage faible.*

Such a set U is constructed in [28] by successively removing countably many Lebesgue spines from, say, a ball. This procedure can be made to work using the complete information on cluster sets of harmonic measures given in Sec. 2, and the following "perturbation lemma":

Lemma 1. *Let $M \subset X, x \in X$ and $x_n \in X, n \in \mathbb{N}$, satisfy $x_n \to x$ and $\varepsilon_{x_n}^M \to \alpha\varepsilon_x + (1-\alpha)\varepsilon_x^M$ for some $\alpha \in [0,1]$. Let $P \subset X$ be such that $\hat{R}_1^{P\cap V}$ tends to 0 uniformly on the set $\{x_n; n \in \mathbb{N}\}$ as the neighbourhood V of x decreases to $\{x\}$. Then $\varepsilon_{x_n}^{M\cup P} \to \alpha\varepsilon_x + (1-\alpha)\varepsilon_x^{M\cup P}$.*

In the example given in [28], U_{irr}^* splits into two disjoint parts: $\{z_n; n \in \mathbb{N}\cup\{0\}\}$ and $A = \{y_{mn}; m, n \in \mathbb{N}\}$, each of them consisting of different points. Here are the properties of the irregular points: $z_n \to z_0$; for every n, $y_{mn} \to z_n$ as $m \to \infty$; $\varepsilon_{z_n}^{\complement U} \to \varepsilon_{z_0}^{\complement U}$; for every n,

$$\varepsilon_{y_{mn}}^{\complement U} \to (1-\frac{1}{n})\varepsilon_{z_n} + \frac{1}{n}\varepsilon_{z_n}^{\complement U} \quad \text{as} \quad m \to \infty,$$

and $\varepsilon_y^{\complement U} \to \varepsilon_{z_0}$ as $y \in A$ tends to z_0.

Then $\lambda_0(A) = A \cup \{z_n; n \in \mathbb{N}\} \neq U_{\text{irr}}^*$ and $\lambda_0(\lambda_0(A)) = \lambda_0(A) \cup \{z_0\} = U_{\text{irr}}^*$.

The example suggests that an important feature of regularizing sets could be that successive iterations of λ_0 should fill in the whole U_{irr}^*. This vague idea is made precise by introducing a new topology on U_{irr}^*. (In topological terminology [18, pp. 237, 272], λ_0 is a closure operation and λ is the topological modification of λ_0.)

Definition 5. For every subset $A \subset U_{\text{irr}}^*$, let $\lambda(A)$ denote the smallest subset B of U_{irr}^* containing A and satisfying $\lambda_0(B) \subset B$.

The mapping $A \mapsto \lambda(A)$ is the *closure operation* for a topology on U_{irr}^* which is finer that the original topology. A set $A \subset U_{\text{irr}}^*$ is said to be *λ-closed* (*λ-dense*) in U_{irr}^*, if $\lambda(A) = A$ (respectively, $\lambda(A) = U_{\text{irr}}^*$). Observe that the *isolated points in the new topology are exactly quasi-isolated points*. It is shown in [28] that, for a λ-closed set $A \subset U_{\text{irr}}^*$, both A and $U_{\text{irr}}^* \setminus A$ are K_σ-sets. The operation λ is, in the following sense, more powerful than λ_0.

Proposition 7. *Let $A \subset U_{\text{irr}}^*$. If $f \in C(U^*)$ satisfies $H_U f = f$ on A, then $H_U f = f$ on $\lambda(A)$.*

Indeed, if $B = \{H_U f = f\} \cap U_{\text{irr}}^*$, then $A \subset B$ and $\lambda_0(B) \subset B$ by Proposition 5. Thus $\lambda(A) \subset B$.

Corollary 4. *Every λ-dense subset of U_{irr}^* is regularizing.*

As we shall see, the converse holds, for example, in classical potential theory. On the other hand, this is not true in general; see [28]. We indicate how a *counterex-*

ample was constructed there for the harmonic space associated to the heat equation in $\mathbb{R} \times \mathbb{R}$.

Let V be a regular set and $z_0 = (x_0, t_0) \in V$. Then closed intervals $I_n \subset \mathbb{R}$ and a strictly increasing sequence (t_n) tending to t_0 can be chosen so that the set $S = \bigcup_{n=1}^{\infty}(I_n \times \{t_n\})$ has the following properties: $\overline{S} \subset V$, $\overline{S} \setminus S = \{z_0\}$, S is thin at z_0 and, for a strict potential q,

$$\lim_{z \to z_0,\, z \in S} \hat{R}_q^S(z) = q(z_0).$$

Then $U = V \setminus \overline{S}$ is open, $U^*_{\text{irr}} = \overline{S}$ and z_0 is quasi-isolated in view of the behaviour of q on S. Thus S is a *proper* λ-closed subset of U^*_{irr}. It is, however, regularizing. The proof of this fact depends on the following fact, see [36]: *Let W be an open subset of $\mathbb{R} \times \mathbb{R}$ and E be a straight-line parallel to the x-axis. If $w \in C(W)$ and $w|_{W \setminus E}$ fulfills the heat equation on $W \setminus E$, then w is a solution of the heat equation on the whole of W.*

An important (even indispensable) feature of this example is that U^*_{irr} is *not* negligible; see Sec. 2. In fact, the following theorem, proved in [28], provides a full characterization of regularizing sets under the additional hypothesis that U^*_{irr} is *negligible*. (This is no restriction as long as semipolar subsets of X are polar as in the classical case.)

Theorem 9. *Suppose that the set U^*_{irr} is negligible. Then a subset $A \subset U^*_{\text{irr}}$ is regularizing if and only if A is λ-dense in U^*_{irr}.*

The proof depends on a duality argument which we now indicate.

Given a Borel subset A of U^*, denote by $\mathcal{M}_A^+$ the set of all positive Radon measures ν on U^* for which $\nu(\complement A) = 0$. Let $\mathcal{N}_A^+$ denote the set $\{\nu - \nu^{\complement U};\ \nu \in \mathcal{M}_A^+\}$ and $\mathcal{N}_A = \mathcal{N}_A^+ - \mathcal{N}_A^+$.

Proposition 8. *Let A be a Borel subset of U^*_{irr}. Then the following statements are equivalent:*

(i) *A is λ-closed;*
(ii) *$\mathcal{N}_A$ is closed.*

One implication is easy. Indeed, let A fail to be λ-closed. Then there exist $z \in U^*_{\text{irr}} \setminus A$, $0 \leqq \alpha < 1$ and a sequence (x_n) in A such that $x_n \to z$ and $\varepsilon_{x_n}^{\complement U} \to \alpha\varepsilon_z + (1-\alpha)\varepsilon_z^{\complement U}$. Thus

$$\varepsilon_z - \varepsilon_z^{\complement U} = \lim_{n \to \infty} \frac{1}{1-\alpha}(\varepsilon_{x_n} - \varepsilon_{x_n}^{\complement U}) \in \overline{\mathcal{N}}_A,$$

whereas $\varepsilon_z - \varepsilon_z^{\complement U} \notin \mathcal{N}_A$. This proves that (ii)$\Rightarrow$(i).

The proof that (i)$\Rightarrow$(ii) requires several auxiliary results which depend essentially on the fact that U^*_{irr} is negligible. The reader is referred to [28] for the details.

Theorem 9 characterizing regularizing sets follows immediately from the following consequence of Proposition 8: *If $A \subset U^*_{\text{irr}}$ and $z \in U^*_{\text{irr}} \setminus \lambda(A)$, then there exists a function $f \in C(U^*)$ such that $\varepsilon_x^{\complement U}(f) = f(x)$ for every $x \in A$ whereas $\varepsilon_z^{\complement U}(f) \neq f(z)$.*

To see this, remark first that $\lambda(A)$ is λ-closed, so that $\mathcal{N}_{\lambda(A)}$ is closed. Since $\varepsilon_z - \varepsilon_z^{\complement U} \notin \mathcal{N}_{\lambda(A)}$, a function $f \in C(U^*)$ with the above stated properties exists by the Hahn–Banach theorem.

Finally, we remark that a stronger separation property by means of a potential is also proved in [28]: *Let U^*_{irr} be negligible, $K \subset U^*$ be a compact polar set and $U^*_{\text{irr}} \setminus K$ be λ-closed. Then there exists a potential q on X such that $q\,|_{U^* \setminus K}$ admits a continuous extension f on U^* satisfying $H_U f = f$ on $U^*_{\text{irr}} \setminus K$ and $H_U f(z) \neq f(z)$ for every $z \in K$.*

References

[1] Bauer, H., Axiomatische Behandlung des Dirichletschen Problems für elliptische und parabolische Differentialgleichungen. Math. Ann. 146 (1962), 1–59

[2] —, Harmonische Räume und ihre Potentialtheorie. LN in Math. 22, Springer–Verlag, 1966

[3] —, Harmonic spaces: A survey. Conferenze del Seminario di Matematica dell'Università di Bari, Vol. 197, 1984

[4] —, Fine boundary limits of harmonic and caloric functions. Bull. Sci. Math. 109 (1985), 337–361

[5] —, Fine boundary limits and maximal sequences. Potential Theory and its Related Fields, Kyoto University, Kyoto, 1987

[6] Bliedtner, J., Hansen, W., Simplicial cones in potential theory. Invent. Math. 29 (1975), 83–110

[7] —, Cones of hyperharmonic functions. Math. Z. 151 (1976), 71–87

[8] —, Potential Theory - An Analytic and Probabilistic Approach to Balayage. Springer–Verlag, 1986

[9] Boboc, N., Cornea, A., Comportement des balayées des mesures ponctuelles. Comportement des solutions du problème de Dirichlet aux points irréguliers. C.R. Acad. Sci. Paris 264 (1967), 995–997

[10] Brelot, M., Sur les ensembles effilés. Bull. Sci. Math. 68 (1944), 12–36

[11] —, Sur l'allure des fonctions harmoniques à la frontière. C.R. Acad. Sci. Paris 220 (1945), 676–678

[12] —, Etude générale des fonctions harmoniques et surharmoniques positives au voisinage d'un point-frontière irrégulier. Ann. Univ. Grenoble 22 (1946), 205–219

[13] —, On the behavior of harmonic functions in the neighborhood of an irregular boundary point. J. Anal. Math. 4 (1954–56), 209–221

[14] —, Élements de la théorie classique du potentiel, 4e ed. Centre de Documentation Universitaire, Paris 1969

[15] —, On topologies and boundaries in potential theory. LN in Math. 175, Springer–Verlag, 1971

[16] —, Quelques recherches sur allure à la frontière des fonctions harmoniques. LN in Math. 419, Springer–Verlag, 1974; 60–66

[17] —, Sur l'allure à la frontière des familles de fonctions harmoniques ou surharmoniques. Rendiconti di Mat. (2) 8 (1975), 447–456

[18] Čech, E., Topological spaces. Interscience, New York 1966

[19] Choquet, G., Sur un théorème de Keldych concernant le problème de Dirichlet. Ann. Inst. Fourier (Grenoble) 18 (1968), 309–315

[20] Constatinescu, C., Cornea, A.,Potential Theory on Harmonic Spaces. Springer–Verlag, 1972

[21] Dieudonné, J., Sur la convergence des suites des mesures de Radon. Anais Acad. Brasil. Ci. 23 (1951), 21–38
[22] Doob, J.L., Classical potential theory and its probabilistic counterpart. Springer–Verlag, 1984
[23] Effros, E.G., Kazdan, J.L., Applications of Choquet simplexes to elliptic and parabolic boundary value problems. J. Differential Equations 8 (1970), 95–134
[24] Frostman, O., Les points irréguliers dans la théorie du potentiel et le critère de Wiener. Medd. Lunds Univ. Mat. Sem. 4 (1939), 1–10
[25] Hansen, W., Potentialtheorie semi-regulärer harmonischer Kerne. Z. Warscheinlichkeitstheorie u. Verw. Gebiete 18 (1971), 298–304
[26] —, Convergence of balayage measures. Math. Ann. 264 (1983), 437–446
[27] —, Fine boundary limits of finely harmonic functions. Math. Ann. 274 (1986), 643–647
[28] Hansen, W., Netuka, I., Regularizing sets of irregular points. J. Reine Angew. Math. 409 (1990), 205–218
[29] Helms, L.L., Introduction to potential theory. Wiley and Sons, 1969
[30] Hyvönen, J., On the harmonic continuation of bounded harmonic functions. Math. Ann. 245 (1979), 151–157
[31] Ikegami, T., On the regularity of boundary points in a resolutive compactification of a harmonic space. Osaka J. Math. 14 (1977), 271–289
[32] —, On the boundary behavior of the Dirichlet solutions at an irregular boundary point. Osaka J. Math. 21 (1984), 851–858
[33] Keldych, M.V., On the solvability and stability of the Dirichlet problem (Russian). Dokl. Akad. Nauk SSSR 18 (1938), 315–318
[34] —, On the solvability and stability of the Dirichlet problem (Russian). Uspehi Mat. Nauk 8 (1941), 172–231
[35] Kellogg, O.D., Recent progress with the Dirichlet problem. Bull. Amer. Math. Soc. 32 (1926), 601–625
[36] Köhn, J., Sieveking, M., Reguläre und extremale Randpunkte in der Potentialtheorie. Rev. Roumaine Math. Pures Appl. 12 (1967), 1489–1502
[37] Král, J., Problem No. 1 (Czech). Časopis Pěst. Mat. 97 (1972), 334
[38] Landkof, N.S., On the structure of the set of irregular points in the Dirichlet problem (Russian). Dokl. Akad. Nauk SSSR 28 (1940), 291–293
[39] —, On the solvability of the generalized Dirichlet problem (Russian). Izv. Akad. Nauk SSSR, ser. matem. 11 (1947), 181–196
[40] —, Foundations of Modern Potential Theory. Springer–Verlag, 1972
[41] Lukeš, J., Semiregular sets in harmonic spaces (Czech). Časopis Pěst. Mat. 100 (1975), 195–197
[42] —, On the set of semiregular points. Potential theory Copenhagen 1979, LN in Math. 787, Springer–Verlag, 1980; 212–218
[43] Lukeš, J., Malý, J., On the boundary behaviour of the Perron generalized solution. Math. Ann. 257 (1981), 355–366
[44] Lyons, T.J., A theorem in fine potential theory and applications to finely holomorphic functions. J. Funct. Anal. 37 (1980), 19–26
[45] Netuka, I., Remark on semiregular sets (Czech). Časopis Pěst. Mat. 98 (1973), 419–421
[46] —, Solution of the problem No. 1 (author J. Král) from 97 (1972), 334 (Czech). Časopis Pěst. Mat. 99 (1974), 90–93
[47] —, Fine behaviour of solutions of Dirichlet problem near an irregular point. Bull. Sci. Math. 114 (1990), 1–22
[48] —, Cluster sets of harmonic measures and the Dirichlet problem in potential theory. Summer School in Potential Theory, Joensuu 1990 (to appear)
[49] Smyrnélis, E., Allure des fonctions harmoniques au voisinage d'un point-frontière irrégulier. C.R. Acad. Sci. Paris 267 (1968), 157–159
[50] —, Sur les limites fines des fonctions harmoniques et les suites maximales. Bull. Sci. Math. 97 (1973), 161–175
[51] Vasilesco, F., Le problème de Dirichlet dans le cas le plus général. Enseign. Math. 35 (1936), 88–106

[52] Wittmann, R., Nicht-tangentieles Randverhalten von harmonischer Funktionen in irregulären Randpunkten. Arch. Math. 48 (1987), 526–529
[53] Boboc, N., Bucur, Gh., On Frostman property in standard H-cones of functions. Stud. Cerc. Mat. 40 (1988), 125–136
[54] Oja, K., Connectivity properties for complements of exceptional sets. Ann. Acad. Sci. Fenn. (AI) Math. 12 (1987), 171–176

Capacities in function theory

Anthony G. O'Farrell

1. Introduction

Ever since the famous thesis of Frostman, capacities have been important in many areas of function theory. In this talk I shall be concerned only with one–variable function theory on arbitrary open subsets of the complex plane, **C**. It is important to stress that the open sets need not be connected. I will discuss the use of analytic capacities in connection with problems of removable singularities, holomorphic approximation, and boundary smoothness.

A brief reference to the applications is in order. The connections between analytic (and harmonic) functions and the physics of perfect fluids, electrostatics, magnetostatics, classical gravitation and heat are well–known. Much of what I shall say about analytic functions (solutions of the d–bar equation) applies also to solutions of other elliptic equations, and so there are other applications, for instance to elasticity (connected to the bi-Laplacian and the d-bar-squared operator). It should also be noted that students of capacitary problems were among the first to discover and examine many of the weird–looking sets formerly regarded as pathological by most people, but now known as fractals and accepted as natural objects of study for many applications.

Here is a brief outline of the talk. Let U be an open subset of **C**, and $F \subset \mathcal{D}'(\mathbf{C}, \mathbf{C})$ be a topological vector space of distributions (generalised functions) on **C**. We define

$$AF(U) = \{f \in F : \bar{\partial} f = 0 \text{ on } U\}.$$

In other words, $AF(U)$ consists of those distributions $f \in F$ that are analytic on U. Typically, the space F is supposed to be characterised by some real–variable property, such as boundedness, continuity, etc. One expects that the functions $f \in AF(U)$ are in some respects nicer than typical elements of F. To the space F, we shall associate an analytic capacity

$$\bar{\partial} - F - \text{cap} : \quad 2^{\mathbf{C}} \to [0, +\infty].$$

The definition of $\bar{\partial}$–F–cap will be given presently, but for now it is enough to note that it is a monotone nonnegative set function. The main idea of analytic capacity theory is that it is possible to recover many *local* properties of the spaces $AF(U)$ from a knowledge of the $\bar{\partial}$–F capacity. Consider the following problems.

Potential Theory, ed. by M. Kishi

Removability:
A compact $K \subset \mathbf{C}$ is said to be $\bar{\partial}$–F–null if

$$AF(U \sim K) = AF(U), \qquad \forall \text{ open } U \subset \mathbf{C}.$$

The removability question is to characterise the $\bar{\partial}$–F–null sets in some explicit way.

Approximation:
For which compact sets $X \subset \mathbf{C}$ is

$$\bigcup\{AF(U) : X \subset U, U \text{ open}\}$$

dense in $AF(U)$ in the topology of F?

Boundary smoothness:
For which open $U \subset \mathbf{C}$, $a \in \text{bdy}U$, and nonnegative integers k, is the map

$$f \mapsto f^{(k)}(a)$$

reasonably well–defined on $AF(U)$?

2. Example: lipα

To appreciate these problems, it is useful to look at an example. Consider the case $F =$lipα.

Let $0 < \alpha \leq 1$.

Definition. $f \in \text{Lip}\alpha$ means $f : \mathbf{C} \to \mathbf{C}$ and

$$|f(z) - f(w)| \leq \kappa |z - w|^{\alpha} \qquad \forall z, w \in \mathbf{C}$$

for some constant κ, independent of z and w.

Lipα becomes a Banach space when given the norm

$$\|f\|_{\text{Lip}\alpha} = |f(0)| + \text{ least } \kappa.$$

The subspace of Lipα consisting of those f such that

$$\lim_{|z-w|\downarrow 0} \frac{|f(z) - f(w)|}{|z - w|^{\alpha}} = 0$$

and

$$\lim_{|z-w|\uparrow\infty} \frac{|f(z)-f(w)|}{|z-w|^\alpha} = 0$$

is a closed subspace, the closure of the space $\mathcal{D}$ of test functions. We denote it lipα.

The $\bar{\partial}$–Lipα–cap turned out to be equivalent to $(1+\alpha)$*–dimensional Hausdorff content*, $M^{1+\alpha}$ (Dolzhenko, 1964). The Hausdorff content corresponding to a non-negative increasing function h on $[0,\infty)$ is defined by

$$M_h(E) = \inf_{\mathcal{S}} \sum_{B\in\mathcal{S}} h(\mathrm{diam} B), \qquad \forall E \subset \mathbf{C},$$

where $\mathcal{S}$ runs over all countable coverings of E. The $\bar{\partial}$–lipα–cap is equivalent to $(1+\alpha)$–dimensional *lower* Hausdorff content, $M_*^{1+\alpha}$. This content is defined by

$$M_*^\beta(E) = \sup\left\{M_h(E) : h(r) \le r^\beta, r^{-\beta}h(r) \to 0 \text{ as } r\downarrow 0\right\}.$$

The solutions to the three problems for lipα are as follows (Similar results hold for Lipα, but the statements are less straightforward, because in that case reference must be made to the weak–star topology on Lipα).

Theorem 1 (Dolzhenko). *Let $K \subset \mathbf{C}$ be compact. Then K is $\bar{\partial}$–lipα–null if and only if*

$$M_*^{1+\alpha}(K) = 0.$$

Theorem 2. *Let X be a compact subset of* $\mathbf{C}$. *Then*

$$\bigcup\{\mathit{Alip}\alpha(V) : V \text{ is a neighbourhood of } X\}$$

is dense in Alipα(int$X$) if and only if

$$M_*^{1+\alpha}(D\sim X) \ge \mathrm{const}\cdot M_*^{1+\alpha}(D\sim \mathrm{int}X), \qquad \forall \text{ discs } D\subset\mathbf{C}.$$

In fact, in this statement, one may replace the $M_*^{1+\alpha}(D\sim X)$ on the left–hand side by $M^{1+\alpha}(D\sim X)$. The statement as given serves to emphasise the fact that the problem may be solved using only the capacity $M_*^{1+\alpha}$.

To understand the statement intuitively, observe that the approximation problem involves the approximation of functions whose singularities lie in $\mathbf{C}\sim$ int$X$ by functions whose singularities lie in the smaller set $\mathbf{C}\sim X$. It is a question of "pushing" singularities off bdy$X$ into the complement of X. The capacity $\bar{\partial}$–lipα–cap ($\sim M_*^{1+\alpha}$) measures, in fact, the *capacity to carry singularities* of lipα analytic functions. Also, the problem is local. So it is reasonable that the solution should involve the comparison of the capacity of $D \sim X$ and $D \sim \mathrm{int}X$ for all disks D. Naturally, it is not important that disks are used; squares, or arbitrary open sets would do just as well.

The first theorem of this type for analytic functions was for continuous functions and uniform approximation, and was proved by Vitushkin. Earlier, Keldysh proved the corresponding result for harmonic functions and uniform approximation.

Lemma 1. *Let $U \subset \mathbf{C}$ be open and $a \in \mathbf{C}$. Then the set of all $f \in Alip\alpha(U)$ that are analytic on a neighbourhood of a (depending on f) is a dense subset of $Alip\alpha(U)$.*

The uniform approximation version of this lemma was first discovered by Arens.

This lemma enables us to state more precisely the boundary smoothness question for $\bar{\partial}$–lipα. It is: to characterise those U, a, and $k \in \mathbf{N}$ such that the functional $f \mapsto f^{(k)}(a)$, defined on

$$\{f \in \mathrm{Alip}\alpha(U) : f \text{ is analytic near } a\}$$

extends continuously to all of Alip$\alpha(U)$. When such a continuous extension (necessarily unique) exists, we say that Alip$\alpha(U)$ admits a *k–th order continuous point derivation* at a.

Only points a that lie on the boundary of U are of interest, because Alip$\alpha(U)$ admits continuous point derivations of all orders at points of U, and of no order $k \geq 1$ at points outside clos$U$.

Theorem 3. *Let $U \subset \mathbf{C}$ be open, $a \in \mathbf{C}$, and $k \in \mathbf{N}$. Then $Alip\alpha(U)$ admits a k–th order continuous point derivation at a if and only if*

$$\sum_{n=1}^{\infty} 2^{(k+1)n} M_*^{1+\alpha}(A_n \sim U) < +\infty.$$

Here A_n denotes the annulus

$$A_n = \left\{z \in \mathbf{C} : \frac{1}{2^{n+1}} < |z-a| \leq \frac{1}{2^n}\right\}.$$

3. Symmetric concrete spaces

A Symmetric Concrete Space (SCS) on $\mathbf{R}^d$ is a complete LCTVS, F, such that

1. $\mathcal{D} \hookrightarrow F \hookrightarrow \mathcal{D}'$,
2. $f \mapsto \bar{f}$ maps $F \to F$ continuously,

3. $$\left\{\begin{matrix} \mathcal{D}\times F \to F \\ (\phi, f) \mapsto \phi f \end{matrix}\right\}$$

makes F a topological $\mathcal{D}$–module,

4. for each $T\in$Aff,

$$c_T : \left\{\begin{matrix} F \to F \\ f \mapsto f\circ T \end{matrix}\right\} \text{ is continuous}$$

and $T \mapsto c_T$ maps compact subsets of Aff to equicontinuous subsets of End(F).

If an SCS is normable, we call it a symmetric concrete Banach space (SCBS).

An SCS is called *small* if $\mathcal{D}$ is sequentially dense in it.

As examples, L^p, C^k, Lipα, lipα, BMO, VMO, Sobolev spaces, Besov spaces, Bloch space, Zygmund class (ZC) and Zygmund smooth class (ZS) are SCS. The space L^p is small if $p < +\infty$. Other small spaces are C^k, lipα, VMO, and ZS $_{\text{loc}}$.

I now give a brief account of basic SCS theory.

Let F be an SCS. To a closed subset $X \subset \mathbf{R}^d$, we associate spaces $F(X)$ — germs on X, F_X — elements of F that are supported on X, and

$$F_{\text{loc}} = \mathcal{E}\cdot F \qquad ; \qquad F_{\text{cs}} = \mathcal{D}\cdot F.$$

The spaces F_{loc} and F_{cs} inherit natural topologies which make them into SCS's.

The most useful equivalence relation on SCS is *local equivalence*, defined by

$$F_1 \overset{\text{loc}}{=} F_2 \Leftrightarrow F_{1\text{loc}} = F_{2\text{loc}}.$$

The notion of *local inclusion*, defined by

$$F_1 \overset{\text{loc}}{\hookrightarrow} F_2 \Leftrightarrow F_{1\text{loc}} \subset F_{2\text{loc}}$$

gives a partial order on the local equivalence classes. It turns out that for metrisable SCS, $F_1 \subset F_2$ is equivalent to $F_1 \hookrightarrow F_2$, so that

$$F_1 \overset{\text{loc}}{\hookrightarrow} F_2 \Leftrightarrow F_{1\text{loc}} \hookrightarrow F_{2\text{loc}}.$$

The usefulness of this notion is illustrated by the L^p spaces. One never has $L^p \hookrightarrow L^q$ if $p \neq q$, but $L^p \overset{\text{loc}}{\hookrightarrow} L^q$ if and only if $p \geq q$, so $\overset{\text{loc}}{\hookrightarrow}$ gives a linear order on the L^p spaces. In general, $\overset{\text{loc}}{\hookrightarrow}$ is not a total order on the SCS. For instance, in two dimensions, the space $C = C^0$ of continuous functions and the Sobolev space $W^{1,2}$ are unrelated by $\overset{\text{loc}}{\hookrightarrow}$

The following "F_∞ construction" is useful:

The space F_∞ associated to an SCS, F, is the set of all those $f \in F_{\text{loc}}$ such that

$$f(\cdot + a)|\mathbf{B}(0,1) \to 0$$

in $F(\mathbf{B}(0,1))$–topology as $a \to \infty$. This construction produces a new SCS, locally–equivalent to the original. It may be larger or smaller than the original. For instance, C^0_∞ is the space, often denoted C_0, of continuous functions that tend to zero at infinity, and is smaller than C^0, whereas L^2_∞ is the space of measurable functions that have

$$\int_{\mathbf{B}(a,1)} |f|^2 dx \to 0$$

as $a \to \infty$, and is larger than L^2. The space F_∞ has some canonical properties. For instance, if there is an SCBS in the local equivalence class of F, then F_∞ is normable.

For any SCS, F, there is a canonical map $i : F^* \to \mathcal{D}'$, the adjoint of the inclusion map $\mathcal{D} \hookrightarrow F$. This map is injective if and only if $\mathcal{D}$ is dense in F, and then $i(F^*)$ is an SCS, a canonical SCS realisation of F^*. Similarly, if F is a dual space, then there is a canonical SCS realisation of the predual F_* if and only if $\mathcal{D}$ is weak–star dense in F; in that case, the canonical predual is just the canonical dual of (F, weak–star).

There is a duality between restriction spaces and support spaces, given by

$$F(X)^* = (F^*)_X, \qquad (F_X)^* = F^*(X),$$

for all $F \in$SCS and all closed $X \subset \mathbf{R}^d$.

If F_1 and F_2 are SCS, then so are the intersection $F_1 \cap F_2$ and the sum $F_1 + F_2$ (— the inner vector space sum in $\mathcal{D}'$), when endowed with the obvious topologies. The only point that requires a little care in the proof is the completeness of these spaces. A corollary is the fact, quoted above, that inclusion inplies continuous inclusion, for SCS.

I conclude this brief summary of the basic properties of SCS with an important observation about the action of convolution. The convolution $f * g$ makes sense when f is a distribution having compact support and g is any distribution. Thus we may consider it for $f \in F_{\text{cs}}$ and $g \in \mathrm{L}^1{}_{\text{loc}}$.

Theorem 4. *Suppose (a) F is a small SCS,* or *(b) F is the SCS dual of a small SCS. Then the map*

$$(f, g) \mapsto f * g$$

maps

$$F_{\text{cs}} \times L^1{}_{\text{loc}} \to F_{\text{loc}}$$

continuously.

The conditions given on F may be relaxed, but those given cover all spaces of interest to me.

To indicate the proof, one may begin by remarking that the case (b) may be reduced to case (a). The difficult thing is to demonstrate that convolution maps

$$F_{cs} \times \mathrm{L}^1{}_{\mathrm{loc}} \to F_{\mathrm{loc}};$$

once this is known, continuity is not a problem. Thus part (b) is obtained by applying part (a) to the SCS (F, weak–star).

To prove the result on hypothesis (a), fix $f \in F_{cs}$ and $g \in \mathrm{L}^1{}_{\mathrm{loc}}$. Then $f * g$ is defined, as an element of $\mathcal{D}'$, by the usual formula:

$$\langle \psi, f * g \rangle = \langle \langle \psi(x+y), f_x \rangle, g_y \rangle, \quad \forall \psi \in \mathcal{D}. \qquad (*).$$

The first step is to extend the domain of $f * g$ from $\mathcal{D}$ to $(F_{\mathrm{loc}})^*$. It is straightforward to check that (*) makes sense whenever we replace ψ by an element of $(F_{\mathrm{loc}})^* = (F^*)_{cs}$, and defines an element $f * g$ of $(F_{\mathrm{loc}})^{**}$.

The next step is to apply the Banach–Grothendieck Theorem to show that in fact $f * g$ lies in the image of F_{loc} under the canonical injection of F_{loc} into $(F_{\mathrm{loc}})^{**}$. According to that theorem, this amounts to showing that $f * g$ is weak–star continuous on the polar of each neighbourhood of zero in F_{loc}. Because of smallness, this reduces to checking a simple sequential statement. □

One might remark that the intuition behind this result is quite simple. Convolving a distribution with a locally–integrable function is a process of taking limits of averages over translates of reflections (in 0) of the distribution. SCS are nicely preserved under translation and reflection, and they are complete, so that it is reasonable to suppose that they will be essentially preserved by convolutions.

4. Analytic capacities

Now we define the analytic capacities associated to an SCS on the plane. For simplicity, we restrict to the case where the SCS F is locally–equivalent to an SCBS. In that case, F_∞ may be normed. We assume this done, and (as is always possible) that the norm on F_∞ is translation–invariant.

Let K be a compact subset of $\mathbf{C}$. Let $\phi \in \mathcal{D}$ be a test function such that $\phi = 1$ on a neighbourhood of K. We define the analytic–F– capacity

$$\bar{\partial} - F - \mathrm{cap}(K) = \sup \left\{ \frac{1}{\pi} |\langle \bar{\partial}\phi, f \rangle| : f \in \mathrm{Ball}\ \mathrm{A}F_\infty(\mathbf{C} \sim K) \right\}.$$

This capacity is a nonnegative, monotone set function, which carries information about the space F and the analytic functions. The quantity

$$\frac{1}{\pi} \langle \bar{\partial}\phi, f \rangle$$

represents the coefficient a_1 in the Laurent expansion

$$f = a_0 + \frac{a_1}{z} + \cdots$$

of f about ∞.

Given two set functions α and β on a family of subsets of C, we say that they are locally equivalent if for each compact set $X \subset \mathbf{C}$ there is a constant $\kappa > 0$ such that

$$\frac{1}{\kappa}\alpha(K) \leq \beta(K) \leq \kappa\alpha(K)$$

whenever $K \subset X$.

For application to the problems we have mentioned above, the precise values of the capacities are not important: all that matters is the local equivalence class of the appropriate capacity. Many of the interesting analytic capacities have been explicitly identified up to local equivalence. Here is a brief summary of the main results:

The L^∞ capacity (*the* analytic capacity, in ordinary parlance) was the first to be introduced (Ahlfors, 1947), and has at least four distinct descriptions.

The C analytic capacity generates the same outer capacity as the L^∞ analytic capacity, so both are as well (or as little) understood on sets with fat interior. On sets with no interior, the two functions are in general quite different, and not much is known about the C capacity. It is worth mentioning the remarkable result of Browder–Wermer–Carleson–Garnett– Bishop–Jones that arcs with no tangents have positive $\bar{\partial}$–C–capacity.

The L^2 analytic capacity is locally–equivalent to logarithmic capacity, which is very well understood (Hedberg). It may be computed as a kernel capacity, a Chebyshev constant, a transfinite diameter, a condenser capacity associated to the Sobolev space $W^{1,2}$ (Dirichlet space).

For $2 < p < +\infty$, the L^p analytic capacity is equivalent to a condenser capacity, an extremal length, and an iterated potential capacity. (Hedberg, Havin, Mazya, Ziemer)

The C^1 analytic capacity of a compact set is equivalent to the area of its interior. (Nguyen, Hrushchev).

The Lipα capacity ($0 < \alpha < 1$) is equivalent to the Haudorff content $M^{1+\alpha}$, and the lipα capacity is equivalent to the corresponding lower content $M_*^{1+\alpha}$. (Dolzhenko, Gonchar, Mergelyan, Garnett, author).

The Lip1 capacity is equivalent to area (Nguyen, Hrushchev). The space lip1 is not an SCS (too small).

The BMO capacity is equivalent to M^1, and the VMO capacity is equivalent to M_*^1 (Kaufman).

For $p > 1$, the $W^{1,p}$ capacity is equivalent to $|\text{area}|^{1/q}$, where q is the index conjugate to p.

The C^k capacity of a compact K is equivalent to

$$\int_K \text{dist}(z, \mathbf{C} \sim K)^{k-1} dxdy.$$

A few interesting analytic capacities remain to be constructively identified, such as the $W^{1,1}$, $W^{2,p}$, ZC and ZS capacities. Other interesting questions concern the local affine invariance and quasisubadditivity of various capacities. Both questions remain open for the Ahlfors capacity $\bar{\partial}$–L$^\infty$–cap. Quasisubadditivity is not a necessary property of analytic capacities in general (— for instance, it fails for the lip(3/2) capacity), but no example is known of an SCS analytic capacity that is not locally affine–invariant.

There is, naturally, no problem of removable singularities for spaces like C^3: the sets of removable singularities are exactly the sets with no interior. The capacity associated to such a space is however of value is examining boundary smoothness. For instance, at which points on the boundary do we have 4–th order bounded point derivations on AC$^3(U)$? This question requires quantitative estimates on the thinness of the complement of U at boundary points.

5. Operators

Now we consider operators that are associated to analytic function theory.

Definition. Let S be an operator defined on a set of distributions, with values in the set of distributions.

We say that a topological vector space $F \subset \mathcal{D}'$ is S–invariant if F lies in the domain of S and S maps F into F, continuously (with respect to the topology of F). This is standard terminology, but we wish to introduce some more, useful when F is an SCS.

We say that F is locally S–invariant if $S : F \to F_{\text{loc}}$, continuously.

We say that F is co–locally S–invariant if $S : F_{\text{cs}} \to F$, continuously.

We say that F is bi–locally S–invariant if $S : F_{\text{cs}} \to F_{\text{loc}}$, continuously.

The important integral operators for function theory on general plane open sets are:

1) The **Cauchy transform**:

$$\mathfrak{C} : f \mapsto \frac{-1}{\pi z} * f.$$

This inverts the $\bar{\partial}$ operator:

$$\bar{\partial}\mathfrak{C}f = f = \mathfrak{C}\bar{\partial}f, \qquad \forall f \in \mathcal{E}'.$$

2) The **Beurling transform**:

$$\mathfrak{B} : f \mapsto \frac{\mathrm{PV}}{\pi z^2} * f.$$

This has

$$\frac{\partial \mathfrak{B} f}{\partial \bar{z}} = \frac{\partial f}{\partial z} \qquad \forall f \in \mathcal{E}'.$$

3) The **Vitushkin localisation operator**:

$$f \mapsto T_\phi f = \frac{-1}{\pi z} * \left(\phi \frac{\partial f}{\partial \bar{z}} \right),$$

defined for all $f \in \mathcal{D}'$ and all $\phi \in \mathcal{D}$. This satisfies

$$\bar{\partial} T_\phi f = \phi \bar{\partial} f,$$

and so may be used to localise singularities.

We make the convention that T–invariant means T_ϕ–invariant, for each $\phi \in \mathcal{D}$.

Theorem 5. *Let F be a small SCS or the SCS dual of a small SCS. Then F is locally T–invariant, and F_∞ is T–invariant.*

Proof. The function $-1/\pi z$ belongs to $\mathrm{L}^1{}_{\mathrm{loc}}$, so Theorem 4 shows that F is bilocally $\mathfrak{C}$–invariant:

$$\mathfrak{C} : F_{\mathrm{cs}} \to F_{\mathrm{loc}}.$$

Using Leibnitz' rule,

$$T_\phi f = \phi \cdot f + \mathfrak{C} \left(\frac{\partial \phi}{\partial \bar{z}} f \right),$$

so the local T–invariance of F follows easily, using the $\mathcal{D}$–module property of F.

Thus F_{loc} is T–invariant.

Given this, it is easy to see that F itself is T–invariant if and only if it contains all those functions $f \in F_{\mathrm{loc}}$ that are analytic near ∞ and vanish at ∞. Since F_∞ has this property, and $F_{\infty\mathrm{loc}} = F_{\mathrm{loc}}$, we get the last assertion of the Theorem. □

It is noteworthy that previous proofs of the T–invariance of various special SCS, such as C, L^p ($p > 2$), Lipα, and BMO, have involved substantial spadework. This theorem uncovers the essential pattern in these results. Of course, not all the SCS properties are needed for this result: full affine local invariance may be relaxed to local invariance under translations and reflection in a point. The theorem also throws up important new observations, such as the availability of a T–invariant SCBS that is locally–equivalent to L^p, when $p \leq 2$.

To illustrate the use of the ideas introduced here, we give a simple application. This is a result which originated with Ahlfors in the case $F = \mathrm{L}^\infty$, and it solves the removability problem, modulo constructive identification of the $\bar{\partial}$–F–cap null sets (a big modulo).

Corollary. *Let F be a small SCS, or the dual of a small SCS. Let $K \subset \mathbf{C}$ be compact. Then the following are equivalent:*
(1) K is $\bar{\partial}$-F-null;
(2) $AF_\infty(\mathbf{C} \sim K) = \{0\}$.
(3) $\bar{\partial}$-F-cap$(K) = 0$.

Proof. It is easy to see that the family of $\bar{\partial}$–F–null compacts depends only on F_{loc}. So we may take $F = F_\infty$, without loss in generality.

(1) $\Rightarrow$ (2) is easy: Suppose that K is $\bar{\partial}$–F–null. Let $f \in AF(\mathbf{C} \sim K)$. Then $F \in AF(\mathbf{C})$. So f is entire. Since $f \in F_\infty$, we deduce $f(a + \cdot)|\mathbf{B}(0,1) \to 0$ in the topology of $\mathcal{D}'(\mathbf{B}(0,1)$ as $a \to \infty$. Using a smeared Cauchy integral formula, this is readily seen to yield $f(a) \to 0$ as $a \to \infty$, so Liouville's theorem gives $f = 0$ identically.

(2) $\Rightarrow$ (3) is trivial.

(3) $\Rightarrow$ (2): Suppose that $\bar{\partial}$–F–cap$(K) = 0$. Suppose f were a nonzero element of $AF(\mathbf{C} \sim K)$. Choose $n \in \mathbf{N}$ such that

$$f(z) = \frac{a_n}{z^n} + \cdots, \text{ near } \infty,$$

with $a_n \neq 0$. Using the $\mathcal{D}$–module property, and the fact that $F = F_\infty$, one sees that $z^{n-1}f$ also belongs to $AF(\mathbf{C} \sim K)$, hence the $\bar{\partial}$–F–cap of K is nonzero.

(2) $\Rightarrow$ (1): Suppose that $AF(\mathbf{C} \sim K) = \{0\}$. Let $f \in AF(U \sim K)$ for some open set U. We wish to show that $f \in AF(U)$.

It is enough to show that f is holomorphic on each open disc D such that $\text{clos}(D) \subset U$. Fix such a disc D.

Take $\phi \in \mathcal{D}$ with $\phi = 1$ near $K \cap \text{clos}D$ and $\text{spt}\phi \subset U$, and form $T_\phi f$. Since F is T–invariant, and $T_\phi f$ is analytic off $\text{spt}\phi$, we have $T_\phi f \in AF(\mathbf{C} \sim K)$, hence $T_\phi f = 0$. So

$$\bar{\partial} f = \bar{\partial}(f - T_\phi f) = (1 - \phi)\bar{\partial} f,$$

hence f is holomorphic on $D \sim K$ and on a neighbourhood of $K \cap \text{clos}(D)$, and hence on all of D. □

Work in progress offers the prospect of attaining a similarly general solution to the other two main problems posed above. In particular, the author and J. Verdera are working on implementing the Vitushkin "coefficient matching" technique, without using algebra structure or $\mathfrak{B}$–invariance.

6. The one–reduction

"Most" SCS are bi–locally $\mathfrak{B}$–invariant, and these are easier to work with. We illustrate this by describing the most important technique that depends on $\mathfrak{B}$–invariance: 1–reduction.

Before starting, we note that the "most" excludes some very, very interesting spaces.

There is a straightforward way to "differentiate" a SCS, F. You just form

$$\left\{ \frac{\partial f_1}{\partial x} + \frac{\partial f_2}{\partial y} : f_1, f_2 \in F \right\},$$

and give it the naturally–induced topology. This space does not necessarily contain $\mathcal{D}$, so in fact we define the derivative of F by

$$DF = \mathcal{D} + \left\{ \frac{\partial f_1}{\partial x} + \frac{\partial f_2}{\partial y} : f_1, f_2 \in F \right\}.$$

Similarly, we define the integral of F by

$$\int F = \left\{ f \in \mathcal{D}' : \frac{\partial f}{\partial x} \in F \text{ and } \frac{\partial f}{\partial y} \in F \right\}.$$

Again, there is a natural topology to use.

Theorem 6. *If F is a bi-locally $\mathfrak{B}$-invariant SCS, then DF and $\int F$ are also SCS, and*

$$\int DF \overset{\text{loc}}{=} F \overset{\text{loc}}{=} D \int F.$$

This is the Fundamental Theorem of the Integral Calculus, for function spaces. Part of the proof is the formula

$$\bar{\partial} F \overset{\text{loc}}{=} DF,$$

where $\bar{\partial} F$ stands for the collection of $\bar{\partial} f$, where f ranges over F. Now for the SCS of Theorem 5, to say that K is $\bar{\partial}$–F–null is the same as saying that no function $g \in \bar{\partial} F$ has support in K. This yields the 1–reduction for $\bar{\partial}$:

Corollary. *Let F be a bi-locally $\mathfrak{B}$-invariant SCS, such that F or F_* is a small SCS. Let $K \subset \mathbf{C}$ be compact. Then K is $\bar{\partial}$-F-null if and only if K supports no nonzero element of DF.*

This reduces the nullity problem for $\bar{\partial}$–F to a problem about supports, a *real–variable* problem. In many cases, one finds that the real–variable problem has already been studied and elucidated.

As an example, consider the case $F = W^{2,p}$. The interesting range of p is $[1,2)$. For $1 < p < 2$, the Corollary applies, and $DW^{2,p} \overset{\text{loc}}{=} W^{1,p}$, so the $\bar{\partial}$–$W^{2,p}$–null sets are the compact sets that cannot support nonzero $W^{1,p}$ functions. These have been intensively studied (Havin, Hedberg, Bagby), because they are the sets K such that each function $f \in \mathrm{L}^q$ is the L^q limit of functions holomorphic on a neighbourhood of K.

Acknowledgments. The author wishes to express his gratitude to the Organising Committee and Sponsors for enabling him to take part in the meeting. It was a very useful meeting, and an especial pleasure to meet Professor Ohtsuka, whose penetrating work on the capacity and Hausdorff dimension of Cantor sets he has long admired. He would also like to record his debt to Larry Zalcman, whose expository account in the American Mathematical Monthly first aroused his interest in analytic capacities.

Potential theory on non-locally compact space via Dirichlet forms

Michael Röckner

1. Introduction

Let E be a locally convex Hausdorff topological vector space over $I\!R$ which is a *Souslin space*, (i.e. the continuous image of a Polish space and hence in particular separable) with dual E'. Let $\mathcal{B}(E)$ be its Borel σ-algebra and μ a finite positive measure on $(E, \mathcal{B}(E))$. Define for $K \subset E'$ the linear space

$$\mathcal{F}C_b^\infty(K) := \{f(l_1, \cdots, l_m) | m \in I\!N, f \in C_b^\infty(I\!R^m), l_1, \cdots, l_m \in K\} \tag{1.1}$$

and set

$$\mathcal{F}C_b^\infty := \mathcal{F}C_b^\infty(E'). \tag{1.2}$$

Here $C_b^\infty(I\!R^m)$ denotes the set of all infinitely differentiable (real-valued) functions on $I\!R^m$ such that all partial derivatives are bounded. We denote the corresponding set of μ-classes by $\widetilde{\mathcal{F}C_b^\infty}(K)$. (Note that if $\text{supp}\mu \neq E$, two different elements in $\mathcal{F}C_b^\infty(K)$ might belong to the same class in $\widetilde{\mathcal{F}C_b^\infty}(K)$.) $\widetilde{\mathcal{F}C_b^\infty}$ is dense in (real) $L^2(E;\mu)$ (cf.[AR 90, Remark 3.1]). Now let for $k \in E$ and $u \in \mathcal{F}C_b^\infty$

$$\frac{\delta u}{\delta k}(z) := \frac{d}{ds} u(z + sk)_{|_{s=0}} \; , z \in E. \tag{1.3}$$

Let $(H, \langle \; , \; \rangle_H)$ be a separable real Hilbert space densely and continuously embedded in E. Identifying H with its dual H' we have that

$$E' \subset H \subset E \text{ densely and continuously.} \tag{1.4}$$

and the dualisation ${}_{E'}\langle \; , \; \rangle_E$ between E' and E coincides with $\langle \; , \; \rangle_H$ when restricted to $E' \times H$. Here E' is considered to be endowed with the strong topology. From now on we consider H as fixed, so it will not appear in notations of quantities introduced below. H should be thought of as a *tangent space* to E at each point. Observe that for $u \in \mathcal{F}C_b^\infty$ and $z \in E$ fixed, $h \to \frac{\partial u}{\partial h}(z)$ is linear and continuous on H. Define $\nabla u(z) \in H$ by

$$\langle \nabla u(z), h \rangle_H = \frac{\partial u}{\partial h}(z) \; , h \in H, \tag{1.5}$$

Potential Theory, ed. by M. Kishi

and

$$\mathcal{E}_\mu(u,v) := \int \langle \nabla u, \nabla v\rangle_H d\mu \ ; \ u,v \in \mathcal{F}C_b^\infty. \tag{1.6}$$

If

$$\nabla u = \nabla v \ \mu - a.e. \text{ for } u,v \in \mathcal{F}C_b^\infty \text{ with } u = v \ \mu - a.e., \tag{1.7}$$

$(\mathcal{E}_\mu, \widetilde{\mathcal{F}C_b^\infty})$ is a symmetric positive definite bilinear form (abbreviated *form*) on $L^2(E;\mu)$ which is densely defined. If this form is closable we denote its closure by $(\mathcal{E}_\mu, D(\mathcal{E}_\mu))$ and then it can be seen (cf. [AR 88a, Theorem 1.1]) that for every *normal contraction* $T : \mathbb{R} \to \mathbb{R}$ (i.e. $T(0) = 0$ and $|T(x) - T(y)| \leq |x - y|$; $x, y \in \mathbb{R}$), $T(u) \in D(\mathcal{E}_\mu)$ if $u \in D(\mathcal{E}_\mu)$, and $\mathcal{E}_\mu(T(u), T(u)) \leq \mathcal{E}_\mu(u,u)$. In other words $(\mathcal{E}_\mu, D(\mathcal{E}_\mu))$ is a (symmetric) *Dirichlet form* on $L^2(E;\mu)$ (cf. [BD 58, 59], [F 80], [S 74]).

Forms of type (closure of) (1.6) have been studied extensively in recent years (cf. e.g. [AH-K 75, 77a,b], [AH-KS 77], [Pa 78], [F 81, 82, 84a,b], [Ku 82], [Wi 85], [T 85,87,90], [BH 86a,b], [AK 88], [AKR 88], [HPS 88], [AR 88a,b, 89a,b,c, 90], [PR 90], [S 90], [AHP 90a,b], [RZ 90], [ARZ 90], [RS 90]). In part I of this paper we give a brief survey on some of the main results obtained so far. The analytic potential theory of general coercive bilinear forms satisfying certain contraction properties have been studied from an abstract point of view in classical papers by C. Preston and A. Ancona ([P 71], [An 72, 75, 76]) generalizing earlier work by J. Deny, M. Ito and J. Bliedtner ([D 65], [I 67a,b], [B 71]). As in [F 80], however, our survey puts the emphasis on the probabilistic counterpart to forms $(\mathcal{E}_\mu, D(\mathcal{E}_\mu))$, i.e. the associated Markov process, and its applications (e.g. to solving weak stochastic differential equations on E). Note that if $dim E = \infty$, E is not locally compact, so the theory in [F 80] is not directly applicable to $(\mathcal{E}_\mu, D(\mathcal{E}_\mu))$. But by proving that the capacity associated with $(\mathcal{E}_\mu, D(\mathcal{E}_\mu))$ is *tight* (see below and [LR 90]) we provide a substitute for local compactness. Moreover, we develop a suitable compactification method that makes all classical potential theory on locally compact space applicable to $(\mathcal{E}_\mu, D(\mathcal{E}_\mu))$ and, in fact, to more general forms of type closure of

$$\mathcal{E}(u,v) = \int \langle A(z)\nabla u(z), \nabla v(z)\rangle_H \mu(dz) \ ; u,v \in \widetilde{\mathcal{F}C_b^\infty}, \tag{1.8}$$

where A is a measurable map from E to the set of all bounded positive definite, self-adjoint operators on H (cf. [K 82, AR 89b,c] for details). Thus, these forms lead to a rich (and concrete) potential theory on the (in general) non-locally compact space E.

In part II of this paper we place the *generator* of $(\mathcal{E}_\mu, D(\mathcal{E}_\mu))$ (i.e. the unique negative definite self-adjoint operator L_μ with domain $D(L_\mu)$ on $L^2(E;\mu)$ such that $(-L_\mu u, v)_{L^2(E;\mu)} = \mathcal{E}(u,v)$ for all $u \in D(L_\mu), v \in D(\mathcal{E}_\mu)$) in the centre of our consideration. With regard to the length of this article we refer the reader who is interested in results on the *(parabolic) Martin boundaries* of these operators (mentioned in my talk) and the solution of the so-called *maximality problem* to [R 89] resp. [AK 88], [AKR 88] and [T 90]. Here we concentrate instead on the uniqueness results recently obtained in [RZ 90] which are even new if dim $E < \infty$. To give some idea of the proof and for the convenience of the reader who is only interested in the finite dimensional situation, we include a special proof for this case.Finally, we discuss some new applications of these results using the work of Yu. G. Kondratiev ([Ko 85]) and we briefly explain the connection with essential self-adjointness.

I. Survey

2. Closability

From now on we assume that (1.7) holds, so $\mathcal{E}_\mu$ defined by (1.6) with domain $\widetilde{\mathcal{F}C_b^\infty}$ is a densely defined form on $L^2(E;\mu)$. We mentioned in the introduction that if $(\mathcal{E}_\mu, \widetilde{\mathcal{F}C_b^\infty})$ is closable its closure is a Dirichlet form on $L^2(E;\mu)$. We recall that a form $\mathcal{E}$ with domain $D \subset L^2(E;\mu)$ is *closable on* $L^2(E;\mu)$ if the continuous extension of the embedding $D \subset L^2(E;\mu)$ (where $L^2(E;\mu)$ is equipped with the usual inner product $(\ ,\ )_\mu$ and D with $\mathcal{E}_1 := \mathcal{E} + (\ ,\ )_\mu$) mapping the abstract completion $\overline{D}^{\mathcal{E}_1}$ of D w.r.t. $\mathcal{E}_1^{1/2}$ into $L^2(E;\mu)$, is one to one. In this case $\mathcal{E}$ naturally extends to $D(\mathcal{E}) := \overline{D}^{\mathcal{E}_1}$ and $(\mathcal{E}, D(\mathcal{E}))$ is called the *closure* of $(\mathcal{E}, D)$. To decide whether $(\mathcal{E}_\mu, \widetilde{\mathcal{F}C_b^\infty})$ is closable on $L^2(E;\mu)$, observe that for any orthonormal basis K_0 in H

$$\mathcal{E}_\mu(u,v) = \int \langle \nabla u, \nabla v \rangle_H d\mu = \sum_{k \in K_0} \int \frac{\partial u}{\partial k} \frac{\partial v}{\partial k}\, d\mu\ ; u, v \in \widetilde{\mathcal{F}C_b^\infty}. \tag{2.1}$$

Since sums of closable forms are closable it is enough to have that the summands

$$\mathcal{E}_k(u,v) = \int \frac{\partial u}{\partial k} \frac{\partial v}{\partial k}\, d\mu\ ; u, v \in \widetilde{\mathcal{F}C_b^\infty}. \tag{2.2}$$

are closable on $L^2(E;\mu)$ for each k in an orthonormal basis K_0 of H. In [AR 90, Section 3] a characterization of such $k \in E$ was proved (cf. Theorem 2.1 below). We need some preparations.

Fix $k \in E \setminus \{0\}$ and let E_k be a "complementing subspace" , i.e. a closed subspace of E such that $E = E_k \oplus k\mathbb{R}$. For each $z \in E$, $z = x + sk$ where $x \in E_k$, $s \in \mathbb{R}$ are uniquely determined. Let $\pi_k : E \to E_k$ be the canonical projection . Since E and E_k are Souslinean we can disintegrate μ w.r.t. $\pi_k : E \to E_k$, i.e. there exists a kernel $\rho_k : E_k \times \mathcal{B}(\mathbb{R}) \to [0,1]$ such that for all $u : E \to (0,\infty]$, $\mathcal{B}(E)$-measurable

$$\int_E u(z)\ \mu(dz) = \int_{E_k} \int_{\mathbb{R}} u(x+sk)\ \rho_k(x,ds)\ \nu_k(dx) \tag{2.3}$$

where $\nu_k := \mu \cdot \pi_k^{-1}$ (i.e. the image measure of μ under π_k). $\rho_k(\cdot, ds)$ is ν_k-a.e. uniquely determined. Correspondingly, $L^2(E;\mu)$ can be written as a direct integral of L^2-spaces over $\mathbb{R}$

$$L^2(E;\mu) = \int_{E_k}^{\oplus} L^2(\mathbb{R}; \rho_k(x, ds))\nu_k(dx) \tag{2.4}$$

in the sense that each $u \in L^2(E;\mu)$ corresponds to a "field of vectors" $(u_x)_{x\in E_k}$ where $u_x := u(x + \cdot k)$, $x \in E_k$ (cf.[AR 90] for details and references). Below ds denotes one dimensional Lebesgue measure and we introduce the condition

$$\frac{\partial u}{\partial k} = \frac{\partial v}{\partial k}\ \mu - a.e. \text{ if } u, v \in \mathcal{F}C_b^\infty \text{ with } u = v\ \mu - a.e. \tag{2.5}$$

Theorem 2.1. *Let $k \in E \setminus \{0\}$. Then (2.5) holds and $(\mathcal{E}_k, \widetilde{\mathcal{F}C_b^\infty})$ is closable on $L^2(E;\mu)$ if and only if for some (resp. every) complementing subspace E_k (as above) we have that for $\nu_k - a.e.$ $x \in E_k$, $\rho_k(x, ds) = \rho_k(x, s)ds$ for some Borel measurable function $\rho_k(x,\cdot) : \mathbb{R} \to [0,\infty[$ such that for $ds - a.e.$ $t \in \{\rho_k(x,\cdot) > 0\}$ there exists $\epsilon > 0$ such that*

$$\int_{t-\epsilon}^{t+\epsilon} \rho_k(x,s)^{-1} ds < \infty.$$

We call $k \in E$ *μ-admissible* if $k = 0$ or one of the equivalent conditions in 2.1 is fulfilled.

3. Existence of an associated diffusion process

Assume that (1.7) holds and that $(\mathcal{E}_\mu, \widetilde{\mathcal{F}C_b^\infty})$ is closable on $L^2(E;\mu)$. Let $(\mathcal{E}_\mu, D(\mathcal{E}_\mu))$ be its closure and L_μ its generator. Consider the associated semigroup

$$T_t := e^{tL_\mu}, t \geq 0,$$

on $L^2(E;\mu)$. The fact that $(\mathcal{E}_\mu, D(\mathcal{E}_\mu))$ is a Dirichlet form is reflected by the (sub-) Markov property of $(T)_{t\geq 0}$, i.e. $0 \leq u \leq 1$ implies $0 \leq T_t u \leq 1$ μ $-a.e.$ (cf. [F 80]).

Theorem 3.1. *Assume E is a Banach space. Then there exists a diffusion process $\mathbb{M} := (\Omega, \mathcal{F}, (\mathcal{F}_t)_{t\geq 0}, (X_t)_{t\geq 0}, (P_z)_{z\in E})$ (cf. [Dy 65]) with state space E* associated *with $(\mathcal{E}_\mu, D(\mathcal{E}_\mu))$, i.e. for all $\mathcal{B}(E)$-measurable, bounded u and all $t \geq 0$*

$$\int_\Omega u(X_t)dP_z = T_t u(z)\ \mu - a.e.\ z \in E.$$

Theorem 3.1 is in this generality due to B. Schmuland (cf. [S 90]) and extends previous results in [K 82], [AR 89b].

Corollary 3.2. *Assume E is conuclear and that*

$$\int |\ _{E'}\langle l, z\rangle_E|\mu(dz) < \infty \quad \textit{for all } l \in E'. \tag{3.1}$$

Then there exists a diffusion process associated with $(\mathcal{E}_\mu, D(\mathcal{E}_\mu))$.

By (3.1) this case can be reduced to the Banach space case (cf. [AR 89b, Sect. 3c)] for details).

4. Capacity and compactification

Now let us recall the definition of the (1-) *capacity* associated with $(\mathcal{E}_\mu, D(\mathcal{E}_\mu))$:

For $U \subset E$, U open, define

$$Cap(U) := \inf\{\mathcal{E}_\mu(u,u) + (u,u)_\mu | u \in D(\mathcal{E}_\mu),\ u \geq 1 \text{ on } U\ \mu - a.e.\} \tag{4.1}$$

and for an arbitrary subset $A \subset E$ define

$$Cap(A) := \inf\{Cap(U) | A \subset U,\ \ U \text{ open}\}. \tag{4.2}$$

Observe that Cap takes values in [0,1] since $1 \in D(\mathcal{E}_\mu)$, and that $Cap(A) = 0$ implies $\mu(A) = 0$ for any $A \in \mathcal{B}(E)$. As in [F 80, Section 3.1] it follows that Cap is a Choquet capacity, i.e. it has the following two properties:

$$\text{If } A_n \subset E, n \in I\!N, \text{ are increasing then } Cap(\bigcup_{n \in I\!N} A_n) = \sup_{n \in I\!N} Cap(A_n). \tag{4.3}$$

If $K_n \subset E$, $n \in I\!N$, are compact and decreasing then (4.4)

$$Cap(\bigcap_{n \in I\!N} K_n) = \inf_{n \in I\!N} Cap(K_n).$$

Furthermore, we have that

$$Cap(\bigcup_{n \in I\!N} A_n) \leq \sum_{n=1}^{\infty} Cap(A_n) \text{ for } A_n \subset E, n \in I\!N, \tag{4.5}$$

and that

$$Cap(A) = \sup\{Cap(K)|K \subset A, \text{ K compact }\} \text{ for } A \in \mathcal{B}(E) \tag{4.6}$$

(since E is Souslinean, cf. [AR 89b, Proposition 2.2]).
In sufficiently regular cases sets of capacity zero are exactly the sets which are not hit by the diffusion process $\mathbb{M}$ associated with $(\mathcal{E}_\mu, D(\mathcal{E}_\mu))$ (cf. [F 80, Theorem 4.3.1]). Cap is called *tight* if

$$\begin{aligned}&\text{there exist compact sets } K_n \subset E, n \in \mathbb{N},\\ &\text{such that } \lim_{n\to\infty} Cap(E \setminus K_n) = 0.\end{aligned} \tag{4.7}$$

Tightness of Cap is intimately related to the existence of an associated diffusion.

Theorem 4.1. *Assume E is completely metrizable then the following are equivalent:*
(i) There exists a diffusion process $\mathbb{M}$ associated with $(\mathcal{E}_\mu, D(\mathcal{E}_\mu))$.
(ii) Cap is tight.

Proof. (ii) $\Rightarrow$ (i): This is a direct consequence of [AR 89b, Proposition 2.6 and Theorem 2.7] (cf. Remark 4.4 below for some details). (i) $\Rightarrow$ (ii). This follows from the general result in [LR 90] which says that the capacity of an arbitrary conservative Dirichlet form on a polish space is tight if it has associated to it a diffusion process. □

Remark 4.2. (i) In particular, if E is a Banach space, 3.1 and 4.1 imply that Cap is tight. Using the method developed by Schmuland to prove 3.1 it is, however, possible to prove tightness directly without using $\mathbb{M}$ (cf. [RS 90]).
(ii) Also in the situation of 3.2 one can prove directly that Cap is tight and then use 4.1 to ensure the existence of an associated diffusion (cf. [AR 89b, Section 3c)]).
(iii) The correspondence between tightness of Cap and existence of $\mathbb{M}$ has been extended to more general cases in [AM 90].

Let $\widehat{E}$ be a *compactification* of E, i.e. $\widehat{E}$ is a compact Hausdorff space with E continuously and densely embedded into $\widehat{E}$ by some map i. Define the image measure of μ under i by $\widehat{\mu} := \mu \circ i^{-1}$. It can be seen (cf. [AR 89c, Section 1]) that the map $\widehat{u} \to u \circ i, \widehat{u} \in L^2(\widehat{E}; \widehat{\mu})$, defines an isometry between $L^2(\widehat{E}; \widehat{\mu})$ and $L^2(E; \mu)$. Define the image form $(\widehat{\mathcal{E}}_\mu, D(\widehat{\mathcal{E}}_\mu))$ of $(\mathcal{E}_\mu, D(\mathcal{E}_\mu))$ on $L^2(\widehat{E}; \widehat{\mu})$ as follows

$$D(\widehat{\mathcal{E}}_\mu) := \{\widehat{u} \in L^2(\widehat{E}; \widehat{\mu}) | \widehat{u} \circ i \in D(\mathcal{E}_\mu)\} \tag{4.8}$$

$$\widehat{\mathcal{E}}_\mu(\widehat{u}, \widehat{v}) := \mathcal{E}_\mu \ (\widehat{u} \circ i, \widehat{v} \circ i) \ ; \widehat{u}, \widehat{v} \in D(\widehat{\mathcal{E}}_\mu). \tag{4.9}$$

Clearly, $(\widehat{\mathcal{E}}_\mu, D(\widehat{\mathcal{E}}_\mu))$ is a Dirichlet form on $L^2(\widehat{E}; \widehat{\mu})$.

Proposition 4.3. *There exists a compactification $\widehat{E}$ of E such that*
(i) $\widehat{E}$ is a complete separable metric space.
(ii) The image $(\widehat{\mathcal{E}}_\mu, D(\widehat{\mathcal{E}}_\mu))$ of $(\mathcal{E}_\mu, D(\mathcal{E}_\mu))$ on $L^2(\widehat{E};\widehat{\mu})$ is a regular *Dirichlet form on $L^2(\widehat{E};\widehat{\mu})$, i.e. $D(\widehat{\mathcal{E}}_\mu)\cap C(\widehat{E})$ is dense both in $D(\widehat{\mathcal{E}}_\mu)$ w.r.t. $\widehat{\mathcal{E}}_{\mu,1}(:=\widehat{\mathcal{E}}_\mu+(\ ,\ )_{\widehat{\mu}})$ and in $C(\widehat{E})$ (:= all continuous functions on $\widehat{E}$) w.r.t. uniform norm.*

Proof. By [AR 89b, Proposition 2.9] there exists a countable $\mathbb{Q}$-algebra $D \subset \mathcal{F}C_b^\infty$ containing the constant function 1 and separating the points of E, which is dense in $D(\mathcal{E}_\mu)$ w.r.t. $\mathcal{E}_{\mu,1}$. By Gelfand's representation theorem applied to the uniform closure $\overline{D}$ of D there exists a compact Hausdorff space $\widehat{E}$ such that E is densely and continuously embedded into $\widehat{E}$ and the restriction to E gives rise to an isomorphism from $C(\widehat{E})$ to $\overline{D}$ (cf. e.g. [G 69, Chap.1, Section 8]). Since $\overline{D}$ is separable, $C(\widehat{E})$ is a separable metric space, hence second countable. Consequently, $\widehat{E}$ is second countable hence completely metrizable as a compact space. Thus (i) is proved. Since every function in D extends to a continuous function on $\widehat{E}$ whose corresponding μ-class is by definition in $D(\widehat{\mathcal{E}}_\mu)$, assertion (ii) follows. □

Remark 4.4. Now the main idea to prove that the tightness of Cap implies the existence of an associated diffusion $\mathbb{M}$ (i.e. 4.1 (ii) ⇒ (i)) can be easily explained. By [F 80, Sect. 6] there exists a diffusion $\widetilde{\mathbb{M}}$ on $\widehat{E}$ associated with $(\widehat{\mathcal{E}}_\mu, D(\widehat{\mathcal{E}}_\mu))$. Since Cap is tight one obtains that $\widetilde{\mathbb{M}}$ does not leave E if started in E. If $\mathbb{M}$ is defined to be the *restriction* of $\widetilde{\mathbb{M}}$ to E, one can then prove that $\mathbb{M}$ in fact has continuous sample paths w.r.t. the stronger topology on E, and is (obviously) associated to $(\mathcal{E}_\mu, D(\mathcal{E}_\mu))$, i.e. $\mathbb{M}$ is the desired diffusion.

From now on we fix a compactification $\widehat{E}$ of E satisfying 4.3(i) and (ii) and identify E with $i(E)$ i.e. $E \subset \widehat{E}$. We consider any function u on E as a function on $\widehat{E}$ by putting $u \equiv 0$ on $\widehat{E} \setminus E$. We recall that a property of points in E is said to hold *quasi-everywhere* (abbreviated "q.e.") if it holds outside a set $N \subset E$ with $Cap(N) = 0$. A q.e. defined function $u : E \to \mathbb{R}$ is called *quasi-continuous* if for each $\epsilon > 0$ there exists a closed set $F \subset E$ such that $Cap(E \setminus F) < \epsilon$ and the restriction $u_{|F}$ of u to F is (everywhere defined and) continuous. Let $C\widehat{ap}$ denote the capacity associated with $(\widehat{\mathcal{E}}_\mu, D(\widehat{E}_\mu))$ defined analogously to (4.1), (4.2). Below we affix an additional superscript "^" in notations corresponding to $(\widehat{\mathcal{E}}_\mu, D(\widehat{E}_\mu))$ (e.g. $\widehat{q.e.}$,qu$\widehat{a}$si-continuous etc.). Then we have (cf. [AR 89c,1.3,4.1]).

Proposition 4.5. *Suppose Cap is tight, then $C\widehat{ap}(A) = Cap(A \cap E)$ for every $A \subset \widehat{E}$ and a function $u : E \to \mathbb{R}$ is quasi-continuous if and only if it is qu$\widehat{a}$si-continuous. This holds, in particular, whenever E is a Banach or conuclear space.*

For simplicity from now on we assume that E is a separable real Banach space and let $\mathbb{M}$ be a diffusion process associated with $(\mathcal{E}_\mu, D(\mathcal{E}_\mu))$ (which exists by 3.1). Since $E \in \mathcal{B}(\widehat{E})$, it follows by [F 80, Theorem 4.1.3] that $\mathbb{M}$ is the restriction of a Hunt process $\widehat{\mathbb{M}} = \left(\widehat{\Omega}, \widehat{\mathcal{F}}, (\widehat{\mathcal{F}}_t)_{t\geq 0}, (\widehat{X}_t)_{t\geq 0}, (\widehat{P}_z)_{z\in\widehat{E}}\right)$ defined on $\widehat{E}$ in such a way each $z \in \widehat{E}\setminus E$ is a *trap* for $\mathbb{M}$ and E is *$\widehat{\mathbb{M}}$-invariant.* We can now "lift" all questions about $(\mathcal{E}_\mu, D(\mathcal{E}_\mu))$ and $\mathbb{M}$ on the non-locally compact space E to questions on $(\widehat{\mathcal{E}}_\mu, D(\widehat{\mathcal{E}}_\mu))$ and $\widehat{\mathbb{M}}$ on the compact space $\widehat{E}$ where standard potential theory is available. Proposition 4.5 then enables us to transfer the answers back to $(\mathcal{E}_\mu, D(\mathcal{E}_\mu))$ and $\mathbb{M}$ on E. Below for brevity, we will refer to this procedure as "by compactification". For example, by compactification all results on *measures of bounded energy* proved in [F 80] in the locally compact case carry over to our situation. Also, the correspondence of *additive functionals* of $\mathbb{M}$ and *smooth measures* on E is immediate from [F 80] by compactification.

5. (Weak) stochastic differential equations on E

Definition 5.1. $k \in E$ is called *well-μ-admissible* if there exists $\beta_k \in L^2(E;\mu)$ such that for all $u, v \in \mathcal{F}C_b^\infty$

$$\int u\,\frac{\partial v}{\partial k}\,d\mu = -\int \frac{\partial u}{\partial k}\,v d\mu - \int u\;v\;\beta_k d\mu. \tag{5.1}$$

It is easy to see that if k is well-μ-admissible then it is μ-admissible. Assume from now on that the following condition holds.

There exists a dense linear subspace K of $E'(\subset H \subset E)$ consisting of well-μ-admissible elements in E such that $\int {}_{E'}\langle k, z\rangle_E^2\;\mu(dz) < \infty$ for all $k \in K$. (C)

Theorem 5.2. *Suppose that there exists a Brownian motion on E with covariance given by $\langle\;,\;\rangle_H$, then for q.e. $z \in E$*

$$X_t = z + W_t + N_t \quad, t \geq 0, \quad P_z - a.s. \tag{5.2}$$

Here $(W_t)_{t\geq 0}$ is an $(\mathcal{F}_t)_{t\geq 0}$-Brownian motion on E with covariance $2\langle\;,\;\rangle_H$ and $(N_t)_{t\geq 0}$ is an $(\mathcal{F}_t)_{t\geq 0}$-adapted continuous process on E such that for all $k \in K$

$${}_{E'}\langle k, N_t\rangle_E = \int_0^t \beta_k(X_s)ds, t \geq 0, P_z - a.s., \;q.e.\; z \in E. \tag{5.3}$$

The proof is based on the *Fukushima decomposition* (cf. [F 80]) which is applicable by compactification and which leads to a "componentwise" solution of (5.2). The technically hard part is then to "localize" their sample paths. We refer the reader to [AR 89c] for details. If E is "large" compared with H, a Brownian motion on E with covariance $\langle\ ,\ \rangle_H$ always exists (cf. [Gr 65]).

If there exists $\beta : E \to E$ such that for all $k \in K$

$$ {}_{E'}\langle k, \beta\rangle_E = \beta_k \quad \mu - a.e., \tag{5.4} $$

then (5.2) can be rewritten as

$$ \begin{aligned} dX_t &= dW_t + \beta(X_t)dt \\ X_0 &= z \quad (\text{under } P_z) \end{aligned} \tag{5.5} $$

So, Dirichlet forms $(\mathcal{E}_\mu, D(\mathcal{E}_\mu))$ on $L^2(E;\mu)$ provide a method to solve (weak) stochastic differential equations on the infinite dimensional space E. Of course, the main obstacle that has to be overcome in practice for solving (5.5) is to find μ corresponding to the given β. We refer to [AR 89c], [R 90] and [RZ 90] for concrete examples. In the latter reference uniqueness of the solution of (5.5) (in the sense of the martingale problems) has been proved in certain cases.

II. Uniqueness

6. The general case

We fix K as in (C) and $\varphi : E \to \mathbb{R}, \mathcal{B}(E)$-measurable with $\varphi > 0$ $\mu - a.e.$. Set

$$ \bar{\mu} := \varphi^2 \cdot \mu \tag{6.1} $$

Let $(\mathcal{E}_{\bar{\mu}}, D(\mathcal{E}_{\bar{\mu}}))$ be the closure (if it exists) of

$$ \mathcal{E}_{\bar{\mu}}(u,v) = \int \langle \nabla u, \nabla v\rangle_H d\bar{\mu}\ ; u, v \in \widetilde{\mathcal{F}C}_b \tag{6.2} $$

on $L^2(E;\bar{\mu})$. Let $L_{\bar{\mu}}$ with domain $D(L_{\bar{\mu}})$ be the generator of $(\mathcal{E}_{\bar{\mu}}, D(\mathcal{E}_{\bar{\mu}}))$ on $L^2(E;\bar{\mu})$. $L_\mu, L_{\bar{\mu}}$ are ***Dirichlet operators***, i.e. the corresponding semigroups $(e^{tL_\mu})_{t\geq 0}$, $(e^{tL_{\bar{\mu}}})_{t\geq 0}$ on $L^2(E;\mu)$ resp. $L^2(E;\bar{\mu})$ are (sub-) Markovian or equivalently the corresponding forms are Dirichlet forms (cf. section 3). We are interested in the following problem:

Suppose L_μ is uniquely determined by its values on $\widetilde{\mathcal{F}C}_b^\infty(K)$ in the class of all Dirichlet operators on $L^2(E;\mu)$. Is the same true for $L_{\bar{\mu}}$ as an operator on $L^2(E;\bar{\mu})$?

Since (C) is still in force, we have indeed that $\widetilde{\mathcal{F}C}_b^\infty(K) \subset D(L_\mu)$ and that for $u = f(l_1, \cdots, l_m) \in \widetilde{\mathcal{F}C}_b^\infty(K)$

$$L_\mu u = \sum_{k \in K_0} \left[\frac{\partial}{\partial k} \left(\frac{\partial u}{\partial k} \right) + \beta_k \frac{\partial u}{\partial k} \right] \tag{6.3}$$

where $K_0 \subset K$ is any orthonormal basis of H having $l_1, \cdots, l_m$ in its linear span (i.e. the sum in (6.3) is finite).

Theorem 6.1. *Suppose that $\varphi \in D(\mathcal{E}_\mu)$ and that $\varphi \cdot \beta_k \in L^2(E; \mu)$ for all $k \in K$. Then $(\mathcal{E}_{\bar{\mu}}, \widetilde{\mathcal{F}C}_b^\infty)$ is closable on $L^2(E; \bar{\mu}), \widetilde{\mathcal{F}C}_b^\infty(K) \subset D(L_{\bar{\mu}})$ and*

$$L_{\bar{\mu}} u = L_\mu u + 2\varphi^{-1} \langle \nabla \varphi, \nabla u \rangle_H \text{ for all } u \in \widetilde{\mathcal{F}C}_b^\infty(K). \tag{6.4}$$

If in addition, $\varphi \in D(\mathcal{E}_{\bar{\mu}})$ and L_μ is the unique Dirichlet operator on $L^2(E; \mu)$ satisfying (6.3) for all $u \in \widetilde{\mathcal{F}C}_b^\infty(K)$ then $L_{\bar{\mu}}$ is the only Dirichlet operator on $L^2(E; \bar{\mu})$ satisfying (6.4).

This result was obtained in [RZ 90] to which we refer for its proof and also for several applications (like e.g. to quantum fields). We also like to mention that as a consequence one can prove a Cameron-Martin-Girsanov formula for diffusions with infinite dimensional state space and singular drifts (cf. [ARZ 90]). For further applications in the case $\dim E \leq \infty$ see section 8 below. We now consider the case $\dim E < \infty$.

7. The finite dimensional case

Consider the case $E = H = K = \mathbb{R}^d$ and μ=Lebesgue measure λ^d on $\mathbb{R}^d$. Though λ^d is not a finite measure we adopt the notation introduced before but replace $\mathcal{F}C_b^\infty(K) = C_b^\infty(\mathbb{R}^d)$ by $C_0^\infty(\mathbb{R}^d)(:= \{u \in C_b^\infty(\mathbb{R}^d) |$ supp u compact $\})$. Let $\varphi : \mathbb{R}^d \to \mathbb{R}$ be $\mathcal{B}(\mathbb{R}^d)$-measurable with $\varphi > 0$ λ^d - a.e. and $\bar{\mu} := \varphi^2 \cdot \lambda^d$.

Theorem 7.1. *Suppose there exist $p, q \in [2, \infty]$ with $\frac{1}{p} + \frac{1}{q} = \frac{1}{2}$ such that $\varphi \in L^p_{loc}(\mathbb{R}^d; \lambda^d)$ and $\|\nabla \varphi\|_{\mathbb{R}^d} \in L^q_{loc}(\mathbb{R}^d; \lambda^d)$ then there exists exactly one Dirichlet operator L_φ on $L^2(\mathbb{R}^d; \varphi^2 \cdot \lambda^d)$ such that $C_0^\infty(\mathbb{R}^d) \subset D(L_\varphi)$ and*

$$L_\varphi u = \Delta u + 2\varphi^{-1} \langle \nabla \varphi, \nabla u \rangle_{\mathbb{R}^d} \text{ for all } u \in C_0^\infty(\mathbb{R}^d). \tag{7.0}$$

Remark 7.2. (i) Observe if there exists an operator on $L^2(\mathbb{R}^d; \varphi^2 \cdot \lambda^d)$ with $C_0^\infty(\mathbb{R}^d)$ in its domain and of the form (7.0) then

$$\varphi, \|\nabla\varphi\|_{\mathbb{R}^d} \in L^2_{loc}(\mathbb{R}^d;\lambda^d). \tag{7.1}$$

i.e. (7.1) is a necessary condition.
(ii) Assume that (7.1) holds and that moreover, $\varphi \in L^\infty_{loc}(\mathbb{R}^d;\lambda^d)$, then Theorem 7.1 applies (with $p=\infty, q=2$). In particular, since in the case $d=1$, $\nabla\varphi \in L^2_{loc}(\mathbb{R};\lambda^1)$ implies that φ has an absolutely continuous version , we always have uniqueness in this case. This suggests that this should also be true for arbitrary $d \geq 1$ if φ is rotational invariant.

Theorem 7.1 was proved in [RZ 90] by using 6.1. Below we give a direct proof, but we confine ourselves to the case where $\varphi \in L^p(\mathbb{R}^d;\lambda^d)$ and $\|\nabla\varphi\|_{\mathbb{R}^d} \in L^q(\mathbb{R}^d;\lambda^d)$. The general case can be deduced from this by a simple localization method and [T 90] (cf.[RZ 90]). We need a few preparations. Let $H^{1,q}(\mathbb{R}^d), q \geq 2$, denote the classical Sobolev space

$$H^{1,q}(\mathbb{R}^d) := \{u \in L^q(\mathbb{R}^d;\lambda^d) | \frac{\partial}{\partial z_i} u \in L^q(\mathbb{R}^d;\lambda^d), 1 \leq i \leq d\} \tag{7.2}$$

(where the partial derivatives are in the sense of Schwartz distributions) with norm

$$\|u\|_{1,q} := \left[\int (\|\nabla u\|^q_{\mathbb{R}^d} + |u|^q) d\lambda^d\right]^{1/q} \tag{7.3}$$

$\left(\nabla u := \left(\frac{\partial u}{\partial z_1}, \cdots, \frac{\partial u}{\partial z_d}\right)\right)$. The following fact will be essential below.

$$H^{1,q}(\mathbb{R}^d) = \text{ closure of } C_0^\infty(\mathbb{R}^d) \text{ w.r.t. } \| \ \|_{1,q}. \tag{7.4}$$

Remark. In the infinite dimensional case for μ replacing λ^d, of course, we do not have (7.4) which is therefore taken into the assumptions of Theorem 6.1 (cf. [RZ 90, Section 2] for details).

We shall also use the following characterization (cf. e.g. [Mi 73, Chap.2])

$u \in H^{1,2}(\mathbb{R}^d)$ if and only if $u \in L^2(\mathbb{R}^d;\lambda^d)$ and for each $1 \leq i \leq d$ there exists a (λ^d-)version u_i of u such that u_i is absolutely continuous on λ^{d-1} – a.e. lines parallel to the i-th axis and its ordinary derivative $\frac{\partial u_i}{\partial z_i}$ belongs to $L^2(\mathbb{R}^d;\lambda^d)$. (7.5)

Observe that by the assumptions of the theorem any $k \in \mathbb{R}^d$ is well-$\bar{\mu}$-admissible. Hence the form

$$\mathcal{E}_\varphi(u,v) := \int \langle\nabla u, \nabla v\rangle_{\mathbb{R}^d} \varphi^2 d\lambda^d; \quad u,v \in C_0^\infty(\mathbb{R}^d), \tag{7.6}$$

is closable on $L^2(\mathbb{R}^d, \bar{\mu})$ and for the generator $(L_\varphi, D(L_\varphi))$ of its closure $(\mathcal{E}_\varphi, D(\mathcal{E}_\varphi))$ we have $C_0^\infty(\mathbb{R}^d) \subset D(L_\varphi)$ and (7.0) holds. Thus existence is proved. Define

$$\begin{aligned} \mathcal{A}_D := \{L | L &\text{ Dirichlet operator on } L^2(\mathbb{R}^d; \bar{\mu}) \text{ such} \\ &\text{that } C_0^\infty(\mathbb{R}^d) \subset D(L) \text{ and (7.0) holds for } L\}. \end{aligned} \tag{7.7}$$

Then $L_\varphi \in \mathcal{A}_D$ and it remains to prove that $\#\mathcal{A}_D = 1$. We define a partial order "$\prec$" on $\mathcal{A}_D$ by setting $L_1 \prec L_2$ if for the corresponding Dirichlet forms $(\mathcal{E}_1, D(\mathcal{E}_1)), (\mathcal{E}_2, D(\mathcal{E}_2))$

$$D(\mathcal{E}_1) \subset D(\mathcal{E}_2) \text{ and } \mathcal{E}_1(u,u) \geq \mathcal{E}_2(u,u) \text{ for all } u \in D(\mathcal{E}_1).$$

Given $u : \mathbb{R}^d \to \mathbb{R}$ define for $1 \leq i \leq d$ and $x = (x_1, \cdots, x_{d-1}) \in \mathbb{R}^{d-1}$

$$u_x^{(i)}(s) := u(x_1, \cdots, x_{i-1}, s, x_i, \cdots, x_d), s \in \mathbb{R}$$

and

$$R_{i,x}(\varphi^2) := \left\{ t \in \mathbb{R} \middle| \int_{t-\epsilon}^{t+\epsilon} \varphi_x^{(i)}(s)^{-2} ds < \infty \text{ for some } \epsilon > 0 \right\}.$$

Then $R_{i,x}$ is an open subset of $\mathbb{R}$ and it follows by 2.1 that for λ^{d-1} – *a.e.* $x \in \mathbb{R}^{d-1}, \varphi_x^{(i)} = 0$ ds - a.e. on $\mathbb{R} \setminus R_{i,x}(\varphi^2)$. Hence for each i we can define a form $(\mathcal{E}^{(i)}, D(\mathcal{E}^{(i)}))$ by

$$\begin{aligned} D(\mathcal{E}^{(i)}) := \{u \in L^2(\mathbb{R}^d; \bar{\mu}) | &\text{ for } \lambda^{d-1} \text{ - a.e. } x \in \mathbb{R}^{d-1}, u_x^{(i)} \text{ has an} \\ &\text{absolutely continuous } (ds-) \text{ version } \widetilde{u}_x^{(i)} \text{ on } R_{i,x}(\varphi^2) \text{ such that} \\ D_i u := \left(\frac{d\widetilde{u}_x^{(i)}}{ds} \right)_{x \in \mathbb{R}^{d-1}} &\in \int_{\mathbb{R}^{d-1}}^{\oplus} L^2(\mathbb{R}; (\varphi_x^{(i)})^2 ds) \lambda^{d-1}(dx) (= L^2(E; \bar{\mu}))\} \end{aligned} \tag{7.8}$$

and

$$\begin{aligned} \mathcal{E}^{(i)}(u,v) := &\int_{\mathbb{R}^{d-1}} \int_{\mathbb{R}} \frac{d}{ds}\widetilde{u}_x^{(i)} \frac{d}{ds}\widetilde{v}_x^{(i)} \varphi_x^{(i)}(s)^2 \, ds \, \lambda^{d-1}(dx) \\ &(= \int D_i u D_i u \, d\bar{\mu}) \, , u, v \in D(\mathcal{E}^{(i)}). \end{aligned} \tag{7.9}$$

Now define

$$\mathcal{E}_\varphi^+(u,v) := \sum_{i=1}^d \mathcal{E}^{(i)}(u,v); \; u, v \in D(\mathcal{E}_\varphi^+) := \bigcap_{i=1}^d D(\mathcal{E}^{(i)}).$$

Then $(\mathcal{E}_\varphi^+, D(\mathcal{E}_\varphi^+))$ is a Dirichlet form on $L^2(\mathbb{R}^d; \bar{\mu})$ (cf.[AR 90, Section 3]). Clearly, $D(\mathcal{E}_\varphi) \subset D(\mathcal{E}_\varphi^+)$ and $\mathcal{E}^+ = \mathcal{E}_\varphi$ on $D(\mathcal{E}_\varphi)$. Furthermore, it is easy to see that the associated generator L_φ^+ belongs to $\mathcal{A}_D$ (cf. [AKR 88, Theorem 2.5]). Moreover, L_φ^+ is the maximal element in $\mathcal{A}_D$ w.r.t. $\prec$. This follows from the solution of the maximality problem in [AKR 88] mentioned in the introduction (cf. [AKR 88, Theorems 3.1, 3.4]). This means in order to prove that $\#\mathcal{A}_D = 1$ we have to show that $L_\varphi^+ = L_\varphi$ or equivalently $D(\mathcal{E}_\varphi^+) = D(\mathcal{E}_\varphi)$. Hence it remains to prove that

$$C_0^\infty(\mathbb{R}^d) \text{ is dense in } D(\mathcal{E}_\varphi^+) \; w.r.t. \; \mathcal{E}_{\varphi,1}^+ := \mathcal{E}_\varphi^+ + (\;,\;)_{\bar{\mu}} \tag{7.10}$$

So, let $u \in D(\mathcal{E}_\varphi^+)$. Since $(u \vee (-n)) \wedge n \underset{n\to\infty}{\longrightarrow} u$ w.r.t. $\mathcal{E}_{\varphi,1}^+$, we can assume that u is bounded. For $l \in \mathbb{N}$, let $b_l \in C_0^\infty(\mathbb{R})$ such that

$$1_{[-l,l]} \leq b_l \leq 1_{]-l-1,l+1[}\;, |b_l'| \leq 2 \tag{7.11}$$

and define

$$a_l := b_l \circ ln \text{ and } \varphi_l := a_l \circ \varphi, \quad l \in \mathbb{N}. \tag{7.12}$$

Since $\varphi \in L^p(\mathbb{R}^d; \lambda^d)$ and $\|\nabla\varphi\|_{\mathbb{R}^d} \in L^q(\mathbb{R}^d; \lambda^d)$, (7.4) (and Hölder's inequality) implies that $H^{1,q}(\mathbb{R}^d) \subset D(\mathcal{E}_\varphi)$. But $\varphi_l \in H^{1,q}(\mathbb{R}^d)$, hence

$$\varphi_l \in D(\mathcal{E}_\varphi), l \in \mathbb{N}. \tag{7.13}$$

Since, φ_l is bounded, we also have that

$$u\varphi_l \in D(\mathcal{E}_\varphi^+). \tag{7.14}$$

Now (7.10) follows by the following four claims (of which the second is the crucial one):

Claim 1. $\varphi_l u \underset{l\to\infty}{\longrightarrow} u$ w.r.t. $\mathcal{E}_{\varphi,1}^+$.

Clearly, $\varphi_l u \underset{l\to\infty}{\longrightarrow} u$ in $L^2(\mathbb{R}^d; \bar{\mu})$ and for all $1 \leq i \leq d$ by the product and chain rule for D_i

$$\begin{aligned}
&\int_{\mathbb{R}^d} |D_i u - D_i(\varphi_l u)|^2 d\bar{\mu} \\
&\leq 2\Big[\int_{\mathbb{R}^d} |D_i u|^2 (1-\varphi_l)^2 d\bar{\mu} + \int_{\mathbb{R}^d} u^2 |D_i \varphi_l|^2 d\bar{\mu}\Big] \\
&\leq 2\Big[\int_{\mathbb{R}^d} |D_i u|^2 (1-\varphi_l)^2 d\bar{\mu} + \int_{[\varphi > e^l]} u^2 (b_l' \circ ln\varphi)^2 |D_i \varphi|^2 d\lambda^d \\
&\longrightarrow 0 \text{ as } l \to \infty.
\end{aligned}$$

Claim 2. $\varphi_l u \in H^{1,2}(\mathbb{R}^d)$ for each $l \in \mathbb{N}$.

Let $l \in \mathbb{N}, 1 \le i \le d$. Since $\varphi \in H^{1,2}(\mathbb{R}^d)$, by (7.5) we can construct a λ^d-version, again denoted by φ, such that $\varphi_x^{(i)}$ is absolutely continuous on $\mathbb{R}$ for every $x \in \mathbb{R}^{d-1}$. Since $u \in D(\mathcal{E}_\varphi^+) \subset D(\mathcal{E}^{(i)})$, we can find $E_0 \in \mathcal{B}(\mathbb{R}^{d-1})$ with $\lambda^{d-1}(\mathbb{R}^{d-1} \setminus E_0) = 0$ and for all $x \in E_0$, $u_x^{(i)}$ has a $(ds-)$version $\widetilde{u}_x^{(i)}$ which is absolutely continuous on $R_{i,x}(\varphi^2)$. Define

$$\widetilde{\Psi}_x := \begin{cases} (a_l \circ \varphi_x^{(i)}) \cdot \widetilde{u}_x^{(i)} & , \quad x \in E_0 \\ 0 & \quad \text{else} \end{cases}$$

Then

$$\widetilde{\Psi}_x \text{ is a (ds-)version of } (\varphi_l u)_x^{(i)} \text{ for each } x \in E_0. \tag{7.15}$$

Set for $x \in \mathbb{R}^{d-1}, U := \{\varphi_x^{(i)} > e^{-(l+2)}\}$ and $V := \{\varphi_x^{(i)} < e^{-(l+1)}\}$. Then $\widetilde{\Psi}_x \equiv 0$ on V and $\widetilde{\Psi}_x$ is absolutely continuous on U since $U \subset R_{i,x}(\varphi^2)$. Therfore, $\widetilde{\Psi}_x$ is absolutely continuous on $\mathbb{R} = U \cup V$. Since for every $f \in C_0^\infty(\mathbb{R})$ by (7.15)

$$\int f \widetilde{\Psi}_x ds = \int f \, (\varphi_l u)_x^{(i)} ds \quad , x \in E_0,$$

it follows that $x \mapsto \widetilde{\Psi}_x(t)$ is $\mathcal{B}(\mathbb{R}^{d-1})$-measurable for all $t \in \mathbb{R}$. Hence $(x,t) \mapsto \widetilde{\Psi}_x(t)$ is $\mathcal{B}(\mathbb{R}^{d-1}) \otimes \mathcal{B}(\mathbb{R})$-measurable by the continuity in t; consequently the function

$$\Psi_i(z) := \widetilde{\Psi}_{(z_1,\cdots,z_{i-1},z_{i+1},\cdots,z_d)}(z_i), \quad z = (z_1, \cdots, z_d) \in \mathbb{R}^d$$

is $\mathcal{B}(\mathbb{R}^d)$-measurable, as well as its ordinary derivative $\frac{\partial \Psi_i}{\partial z_i}$. But

$$\begin{aligned} \int \left(\frac{\partial \Psi_i}{\partial z_i}\right)^2 \lambda^d(dz) &= \int \int_{[\varphi_x^{(i)} \ge e^{-(l+1)}]} \left(\frac{d}{ds}\widetilde{\Psi}_x\right)^2 ds \lambda^{d-1}(dx) \\ &\le e^{2l+2} \int_{\mathbb{R}^{d-1}} \int_{\mathbb{R}} \left(\frac{d}{ds}(\widetilde{\varphi_l u})_x^{(i)}\right)^2 \varphi_x^{(i)}(s)^2 ds \lambda^{d-1}(dx) \\ &= e^{2l+2} \int_{\mathbb{R}^d} D_i(\varphi_l u)^2 d\bar{\mu} < \infty. \end{aligned}$$

Hence $\varphi_l u \in H^{1,2}(\mathbb{R}^d)$, by (7.5).

Claim 3. Let $l \in \mathbb{N}$, then there exist $u_n \in \mathcal{F}C_b^\infty$ such that $\varphi_{l+1} u_n \underset{n\to\infty}{\longrightarrow} \varphi_l u$ w.r.t. $\mathcal{E}_{\varphi,1}^+$.

Since $\varphi_l u \in H^{1,2}(\mathbb{R}^d)$ (cf. Claim 2) and $\varphi_l u$ is bounded, by (7.4) there exist $u_n \in C_0^\infty(\mathbb{R}^d), n \in \mathbb{N}$, such that $u_n \underset{n\to\infty}{\longrightarrow} \varphi_l u$ in $H^{1,2}(\mathbb{R}^d)$ and $(\|u_n\|)_{n\in\mathbb{N}}$ is bounded. Hence , since $\varphi_{l+1}\varphi_l = \varphi_l$ and by the product/chain rule, we obtain for all $1 \le i \le d$

$$\begin{aligned}
&\int |D_i(\varphi_{l+1}u_n - \varphi_l u)|^2 d\bar{\mu} + \int (\varphi_{l+1}u_n - \varphi_l u)^2 \, d\bar{\mu} \\
\le 2 &\int_{[\varphi \le e^{l+2}]} \left[|D_i(u_n - \varphi_l u)|^2 + (u_n - \varphi_l u)^2 |D_i \varphi_{l+1}|^2 + (u_n - \varphi_l u)^2\right] \varphi^2 d\lambda^d \\
\le 2e^{2l+4} &\int \left[|D_i(u_n - \varphi_l u)|^2 + (u_n - \varphi_l u)^2\right] d\lambda^d + 2 \int (u_n - \varphi_l u)^2 |D_i \varphi|^2 d\lambda^d \\
&\longrightarrow 0 \text{ as } n \to \infty.
\end{aligned}$$

Claim 4. Let $v \in C_0^\infty(\mathbb{R}^d)$, $l \in \mathbb{N}$, then $\varphi_l v \in D(\mathcal{E}_\varphi)$.
This is obvious since $\varphi_l \in D(\mathcal{E}_\varphi)$ by (7.13) and both φ_l and v are bounded.

Thus the proof is complete. □

Remark 7.3. Observe that in the above proof the assumption $\varphi \in L^p(\mathbb{R}^d; \lambda^d)$, $\|\nabla\varphi\|_{\mathbb{R}^d} \in L^q(\mathbb{R}^d; \lambda^d)$ for some $p, q \in [2, \infty]$ with $\frac{1}{p} + \frac{1}{q} = \frac{1}{2}$, was only used to derive (7.13). Hence 7.1 can be generalized by just assuming (7.13) (cf. the infinite dimensional case in Theorem 6.1).

As a consequence of 7.1 one obtains

Corollary 7.4. *Suppose $\varphi, \|\nabla\varphi\|_{\mathbb{R}^d} \in L^2_{loc}(\mathbb{R}^d; \lambda^d)$ and there exists a closed set $N \subset \mathbb{R}^d$ of $\mathcal{E}_\varphi$-capacity zero (i.e. the capacity associated with $(\mathcal{E}_\varphi, D(\mathcal{E}_\varphi))$) such that $\varphi \in L^p_{loc}(\mathbb{R}^d \setminus N; \lambda^d)$, $\|\nabla\varphi\|_{\mathbb{R}^d} \in L^q_{loc}(\mathbb{R}^d \setminus N; \lambda^d)$ for some $p, q \in [2, \infty]$ with $\frac{1}{p} + \frac{1}{q} = \frac{1}{2}$. Then the assertion of Theorem 7.1 still holds.*

Proof. See [RZ 90a]. □

7.4 generalizes [T 90, Remark 2].

8. Applications and final remarks

In order to apply Theorem 6.1 one first has to check whether for the given measure μ the operator L_μ is the unique Dirichlet operator satisfying (6.3) for all $\widetilde{\mathcal{F}C}_b^\infty(K)$. In [RZ 90] this has been done for μ's which are Gaussian, but with reproducing kernel Hilbert space not necessarily equal to the tangent space H. Below we want to describe a class of (non-Gaussian) measures which can also serve as μ using the work of Yu.G. Kondratiev.

Assume E itself is a real Hilbert space and that the embeddings

$$E' \subset H \subset E \tag{8.1}$$

are nuclear. Assume that μ is a finite positive measure on $\mathcal{B}(E)$ which is quasi-invariant w.r.t. E', i.e. for all $k \in E'$ the image measure $\mu \circ \tau_k^{-1}$ under $\tau_k(z) := z + k, z \in E$, is absolutely continuous w.r.t. μ. Let $a_k(z) := \frac{d\mu\circ\tau_k^{-1}}{d\mu}(z)$ and suppose that for all $k \in E'$

$$\beta_k := L^2(E;\mu) - \lim_{s\to 0} \frac{1}{s}(\sqrt{a_{-sk}} - 1) \tag{8.2}$$

exists. Then each $k \in E'$ is well-μ-admissible in the sense of 5.1 with corresponding β_k given by (8.2) (cf. [AKR 88, Example 2.7]). Then by [AH-K 75] there exists $\beta : E \to E, \mathcal{B}(E)$-measurable such that for all $k \in E'$

$$_{E'}\langle k, \beta\rangle_E = \beta_k \quad \mu - a.e. \tag{8.3}$$

Assume furthermore that

$$\int \|\beta\|_E^2 d\mu < \infty \tag{8.4}$$

and

$$\beta \in C^1_{b,loc}(E, E') \tag{8.5}$$

i.e. β is Frechet differentiable such that its derivative $d\beta(z)$ restricted to E' belongs to $\mathcal{L}_\infty(E')$, the bounded linear operators on E', and

$$z \longmapsto \|\beta(z)\|_E + \|d\beta(z)\|_{\mathcal{L}_\infty(E')}, z \in E,$$

is bounded on bounded subsets of E. Then Kondratiev proved that the generator L_μ of the corresponding Dirichlet form $(\mathcal{E}_\mu, D(\mathcal{E}_\mu))$ (cf. section 6) restricted to the domain $\mathcal{F}C_b^\infty$ is essentially self-adjoint on $L^2(E;\mu)$. Hence $(L_\mu, D(L_\mu))$ is the unique extension of $(L_\mu, \mathcal{F}C_b^\infty)$ among all self-adjoint operators on $L^2(E;\mu)$. This result was privately communicated by Kondratiev. (For a proof of the case where $\beta \in C^2_{b,loc}(E, E')$ see [Ko 85].) Note that this is a stronger "uniqueness" than the one required in 6.1.However, (8.5) is very restrictive and even in the "Gaussian cases" mentioned above, it is not fulfilled in general, since there, β might even be not continuous.

Further results on essential self-adjointness were obtained recently by V.A. Liskevich and Yu.A. Semenov. In the case $E = \mathbb{R}^d$ one of their theorems in [LS 90, Section 3] states that if $\int \varphi^2 d\lambda^d = 1$ and

$$\varphi^{-1}\|\nabla\varphi\|_{\mathbb{R}^d} \in L^4(\mathbb{R}^d; \varphi^2 \cdot \lambda^d) \tag{8.6}$$

then $(\Delta + 2\varphi^{-1}\langle\nabla\varphi, \nabla\cdot\rangle_{\mathbb{R}^d}, C_0^\infty(\mathbb{R}^d))$ has a unique self-adjoint extension on $L^2(\mathbb{R}^d; \varphi^2 \cdot \lambda^d)$. If $d \leq 8$, we have shown that (8.6) implies our condition (7.13) (cf.

[RZ 90a]). Obviously for all $d \geq 1$ there exist functions φ satisfying (7.13), but not (8.6).

Finally, we briefly illustrate the difference between "having a unique Dirichlet extension" and "having a unique self-adjoint extension". Consider the example $\varphi(x) = |x|^{2\alpha}, x \in \mathbb{R}^d$. In order to satisfy the necessary condition (7.1) we need

$$\alpha > -\frac{d}{2} + 1. \tag{8.7}$$

By 7.4 (or 7.3) we have that that $(\Delta + 2|x|^{-1}\langle x, \nabla\cdot\rangle_{\mathbb{R}^d}, C_0^\infty(\mathbb{R}^d))$ always has a unique Dirichlet extension on $L^2(\mathbb{R}^d; \varphi^2 \cdot \lambda^d)$. By the final example in [LS 90], however, it has more than one self-adjoint extension if $-\frac{d}{2} + 1 < \alpha < -\frac{d}{2} + 2$. (In particular, $C_0^\infty(\mathbb{R}^d)$ is not an operator core in this case.) On the other hand if $\alpha > -\frac{d}{2} + 2$, by the Kalf-Walter-Schmincke-Simon theorem it follows that even $(\Delta + 2|x|^{-1}\langle x, \nabla\cdot\rangle_{\mathbb{R}^d}, C_0^\infty(\mathbb{R}^d \backslash \{0\}))$ has a unique self-adjoint extension on $L^2(\mathbb{R}^d; \varphi^2 \cdot \lambda^d)$ (cf. [T 90] and [ReS 75], Theorem X.30]). For further discussions in these directions we refer to [F 80, §2.3], [Wi 85] and [T 90].

Acknowledgement

I would like to thank the organizers of the ICPT 90 for the invitation and their hospitality during this excellently organized and very stimulating conference. I would also like to thank my various collaborators on topics that were presented in my talk and this paper, that is S. Albeverio, S. Kusuoka, T. Lyons, B. Schmuland and T.S. Zhang.

References

[AHP 90a] Albeverio S., Hida, T., Potthoff, J., Röckner, M., Streit, L., Dirichlet forms in terms of white noise analysis I: Construction and QFT examples. Reviews in Math. Phys. 1 (1990), 291-312

[AHP 90b] Albeverio S., Hida, T., Potthoff, J., Röckner, M., Streit, L., Dirichlet forms in terms of white noise analysis I: Closability and diffusion processes. Reviews in Math. Phys. 1 (1990), 313-323

[AH-K 75] Albeverio, S., Høegh-Krohn, R., Quasi-invariant measures, symmetric diffusion processes and quantum fields. In: Les méthodes mathématiques de la théorie quantique des champs, Colloques Internationaux du C.N.R.S., no. 248, Marseille,23-27 Juin 1975, C.N.R.S. (1976)

[AH-K 77a] Albeverio, S., Høegh-Krohn, R., Dirichlet forms and diffusion processes on rigged Hilbert spaces. Z. Wahrscheinlichkeitstheorie verw. Gebiete 40 (1977), 1-57

[AH-K 77b] Albeverio, S., Høegh-Krohn, R., Hunt processes and analytic potential theory on rigged Hilbert spaces. Ann. Inst. Henri Poincaré Sect. B 13 (1977), 269-291

[AH-KSt 77] Albeverio, S., Høegh-Krohn, R., Streit, L., Energy forms, Hamiltonians and distorted Brownian paths. J. Math. Phys. 18 (1977), 907-917

[AK 88] Albeverio, S., Kusuoka, S., Maximality of infinite dimensional Dirichlet forms and Høegh-Krohn's model of quantum fields. Kyoto-Bochum Preprint (1988)

[AKR 88] Albeverio, S., Kusuoka, S., Röckner, M., On partial integration in infinite dimensional space and applications to Dirichlet forms. Preprint (1988). To appear in J. London Math. Soc.

[AM 90] Albeverio, S., Ma, Z.M., Necessary and sufficient conditions for the existence of m-perfect processes associated with Dirichlet forms. BiBoS-preprint (1990)

[AR 88a] Albeverio, S., Röckner, M., New developments in theory and applications of Dirichlet forms. In: Stochastic processes, physics and geometry, 27-76, Ascona/Locarno, Switzerland, 4-9 July 1988, Editors: S. Albeverio et.al., World Scientific 1990, Singapore

[AR 88b] Albeverio, S., Röckner, M., On the maximality problem for classical Dirichlet forms on topological vector spaces. In: Proc. Bad-Honnef Conference, June 1988, Lect. Notes Inform. Control 126 (1989), 14-31, Editors: N. Christopeit et. al., Springer, Berlin

[AR 89a] Albeverio, S., Röckner, M., Dirichlet forms, quantum fields and stochastic quantization. Stochastic analysis, path integration and dynamics. Research Notes in Math. 200, (1989), 1-21. Editors: K.D. Elworthy, J.C.Zambrini

[AR 89b] Albeverio, S., Röckner, M., Classical Dirichlet forms on topological vector spaces - construction of an associated diffusion process. Prob. Th. Rel. Fields 83, (1989), 405-434

[AR 89c] Albeverio, S., Röckner, M., Stochastic differential equations in infinite dimensions: solutions via Dirichlet forms. Preprint Edinburgh (1989). Publication in preparation

[AR 90] Albeverio, S., Röckner, M., Classical Dirichlet forms on topological vector spaces - closability and a Cameron-Martin formula. J. Funct. Anal. 88 (1990), 395-436

[ARZ 90] Albeverio, S., Röckner, M., Zhang, T., Girsanov transform of symmetric diffusions with infinite dimensional state space. Preprint Edinburgh (1990). Publication in preparation

[An 72] Ancona, A., Continuité des contractions dans les espaces de Dirichlet. C.R.A.S., t. 282 (1976), 871

[An 75] Ancona, A., Sur les espaces de Dirichlet: principes, fonctions de Green. J. Maths pures et appl. 54 (1975), 75-124

[An 76] Ancona, A., Continuité des constractions dans les espaces de Dirichlet. Sem. Th. Potentiel. Paris no.2. Lecture Notes in Math. 563 (1976), 1-26, Springer, Berlin

[BD 58] Beurling, A., Deny J., Espaces de Dirichlet. Acta Math. 99 (1958), 203-224

[BD 59] Beurling, A., Deny J., Dirichlet spaces. Proc. Nat. Acad. Sci. 45 (1959), 208-215

[B 71] Bliedtner, J., Dirichlet forms on regular fuctional spaces. Seminar on Potential Theory 11. Lecture Notes in Math. 226 (1971), 1-61, Springer, Berlin

[BH 86a] Bouleau, N., Hirsch, F., Propriétés d'absolue continuité dans les espaces de Dirichlet et application aux équations differentielles stochastiques. Séminaire de Probabilités, no. XX

[BH 86b] Bouleau, N., Hirsch, F., Formes de Dirichlet générales et densité des variables aléatoires réelles sur l'espace de Wiener. J. Funct. Anal. 69 (1986), 229-259

[D 65] Deny, J., Princip complet du maximum et contractions. Ann. Inst. Fourier 15 no.1 (1965), 259-272

[Dy 65] Dynkin, E.B., Markov processes Vol. I and II. Berlin-Heidelberg-New York, Springer (1965)

[F 80] Fukushima, M., Dirichlet forms and Markov processes. Amsterdam-Oxford-New York: North Holland (1980)

[F 81] Fukushima, M., On a stochastic calculus related to Dirichlet forms and distorted Brownian motion. Physical Reports 77 (1981), 255-262

[F 82] Fukushima, M., On absolute continuity of multidimensional symmetrizable diffusions. Lecture Notes in Math. 923 (1982), 146-176, Springer, Berlin

[F 84a] Fukushima, M., Basic properties of Brownian motion and a capacity on the Wiener space. J. Math. Soc. Japan 36, (1984), 161-175

[F 84b] Fukushima, M., Energy forms and diffusion process. In: Mathematics and Physics, Lectures on recent results. Ed. Streit, L. Singapore: World Scientific Publishing Co.

[G 69] Gamelin, T.W., Uniform algebras. Englewood Cliffs, Prentic Hall (1969)

[Gr 65] Gross, L., Abstract Wiener spaces. Proc. 5th Berkeley Symp. Math. Stat. Prob. 2 (1965), 31-42

[HPS 88] Hida, T., Potthoff,J., Streit, L., Dirichlet forms and white noise analysis. Commun. Math. Phys. 116 (1988), 235-245

[I 67a] Itô, M., Note sur contructions et principes du maximum. Osaka J. Math. 4 (1967), 217-226

[I 67b] Itô, M., A note on extended regular functional spaces. Proc. Jap. Acad. 43 (1967), 435-440

[Ko 85] Kondratiev, Yu.G., Dirichlet operators and smoothness of the solutions of infinite-dimensional elliptic equations. Soviet Math Dokl. 31, no.3 (1985), 461-464

[LS 90] Liskevich, V.A., Semenov, Yu.A., Dirichlet operators. A priori estimates. Uniqueness problem. Preprint (1990)

[LR 90] Lyons, T., Röckner, M., A note on tightness of capacities associated with Dirichlet forms. Preprint Edinburgh (1990), to appear in Bull. London Math. Soc.

[Mi 73] Mizohata, S., The theory of partial differential equations. Cambridge Univ. Press, London (1973)

[Pa 78] Paclet, P., Espaces de Dirichlet et capacités fonctionelles sur triplet de Hilbert-Schmidt.Séminaire P. Krée, no.5 (1978)

[PR 90] Potthoff, J., Röckner, M., On the contraction property of energy forms on infinite dimensional space. J. Funct. Anal. 92 (1990), 155-165

[P 71] Preston, C., Maximum principles and contractions in the potential theory of a functional Hilbert space. Preprint (1971), unpublished

[ReS 75] Reed, M., Simon, B., Methods of modern mathematical physics II. Fourier Analysis, self-adjointness. New York-San Francisco-London, Academic Press (1975)

[R 89] Röckner, M., On the parabolic Martin boundary of the Ornstein-Uhlenbeck operator on Wiener space. Preprint Edinburgh (1989), to appear in Ann. Prob.

[R 90] Röckner, M., Dirichlet forms on infinite dimensional state space and applications. Lectures held at Silivri Summer School (1990). To appear

[RS 90] Röckner, M., Schmuland, B., Tightness of capacities (1990). In preparation

[RZ 90] Röckner, M., Zhang, T.S., On uniqueness of generalized Schrödinger operators and applications. Preprint Edinburgh, June (1990). Publication in preparation

[RZ 90a] Röckner, M., Zhang, T.S., In preparation

[S 90] Schmuland, B., An alternative compactification for classical Dirichlet forms on topological vector spaces. Preprint (1989). To appear in Stochastics

[Si 74] Silverstein, M.L., Symmetric Markov Processes. Lecture Notes in Math. 426. Berlin-Heidelberg-New York, Springer (1974)

[T 85] Takeda, M., On the uniqueness of Markovian self-adjoint extension of diffusion operators on infinite dimensional space. Osaka J. Math. 22 (1985), 733-742

[T 87] Takeda, M., On the uniqueness of Markovian self-adjoint extension. Lecture Notes in Math. 1250 (stochastic processes - Mathematics and Physics), (1985), 319-325, Springer, Berlin

[T 90] Takeda, M., The maximum Markovian self-adjoint extensions of generalized Schrödinger operators. Preprint (1990)

[Wi 85] Wielens, N., On the essential self-adjointness of generalized Schrödinger operators. J. Funct. Anal. 61 (1985), 98-115.

The Martin compactification of a symmetric space of non-compact type at the bottom of the positive spectrum: An introduction

John C. Taylor *

Abstract

This is an introduction to joint work with Y. Guivarc'h on the Martin compactification of a symmetric space of non-compact type, in particular of the polydisc.

Key words. lowest eigenvalue, polydisc, Martin compactification, Satake compactification
1980 Mathematics Subject Classification(1985 Revision): 31C35,43A85,60J50.

Introduction

This report is an introduction to some joint work with Y. Guivarc'h on the Martin compactifications of symmetric spaces of non-compact type. While results on this topic have been announced by Ol'shanetskii in [26] no proofs have been published and in light of the review (MR volume 42 #427) there is perhaps reason to try to clarify the situation.

More specifically, in [26] nothing very explicit was said about the Martin compactification at the bottom of the positive spectrum. Explicit descriptions are given in [16] and [17] which identify this compactification with the maximal Furstenberg-Satake compactification of the space [14],[27]. For the case of the polydisc, this compactification is the closure of the polydisc in $\mathbb{C}^{\ell}$.

Here I propose to explain the problem for the case of the polydisc and to give the necessary background needed to understand [16]. The result for the polydisc case can be understood without knowing the general machinery of semi-simple Lie groups, in particular the associated root systems, and may serve to interest readers in the general case [17].

1. The Martin compactification

Suppose that M is a smooth manifold on which a second order elliptic operator with smooth coefficients L is defined. A function u on an open subset U of M will

*Materially supported by NSERC Operating Grant #A3108

Potential Theory, ed. by M. Kishi

be said to be *harmonic* if Lu = 0 on U. Assume that there is an associated Green function G(x,y). By this is meant a continuous function $G : M \times M \mapsto \mathbb{R}^+ \cup \{+\infty\}$ finite off the diagonal such that

(1) for each y, $x \mapsto G(x,y)$ is harmonic on $M \backslash \{y\}$;
(2) if $0 \leq h \leq G(\cdot, y)$ and h is harmonic on M, then h = 0.

Typically, this will mean that $G(\cdot, y)$ vanishes at infinity and $LG(\cdot, y) = -\varepsilon_y$ as a distribution, relative to some " natural" measure σ on M. This is the case for the Newtonian kernel $G(x,y) = c_d||x-y||^{-(d-2)}$ on $\mathbb{R}^d$, the Green kernel for the Laplacian on $\mathbb{R}^d$ where σ is Lebesgue measure. The *Martin kernel* is obtained by normalizing the Green function. Pick a base point x_0 and define K(x,y) as follows:

$$K(x,y) = \begin{cases} \frac{G(x,y)}{G(x_0,y)}, & \text{if defined} \\ 1 & \text{otherwise.} \end{cases}$$

It follows from Harnack's inequality, that every non-convergent sequence $(y_n)_{n\geq 1}$ in M contains a subsequence $(y_{n_k})_{k\geq 1}$such that $\lim_{k\to\infty} K(\cdot, y_{n_k})$ converges uniformly on compact subsets of M to a harmonic function h.

Following R.S. Martin [25] non-convergent sequences $(y_n)_{n\geq 1}$ in M may then be classified by the resulting limit harmonic functions and ideal points added at infinity to correspond to them. A compactification of M results which is called the *Martin compactification* of M relative to L. More generally, given any collection $\mathcal{Q}$ of continuous functions f on M, we have the following result of Constantinescu and Cornea [9].

Proposition 1. *There is a unique compactification $\tilde{M}$ of M such that*

(1) *each function f extends continuously to $\tilde{M}$; and*
(2) *the extended functions separate the points of the ideal boundary $\tilde{M} \backslash M$.*

The Martin compactification is obtained by taking for $\mathcal{Q}$ the set of functions $K(x, \cdot)_{x\in M}$. The possible discontinuity at x of $K(x, \cdot)$ has no effect on the Proposition 1 (cf. [30]). While this concept makes sense in many contexts, it was Šur[29] who first introduced it for differential operators.

There is by now a large number of examples of Martin compactifications. Here are some of them.

(1) $M = B(0,R) \subset \mathbb{R}^d, L = \Delta$; here the Martin compactification is the closed ball and the limit functions determine the Poisson kernel $P(x,b) = R^{d-2}\frac{r^2 - ||x||^2}{||b-x||^d}$, $||b|| = R$ (cf. Bouligand [7] for a direct calculation).
(2) M a Lipschitz domain D in $\mathbb{R}^d$, L = Δ; again the Martin compactification is the closure $\overline{D}$ (Hunt and Wheeden [20]), but in general there are no explicit formulas for the limit functions.
(3) M a simply connected Riemannian manifold of dimension d whose sectional curvatures lie between $-b^2$ and $-a^2$, L = Δ(the Laplace-Beltrami operator

[5]) or a fairly general elliptic operator [3]; in this case the compactification is the so-called " conical" compactification whose ideal boundary is a sphere of dimension $d-1$, each point corresponding to a unique geodesic ray from a fixed base point (Anderson and Schoen [5], Ancona [3]).

(4) M a symmetric space of non-compact type and L the Laplace-Beltrami operator Δ; here the general results are given by Ol'shanetskii in [26] but are relatively complicated to describe: for the so-called rank one spaces (real, complex, quaternionic hyperbolic space of all dimensions plus the space involving the octonions) the sectional curvatures are all "pinched" as in the last example and so the Martin compactification involves adding a sphere at infinity (cf.[23] for a simple argument, also [28] for the case of real hyperbolic space).

2. Symmetric spaces: generalities

A symmetric space is a Riemannian manifold with the property that for each point $x \in M$ there is an isometry I_x of the space such that I_x^2 is the identity and x is an isolated fixed point. Every Euclidean space has this property: $x \mapsto -x$ gives I_0. So too does every sphere: for the 2-sphere, reversing directions along all the great circles through a fixed point p gives I_p. Finally, the hyperbolic disc — $\{z \mid |z| < 1\}$ with the Riemannian metric $ds^2 = (1-|z|^2)^{-2}|dz|^2$ — is a symmetric space: let $\mathrm{I}_0(z) = -z$.

It is known that the isometry group of a symmetric space (even of a Riemannian manifold) is a Lie group G (cf. [18] p205). In the case of a symmetric space, its Lie algebra $\mathfrak{g}$ determines the *type* of the symmetric space. Let p be an arbitrary point of M. The isotropy group K of p — the isometries that fix p — has a Lie algebra $\mathfrak{k}$ and the tangent space $T_{\mathrm{p}}(M)$ may be identified with a direct summand $\mathfrak{p}$ of $\mathfrak{k}$. If $\mathfrak{g}$ is semisimple and this decomposition is a Cartan decomposition, the symmetric space is said to be of *non-compact type* (see the Appendix and [18] Ch V p244). The simplest symmetric space of non-compact type is the hyperbolic disc. It is clear that the product of symmetric spaces is itself a symmetric space since its isometry group is the product of the isometry groups of the factors. Also, it turns out to be of non-compact type if all the factors are of non-compact type. The polydisc (as a symmetric space) is obtained as the product of a number of copies of the hyperbolic disc and hence, is of non-compact type.

Another important symmetric space is the space M of positive definite symmetric $n \times n$ real matrices. The mapping $g \mapsto gg^*$ maps the group $\mathrm{G} = \mathrm{GL}(n, \mathbb{R})$ of $n \times n$ real invertible matrices onto M and it is easy to see therefore that the left cosets of $\mathrm{K} = \mathrm{O}(n)$ are in one to one correspondence with the points of M. Here it is often convenient to factor out the determinant and to view M as $\mathbb{R}^+ \times \mathrm{M}_1$, where M_1 denotes the points of M of determinant one. Then M_1 can be identified with $\mathrm{SL}(n,\mathbb{R})/\mathrm{SO}(n)$, where the letter S stands for "special" and indicates the appropriate subgroup where the determinant is equal to one. The space SL(2,$\mathbb{R}$)/SO(2) is isomorphic to the hyperbolic disc as will be shown below.

3. The hyperbolic disc and the polydisc (1)

It is well known that a Möbius transformation $z \mapsto \dfrac{az+b}{cz+d}$ maps the unit disc into itself if and only if the associated matrix $g = \begin{bmatrix} a & b \\ c & d \end{bmatrix} = \begin{bmatrix} \alpha & \overline{\beta} \\ \beta & \overline{\alpha} \end{bmatrix}$. Möbius transformations of this type act transitively on the unit disc. Clearly, such a transformation is unchanged if the determinant of the associated matrix is assumed to be one. In other words, the group G — denoted by SU(1,1) — of 2×2 complex matrices g of the form

$$g = \begin{bmatrix} \alpha & \overline{\beta} \\ \beta & \overline{\alpha} \end{bmatrix},$$

acts transitively on the unit disc. Up to a constant, the Poincaré metric $ds^2 = (1 - |z|^2)^{-2}|dz|^2$ is the only Riemannian metric on the unit disc that is invariant under SU(1,1).

The isotropy group K of o is SU(1), those matrices for which $\beta = 0$ and $\alpha = e^{it}$. Hence the unit disc can be identified with SU(1,1) /SU(1) = G/K. The Lie algebra $\mathfrak{su}(1,1)$ of SU(1,1) consists of the 2×2 complex matrices X such that the curve $t \mapsto \exp tX$ in
GL(2, $\mathbb{C}$) lies in SU(1,1) (cf. [18] p113). These are the matrices

$$X = \begin{bmatrix} u & h+iy \\ h-iy & -u \end{bmatrix}.$$

The Lie algebra can be written as $\mathfrak{k} \oplus p$, where $\mathfrak{k} = su(1)$ is generated over $\mathbb{R}$ by

$$U = \begin{bmatrix} 1 & 0 \\ 0 & -1 \end{bmatrix}$$

and $\mathfrak{p}$ is generated by the matrices

$$H= \begin{bmatrix} 0 & 1 \\ 1 & 0 \end{bmatrix}, \text{ and } \quad Y = \begin{bmatrix} 0 & -i \\ i & 0 \end{bmatrix}.$$

As shown in the Appendix, this is a Cartan decomposition of $\mathfrak{su}(1,1)$ and $\theta(X) = -X^*$ is a Cartan involution. This symmetric space is therefore of non-compact type.

The Laplacian on the hyperbolic disc is $\mathrm{L} = (1 - |z|^2)^2\Delta$, where $\Delta = \dfrac{\partial^2}{\partial x^2} + \dfrac{\partial^2}{\partial y^2}$ is the usual Laplacian. Writing the Poincaré metric with a 2×2 matrix (g_{ij}), we have $(g_{ij}(z)) = (1-|z|^2)^{-2}(\delta_{ij})$ and the formula for L is easily calculated from the general formula in local coordinates

$$(*) \qquad Lu(z) = \frac{1}{\sqrt{\overline{g}}}\sum_i \frac{\partial u}{\partial x_i}\{\sum_j \sqrt{\overline{g}} g^{ij}\frac{\partial u}{\partial x_j}\},$$

where $\overline{g}$ is the determinant of the matrix (g_{ij}) and (g^{ij}) is its inverse.

Let D denote the product of ℓ hyperbolic discs D_i. Then the metric is the product of the metrics and so the matrix (g_{ij}) has ℓ blocks of size 2×2 along the diagonal each given by the corresponding hyperbolic metric. It follows from (*) that the Laplacian for D is given by the formula

$$Lf(\mathbf{z}) = \{L_1 f(\cdot, z_1, \ldots, z_\ell)\}(z_1) + \ldots \{L_\ell f(z_1, z_2, \ldots, \cdot)\}(z_\ell),$$

where L_i denotes the Laplace-Beltrami operator on the disc D_i and $\mathbf{z} = (z_1, z_2, \ldots, z_\ell)$ is a generic point of D.

Furthermore, the isometry group G is the product of ℓ-copies of SU(1,1) and the isotropy group K of $\mathbf{o} = (o, o, \ldots, o)$ is the product of ℓ-copies of SU(1). As before, the polydisc D can be identified with G/K.

4. The positive spectrum

Let M be a Riemannian manifold and let L denote the Laplace-Beltrami operator. The *positive spectrum* $\sigma^+(L)$ of L on M will be defined as

$$\{\lambda \mid \exists f \geq 0 \text{ and } f \not\equiv 0 \text{ on M such that } Lf = \lambda f\}.$$

As shown in [28], L restricted to a relatively compact submanifold O with smooth boundary, defined initially on the smooth functions with compact support in O can be extended as a self-adjoint operator (relative to the volume measure $d\mu$) to $L^2(O)$ and hence to $L^2(M)$. The supremum $\lambda_0(M)$ of the $L^2(M)$-spectrum is the supremum of

$$\left\{\int \|\nabla\varphi\|^2 d\mu \Big/ \int \varphi^2 d\mu \,\middle|\, \varphi \in C_0^\infty(M)\right\}.$$

Proposition 2. $\sigma^+(L) = [\lambda_0(M), \infty)$.

Proof. Suppose $\lambda > \lambda_0$, then $(L - \lambda \mathrm{Id})^{-1} = G_\lambda$ is a positive bounded operator on $L^2(M)$. Pick a point $x_0 \in M$ and let $\varphi \in C_0^\infty(M)$ with support not containing x_0. Set $u = G_\lambda\varphi$. Then $Lu - \lambda u = 0$ off the support of φ. Normalize so that $u(x_0) = 1$. By letting the support of φ tend to infinity and using Harnack's inequality one sees that $\lambda \in \sigma^+(L)$. Hence, $\sigma^+(L)$ is an interval unbounded above.

Suppose that $L\phi = \lambda\phi$, $\phi > 0$. Define the operator M by $Mu = (1/\phi)L(u\phi) - \lambda u$. Then, relative to the measure $\phi^2 d\mu$, M is symmetric. Since $M1 = 0$, it follows that $(M\varphi, \varphi) = \int (M\varphi)\varphi\phi^2 d\mu = -\int ||\nabla\varphi||^2\phi^2 d\mu \leq 0$, which implies that $\lambda_0 \leq \lambda$.

It remains to show that $\lambda_0 \in \sigma^+(L)$. Assume that $\lambda_0(O) < \lambda_0(M)$ for any open relatively compact subset O of M with smooth boundary. Then the first part of the proof shows that there is a positive solution to $Lu = \lambda_0 u$ on O. The usual argument using normalization and Harnack's inequality shows that there is a global solution i.e. $\lambda_0 \in \sigma^+(M)$. Alternatively, one may use Schauder estimates and Harnack's inequality to extract from a sequence u_n, such that $Lu_n = \lambda_n u_n, \lim_{n\to\infty} \lambda_n = \lambda_0$ a subsequence converging uniformly on compact subsets to a solution of $Lu = \lambda_0 u$. □

Therefore the supremum $\lambda_0 = \lambda_0(M)$ of the $L^2(M)$-spectrum is also *the bottom of the positive spectrum.*

Remarks. This proposition was proved by Cheng and Yau [11] for complete Riemannian manifolds. Sullivan [28] gives a very clear simple proof of the above result, as well as some very interesting computations involving $\lambda_0(M)$ for real hyperbolic space and quotients of it by a discrete subgroup. The above argument was shown to me by N. Varopoulos.

For each λ in the positive spectrum, if the operator $L - \lambda Id$ has a Green function, a Martin compactification $\tilde{M}^\lambda$ of M is determined. As long as $\lambda > \lambda_0$, this is always the case, but when $\lambda = \lambda_0$ this is not necessarily the case (cf. Ancona [4] and Sullivan [28]). A sufficient criterion that a Green's function exist is that there are two non-proportional $L - \lambda Id$-harmonic functions (cf. Brelot [8]).

There seems to be no general theory about how these Martin compactifications depend upon the parameter λ, although Agmon [1] considered a class of uniformly elliptic operators in divergence form with periodic coefficients and identified the Martin boundaries with the boundaries of strictly convex sets in $\mathbb{R}^d$ above the bottom of the positive spectrum and as one point at that number. It is not even clear how to relate these compactifications in general. We might suspect that if $\lambda_2 > \lambda_1$, then there is a continuous map from $\tilde{M}^{\lambda_2}$ to $\tilde{M}^{\lambda_1}$ that agrees with the identity on M, i.e. the compactification increases with λ. This is the case for several examples:

(1) M is a Lipschitz domain D in $\mathbb{R}^d$ and $L = \Delta$; here $\lambda_0 = \lambda_0(D) > 0$ and all the Martin compactifications coincide with the closure of D (cf. Ancona [2]).

(2) M= $\mathbb{R}^d$, $L = \Delta$; here $\lambda_0 = 0$ and we have

$$\tilde{M}^\lambda = \begin{cases} \mathbb{R}^d \cup S^{d-1} & \text{the conical compactification, } \lambda > 0 \\ \mathbb{R}^d \cup \{\infty\} & \text{the one point compactification, } \lambda = 0. \end{cases}$$

(3) M a simply connected Riemannian manifold whose sectional curvatures lie between $-b^2$ and $-a^2$, $L = \Delta$ the Laplace-Beltrami operator or a fairly

general elliptic operator; in this case the compactifications are all the " conical" compactification for $\lambda > \lambda_0$ (Ancona [3]). At the bottom of the positive spectrum the Martin compactification may be the "conical" compactification or the one point compactification (Ancona [4]).

(4) M a symmetric space of non-compact type and L the Laplace-Beltrami operator Δ: here according to Ol'shanetskii [26] the situation is the same as for the Euclidean case; the compactifications all agree for $\lambda > \lambda_0$ and there is a continuous map from any of the Martin compactifications onto the so-called maximal Furstenberg-Satake compactification.

Y. Guivarc'h and the author [17] have recently shown that the Martin compactification at the bottom of the spectrum for all symmetric spaces of non-compact type coincides with the maximal Furstenberg-Satake compactification. In the case of the polydisc D consisting of ℓ-copies of the hyperbolic disc, this compactification is the closure of D in $\mathbb{C}^\ell$ [16].

In what follows I will show how to calculate the positive spectrum for the hyerbolic disc and the polydisc. It turns out that $\lambda_0 = -\ell$ for the polydisc D and that $L + \ell \mathrm{Id}$ has a Green function. In [16] the Martin compactificaton for the operator $L + \ell \mathrm{Id}$ on the polydisc is calculated by relatively elementary means.

5. Hyperbolic space: determination of λ_0

It is well known that the unit disc D may be mapped conformally onto the upper half plane H by a Möbius transformation. One such transformation is given by $w = \gamma(z) = \dfrac{i(1+z)}{(1-z)}$. It maps the line segment from -1 to 1 onto the positive imaginary axis in the w-plane in an increasing manner with o going to i. Let $\gamma = \begin{bmatrix} i & i \\ -1 & 1 \end{bmatrix}$. Then $\gamma \mathrm{SU}(1,1)\gamma^{-1} = \mathrm{SL}(2,\mathbb{R})$ is the subgroup of SL(2,$\mathbb{C}$) corresponding to the Möbius transformations of the upper half plane into itself. Recall that a matrix $\begin{bmatrix} a & b \\ c & d \end{bmatrix} \in$ SL(2,$\mathbb{R}$) if and only if it has real coefficients and determinant equal to one. The Poincaré metric may be transported to the upper halfplane by γ. The resulting metric is $ds^2 = (4v)^{-1}|dw|^2$, if $w = u + iv$. As a result by (*), the Laplacian is given by $Lf(w) = 4v^2 \Delta f(w)$. The constant 4 may be droppd at the price of multiplying the formula for the Laplacian in the unit disc by (1/4).

Consequently, if $F(w) = f(z)$ i.e. $F \circ \gamma = f$, then $(1-|z|^2)^2 \Delta f(z) = 4v^2 \Delta F(w) = 4v^2 \{\frac{\partial^2 f}{\partial u^2} + \frac{\partial^2 f}{\partial v^2}\}$. In the upper half plane the second of the two variables plays a key role. Assume that $F(w) = F(u+iv) = F(v)$. Let $v = e^{2t}$ and $F(v) = \varphi(t) = e^t \psi(t)$. Then

$$e^t\{\psi(t) + \psi'(t)\} = \varphi'(t) = 2vF'(v) \text{ and}$$
$$e^t\{\psi(t) + 2\psi'(t) + \psi''(t)\} = \varphi''(t) = 4v^2 F''(v) + 4vF'(v).$$

Hence, for a function F that depends only on v,

$$LF(w) = 4v^2F''(v) = \varphi''(t) - 2\varphi'(t) = e^t\{\psi''(t) - \psi(t)\}.$$

Consequently, for such a function F that is also ≥ 0,

$$(**) \qquad LF = \lambda F \Leftrightarrow \psi''(t) = (\lambda + 1)\psi(t) \Leftrightarrow \lambda \geq -1.$$

This shows that the bottom of the positive spectrum for the hyperbolic spectrum is less than or equal -1. In fact, the Schauder-Tychonov fixed point theorem implies that it is exactly equal to -1.

Proposition 3. *If L is the hyperbolic Laplacian, then $\sigma^+(L) = [-1, \infty)$.*

Proof. Consider the convex set C of positive solutions F of the equation $LF = \lambda F, \lambda \in \sigma^+(\mathrm{L}), F(i) = 1$. This is a compact convex set with respect to the topology of uniform convergence on compact subsets of H. The group G = SL(2,$\mathbb{R}$) acts on C in an affine way: define $S_gF(w) = F(g.w)/F(g.i)$, where $g.w = \dfrac{aw+b}{cw+d}$ if $g = \begin{bmatrix} a & b \\ c & d \end{bmatrix}$. Then $S_{g_2} \circ S_{g_1} = S_{g_1g_2}$.

There is a very natural subgroup N of SL(2,$\mathbb{R}$) that corresponds to translations parallel to the real axis: $\mathrm{N} = \{n_u | u \in \mathbb{R}\}$ where $n_u = \begin{bmatrix} 1 & u \\ 0 & 1 \end{bmatrix}$. A function F(w) $\in$ C depends only on v if and only if $S_nF = F \; \forall n \in \mathrm{N}$. The Schauder Tychonov theorem (cf. [12] p456) implies that for each n, S_n has a non void compact convex set of fixed points $\mathrm{C}_n \subset \mathrm{C}$. Since N is a commutative subgroup of SL(2, $\mathbb{R}$) any finite number of the convex sets C_n has a finite intersection and so C has a point fixed by N.

Therefore, if $\lambda \in \sigma^+(\mathrm{L})$ there is a positive function F that depends only on v such that $LF = \lambda F$. □

Remark 1. The Schauder-Tychonov Theorem was used in this context by Margulis [24]. Conze and Guivarc'h [10] extended it to nilpotent groups and it is used in a similar way by Guivarc'h in his simplified determination of the minimal functions for L − λId on any symmetric space of non-compact type [15].

Remark 2. The formula (**) also shows that the function F on H that satisfies $LF = \lambda F$ and depends on v alone is of the form

$$F(v) = c_1e^{(1+\sqrt{\lambda+1})t} + c_2e^{(1-\sqrt{\lambda+1})t} = c_1v^{\beta_1(\lambda)} + c_2v^{\beta_2(\lambda)},$$

where the β_i are the roots of the equation $4\beta(\beta - 1) = \lambda$
i.e.

$$2\beta_1(\lambda) = 1 + \sqrt{\lambda+1} \text{ and } 2\beta_2(\lambda) = 1 - \sqrt{\lambda+1}.$$

The larger value of β determines a minimal function while the other function is non-minimal. Notice that for any $\beta, F(v) = v^\beta$ satisfies $4v^2F''(v) = \beta(\beta-1)F$.

6. The bottom of the positive spectrum for the polydisc

It is clear that we can look at the question on a " polyplane"— the product of ℓ copies of the hyperbolic upper half plane — rather than on the polydisc. Let $\mathbf{w} = (w_1, w_2, \dots, w_\ell)$ be a point of the polyplane, $\mathbf{w} = \mathbf{u} + i\mathbf{v}$. Then $\mathbf{v} = (v_1, v_2, \dots v_\ell)$, $v_i = e^{2t_i}$. Define $\rho(\mathbf{t}) = \sum_{i=1}^{\ell} t_i$ if $\mathbf{t} = (t_1, t_2, \dots t_\ell)$. If $F(\mathbf{w}) = F(\mathbf{v}) = e^{\rho(\mathbf{t})}\psi(\mathbf{t})$, then the computation for the hyperbolic disc implies that

$$\mathrm{L}F(\mathbf{w}) = e^{\rho(\mathbf{t})}\{\Delta\psi(\mathbf{t}) - \ell\psi(\mathbf{t})\},$$

where L denotes the Laplace-Beltrami operator on the polyplane and Δ is the usual Laplacian on $\mathbb{R}^\ell$. It follows that for positive functions F depending on $\mathbf{v}$ alone, $Lf = \lambda F \Leftrightarrow$
$\lambda \geq -\ell$. As for the hyperbolic disc, this implies that the bottom of the positive spectrum for the polydisc is $-\ell$. The argument is the same with the group N replaced by the product of ℓ copies of the subgroup N $\subset$ SL(2,$\mathbb{R}$).

Main Theorem. *The Martin compactification $\tilde{D}$ of the ℓ-polydisc D at the bottom of the positive spectrum is its closure $\overline{D}$ in $\mathbb{C}^\ell$. The distinguished boundary T is the set of minimal points. The minimal function associated to* $\mathrm{b} \in T$ *is*

$$K(\mathbf{z}, \mathbf{b}) = \sqrt{P_{\mathbf{b}}(\mathbf{z})} = \prod_{i=1}^{\ell} \sqrt{P_{b_i}(z_i)},$$

where $P_{b_i}(z_i) = \dfrac{1-|z_i|^2}{|b_i - z_i|^2}$ *is the Poisson kernel on the disc* D_i.

Furthermore, the limit functions associated with the non-minimal points are given by explicit formulas (cf. [16])

To prove this theorem it is necessary to understand (i) the Iwasawa decomposition of SU(1,1), equivalently of SL(2,$\mathbb{R}$) and (ii) the Martin compactifications of the hyperbolic disc — at least at the bottom of the positive spectrum. The actual proof is given in [16].

7. The Iwasawa decomposition of SL(2, $\mathbb{R}$)

Let $g = [\mathbf{g}_1 \ \ \mathbf{g}_2] = \begin{bmatrix} a & b \\ c & d \end{bmatrix} \in$ SL(2,$\mathbb{R}$). Then there is an orthogonal matrix k such that $g = ks$ where s is upper triangular. The columns $\mathbf{g}_1, \mathbf{g}_2$

of g form a basis of $\mathbb{R}^2$. Apply the Gram-Schmidt orthonormalization procedure to it. This determines $k = [\mathbf{k}_1 \quad \mathbf{k}_2] = \begin{bmatrix} a/\sqrt{a^2+c^2} & -c\sqrt{a^2+c^2} \\ c\sqrt{a^2+c^2} & a\sqrt{a^2+c^2} \end{bmatrix}$ and $s = \begin{bmatrix} \sqrt{a^2+c^2} & ab+cd/\sqrt{a^2+c^2} \\ 0 & 1/\sqrt{a^2+c^2} \end{bmatrix}$.

Let A denote the following subgroup of matrices $a_t, t \in \mathbb{R}$, where $a_t = \begin{bmatrix} e^t & 0 \\ 0 & e^{-t} \end{bmatrix}$. Then the upper triangular matrice s can be written uniquely as $s = n_u a_t, u = e^t(ab+cd)/\sqrt{a^2+c^2}$ and $t = \log\sqrt{a^2+c^2}$.

Note that

$$(*) \qquad a_t^{-1} n_u a_t = n_{e^{-2t}u}.$$

This implies that the set S of upper triangular matrices is a group and that each $g \in$ SL(2,$\mathbb{R}$) has a unique decomposition as $g = kan$, $k \in$ K $=$ SO(2), $a \in$ A and $n \in$ N, where N was defined earlier. This is the *Iwasawa decomposition* of g. So G= KAN = KS = S^{-1}K^{-1} = SK = NAK — the Iwasawa decomposition of G — and so the upper half plane /K can be identified with the group S = NA.

It is also very important to note that (*) implies

$$\lim_{t\to\infty} a_t^{-1} n a_t = e \text{ for any } n \in \text{N}.$$

This is a special case of a general phenomenon which is the key to Dynkin's method [13] for computing Martin compactifications of symmetric spaces of non-compact type which was used by Karpelevič in [21].

8. The Martin compactifications of the hyperbolic disc

It is of course obvious that the Martin compactification of the unit disc with respect to the hyperbolic Laplacian is the closed disc since the harmonic functions are the same as the usual harmonic functions.

However, it is not so obvious that this is also the Martin compactification for the harmonic space given by the solutions of the equation $\mathrm{L}f(z) = (1-|z|^2)^2\Delta f(z) = \lambda f(z), \lambda \geq -1$ is also the closed disc. It turns out, by a simple argument, that the Poisson kernel for this harmonic space is just a power of the usual Poisson kernel, the power being $1+\sqrt{1+\lambda}$.

The Green kernel G_λ(z,w) depends only on the hyperbolic distance of z from w. To obtain its asymptotics, it suffices to examine the radial form of the operator L $-$ λId. It is well known (cf. Helgason [19]) that the geodesic distance of z from o is t, where $\tanh t = |z|$ and that the rays from the origin are geodesics with the unit speed geodesic given by $t \mapsto e^{i\theta}\tanh t$.

Proposition 4. *If* $f(z) = f(|z|) = w(t)$, *where* $\tanh t = |z|$, *then*

$$(1-|z|^2)^2 \Delta f(z) = w''(t) + 2\coth 2t\, w'(t).$$

Proof. This is a direct calculation or one may use the general formula on p313 of [19] after computing the hyperbolic volume $V(T) = 2\pi \sinh^2 T = 2\pi \int_0^R \frac{r dr}{1-r^2}$ of the euclidean ball of radius $R = \tanh T$. □

Making the substitution $s = e^{-t}$ and setting $w(t) = \varphi(s)$, it follows that

$$w''(t) + 2\coth 2t\, w'(t) - \lambda w(t) = s^2\varphi''(s) + \frac{s(3s^4+1)}{(s^4-1)}\varphi'(s) - \lambda\varphi(s) = \mathrm{M}\varphi(s).$$

Applying the classical method of Frobenius, as in [22], the indicial equation for $M\varphi(s) = 0$ is $r^2 - 2r - \lambda = 0$ and the largest root $r(\lambda) = 1 + \sqrt{1+\lambda}$ gives the asymptotics of G_λ(z,w) as $\exp(-r(\lambda)t)$, where t is the distance from z to w (see [22]).

Proposition 5. *The Martin compactification of the unit disc given by the operator* $L - \lambda Id$ *is the closed unit disc. All points are minimal and the minimal function associated with* $b = e^{i\theta}$ *is* $[P_b(z)]^{\frac{(1+\sqrt{1+\lambda})}{2}}$.

Proof. Let G(z,w) denote G_λ(z,w). Consider a sequence of points $w_n = e^{i\theta_n} a_{t_n}.o = e^{i\theta_n}\tanh t_n$ for which the Martin kernel K has a limit. One may assume the angles converge to 0. Then the limit function equals $h(z) = \lim_{n\to\infty} G(z, \tanh t_n)/G(o, \tanh t_n) = \lim_{n\to\infty} \exp r(\lambda)\{d(o, \tanh t_n) - d(z, \tanh t_n)\} = \lim_{n\to\infty} \exp r(\lambda)\{t_n - d(z, \tanh t_n)\}$.

To compute the so-called Busemann function $\beta(z) = \lim_{t\to\infty}\{d(o, \tanh t_n) - d(z, \tanh t_n)\}$ (see [6]), it is easiest to consider the situation on the upper half plane, replacingo by i, z by w = $\gamma(z)$, and $\tanh t$ by $e^{2t}i = \gamma(\tanh t)$. If $w = u + iv$, let Q(t) = $u + ie^{2t}$ and P(t) = $e^{2t}i$. Then $d(Q(t), P(t)) \to 0$ as $t \to \infty$. Hence, if $v = e^{2t_0}$, $\lim_{t\to\infty}\{d(i, P(t)) - d(w, P(t))\} = \lim_{t\to\infty}\{d(i, P(t)) - d(w, Q(t))\} = t_0 = 1/2 \log v$.

From the explicit form of the Busemann function it follows that the limiting angle for a sequence of points $w_n = e^{i\theta_n}\tanh t_n$ for which the Martin kernel has a limit is unique. □

9. The Furstenberg and Satake compactifications

Let G be a semisimple Lie group with a maximal compact subgroup K — for example G = SL(n,$\mathbb{R}$) and K = SO(n). There is a "natural" measure ν on K which is left K-invariant and obtained by averaging Haar measure over the orbits of a certain subgroup of K. For SO(n) this is the subgroup of diagonal matrices. In the case of SO(2), $\int f d\nu = \int (1/2)[f(k) + f(-k)]dk$. It turns out that the cosets gK in M = G/K can be identified with the set $G\cdot\nu = \{g \cdot \nu \mid g \in G\}$ of probabilty measures on G (recall that $\int f d(g\cdot\nu) = \int f(gx)\nu(dx)$). The maximal Furstenberg compactification of M is the closure of this set of probabilities in the topology of weak convergence [14].

Satake constructed compactifications of M by using finite dimensional irreducible representations of G to imbed M in the projective space associated with a vector space of Hermitian symmetric matrices. The introduction to his article [27] explains the procedure.

Appendix: Lie algebras ; some remarks. Let $\mathfrak{g}$ be a real Lie algebra. Then $\mathfrak{g}$ acts on itself by $X \mapsto \mathrm{ad}X$, where $\mathrm{ad}X(Y) = [X,Y]$. The Jacobi identity [6] p99 ensures that this map is a Lie algebra homomorphism of $\mathfrak{g}$ into the Lie algebra gl($\mathfrak{g}$) of linear transformations of $\mathfrak{g}$ into itself.

Associated with these linear transformations there is an intrinsic bilinear form $B(X,Y)$ on $\mathfrak{g}$, the *Killing form,* defined by

$$B(X,Y) = trace(adX \circ adY).$$

If this bilinear form is non-degenerate then $\mathfrak{g}$ is said to be *semisimple.* For example, $\mathfrak{gl}(n,R)$ the set of $n\times n$ real matrices is not semisimple because the scalar multiples of the identity matrix are in the center. On the other hand the subalgebra $\mathfrak{sl}(n,R)$ given by the matrices of trace zero is semisimple because $B(X,Y) = 2(n+1)trace(XY)$ —[6] p187 and Ch III Lemma 6.1.p180— and $trace XY = 0$ for all Y implies that $X = 0$.

Every semisimple Lie algebra $\mathfrak{g}$ admits a so-called *Cartan decomposition*

$$\mathfrak{g} = \mathfrak{k} \oplus \mathfrak{p}$$

where $\mathfrak{k}$ is a subalgebra on which the Killing form of g is negative definite and $\mathfrak{p}$ is a direct summand on which the Killing form B is positive definite. For example, in case $\mathfrak{g} = sl(n,R)$ one such decomposition is given by $\mathfrak{k} = so(n,R)$ and $\mathfrak{p}$ the set of symmetric matrices of trace zero. These two subspaces are also determined by the involutive automorphism $\theta(X) = -X'$ with $\mathfrak{k} = \{X \mid \theta(X) = X\}$ and $\mathfrak{p} = \{X \mid \theta(X) = -X\}$.

As a result, the form $B_\theta(X,Y) = -B(X,\theta(Y))$ on $\mathfrak{g}$ is positive definite and the subspaces $\mathfrak{k}$ and $\mathfrak{p}$ are orthogonal with respect to it.

In the case of $\mathfrak{su}(1,1)$, which is isomorphic to $\mathfrak{sl}(2,\mathbb{R})$, with generators U,H, and Y the following commutation relations are satisfied:

$$[\mathrm{U},\mathrm{H}] = 2\mathrm{Y},\ [\mathrm{U},\mathrm{Y}] = -2\mathrm{H}, \quad \text{and } [\mathrm{H},\mathrm{Y}] = -2\mathrm{U}.$$

It follows easily from this that

$$B(U,H) = B(U,Y) = B(H,Y) = 0$$

and

$$-B(U,U) = B(H,H) = B(Y,Y) = 8.$$

Therefore, $\mathfrak{k} = \mathbb{R}U$ and $\mathfrak{p} = \mathbb{R}H + \mathbb{R}Y$ is a Cartan decomposition of $\mathfrak{su}(1,1)$. Furthermore, relative to B_θ, the three generators are mutually orthogonal and of length $\sqrt{8}$. The Poincaré metric on $D = \{z \mid |z| < 1\}$ corresponds to scaling B by 1/8. The metric on D for which $Lf(z) = (1/4)(1-|z|^2)^2\Delta f(z)$, corresponds to scaling B by 1/2 (cf. [19] p405).

Bibliography

[1] S. Agmon, *On positive solutions of elliptic equations with periodic coefficients in $\mathbb{R}^n$, spectral results and extensions to elliptic operators on Riemannian manifolds,* "Differential Equations," edited by I. Knowles and R.T.Lewis, Elsevier Science, Publishers B.V. (North-Holland), Amsterdam, 1984, pp. 7-17

[2] A. Ancona, *Principe de Harnack à la frontière et théorme de Fatou pour un opérateur dans un domaine lipschitzien,* Ann. Inst. Fourier **XXVIII** (1978), 169-213.

[3] A. Ancona, *Negatively curved manifolds, elliptic operators, and the Martin boundary,* Ann. of Math. **125** (1987), 495-536.

[4] A. Ancona, *Sur les fonctions propres positives des variétés de Cartan-Hadamard,* Comm. Math. Helv. **64** (1989), 62-83.

[5] M. Anderson and R. Schoen, *Positive harmonic functions on complete manifolds of negative curvature,* Ann. of Math. **121** (1985), 429-461.

[6] W. Ballmann, M. Gromov, and V. Shroeder, "Manifolds of Nonpositive Curvature," Birkhäuser, Boston, 1985.

[7] G. Bouligand, *Démonstration élémentaire d'un théorème de détermination à un facteur près d'une fonction harmonique,* Mathematica **6** (1932), 80-85.

[8] M. Brelot, "Lectures on Potential Theory," Tata Institute of Fundamental Research, Bombay, 1967.

[9] C. Constantinescu and A. Cornea, "Ideale Ränder Riemannscher Flächen," Ergebnisse der Mathematik und ihrer Grenzgebiete," Bd. 32, Springer-Verlag, Heidelberg–Berlin, 1963.

Level sets and the Green function

Jang-Mei Wu

1. Level sets of conformal mappings

Let Ω be a simply–connected domain in $\mathbb{R}^2(\Omega \neq \mathbb{R}^2)$, f be a conformal mapping from Ω onto the unit disk $|z| < 1$, and Γ be any line or circle on the plane. It is proved in [8],[10], [12] and [13], that the length of the level curve

$$\ell(f(\Gamma \cap \Omega)) = \int_{\Gamma\cap\Omega} |f'(z)||dz| \leq 4\pi^2. \tag{1}$$

We shall discuss various generalizations of the above theorem and some open questions.

In [8], Fernández, Heinonen and Martio proved the L_p version: there exists an absolute constant $p_0 (1 < p_0 < 2)$ so that

$$\int_{\Gamma\cap\Omega} |f'(z)|^p |dz| < C(p, \Omega) < \infty, \tag{2}$$

for $1 \leq p \leq p_0$. More recently, Baernstein [2] constructed Ω and f as above, so that

$$\int_{\Gamma\cap\Omega} |f'(z)|^{2-\epsilon} |dz| = \infty$$

for some line Γ and $\epsilon > 0$.

The largest possible p_0 in (2) is unknown. The smallest upper bound for the lengths of all level curves in (1) is also to be determined.

The question of finding curves Γ so that

$$\ell(f(\Gamma \cap \Omega)) < C(\Gamma) < \infty, \tag{3}$$

for all Riemann pairs (Ω, f) has been studied by many authors [7], [9] and [12]. In [3], Bishop and Jones proved that (3) holds if and only if Γ satisfies the Ahlfors regularity condition, i.e. there is a constant $C > 0$ such that $\ell(\Gamma \cap \{|z-a| < r\}) < Cr$ for all $a \in \Gamma$ and $r > 0$.

Potential Theory, ed. by M. Kishi

2. Gradient of the Green function

Let (Ω, f) be a Riemann pair in $\mathbb{R}^2$ and G be the Green function on Ω with pole at $f^{-1}(0) \notin \Gamma$. Because $|\nabla G(z)| = |f'(z)/f(z)|$, (2) becomes

$$\int_{\Gamma\cap\Omega} |\nabla G(z)|^p |dz| < \infty. \tag{4}$$

Thus it is natural to study the integrability of $|\nabla G|$ for general domains in $\mathbb{R}^m (m \geq 2)$. The following is proved in [19].

Suppose that Ω is a domain in $\mathbb{R}^m$, $m \geq 2$, that satisfies the $(m-1)$–dimensional density condition ($(m-1)DC$). Let P be a fixed point in Ω, G the Green function of Ω with pole at P, and Γ an $m-1$ dimensional hyperplane with $P \notin \Gamma$. Then there exists a constant $p_0 > 1$ depending on the $(m-1)DC$ constant, so that if $1 \leq p \leq p_0$,

$$\int_{\Gamma\cap\Omega} |\nabla G(x)|^p d\sigma(x) < C, \tag{5}$$

where $d\sigma$ is the $m-1$ dimensional measure on Ω, C is a constant depending on p, the $(m-1)DC$ constant, $dist(P, \partial\Omega)$ and $dist(P, \Gamma)$.

Denote by Λ^α the α–dimensional content of a set, that is $\Lambda^\alpha(S) = \inf \sum_n r_n^\alpha$ where the infimum is taken over all coverings of S with countably many balls of radii r_n. Ω satisfies the α–dimensional density condition (αDC) for some $\alpha \leq m-1$, if there is a fixed $\zeta > 0$ such that, for any $x \in \Omega$,

$$\Lambda^\alpha\left(\frac{\Omega^c \cap B(x, 2d(x))}{10d(x)}\right) > \zeta,$$

where $d(x) = dist(x, \partial\Omega)$, $B(x, r)$ is the ball with center x, radius r, and $\frac{S}{a}$ is the set $\{\frac{x}{a} : x \in S\}$. Ω satisfies the capacity density condition (CDC) if there is a fixed $\eta > 0$ such that whenever $x \in \Omega$,

$$capacity\left(\frac{\Omega^c \cap B(x, 2d(x))}{10d(x)}\right) > \eta;$$

here capacities are associated with the kernel $-\log r$ when $m = 2$, and r^{-m+2} when $m > 2$.

Simply–connected planar domains satisfy $1DC$; and αDC, $\alpha > m-2$, implies CDC by the Frostman Theorem [4,p.28].

Although $(m-1)DC$ is not necessary for (5), examples exist to show that $(m-1)DC$ cannot be replaced by CDC or αDC for any $\alpha < m-1$. In $\mathbb{R}^3$, let $1 < \alpha < 2$, and γ be a von Koch snowflake curve of Hausdorff dimension α lying on

$\{x : x_1^2 + x_2^2 < 1 \text{ and } x_3 = 0\}$. The domain $\Omega \equiv \{x : |x| < 10\}\backslash\gamma$ satisfies αDC; on the other hand,

$$\int_{\{x_3=0\}\cap\Omega} |\nabla G| d\sigma(x) = \infty$$

for Green function G with pole anywhere in Ω.

It remains interesting to find necessary and sufficient conditions on Ω for (5).

The main step of the proof is to show that if Ω satisfies $(m-1)DC$ and $\Omega^c \subseteq \{x_m = 0\}$, then any harmonic measure $d\omega(x)$ on $\partial\Omega$ can be extended to a measure on the entire hyperplane $\{x_m = 0\}$ which is A_∞ with respect to the surface measure $d\sigma(x)$. See [5] for the definition of A_∞. To prove this, one needs the following theorem, which, in the author's opinion, is extremely useful and deserves more attention than it has obtained.

Let Ω be a domain in $\mathbb{R}^m$, and $\{D_j\}$ be a sequence of closed sets contained in Ω with $dist(D_i, D_j) > 0$ whenever $i \neq j$. Set $\Omega_j = \Omega\backslash \bigcup_{k\neq j} D_k$. If $\inf_j \inf_{z\in D_j} \omega(z, \partial\Omega, \Omega_j) = a > 0$, then for any $x \in \Omega\backslash \bigcup_j D_j$,

$$\sum_j \omega(x, D_j, \Omega\backslash D_j) \leq \frac{1}{a}\omega(x, \cup D_j, \Omega\backslash \cup D_j),$$

where ω is the harmonic measure.

This says that if sets are uniformly separated then their harmonic measures are summable. When Ω and D_j's are disks, this is proved by Garnett, Gehring and Jones [12]. Other versions can be found in [16] and [17]. The theorem as stated in its greatest generality is due to B. Davis by a Brownian motion proof [11,Chap.2]. The inverse of this theorem is also true if $\{D_j\}$ are quasi-hyperbolically separated, see [16].

3. Covering mappings

Let Ω be a hyperbolic planar domain, π a universal covering map from $\Delta \equiv \{|z| < 1\}$ onto Ω, $\mathcal{G}$ the covering group. Let δ be the order of convergence, i.e. $\delta = \inf\{a > 0 : \sum_{T\in\mathcal{G}} (1 - |T(0)|)^a < \infty\}$. Fernández and Hamilton [7] proved that if $\delta < \frac{1}{2}$, then

$$\ell(\pi^{-1}(\Omega \cap \Gamma)) < C(\Omega) < \infty \tag{6}$$

for any line Γ. Fernández proved in [6] that if (6) holds then Ω satisfies the capacity density condition (CDC) which implies $\delta < 1$; and that if Ω is a Denjoy domain (Ω^c lies on a line), then (6) holds if and only if Ω^c is homogeneous (equivalent to $1DC$).

It is interesting to note that

$$\ell(\pi^{-1}(\Omega \cap \Gamma)) = \int_{\Gamma\cap\Omega} |\sum_{\mathcal{G}} \frac{(T\pi^{-1}(w))'}{T\pi^{-1}(w)}||dw|,$$

and that for the Green function with pole at $\pi(0)$,

$$\int_{\Gamma\cap\Omega} |\nabla G(w)||dw| = \int_{\Gamma\cap\Omega} |\sum_{\mathcal{G}} \frac{(T\pi^{-1}(w))'}{T\pi^{-1}(w)}||dw|.$$

4. Diameter problem for quasicircles

Let Γ be a K-quasicircle, one may ask whether

$$\sum_{\gamma\in J} diam f(\gamma) < C(K) < \infty \tag{7}$$

for all Riemann pairs (Ω, f), with J being the collection of components of $\Gamma \cap \Omega$. An analogous question in $\mathbb{R}^m$ is the following: if f is a K-quasiconformal mapping of Ω onto the unit ball and Γ is any line, is it true that

$$\sum_{\gamma\in J} diam f(\gamma)^{m-1} < C(K,m)? \tag{8}$$

These questions are equivalent when $m = 2$; Astala and Rohde have just answered them negatively [1]. (8) is still open for $m > 2$.

Väisälä [18] recently proved a weaker version of (8): $\sum_{\gamma\in J} diam f(\gamma)^{\alpha} < C(K,m,\alpha)$ for all $\alpha > m-1$. He also proved (8) if f has a quasiconformal extension to the whole space.

It is challenging to characterize those quasicirdes Γ so that (7) holds.

5. A_∞ and the Green function

The A_∞ property of certain harmonic measure was studied in §2. We shall study the A_∞ property of $|f'|$ and $|\nabla G|$ in the interior. In [14], Heinonen proved that if (Ω, f) is a Riemann pair and L is a line segment in Ω, then $|f'| \in A_\infty(|dz|)$ on L if and only if there exists a quasidisk $D \subseteq \Omega$ so that $L \subseteq \bar{D}$. There are three conditions involved: thickness of Ω^c, shape of L, and a surrounding domain D.

Let Ω be a domain in $\mathbb{R}^m$, P a fixed point in Ω, G the Green function with pole P and Γ a $(m-1)$ dimensional hyperplane that does not contain P. The following is proved in [20].

If Ω is a Lipschitz domain, then $\frac{G(x)}{dist(x,\partial\Omega)}|_{\Gamma\cap\Omega}$ can be extended to an A_∞ function on Γ with respect to the surface measure $d\sigma$.

If Ω is a nontangentially accessible (NTA) domain, L is a uniform domain on Γ and $L \subseteq \bar{D}$ for some quasiball $D \subseteq \Omega$, then $\frac{G(x)}{dist(x, \partial\Omega)}|_L$ can be extended to become a function $A_\infty(d\sigma)$ on Γ. See [15] for the definition of NTA.

When Ω is Lipschitz, no additional condition is imposed on the shape of $\Gamma \cap \Omega$. However there exists an NTA domain Ω, so that $\Omega \cap \Gamma$ is a square, but $\frac{G(x)}{dist(x, \partial\Omega)}$ is not in $A_\infty(d\sigma)$ on $\Omega \cap \Gamma$.

The question on necessary and sufficient conditions remains open.

References

[1] K. Astala and S. Rohde, preliminary.
[2] A. Baernstein, A counterexample concerning integrability of derivative of conformal mappings, J. d'Analyse Math., 53 (1989), 253–268.
[3] C. Bishop and P. Jones, Harmonic measure and arc length., Ann. Math. (to appear).
[4] L. Carleson, "Selected Problems on Exceptional Sets," Wadsworth, 1983.
[5] R. R. Coifman and C. Fefferman, Weighted norm inequalities for maximal functions and singular integrals, Studia Math. 51 (1974), 241–250.
[6] J. Fernández, Domains with strong barrier, Revista Math. Ibernoamericana (to appear).
[7] J. Fernández and D. Hamilton, Length of curves under conformal mappings, Comm. Math. Helv. 62 (1987), 122–134.
[8] J. Fernández, J. Heinonen and O. Martio, Quasilines and conformal mappings, J. d'Analyse Math., 52 (1989), 117–132.
[9] J. Fernández and M. Zinsmeister, Ensembles de niveau des representations conformes, C. R. Acad. Sci. Paris, 305 (1987), 449–452.
[10] B. B. Flinn, Hyperbolic convexity and level sets of analytic functions, Indiana Univ. Math. J. 32 (1983), 831–841.
[11] J. Garnett, "Applications of Harmonic Measure" Wiley, 1986.
[12] J. Garnett, F. Gehring and P. Jones, Conformally invariant length sums, Indiana Univ. Math. J., 32 (1983), 809–829.
[13] W. Hayman and J.-M. Wu, Level sets of univalent functions, Comment. Mat. Helv., 56 (1981), 366–403.
[14] J. Heinonen, preliminary.
[15] D. Jerison and C. Kenig, Boundary behavior of harmonic functions in nontangentially accessible domains, Adv. in Math. 46 (1982), 80–147.
[16] P. Jones and D. Marshall, Critical points of Green's function, harmonic measure, and the corona problem, Ark. Math., 23 (1985), 281–314.
[17] Ch. Pommerenke, On uniformly perfect sets and Fuchsian groups, Analysis, 4 (1984), 299–321.
[18] J. Väisälä, Bounded turning and quasiconformal maps, preprint.
[19] Jang-Mei Wu, Boundary density and the Green function, Michigan J. Math. (to appear).
[20] Jang-Mei Wu, A_∞ and the Green function, preprint.

Local properties of solutions to equations involving square Hörmander's operators

Marco Biroli

Abstract
We study estimates on oscillation and energy decay for solutions to equations involving square Hörmander's operators.

Mathematics Subject classification: 31B05, 35D10.

1. Introduction and results

We will point out at first that most of the results in the following have been obtained jointly with U. Mosco.

Let $a_{il}(x), b_i(x), c_{ij}(x)$, $i, j = 1, \dots, m$, $l = 1, \dots, N$, be $C^\infty(\mathbf{R}^N)$ functions; we consider the vector fields

$$X_i = \sum_{l=1}^{N} a_{il} D_{x_l}$$

and we suppose in the following that a uniform Hörmander's condition holds for the fields X_i (i.e. X_i and the commutators up to a fixed order K span the tangent space).

A natural distance associated to vector fields can be defined as

$$d(x,y) = inf\{b; \gamma : [0,b] \to \mathbf{R}^N \ admissible\ path,\ \gamma(0) = x, \quad \gamma(b) = y\}; \qquad (1.1.)$$

an admissible path is a Lipschitz curve such that

$$\gamma'(s) = \sum_{i=1}^{m} d_i(s) X_i(\gamma(s))$$

a.e., where d_i are bounded measurable functions with $\sum_{i=1}^m d_i^2(s) \le 1$, [3][6][7]. We denote

$$B(r,x) = \{y | d(x,y) \le r\} \qquad (1.2.)$$

and we say $B(r,x)$ are the *intrinsic balls* (for our vector fields).

We recall that for intrinsic balls a Poincaré's type inequality holds, [3]:

Potential Theory, ed. by M. Kishi

Proposition 1. *Let f be in $C^\infty\overline{(B(2r,x))}$; then*

$$\int_{B(r,x)} |f-f_r|^2 dx \le Cr^2 \sum_{i=1}^{m} \int_{B(r,x)} |X_i(u)|^2 dx$$

where f_r is the average of f over $B(r,x)$.

Given an open set Ω in $\mathbf{R}^N$ we denote

$$E(u,v;\Omega) = \sum_{i=1}^{m} \int_\Omega X_i(u)X_i(v)dx + \int_\Omega uvdx.$$

The space $H^1(\Omega;E)$ $(H^1_0(\Omega;E))$ will be the closure of $C^\infty(\Omega)$ $(C_0^\infty(\Omega))$ to the norm $E(u,v;\Omega)^{\frac{1}{2}}$.

We can give an extension of Proposition 1.

Proposition 1′. *Let f be in $H^1(B(r,x);E)$ $(H^1_0(B(r,x);E))$ then we have*

$$\int_{B(r,x)} |f-f_r|^2 dx \le Cr^2 \sum_{i=1}^{m} \int_{B(r,x)} |X_i(u)|^2 dx$$

$$\left(\int_{B(r,x)} |f|^2 dx \le Cr^2 \sum_{i=1}^{m} \int_{B(r,x)} |X_i(u)|^2 dx\right).$$

We consider the operator

$$L = -\sum_{i=1}^{m} X_i^* X_i + \sum_{i=1}^{m} b_i X_i + \sum_{ij=1}^{m} c_{ij}[X_i, X_j] + c_0 \qquad (1.3.)$$

where X_i^* is the formal adjoint of X_i.

We can associate to L the bilinear form $a(u,v;\Omega)$ on $H^1(\Omega;E)XH^1_0(\Omega;E)$ defined by

$$a(u,v;\Omega) = \sum_{i=1}^{m} \int_\Omega X_i(u)X_i(v)dx + \sum_{i=1}^{m} \int_\Omega b_i X_i(u)vdx +$$
$$+\sum_{ij=1}^{m} \int_\Omega c_{ij}[X_i,X_j](u)vdx + \int_\Omega c_0 uvdx. \qquad (1.4.)$$

We observe that, taking into account proposition 1′, the bilinear form $a(u,v;B(r,x))$ is coercive on $H^1_0(B(r,x);E)$ for $r \le \overline{R_0}$, $\overline{R_0}$ positive suitable.

This last property allow us to define the *potential* ϕ of $B(sr,x)$ $(0 < s < 1)$ with respect to $B(r,x)$, $r \le \overline{R_0}$, as the solution of the problem

$$a(\phi, v-\phi; B(r,x)) \ge 0$$

$$\forall v \in H_0^1(B(r,x);E), \qquad v \geq 1 \qquad on \qquad B(sr,x)$$
$$\phi \in H_0^1(B(r,x);E), \qquad \phi \geq 1 \qquad on \qquad B(sr,x). \tag{1.5.}$$

We observe that $L\phi$ is a measure on $B(r,x)$ which belongs to $H^{-1}(B(r,x);E) = (H_0^1(B(r,x);E))'$; moreover the support of $L\phi$ is contained in $\partial B(sr,x)$.

Consider now $f \in L^\infty(\Omega)$; we say that u is a *weak local solution* in Ω of

$$Lu = f \tag{1.6.}$$

iff $u \in H^1(\Omega;E)$ and

$$a(u,v;\Omega) = \int_\Omega fv dx \quad , \quad \forall v \in H_0^1(\Omega;E).$$

In this paper we will investigate the properties of weak local solutions of (1.6.); we obtain the following result:

Theorem 1. *Let u be a weak bounded local solution of (1.6.) in $B(\overline{R}_0;x_0)$; we have*

$$V(r) \leq C_1\left(\frac{r}{R}\right)^\beta V(R) + C_2 R^\alpha \quad , \quad r \leq qR \leq \frac{q^2}{2}\overline{R}_0 \tag{1.7.}$$

where

$$V(r) = (osc_{B(r,x_0)}u)^2 + \sum_{i=1}^m \int_{B(r,x_0)} |X_i(u)|^2 G^{x_0}_{B(2q^{-1}r;x_0)} dx;$$

$G^{x_0}_{B(2q^{-1}r;x_0)}$ is the Green function of L relative to $B(2q^{-1}r;x_0)$ with singularity at x_0 (see Section 2.), α, $\beta \in (0,1)$, $q \in (0,q_0)$ $q_0 \in (0,\frac{1}{4})$ suitable, $C_1 > 0$ depends on the coefficients of L and $C_2 \geq 0$ on $sup_{B(\overline{R}_0;x_0)}|f|$ and $sup_{B(\overline{R}_0;x_0)}|u|$; moreover if $f = c_0 = 0$ we have $C_2 = 0$.

Then u is locally Hölder continuous in $B(\overline{R}_0;x_0)$ and

$$\sum_{i=1}^m \int_{B(r,x_0)} |X_i(u)|^2 G^{x_0}_{B(2q^{-1}r;x_0)} dx \leq$$
$$\leq C_3\left(\frac{r}{R}\right)^\alpha \sum_{i=1}^m \int_{B(R,x_0)} |X_i(u)|^2 G^{x_0}_{B(2q^{-1}R;x_0)} dx \tag{1.8.}$$

$r \leq q^2 R \leq \frac{q^3}{2}\overline{R}_0$.

We observe that estimates on the oscillation of u have been proved in [4] for continuous solutions and $f = 0$ in the parabolic case. To our knowledge the estimate (1.8.) on the "energy decay" is completely new.

To prove Theorem 1. we use a method developped in [1] to study analogous estimates in the case of an obstacle problem relative to a degenerate elliptic operator with a weight in the A_2 Muckenhoupt's class. This method allows us to overcome the difficulty due to the fact that we have a Poincaré's inequality on intrinsic balls, but not on intrinsic annuli.

The method of [1] has to be modified using suitable cut–off functions, which are related to the level sets of the Green function and this makes the method more intrinsic.

Finally we observe that we deal in this paper with weak local solutions of (1.6.), but the method can be used also to prove Wiener estimates at the boundary and for local solutions of obstacle problems relative to the bilinear form (1.4.).

2. The Green function

The existence of a Green function for the operator (1.3.) relative to $\mathbf{R}^N$ has been proved in [7] and the following estimates are given

$$\Lambda_1 \frac{r^2}{|B(r,x)|} \leq G^x(y) \leq \Lambda_2 \frac{r^2}{|B(r,x)|} \tag{2.1.}$$

where $G^x(y)$ is the Green function relative to $\mathbf{R}^N$ with singularity at x and $y \in \partial B(r,x)$, $r \leq \overline{R}_0 \leq \overline{R}$ suitable.

We observe that $G^x(y)$ is in $H^1(\mathbf{R}^N - \{x\}; E)$ and

$$L^* G^x(y) = \delta_x$$

in the distribution sense, where δ_x is the Dirac distribution with singularity at x and L^* denotes the adjoint of the operator L.

The form $a(u,v;B(r,x))$ is coercive on $H_0^1(B(r,x);E)$ for $r \leq \overline{R}_0$; this property allows us to define the Green function relative to $B(r,x)$ with singularity at x; moreover the estimates (2.1.) and the properties of the measure of intrinsic balls give easily

$$\Lambda_3 \frac{r^2}{|B(r,x)|} \leq G^x{}_{B(r,x)}(y) \leq \Lambda_4 \frac{r^2}{|B(r,x)|} \tag{2.2.}$$

for $y \in \partial B(qr,x)$, $0 < q < \overline{q}_0 < 1$, $\overline{q}_0$ suitable, where Λ_3 and Λ_4 depend on q but don't depend on x.

Using the same methods as in [2] we can prove that

$$G^x_{B(R,x)}(y) \cong \int_r^R \frac{s^2}{|B(s,x)|}\frac{ds}{s} \quad , \qquad r \leq qR. \tag{2.3.}$$

Proposition 2. We have $G^x_{B(\overline{R}_0,x)} \in L^{1+\epsilon}(B(\overline{R}_0;x))$, with $\epsilon > 0$ suitable.

Proof. We have

$$\int_{B(\overline{R}_0,x)} (G^x_{B(\overline{R}_0,x)})^{1+\epsilon} dy \leq \int_0^\infty t^\epsilon |\{x | G^x_{B(\overline{R}_0,x)}(y) > t\}| dt + Cst \leq$$

$$\leq C_1 \int_0^\infty \left(\int_r^{\overline{R}_0} \frac{s^2}{|B(s,x)|} \frac{ds}{s} \right)^\epsilon |B(r,x)| \frac{r}{|B(r,x)|} dr + Cst \leq Cst$$

if $\epsilon < (N + K - 1)^{-1}$, where K is the greatest order of the commutators in Hörmander's condition (here we use the estimates on the volume of intrinsic balls, [6]). ◇

From the result in Proposition 1 we easily obtain:

Corollary 1. *$G^x \in L^{1+\epsilon}(B(R_0; x_0))$ and $G^x_{B(r,x)} \in L^{1+\epsilon}(B(R_0; x_0))$.*

3. Proof of Theorem 1

At first we prove a Caccioppoli's inequality for local bounded solutions of the equation (1.6.) in $B(\overline{R}_0; x_0)$.

Lemma 1. *Let u be a bounded weak solution of (1.6.) in $B(R_0; x_0)$; we have*

$$\sum_{i=1}^m \int_{B(qr,x_0)} |X_i(u)|^2 G^{x_0}_{B(sr;x_0)} dx + sup_{B(qr,x_0)} |u-k|^2 \leq$$

$$\leq \frac{C_1}{|B(r,x_0)|} \int_{B(r,x_0)-B(qr,x_0)} |u-k|^2 dx + C_2 r^\alpha \qquad (3.1.);$$

where $0 < r < q\overline{R}_0$, $0 < q < q_0$, with $q_0 \in (0, \frac{1}{4})$ and $\overline{R}_0 > 0$ suitable independent of x_0, $s \geq 1$, k in R, $C_1(C_2)$ is a constant dependent on q, f, $sup_{B(\overline{R}_0;x_0)}|u|^2$, k and α depends on N and on the order of the Hörmander condition.

Sketch of the Proof. First we observe that to prove (3.1.) it is enough to prove the same inequality with $G^{x_0}_{B(sr;x_0)}$ replaced by G^{x_0} (the Green function relative to $\mathbf{R}^N$); this last inequality can be proved using as test function $v = (u-k)^2 \phi^2((G^z_\rho)^{-1}) G^z_\rho$, $z \in B(qr, x_0)$, where G^z_ρ is the regularized Green function defined as in [1] and $\phi \in C_0^\infty(\mathbf{R})$, $\phi(t) = 1$ for $|t| \leq \lambda$, $\phi(t) = 0$ for $|t| \geq c_0\lambda$, $\phi'(t) \leq \frac{2}{(c_0-1)\lambda}$, λ and c_0 suitable. ◇

Remark 1. (1) Every weak solution of a boundary value Dirichlet problem in a bounded set Ω with bounded data is bounded.

(2) If $f = 0$ and we choose $k = 0$ the same proof as in Lemma 1 allow us to prove that every weak local solution of (1.6.) is locally bounded, then the result of Lemma 1 holds for every local solution, with C_2 depending only on the L^2-norm of u in $B(\overline{R}; x_0)$.

We now prove the estimate (1.7.) in Theorem 1.

We use as test function $v = (u - \overline{u}) G^z_\rho \phi$, $\rho < \frac{1}{2} sr$, where $\overline{u}$ is the average of u on $B(qr, x_0)$, $G^z_\rho = G^z_{B(tr,z),\rho}$ is the regularized Green function (as defined in [1]) of

$G^z_{B(tR,z)}$, which we denote in the following by G^z, and ϕ is the potential of $B(sr,z)$ in $B(tR,z)$, $s<t$.

We recall that $\langle L\phi, G^z_\rho\rangle = f_{B(tr,z)}\phi dx = 1$ and that $L\phi$ is a positive measure with support in $\partial B(sr,z)$.

We have

$$\sum_{i=1}^m \int_{B(sr,z)} |X_i(u)|^2 \phi G^z_\rho dx + \frac{1}{2} f_{B(\rho,z)} |u-\overline{u}|^2 dx \leq$$

$$\leq -2\sum_{i=1}^m \int_{B(tr,z)} X_i(u)X_i(\phi)(u-\overline{u})G^z_\rho dx + \frac{1}{2}\langle L\phi, |u-\overline{u}|^2 G^z_\rho\rangle + C_1 \int_{B(tr,z)} G^z_\rho dx. \tag{3.2.}$$

Passing to the limit as $\rho \to 0$ we obtain

$$2\sum_{i=1}^m \int_{B(sr,z)} |X_i(u)|^2 \phi G^z dx + |u-\overline{u}|^2(z) \leq$$

$$\leq sup_{B(tr,z)}|u-\overline{u}|^2 - 4\sum_{i=1}^m \int_{B(tr,z)} X_i(u)X_i(\phi)(u-\overline{u})G^z dx + C_1 \int_{B(tr,z)} G^z dx. \tag{3.3.}$$

Taking into account the result of Proposition 2 and the estimates on the measures of intrinsic balls, [6], we can estimate the last term in the right hand side by $C_2 r^\alpha$.

We also have

$$\sum_{i=1}^m \int_{B(tr,z)} |X_i(\phi)|^2 dx \leq \langle L\phi, \phi\rangle \leq \frac{C_3 \langle L\phi, G^z\phi\rangle}{inf_{B(sr,z)} G^z} \leq \frac{C_4}{inf_{B(sr,z)} G^z};$$

then from (3.3.) we easily obtain

$$\sum_{i=1}^m \int_{B(sr,z)} |X_i(u)|^2 G^z dx + |u-\overline{u}|^2(z) \leq$$

$$\leq (1+C_5\eta) sup_{B(tr,z)} |u-\overline{u}|^2 + \frac{C_6}{\eta} \sum_{i=1}^m \int_{B(tr,z)-B(sr,z)} |X_i(u)|^2 G^z dx + C_7 r^\alpha \tag{3.4.}$$

Taking now the supremum for z in $B(qr,x_0)$, choosing $t = 1-2q$, $s = 2q$, $q < min(\frac{1}{5}; q_0)$, and neglecting the first term in the right hand side we obtain

$$sup_{B(2qr,x_0)}|u-\overline{u}|^2 \leq (1+C_5\eta) sup_{B(r,x_0)} |u-\overline{u}|^2 +$$

$$+\frac{C_8}{\eta} sup_{B(qr,x_0)}\left(\frac{r^2}{|B(sr,z)|}\right) \sum_{i=1}^m \int_{B(r,x_0)-B(qr,x_0)} |X_i(u)|^2 dx + C_7 r^\alpha \tag{3.5.}$$

Taking into account the properties of the measures of intrinsic balls, [6], and the estimates for the Green function, we obtain

$$sup_{B(2qr,x_0)}|u-\overline{u}|^2 \leq (1+C_5\eta) sup_{B(r,x_0)} |u-\overline{u}|^2 +$$

$$+\frac{C_8}{\eta}\sum_{i=1}^{m}\int_{B(r,x_0)-B(qr,x_0)}|X_i(u)|^2 G^{x_0}_{B(2r;x_0)}dx + C_7 r^{\alpha}. \tag{3.6.}$$

From Lemma 1 we have

$$C_9\sum_{i=1}^{m}\int_{B(r,x_0)}|X_i(u)|^2 G^{x_0}_{B(2r;x_0)}dx + sup_{B(2qr,x_0)}|u-\overline{u}|^2 \leq$$

$$\leq (2+C_5\eta)sup_{B(r,x_0)}|u-\overline{u}|^2 +$$

$$+\frac{C_8}{\eta}\sum_{i=1}^{m}\int_{B(r,x_0)-B(qr,x_0)}|X_i(u)|^2 G^{x_0}_{B(2r;x_0)}dx + C_7 r^{\alpha}. \tag{3.7.}$$

By the usual "hole filling" argument we obtain

$$(C_8+C_9\eta)\sum_{i=1}^{m}\int_{B(qr,x_0)}|X_i(u)|^2 G^{x_0}_{B(2r;x_0)}dx + \eta sup_{B(qr,x_0)}|u-\overline{u}|^2 \leq$$

$$\leq \eta(2+C_5\eta)sup_{B(r,x_0)}|u-\overline{u}|^2 +$$

$$+C_8\sum_{i=1}^{m}\int_{B(r,x_0)}|X_i(u)|^2 G^{x_0}_{B(2r;x_0)}dx + C_7\eta r^{\alpha}. \tag{3.8.}$$

We now estimate the second term on the right hand side, taking into account that, for q_0 suitable, we have

$$inf_{\partial B(r,x_0)}\left(G^{x_0}_{B(2q^{-1}r;x_0)} - G^{x_0}_{B(2r;x_0)}\right) \geq C_{10}\left(\frac{r^2}{|B(sr,x_0)|}\right).$$

Therefore

$$(C_8+C_9\eta)\sum_{i=1}^{m}\int_{B(qr,x_0)}|X_i(u)|^2 G^{x_0}_{B(2r;x_0)}dx + \eta sup_{B(qr,x_0)}|u-\overline{u}|^2 \leq$$

$$\leq \eta(2+C_5\eta)sup_{B(r,x_0)}|u-\overline{u}|^2 + C_8\sum_{i=1}^{m}\int_{B(r,x_0)}|X_i(u)|^2 G^{x_0}_{B(2q^{-1}r;x_0)}dx +$$

$$+C_{11}\left(\frac{r^2}{|B(r,x_0)|}\right)\sum_{i=1}^{m}\int_{B(r,x_0)}|X_i(u)|^2 dx + C_7\eta r^{\alpha}. \tag{3.9.}$$

Using the Poincaré's and the Caccioppoli's inequalities we obtain

$$(C_8+C_9\eta)\sum_{i=1}^{m}\int_{B(qr,x_0)}|X_i(u)|^2 G^{x_0}_{B(2r;x_0)}dx + (C_{12}+\eta)sup_{B(qr,x_0)}|u-\overline{u}|^2 \leq$$

$$\leq \eta(2+C_5\eta)sup_{B(r,x_0)}|u-\overline{u}|^2 + C_8\sum_{i=1}^{m}\int_{B(r,x_0)}|X_i(u)|^2 G^{x_0}_{B(2q^{-1}r;x_0)}dx + C_7\eta r^{\alpha}.$$

then

$$(C_8+C_9\eta)\sum_{i=1}^{m}\int_{B(qr,x_0)}|X_i(u)|^2G^{x_0}_{B(2r;x_0)}dx+\frac{(C_{12}+\eta)}{4}(osc_{B(qr,x_0)}u)^2\leq$$

$$\leq\eta(2+C_5\eta)(osc_{B(r,x_0)}u)^2+C_8\sum_{i=1}^{m}\int_{B(r,x_0)}|X_i(u)|^2G^{x_0}_{B(2q^{-1}r;x_0)}dx+C_7\eta r^{\alpha}. \quad (3.10.)$$

Choosing now η suitable we have

$$V(qr)\leq\eta V(r)+C_{13}r^{\alpha}$$

where

$$V(r)=(osc_{B(r,x_0)}u)^2+\sum_{i=1}^{m}\int_{B(r,x_0)}|X_i(u)|^2G^{x_0}_{B(2q^{-1}r;x_0)}dx.$$

The integration Lemma in [5] gives the inequality (1.7.) in Theorem 1.

We now prove inequality (1.8.).

From Lemma 1 we have

$$(osc_{B(r,x_0)}u)^2\leq\frac{C_{14}}{|B(R,x_0)|}\int_{B(q^{-1}r,x_0)}|u-\overline{\overline{u}}|^2dx\leq$$

$$\leq\sum_{i=1}^{m}\int_{B(q^{-1}r,x_0)}|X_i(u)|^2G^{x_0}_{B(2q^{-2}r;x_0)}dx.$$

From (1.7.) and (3.11.) we obtain (1.8.). ◇

Remark 2. If $f=c_0=0$ then $C_{13}=0$.

References

[1] Biroli M., Mosco U. - Wiener Criterion and potential estimates for obstacle problems relative to degenerate elliptic operators - Ann. Mat. Pura Appl., in print.

[2] Fabes E., Jerison D., Kenig C. - The Wiener test for degenerate elliptic operators - Ann. Inst. Fourier, 3, 1982 , 151-183.

[3] Jerison D. - The Poincaré inequality for vectors fields satisying Hörmander condition - Duke Math. J., 53, 1986, 503-526.

[4] Kusuoka S., Stroock D. - Applications of the Malliavin calculus , II - J. Fac. Sc. Tokyo, IA, 32, 1985, 1-76.

[5] Mosco U.- Wiener criterion and potential estimates for elliptic obstacle problem - Indiana Math. J., 36, 1987, 455-494.

[6] Nagel A., Stein E.M., Wainger S. - Balls and metrics defined by vector fields: basic properties - Acta Math., 137, 1986, 249-320.

[7] Sanchez Calle A. - Fundamental solutions and geometry of squares of vector fields - Invent. Math., 78, 1984, 143-160.

On the generalized Neumann problem

Andrej V. Bitsadze

Let D be a bounded domain in the Euclidean space $\mathbf{E}_n$, $n \geq 2$, of points x with the Cartesian orthogonal coordinates $(x_1, \cdots, x_n)$. Let us consider the linear uniformly elliptic differential equation of the second order

$$\sum_{i,j=1}^{n} A_{ij}(x)u_{x_i x_j} + \sum_{i=1}^{n} B_i(x)u_{x_i} + c(x)u = 0. \tag{1}$$

The boundary condition for the equation (1) is of the form

$$\sum_{k=0}^{m} B_k^*(p_{i_1,\cdots,i_n}) = h(x), \quad p_{i_1,\cdots,i_n} \equiv \frac{\partial^k u}{\partial x_1^{i_1} \cdots \partial x_n^{i_n}}, \quad \sum_{j=1}^{n} i_j = k, \tag{2}$$

on the boundary S of D, where B_k^*, $k = 0, \cdots, m$, are real linear homogeneous combinations of the variables $p_{i_1,\cdots,i_n}$.

An extensive class of problems for the equation (1) is covered by the following linear boundary value problem: Determine a regular solution $u(x)$ of the equation (1) in D which satisfies the boundary condition (2).

In the case of $n = 2$ the equation (1) takes the form of

$$u_{z\overline{z}} + a(z,\overline{z})u_z + \overline{a(z,\overline{z})}u_{\overline{z}} + c(z,\overline{z})u = 0, \tag{3}$$

where $z = x_1 + ix_2$, $\overline{z} = x_1 - ix_2$.

When a and c are real analytic functions of x_1, x_2 in D, any regular solution $u(x_1, x_2)$ of the equation (3) is analytic in D.

In a simply connected domain D with $z_0 = 0 \in D$, any real regular solution $u(x_1, x_2)$ of the equation (3) is represented by the formula

$$u(x_1, x_2) = \Re\{\alpha(z,\overline{z})\varphi(z) + \int_0^z \beta(z,\overline{z},t)\varphi(t)dt\}, \tag{4}$$

with an analytic function φ in D, where

$$\alpha(z,\overline{z}) = R(z,0;z,\overline{z}),$$
$$\beta(z,\overline{z},t) = R(t,0;z,\overline{z})\overline{\alpha(t,0)} - \frac{\partial}{\partial t}R(t,0;z,\overline{z}).$$

Potential Theory, ed. by M. Kishi

In this case $\varphi(z)$ is of the form

$$\varphi(z) = u(\frac{z}{2}, \frac{z}{2i}) - \frac{1}{2}R(0,0;z,0)u(0,0) \tag{5}$$

whenever

$$\Im\varphi(0) = 0. \tag{6}$$

Here $R(z,\overline{z};t,\overline{t})$ is the Riemann function, that is; 1) $R(z,\overline{z};t,\overline{t})$ is a solution of (3) with respect to the variables z, $\overline{z}$. 2) The equalities

$$\begin{aligned} R(t,\overline{z};t,\overline{t})_{\overline{z}} - R(t,\overline{z};t,\overline{t})a(t,\overline{z}) &= 0,\\ R(z,\overline{t};t,\overline{t})_{z} - R(z,\overline{t};t,\overline{t})\overline{a(t,\overline{z})} &= 0,\\ R(t,\overline{t};t,\overline{t}) &= 1 \end{aligned}$$

hold.

Thus, the formulas (4) and (5) establish a one-to-one correspondence between the set of all real regular solutions $u(x_1,x_2)$ of the equation (3) in D and the set of analytic functions $\varphi(z)$ in D which satisfy condition (6) ([1],[2]).

In the case of a simply connected domain D of class $A^{m,h}$, we assume that a solution $u(x_1,x_2)$ of the problem (2), (3) is a function of class $C^{m,h}(D\cup S)$. By virtue of (5) the function $\varphi(z)$ also belongs to the class $C^{m,h}(D\cup S)$.

Between the set of analytic functions $\varphi(z)\in C^{m,h}(D\cup S)$ satisfying the condition (6) and the set of real functions $\mu\in C^{m,h}(S)$ there exists a one-to-one correspondence given by the formula

$$\varphi(z) = \int_S \mu(t)(1-\frac{z}{t})^{m-1}\log(1-\frac{z}{t})ds_t + \int_S \mu ds + ic, \tag{7}$$

with a real constant c ([3]).

Having substituted the expression (7) of $\varphi(z)$ into the formula (4), we obtain the singular integral equation

$$\alpha(t_0)\mu(t_0) + \beta(t_0)\int_S \frac{\mu(t)}{t-t_0}dt + \int_S k(t,t_0)\mu(t)dt = g(t_0), \quad t_0\in S, \tag{8}$$

where $\mu(t)$ is an unknown function.

The theory of one-dimensional singular integral equations of the form (8) allows one to investigate the boundary value problem (1), (2) in two independent variables. But in studying this problem in the case when the number of independent variables n is not smaller than 3, one encounters considerable difficulties caused by the lack of a sufficiently complete theory of multi-dimensional singular integral equations.

The basic content of our report is an assertion about the normal solvability of the problem (1), (2) in the sense of Fredholm, Noether or Hausdorff when we eliminate the degeneracy of $B^*_m(p_{i_1,\cdots,i_n})$, $\sum_{j=1}^n i_j = m$ in (2).

The following boundary value problem will be called the generalized Neumann problem: Find a function $u(x)$ which is harmonic in D, belongs to the class $C^{m,h}(D\cup S)$ and satisfies the condition

$$\frac{d^m u}{d\nu_y^m} = g_m(y), \quad y \in S, \tag{9}$$

where ν_y is the outer normal to S at the point y and g_m is a given function of class $C^{0,h}(S)$.

When $m = 1$, this is the well-known Neumann problem.

A function $N_m(x, y)$ will be called the generalized Neumann kernel if the formula

$$u(x) = \int_S N_m(x, y) g_m(y) ds_y \tag{10}$$

gives a solution of the generalized Neumann problem.

In [4] we get the kernel $N_m(x, y)$ when D is the ball $|x| < 1$. In fact a solution of this problem has the expression

$$u(x) = u^*(x) + \sum A_{\alpha_1,\cdots,\alpha_{n-1}} P_{\alpha_1,\cdots,\alpha_{n-1}}(x) + \sum B_{\beta_1,\cdots,\beta_{n-1}} Q_{\beta_1,\cdots,\beta_{n-1}}(x). \tag{11}$$

Here

$$u^*(x) = \frac{1}{(m-1)!} \int_0^1 (\frac{1}{\xi} - 1)^{m-1} \frac{v(\xi x)}{\xi} d\xi, \tag{12}$$

$$P_{\alpha_1,\cdots,\alpha_{n-1}}(x), \quad \sum_{i=1}^{n-1} \alpha_i = k, \quad k = 0, \cdots, m-1,$$

and

$$Q_{\beta_1,\cdots,\beta_{n-1}}(x), \quad \sum_{i=1}^{n-1} \beta_i = k - 1, \quad k = 1, \cdots, m-1$$

are linearly independent homogeneous harmonic polynomials of variables $x_1, \cdots, x_n$, $A_{\alpha_1,\cdots,\alpha_{n-1}}$ and $B_{\beta_1,\cdots,\beta_{n-1}}$ are real constants and

$$v(\xi x) = \frac{1}{\omega_n} \int_S \frac{1 - |\xi x|^2}{|y - \xi x|^n} g_m(y) ds_y. \tag{13}$$

For the solvability of the generalized Neumann problem, it is necessary and sufficient that the conditions

$$\int_S P_{\alpha_1,\cdots,\alpha_{n-1}}(y)g_m(y)ds_y = 0$$

and

$$\int_S Q_{\beta_1,\cdots,\beta_{n-1}}(y)g_m(y)ds_y = 0$$

are satisfied.

By virtue of (10), (12) and (13) we have the expression

$$(14)\qquad N_m(x,y) = \frac{1}{(m-1)!\omega_n}\int_0^1 (\frac{1}{\xi}-1)^{m-1}\frac{1}{\xi}\frac{1-|\xi x|^2}{|y-\xi x|^n}d\xi \ (m\geq 1),$$
$$N_0(x,y) = \frac{1-|x|^2}{\omega_n|x-y|^n}.$$

In the case of the Neumann problem ($m=1$), the formula (14) gives us

$$2\pi N_1(x,y) = -2\log|y-x|,\ n=2\ \text{(Dini)},$$
$$4\pi N_1(x,y) = \frac{2}{|y-x|} + \log\frac{2}{1-xy+|y-x|},\ n=3\ \text{(Kellogg)},$$
$$2\pi^2 N_1(x,y) = \frac{1}{|y-x|^2} - \log|y-x| + \frac{x\cdot y}{\delta^{1/2}}\arctan\frac{\delta^{1/2}}{1-x\cdot y},\ n=4,$$

where $\delta = |x|^2 - x\cdot y$, and $x\cdot y = \sum_{i=1}^n x_i y_i$ ([4]).

The formula (11) allows us to find the inversion formula for the singular integral equation of the first kind

$$\int_S N_m(x,y)g(y)ds_y = f(x),\ x\in S$$

with the kernel $N_m(x,y)$ defined by the formula (14).

When $m=0$ or $m=1$, by (9) we have the Dirichlet problem and the Neumann problem

$$(15)\qquad u_0(x) = g_0(x),\ x\in S$$

and

$$(16)\qquad u_1(x) = g_1(x),\ x\in S$$

respectively.

The solution $u_0(x)$ of the Dirichlet problem is given by

(17) $$u_0(x) = \int_S N_0(x, y) g_0(y) ds_y$$

and it is unique. Under the condition

(18) $$\int_S g(y) ds_y = 0,$$

the solution $u_1(x)$ of the Neumann problem is given by

(19) $$u_1(x) = \int_S N_1(x, y) g_1(y) ds_y + c,$$

where c is a real constant.

The harmonic solution of the Cauchy problem with the boundary conditions

$$u(x_0) = g_0(x_0), \quad \frac{du}{d\nu_{x_0}} = g_1(x_0), \quad x_0 \in S$$

exists if and only if the equalities

(20) $$u(x) = u_0(x) = u_1(x)$$

hold for all $x \in D$.

It is obvious that under (18), the condition (20) is equivalent to each of the equalities

(21) $$\lim_{x \to x_0} u_1(x) = \int_S N_1(x_0, y) g_1(y) ds_y + c = g_0(x_0)$$

and

(22) $$\lim_{x \to x_0} \nu_{x_0} \cdot \operatorname{grad} u_0 = \lim_{x \to x_0} \nu_{x_0} \cdot \operatorname{grad} \int_S N_0(x, y) g_0(y) ds_y = g_1(x_0)$$

for all $x_0 \in S$.

The equalities (21) and (22) are the integral equations of the first kind for the functions $g_1(x)$ and $g_0(x)$.

Under (18) the integral equations (21) and (22) are mutually inverse.

Let $n = 2$. In the disk $D : |x| < 1$, the equalities (21) and (22) are the Hilbert inversion formulas

(23) $$\frac{1}{2\pi} \int_0^{2\pi} \operatorname{ctg} \frac{s - \varphi}{2} g_1(s) ds = g_0'(\varphi), \quad -\frac{1}{2\pi} \int \operatorname{ctg} \frac{s - \varphi}{2} g_0'(s) ds = g_1(\varphi).$$

In the case of the half-plane $D : x_2 > 0$, they pass into the Hardy inversion formulas

$$\frac{1}{\pi}\int_{-\infty}^{\infty}\frac{g_1(t)}{t-x}dt = g_0'(x), \quad -\frac{1}{\pi}\int_{-\infty}^{\infty}\frac{g_0'(t)}{t-x}dt = g_1(x). \tag{24}$$

When $S = \mathbf{E}_{n-1}$ is the hyperplane $x_n = 0$, the equalities (21) and (22) give us the multi-dimensional analogs of (24)

$$\frac{\Gamma(n/2)}{\pi^{n/2}}\int_{\mathbf{E}_{n-1}}\frac{y-x}{|y-x|^n}g_1(y)ds_y = \operatorname{grad} g_0(x),$$

$$\frac{\Gamma(n/2)}{\pi^{n/2}}\int_{\mathbf{E}_{n-1}}\frac{(y-x)\operatorname{grad} g_0(y)}{|y-x|^n}ds_y = -g_1(x).$$

In the case of the n-dimensional ball $D : |x| < 1$ it is natural to call (21) and (22) multi-dimensional analogs of the Hilbert formulas (23).

References

[1] Vekua, I.N., New Methods of Solving Elliptic Equations. Gostekhizdat, Moscow-Leningrad 1948

[2] Bitsadze, A.V., Boundary Value Problems for Second-Order Elliptic Equations. Nauka, Moscow 1966

[3] Muskhelishvili, N.I., Singular Integral Equations. Nauka, Moscow 1968

[4] Bitsadze, A.V., On the Neumann problem for harmonic functions. Dokl. Akad. Nauk SSSR 311 (1990), 11-13

Principe de Picard pour les mesures invariantes par rotation et applications*

Abderrahman Boukricha

0. Introduction

Dans cet article nous considérons l'équation de Schrödinger généralisée

(1) $$\Delta u - u\mu = 0 \text{ au sens des distributions}$$

sur $U = \{x \in \mathbb{R}^n | 0 < \|x\| < 1\}$, $n \geq 2$, avec μ une mesure invariante par rotation dans la classe de Kato locale associée à $\overline{U} \setminus \{0\}$. Nous prouvons des tests pour le principe de Picard relatif à μ et à l'ouvert U.

Au paragraphe 1 nous rappelons la notion de principe de Picard pour l'équation (1), l'existence de solutions bornées et non bornées et leur comportement au voisinage de la singularité éventuelle de μ.

Au paragraphe 2, nous rappelons les caractérisations connues de la validité du principe de Picard dans le cas des mesures radiales et nous prouvons entre autre qu'il est valable pour μ si et seulement si, il existe $r \in]0,1[$ et deux solutions continues positives u et v respectivement bornée et non-bornée au voisinage de l'origine telles que

$$\int_{B(0,r)} u(t)v(t)\frac{1}{\|t\|^2}\lambda(dt) = +\infty \quad (\lambda \text{ mesure de Lebesgue sur } \mathbb{R}^n).$$

Au paragraphe 3 nous considérons $\mu = c\lambda$ avec c strictement positive et deux fois differentiable sur $\overline{U} \setminus \{0\}$. Nous posons $\tilde{c} = c + \frac{\Delta[(r^{2(n-1)}c)]^{-\frac{1}{4}}}{(r^{2(n-1)}c)^{-\frac{1}{4}}}$ et nous supposons que $|\tilde{c}|\lambda$ est dans la classe de Kato locale $M(U)$ définie au paragraphe 1. Nous montrons que si $\tilde{c} \geq 0$ sur un voisinage V de zéro et si $\int_0^1 \sqrt{c(t)}dt = +\infty$ alors $\tilde{c}$ verifie le principe de Picard sur V si et seulement si il existe $l \in]0,1]$ tel que

$$\int_0^l \frac{1}{r^2\sqrt{c(r)}}dr = +\infty .$$

Au paragraphe 4 nous montrons que la fonction b_α (resp. c_α) définie par $b_\alpha(x) = \frac{1}{\|x\|^\alpha}$ (resp. $c_\alpha(x) = \frac{(-\log\|x\|)^\alpha}{\|x\|^2}$ sur $\{x \in \mathbb{R}^n : 0 < \|x\| < 1/2\}$) vérifie le principe

*Ce travail est soutenu par "Commemorative Association for the Japan World Exposition (1970)" et par la Fondation Nationale pour la Recherche Scientifique (projet MA_3 89).

Potential Theory, ed. by M. Kishi

de Picard si et seulement si $\alpha \leq 2$. Les résultats de ce paragraphe sont bien connus par M. Nakai [9,10] dans le cas $n = 2$, mais les méthodes sont différentes.

1. Notations et préliminaires

Soit U un ouvert relativement compact de $\mathbb{R}^n$ $(n \geq 2)$ tel que $U = W \setminus \{0\}$ avec W un ouvert qui contient l'origine. Soit $M(U)$ l'ensemble des mesures positives μ sur $\mathbb{R}^n$ portées par $\overline{U}$ qui vérifient la propriété suivante: Pour tout ouvert V tel que $V \subset \overline{V} \subset \complement\{0\}$ la fonction

$$p_V = \int G_t^V \mu(dt)$$

est continue et réelle sur V . (G^V est la fonction de Green associeé au Laplacien et à V).
Pour $\mu \in M(U)$ et $V \subset U$, nous considérons les ensembles suivants:

$${}^\mu H(V) = \{u \in C(V) : \Delta u = u\mu \text{ au sens des distributions}\},$$

$${}^\mu H_0 = \{h \in {}^\mu H^+(U) : \lim_{x \to z} h(x) = 0 \quad \forall z \in \partial\overline{U} \setminus \{0\}\}.$$

1.1. Définition. Soit $\mu \in M(U)$. Nous dirons que le principe de Picard est valable pour μ sur U si le cône ${}^\mu H_0$ est engendré par une seule fonction.
Notons que la notion de principe de Picard a été introduite par G. Bouligand et reprise par M. Nakai.

1.2. Remarque. Soit $\mu \in M(U)$ et V un ouvert dans U tel que $V \cup \{0\}$ est un voisinage de 0 . Alors le principe de Picard est valable pour μ sur U si et seulement si il l'est pour μ sur V .

1.3. Lemme. *Soit* $\mu \in M(U)$. *Alors:*
a) Il existe une unique fonction continue bornée e_μ *dans* ${}^\mu H(U)$ *telle que* $\lim_{|x| \to 1} e_\mu(x) = 1$.
Si μ *et* U *sont invariantes par rotation alors:*
b) Il existe une unique fonction h_μ *invariante par rotation telle que* $h_\mu \in {}^\mu H_0$ *et* $h_\mu(x_0) = 1$ *pour* $x_0 \in U$. *De plus* $\lim_{x \to 0} h_\mu(x) = +\infty$.
c) Si μ' *est une autre mesure invariante par rotation dans* $M(U)$ *telle* $\mu \leq \mu'$ *alors* $\lim_{x \to 0} \frac{e_{\mu'}}{e_\mu}(x)$ *et* $\lim_{x \to 0} \frac{h_\mu}{h_{\mu'}}(x)$ *existent et sont toutes les deux dans* $[0, 1]$.

Preuve. Pour a) voir [2] et pour b) et c) voir [3].
Dans la suite $C(U)$ désignera l'ensemble des fonctions continues et réelles sur U et,

pour $x_0 \in \mathbb{R}^n$ et $r > 0$, $B(x_0, r)$ désignera la boule ouverte de centre x_0 et de rayon r et $U_r = B(0,r) \setminus \{0\}$. □

2. Principe de Picard pour les measures invariantes par rotation

Dans ce paragraphe nous considérons $U = \{x \in \mathbb{R}^n | 0 < \|x\| < 1\}$ et $\mu \in M(U)$ une mesure invariante par rotation.

2.1. Théorème. *Les proprietés suivantes sont équivalentes:*
1) Le principe de Picard est valable pour μ.
2) $\lim_{x \to 0} \frac{h_\mu}{h_{(\mu + \frac{\alpha\lambda}{|t|^2})}}(x) = 0$ *pour un réel* $\alpha > 0$.
3) $\lim_{x \to 0} \frac{e_{(\mu + \frac{\alpha\lambda}{|t|^2})}}{e_\mu}(x) = 0$ *pour un réel* $\alpha > 0$.
4) $\int_{B(0,1)} e_\mu(t) h_\mu(t) \frac{1}{|t|^2} \lambda(dt) = +\infty$.

Preuve. Pour 1) ⇔ 2), nous reprenons la même que celle du cas où μ est absolument continue par rapport à la mesure de Lebesgue (voir [5]). Pour 2) ⇔ 3) ⇔ 4) (voir [3], Thm 4.5.). □

2.2. Corollaire. *Les proprietés suivantes sont équivalentes:*
1) Le principe de Picard est valable pour μ.
2) Il existe $r \in]0,1[$ *et deux solutions strictement positives continues* u *et* v *dans* ${}^\mu H(U)$ *telles que* u *est bornée et* v *est non bornée sur la boule de centre* 0 *et de rayon* r *et:*

$$\int_{B(0,r)} u(t) v(t) \frac{1}{|t|^2} \lambda(dt) = +\infty.$$

Démonstration. D'après le théorème 1, il suffit de prouver " ⇐ ". u est une solution continue bornée de $\Delta h = h\mu$ sur U. D'après le principe du minimum ([2],II) il existe une constante $M > 0$ telle que $u \leq M e_\mu$ sur $B(0,r) \setminus \{0\}$. Soit h la fonction definie sur U comme suit:

$$h(x) = \int v(\|x\|, \Theta) d\sigma(\Theta) ,$$

où $(\|x\|, \Theta)$ sont les coordonnées polaires et σ la mesure de Haar sur la sphère $S^{n-1} = \{t \in \mathbb{R}^n : \|t\| = 1\}$. h est invariante par rotation sur U, et vérifie:

$$\Delta h = h\mu \text{ sur } U .$$

De nouveau d'après le principe du minimum, il existe $M_1 > 0$ telle que $h \leq M_1 h_\mu$ sur $B(0,r) \setminus \{0\}$. Il en resulte de ce qui précède et de 2) que

$$\int_{B(0,r)} e_\mu(t) h_\mu(t) \tfrac{1}{|t|^2} \lambda(dt) = +\infty .$$

Par suite $\int_{B(0,1)} e_\mu(x) h_\mu(t) \frac{1}{|t|^2} \lambda(dt) = +\infty$. D'après le théorème précédent μ vérifie le principe de Picard.
Si $\mu = c\lambda$ avec c localement Höldérienne sur $\{x \in \mathbb{R}^n | 0 < \|x\| \leq 1\}$ on a d'après [9] et [3]:

2.3. Théorème. *Le principe de Picard est valable pour μ si et seulement si, il existe $\alpha > 0$ telle que le principe de Picard est valable pour $\mu + \frac{\alpha}{|t|^2}\lambda$.*

2.4. Remarque. Les résultats précédents sont valables si on remplace U par $U_r := \{x \in \mathbb{R}^n | 0 < \|x\| < r\}$ avec $r \in \mathbb{R}_+^*$.

3. Test pour le principe de Picard

Dans la suite nous considérons une fonction c invariante par rotation et strictement positive sur $\{x \in \mathbb{R}^n | 0 < \|x\| \leq 1\}$. Nous écrirons $c(r) = c(x)$ pour $\|x\| = r$. Nous supposons que c possède une fonction dérivée seconde (au sens des distributions) telle que $|c_1|\lambda \in M(U)$, où $c_1 = \frac{\Delta[(r^{2(n-1)}c)^{-\frac{1}{4}}]}{(r^{2(n-1)}c)^{-\frac{1}{4}}}$. Nous posons $\tilde{c} = c_1 + c$ et $u_i(x) = c_1(x) e^{\varepsilon_i \int_{\|x\|}^1 \sqrt{c(t)}dt}$, $i \in \{1,2\}$, ou $\varepsilon_1 = -1$, $\varepsilon_2 = +1$.

3.1. Lemme. *On a $\Delta u_i = u_i \tilde{c}$ sur U pour $i \in \{1,2\}$.*

Démonstration. Nous posons $v = f(r) e^{\varepsilon_i \int_r^1 \sqrt{c(t)}dt}$ avec f deux fois différentiable. Un calcul simple donne $\Delta v = gv$ avec

$$g = c + \frac{\Delta f}{f} + \frac{\varepsilon_i}{2\sqrt{c}} (\log c f^4 r^{2(n-1)})' .$$

En remplaçant f par c_1 on obtient le résultat demandé.

3.2. Lemme. *Si $\tilde{c} \geq 0$ alors u_2 est non-bornée sur U.*

Preuve. Supposons que u_2 est bornée. u_1 et u_2 sont alors deux solutions bornées de $\Delta h = h\tilde{c}$ sur U. De plus elles possèdent la même limite sur la frontière de

$\overline{U}$. Il en résulte du principe du minimum que $u_1 = u_2$ et par suite $e^{-\int_r^1 \sqrt{c(t)}dt} = e^{+\int_r^1 \sqrt{c(t)}dt}$. On en deduit que $\int_r^1 \sqrt{c(t)}dt = 0$ et alors $c = 0$ ce qui est absurde car $c > 0$. □

3.3. Lemme. *Si* $\tilde{c} \geq 0$ *et* $\int_0^1 \sqrt{c(t)}dt = +\infty$ *alors* u_1 *est bornée et* u_2 *est non bornée.*

Preuve. D'après le lemme précédent, il suffit de montrer que u_1 est bornée. Supposons que u_1 est non-bornée. Soit $v_i = u_i - \alpha e_{\tilde{c}}$ avec $\alpha = c(1)^{-\frac{1}{4}}$ et $i \in \{1,2\}$. D'après le lemme 1.3, nous avons, pour $x_0 \in U$ fixé,

$$\frac{v_1}{v_1(x_0)} = \frac{v_2}{v_2(x_0)} \text{ sur } U .$$

Après un calcul simple nous obtenons:

$$\left| \frac{e^{-\int_r^1 \sqrt{c(t)}dt}}{v_1(x_0)} - \frac{e^{\int_r^1 \sqrt{c(t)}dt}}{v_2(x_0)} \right| \leq M \left(r^{2(n-1)} \cdot c \right)^{\frac{1}{4}} \quad (*) .$$

Aussi d'après le lemme 1.3 on a $\lim_{x \to 0} u_1(x) = +\infty$. D'où il existe $M_1 > 0$ telle que $\frac{1}{u_1} \leq M_1$ sur U , et donc on a $(r^{2(n-1)} \cdot c)^{\frac{1}{4}} \leq M_1$. Il en résulte de ce qui précède que le premier membre de l'inégalité $(*)$ est majoré par MM_1 et par suite on a $\int_0^1 \sqrt{c(t)}dt < +\infty$. □

3.4. Théorème. *Supposons que* $\tilde{c} \geq 0$ *et* $\int_0^1 \sqrt{c(t)}dt = +\infty$ *, alors les proprietés suivantes sont équivalentes:*
1) Le principe de Picard est valable pour $\tilde{c}$ *.*
2) Il existe $l \in]0,1[$ *tel que:*

$$\int_0^l \frac{1}{r^2\sqrt{c(r)}}dr = +\infty .$$

Preuve. 1) $\Rightarrow$ 2) . D'après le principe du minimum et le lemme 3.3., il existe deux constantes strictement positives M_1 et M_2 telles que:

$$e_{\tilde{c}} \leq M_1 u_1 \quad \text{et} \quad h_{\tilde{c}} \leq M_2 u_2 .$$

Donc d'apres le théorème 2.1,4 on a:

$$\int_{B(0,1)} u_1(t)u_2(t)\frac{1}{|t|^2}dt = +\infty$$

et donc 2).

2) $\Rightarrow$ 1) : En utilisant les coordonnées polaires et 2) on obtient la proprieté 2) du corollaire 2.2. □

3.5. Remarques. 1) Les résultats du paragraphe 3 restent vrais si on remplace U par $U_r = \{x \in \mathbb{R}^n : 0 < \|x\| < r\}$, $r \in \mathbb{R}_+^*$.

2) Dans le théorème 2.4, on ne peut pas éliminer l'hypothèse $\int_0^1 \sqrt{c(t)}dt = +\infty$. En effet soit $P(x) = \frac{1}{\|x\|^3}$, $h = h_P + e_P$ et $c = \frac{h^{-4}}{r^{2(n-1)}}$. D'après la proposition 4.1, P ne vérifie pas le principe de Picard, d'après [3] on a $\lim_{x\to 0}\|x\|^\alpha h(x) = +\infty$ pour tout $\alpha \in \mathbb{R}$. Il en suit alors que $\lim_{x\to 0} c(x) = 0$ et $\int_0^1 \sqrt{c(t)}dt < +\infty$. On a $\tilde{c} = \frac{1}{\|t\|^3} + c$, $u_1 = he^{-\int_r^1 \sqrt{c(t)}dt}$. u_1 est alors une solution radiale non bornée de $\Delta u = \tilde{c}u$ telle que $\lim_{|x|\to 1} u_1(x) = 1$. D'où il existe $M > 0$ telle que $h_{\tilde{c}} \leq Mu_1$ et par suite on a:

$$\lim_{x\to 0}\frac{h_P}{h_{\tilde{c}}}(x) > 0$$

et d'après [3] on a aussi

$$\lim_{x\to 0}\frac{e_{\tilde{c}}}{e_P}(x) > 0 \ .$$

Il en résulte de ce qui précède qu'il existe $M_2 > 0$ telle que

$$h_{\tilde{c}}e_{\tilde{c}} \leq M_2 e_P h_P \ .$$

Comme P ne vérifie pas le principe de Picard, alors d'après le théorème 2.1, il en est de même pour $\tilde{c}$. Par contre pour tout $l \in]0,1[$ on a:

$$\int_0^l \frac{1}{r^2\sqrt{c(r)}}dr = \int_0^l r^{n-3}h^2(r)dr = +\infty \ .$$

3) Dans [11] M. Nakai et M. Kawamura ont prouvé ce qui suit:
Si $n = 2$ et c est differentiable sur $\{x \in \mathbb{R}^2 : 0 < \|x\| \leq 1\}$ alors les propriétés suivantes sont équivalentes:

i) Le principe de Picard est valable pour c .

ii) $\int_0^1 \frac{1}{r\sqrt{r^2c(r)+1}}dr = +\infty$.

4. Applications

Pour $n \geq 2$, soit $U = \{x \in \mathbb{R}^n : 0 < \|x\| < 1/2\}$, $c_\alpha(x) = \frac{1}{\|x\|^\alpha}$ et $P_\alpha(x) = \frac{(-\log\|x\|)^\alpha}{\|x\|^2}$. Nous allons prouver le resultat suivant: c_α (resp. P_α) vérifie le principe de Picard si et seulement si $\alpha \leq 2$. Notons que ce résultat est bien connu par M. Nakai dans le cas $n = 2$.

4.1. Proposition. *Soit* $c(x) = \frac{1}{\|x\|^\alpha}$ *pour* $x \in U$, *alors le principe de Picard est valable pour* c *si et seulement si* $\alpha \leq 2$.

Preuve. Soit $\alpha > 2$, on a $\int_0^1 \sqrt{c(t)}dt = +\infty$. Un calcul simple donne que $\tilde{c}(x) = \frac{\beta}{\|x\|^2} + \frac{1}{\|x\|^\alpha}$ avec $\beta \in \mathbb{R}$. Donc il existe $r_0 \in]0, \frac{1}{2}[$ tel que $\tilde{c} \geq 0$ sur $\{x \in \mathbb{R}^n : 0 < \|x\| \leq r_0\} = U_{r_0}$. De plus on a $\int_0^l \frac{1}{r^2\sqrt{c(r)}}dr < +\infty$ pour tout $l \in]0,1]$. D'après remarque 3.5,1 et théorème 3.4, $\tilde{c}$ ne vérifie pas le principe de Picard sur U_{r_0}. Il en résulte de ce qui précède et du théorème 2.3 que c ne verifie pas le principe de Picard. La réciproque est vraie d'après [3], thm. 6.4.

Dans la suite soit $\alpha \in \mathbb{R}$ et $c(x) = \frac{(-\log|x|)^\alpha}{\|x\|^2}$ et

$$c_1(r) = (r^{2(n-1)}c(r))^{-\frac{1}{4}} \quad (c(r) = c(x) \text{ pour } \|x\| = r).$$ □

4.2. Lemme. *On a*

$$\tilde{c}(r) = -\frac{(n-2)^2}{4} \times \frac{1}{r^2} + \frac{\alpha}{4}(\frac{\alpha}{4}+1)\frac{(-\log r)^{-2}}{r^2} + c(r), \quad r \in \left]0, \tfrac{1}{2}\right[.$$

4.3. Proposition. *Soit* $\alpha \geq 0$ *et* $P = \frac{\alpha}{4}(\frac{\alpha}{4}+1)\frac{(-\log r)^{-2}}{r^2} + c$. *Alors le principe de Picard est valable pour* P *si et seulement si* $\alpha \in [0,2]$.

Preuve. Il existe $r_0 \in]0,1[$ tel que

$$\tilde{c} \geq 0 \text{ sur } U_{r_0}.$$

Un calcul simple montre que $\int_0^l \frac{1}{r^2\sqrt{c(r)}}dr = +\infty$ pour tout $l \in]0,1]$ si et seulement si $\alpha \in [0,2]$. Les théorèmes 3.4 et 2.3 donnent le résultat demandé. □

4.4. Théorème. *Le principe de Picard est non valable pour* c *si et seulement si* $\alpha > 2$.

Preuve. D'après [3], il suffit de considerer $\alpha > 0$ " $\Rightarrow$ ". D'après le théorème 2.1 on a

$$\int_{B(0,1/2)} e_c(t)h_c(t)\frac{1}{|t|^2}\lambda(dt) < +\infty .$$

Par suite on a $\int_{B(0,1/2)} e_c(t)h_c(t)\frac{(-\log|t|)^{-2}}{|t|^2}\lambda(dt) < +\infty$. D'après [3] thm. 4.5 on a

$$\lim_{x\to 0}\frac{e_P}{e_c}(x) > 0 \quad \text{et} \quad \lim_{x\to 0}\frac{h_c}{h_P}(x) > 0.$$

Donc il existe $M > 0$ telle que $h_P e_P \leq M e_c h_c$. Il en résulte que le principe de Picard est non valable pour P . D'après la proposition précédente on a $\alpha > 2$.
" $\Leftarrow$ ". Soit $\alpha > 2$. Il existe $\beta > 0$ telle que $P \leq c + \frac{\beta}{r^2}$ sur U . Donc on a $e_{c+\frac{\beta}{r^2}} \leq e_P$. Comme $c \leq P$, on a $c + \frac{\beta}{r^2} \leq P + \frac{\beta}{r^2}$ et d'après le lemme 1.3 on a $h_{c+\frac{\beta}{r^2}} \leq h_{P+\frac{\beta}{r^2}}$. D'après (4.3) P ne vérifie pas le principe de Picard, alors $\lim_{x\to 0}\frac{h_P}{h_{P+\frac{\beta}{r^2}}}(x) > 0$ et donc il existe $M > 0$ telle que $h_{P+\frac{\beta}{r^2}} < M h_P$ sur $U_{\frac{1}{4}}$. Il en résulte de ce qui précède que

$$e_{c+\frac{\beta}{r^2}} \times h_{c+\frac{\beta}{r^2}} \leq M e_P h_P \quad \text{sur } U_{\frac{1}{4}}.$$

D'après le théorème 2.1 on obtient que $c + \frac{\beta}{r^2}$ ne vérifie pas le principe de Picard. D'après le théorème 2.3 on a la réciproque. □

Références

[1] J. Bliedtner, W. Hansen. Potential Theory–An Analytic and Probabilistic Approach to Balayage, Universitext, Berlin–Heidelberg–New York, Springer 1986.

[2] A. Boukricha. Das Picard–Prinzip und verwandte Fragen bei Störung von harmonischen Räumen. Math. Ann. 239 (1979), 247–270.

[3] A. Boukricha. Principe de Picard et comportement des solutions continues de l'équation de Schrödinger, au voisinage d'une singularité isolée. Bibos Publications Nr. 247, Universität Bielefeld 1987.

[4] A. Boukricha. Classe de Kato et équation de Schrödinger generalisée avec une singularité isolée. À paraître.

[5] A. Boukricha, H. Hueber. The Poisson space ${}^cP_x$ for $\Delta u = cu$ with rotation free c . Académie royale de Belgique, Bulletin de la classe des sciences, 5ème série, Tome LXIV, 1978-10, 651–658.

[6] M. Brelot. Etude de l'équation de la chaleur $\Delta u(M) = c(M)u(M)$, $c(M) \geq 0$, au voisinage d'un point singulier du coefficient, Ann. Sci. Ecole Norm. Sup. 48 (1931), 153–246.

[7] M. Brelot. Sur le principe des singularités positives et la notion de source pour l'équation $\Delta u(M) = c(M)u(M)$ $(c \geq 0)$. Ann. Univ. Lyon. Section AXI, 9.

[8] M. Murata. Structure of positive solutions to $(-\Delta + V)u = 0$ in $\mathbb{R}^n$, Duke Math. J. 53 (1986), 869-943.

[9] M. Nakai. Martin boundary over an isolated singularity of rotation–free densities. J. Math. Soc. Japan 26 (1974), 483–507.

[10] M. Nakai. A test of Picard principle for rotation free densities. J. Math. Soc. Japan 27 (1975), 412–431.
[11] M. Nakai, M. Kawamura. A test of Picard principle for rotation free densities. II. J. Math. Soc. Japan 28 (1976), 323–342.

Measure and capacity of exceptional sets arising in estimations of δ-subharmonic functions

Vladimir Ya. Eiderman

Abstract

This paper consists of three parts. In the first part we obtain estimates of exceptional sets outside of which a δ-subharmonic function has a regular asymptotic behaviour. In the second part we add a few results to well-known facts about the comparison of capacity and Hausdorff measure (and h-measure which was introduced by Carleson [3]). In the third part the following problem is considered: to find sufficient conditions for a sequence of points z_n in the unit disk D to satisfy $\sum_{n=1}^{\infty} \Phi(\ln |f(z_n)|) = \infty$, $\forall f \in F$, $f \not\equiv 0$, where $\Phi(t) \downarrow 0$ as $t \to -\infty$ and F is in some class of meromorphic functions in D.

1980 Mathematics Subject Classification (1985 Revision): 30D30, 31B05, 31C15.

0. Notation

Let $U(\rho(r))$ be a class of subharmonic functions $u(x)$, $x \in R^m$, $m \geq 2$, of proximate order $\rho(r)$ ($\rho(r) \to \rho$) and finite type. Let $U^d(\rho(r)) = \{u(x) : u(x) = u_1(x) - u_2(x),\ u_1, u_2 \in U(\rho(r))\}$. By V_t, $t \in (0, \infty)$, we denote a family of rotations of R^m such that $V_{t_1} \cdot V_{t_2} = V_{t_1 \cdot t_2} and$ $V_1 = I$ is the identity transformation. Let $P_t = tV_t$, $u_t(x) = u(P_t x)t^{-\rho(t)}$ and let $\mathcal{D}'$ denote the space of generalized functions connected with the space of infinitely differentiable functions with compact supports.

Theorem A [1, p.148]. *Let $u \in U(\rho(r))$. For every sequence $t_j \to \infty$ there exist a subsequence $t'_j \to \infty$ and a subharmonic function $v(x)$ on R^m such that $u_{t'_j} \to v$ in the topology of $\mathcal{D}'$.*

The set of functions $v(x)$ described in Theorem A is denoted by $Fr[u]$. A function $u(x)$ belongs to the class U_{reg} if the set $Fr[u]$ consists of a single function $v(x)$. From [1] it follows that $u(z)$, $m = 2$, is a function of completely regular growth in the sense of Levin-Pfluger [2] if and only if $u(z) \in U_{reg}$ for some family V_t. Let $x^0(x) = P_t x$ with $t = |x|^{-1}$ and let $h(x, u)$ be the indicator of $u(x) \in U(\rho(r))$ (see [1]). By [1] we have $h(x, u) = |x|^\rho h(x^0(x), u)$. If $m = 2$ and $V_t \equiv I$, then $h(x^0(x), u)$ coincides with the Phragmen-Lindelöf indicator function. Let $B_r = \{x : |x| < r\}$ and let μ be a measure in R^m such that $\overline{\lim_{r\to\infty}}\, \mu(r) r^{-\rho(r)-m+2} < \infty$, where $\mu(r) = \mu(B_r)$. For any Borel set E we introduce $\mu_t(E) = \mu(P_t E)t^{-\rho(t)-m+2}$.

Potential Theory, ed. by M. Kishi

Let $h : [0,\infty) \to [0,\infty)$ be a continuous increasing function such that $h(0) = 0$. The h-measure of a set $E \subset R^m$ was introduced in [3] by $h\text{-mes}\, E = \inf \sum_i h(r_i)$, where the infimum is taken over families of open balls with radii r_i which cover E. The relative h-measure of a set $C \subset R^m$ is defined by

$$h\text{-}\overline{\text{mes}}\, C = \varlimsup_{R\to\infty} [\inf \sum_i h(r_i/R)],$$

where the infimum is taken over families of balls with radii r_i which cover $C \cap B_R$. In the case of $h(r) = r^{\alpha+m-2}$, $\alpha > 0$, the value $h\text{-}\overline{\text{mes}}\, C$ was introduced by V.S. Azarin [1].

Let $H(t)$ be a nonnegative continuous increasing convex function, $-\infty < t < \infty$. Put $\varphi(r) = -\ln r$, $m = 2$; $\varphi(r) = r^{2-m}$, $m > 2$. As in [3] we consider kernels $K(r)$ such that $K(r) = H(\varphi(r))$ and $\int_0 K(r) r^{m-1}\, dr < \infty$. Let $\psi(r) = r^m/h(r)$.

1. Relative h-measures of exceptional sets

Theorem 1.1. *Let $u(x) \in U^d(\rho(r))$ and suppose there exists a family V_t such that $u_t \to 0$ in $\mathcal{D}'$ as $t \to \infty$. If the function $h(r)$ satisfies the condition*

$$\int_0 \varphi(r)\, dh(r) < \infty, \tag{1.1}$$

then there exists a set C such that $h\text{-}\overline{\text{mes}}\, C = 0$ and

$$u(x)|x|^{-\rho(|x|)} \to 0, \quad |x| \to \infty, \quad x \notin C. \tag{1.2}$$

Theorem 1.2. *Let $u(x) \in U_{reg}$ for some family V_t. If the function $h(r)$ satisfies (1.1), then there is a set C such that $h\text{-}\overline{\text{mes}}\, C = 0$ and*

$$u(x)|x|^{-\rho(|x|)} - h(x^0(x), u) \to 0, \quad |x| \to \infty, \quad x \notin C. \tag{1.3}$$

Theorems 1.1, 1.2 improve Azarin's estimations of an exceptional set [1] (see also [4]).

In Theorems 1.1, 1.2 the exceptional set C depends on the function $h(r)$. The following question arises: is there an exceptional set C such that C satisfies (1.2), (1.3) and $h\text{-}\overline{\text{mes}}\, C = 0$ for all functions h satisfying (1.1)?

Theorem 1.3. *Let $u = u_1 - u_2 \in U^d(\rho(r))$ and let μ_1, μ_2 be the Riesz measures of u_1, u_2. If $u_t \to 0$, $\mu_{1t} \to 0$, $\mu_{2t} \to 0$ in $\mathcal{D}'$ for some V_t, then there is a set C satisfying (1.2) and such that $h\text{-}\overline{\text{mes}}\, C = 0$ for every function $h(r)$ satisfying (1.1).*

Theorem 1.4. *Let $u(x) \in U_{reg}$ for some V_t and $\mu_t \to 0$ in $\mathcal{D}'$, where μ is the Riesz measure of u. Then there is a set C satisfying (1.3) and such that $h\text{-}\overline{\text{mes}}\, C = 0$ for every function $h(r)$ satisfying (1.1).*

Without the conditions that the measures are vanishing, Theorems 1.3, 1.4 are not true.

Theorem 1.5. *Let $C \subset R^1$ and $\lim \ln|\sin \pi t|/t = 0$ as $t \to \infty$, $t \in (0,\infty)\backslash C$. Then there is a continuous increasing function $h(r)$ such that $h(0) = 0$, $-\int_0 \ln r\, dh(r) < \infty$ and $h\text{-}\overline{\text{mes}}\, C > 0$.*

Similar results in terms of capacity were obtained by S.Yu. Favorov [4].

Let us discuss now the necessity of the condition (1.1). First we are going to show that the set C with properties as in Theorems 1.1, 1.3 exists for some functions $h(r)$ which does not satisfy (1.1), namely the condition (1.1) may be replaced by the other condition which is independent of condition (1.1).

Lemma 1.1. *There exists a continuous increasing function $h(r)$ such that $h(0) = 0$,*

$$\int_0 K(r)dh(r) = \infty \tag{1.4}$$

and

$$\overline{\lim}\, \psi(r) = \alpha > 0, \quad r \to 0+. \tag{1.5}$$

In particular there exists a function $h(r)$ satisfying (1.5) and

$$\int_0 \varphi(r)dh(r) = \infty. \tag{1.6}$$

Theorem 1.6. *If $u \in U^d(\rho(r))$ and $u_t \to 0$ in $\mathcal{D}'$ for some family V_t then there is a set C satisfying (1.2) such that $h\text{-}\overline{\text{mes}}\, C = 0$ for every function $h(r)$ satisfying (1.5).*

Theorem 1.6 does not contain any conditions about convergence of the integral in (1.6). So, the condition (1.1) in Theorems 1.1, 1.3 is not necessary. On the other hand we show now that the condition (1.1) with additional assumption about the function $\psi(r)$ is necessary.

Let $x, y \in R^m$, $x = (x_1, \dots, x_m)$, $y \neq 0$, and let $\varphi(|x-y|) = \sum_{i=0}^{\infty} b_i(x,y)$ be the expansion of the function $\varphi(|x-y|)$ with the homogeneous polynomials $b_i(x,y)$ of $x_1, x_2, \dots, x_m$ with power i. Let $\varphi_p(x,y) = \varphi(|x-y|) - \sum_{i=0}^{p} b_i(x,y)$ [5, §4.1].

Theorem 1.7. *Let $\rho > 0$ and let $q(r)$ be the function in $(0, r_0)$ such that $0 < \varepsilon \leq q(r) \leq 1-\varepsilon$ with some $r_0 > 0$, $\varepsilon > 0$. Let $h(r)$ satisfy (1.6) and $\psi(rq(r)) < \psi(r)$. Then there exists a sequence of points $a_n \in R^m$ with convergence exponent $m+\rho-2$ and genus p and there is a set $E \subset R^m$ such that*

$$\sum_{i=1}^{\infty} |a_i|^{2-\rho-m} < \infty, \tag{1.7}$$

h-$\overline{\text{mes}}\,E > 0$ and $u(x)|x|^{-\rho} \to -\infty$, $|x| \to \infty$, $x \in E$, where $u(x) = -\sum_{i=1}^{\infty} \varphi_p(x, a_i)$.

From (1.7) it follows that $u(x)$ in Theorem 1.7 is a subharmonic function of order ρ and of minimum type, $u(x) \not\equiv -\infty$ and $u(x) \in U_{reg}$ for every family V_t. Lemma 1.1 and Theorem 1.6 show that we cannot remove the condition about $\psi(r)$ from Theorem 1.7. Moreover this condition is best possible in the following sense.

Theorem 1.8. *Let $q(r)$ be an increasing function in $(0, r_0)$, $r_0 > 0$, such that $q(r) \to 0$ as $r \to 0+$. Let $m \geq m_0$ where m_0 is some absolute constant. Then there exists a continuous increasing function $h(r)$ with the following properties: $h(0) = 0$; $h(r)$ satisfies (1.6); $\psi(rq(r)) < \psi(r)$; for every $u(x)$ such that $u(x) \in U^d(\rho(r))$, $x \in R^m$, $u_t \to 0$ in $\mathcal{D}'$ for some V_t, there is a set C satisfying (1.2) and h-$\overline{\text{mes}}\,C = 0$.*

The results of this section were published partly in [6].

2. On the connexion between *h*-measure and capacity

In this section we add a few results to well-known facts about the comparison of capacity and h-measure. We consider the "classical" capacity (see for example [3]) and denote the capacity of bounded analytic set E by $C_K(E)$ (K is the kernel of the potential).

Theorem B [3]. *Let E be an analytic set, let h-mes $E > 0$ and*

$$\int_0 K(r)\,dh(r) < \infty. \tag{2.1}$$

Then $C_K(E) > 0$.

We now discuss the necessity of condition (2.1).

Lemma 2.1. *If* $h\text{-mes}\, E > 0$ *and* $h(r)$ *satisfies (1.5) then* $C_K(E) > 0$.

The conclusion of Lemma 2.1 does not depend on convergence or divergence of the integral in (2.1). Hence the implication ($h\text{-mes}\, E > 0 \Rightarrow C_K(E) > 0$) holds not only for functions $h(r)$ satisfying (2.1). But the condition (2.1) with additional assumption about $\psi(r)$ is necessary.

Theorem 2.1. *Let* $q(r)$ *be a function in* $(0, r_0)$ *such that* $0 < \varepsilon \le q(r) \le 1 - \varepsilon$ *with some* $r_0 > 0$, $\varepsilon > 0$. *Let* $h(r)$ *satisfy (1.4) and* $\psi(rq(r)) < \psi(r)$. *Then there exists an* m*-dimensional symmetric generalized Cantor set* E *such that* $h\text{-mes}\, E > 0$ *but* $C_K(E) = 0$.

This theorem improves Lemma 2 in [7]. Theorem 2.1 is the counterpart of [8, Theorem 4] about L^p-capacity. The condition $h(2r) < 2^m h(r)$ in [8] coincides with our condition $\psi(rq(r)) < \psi(r)$ when $q(r) \equiv 1/2$.

Now we show that the condition about $\psi(r)$ is best possible in the following sense.

Theorem 2.2. *Let* $q(r)$ *be an increasing function in* $(0, r_0)$ *such that* $q(r) \to 0$ *as* $r \to 0+$. *Let* $m \ge m_0$ *where* m_0 *is some absolute constant. Then there exists a continuous increasing function* $h(r)$ *with the following properties:* $h(0) = 0$; $h(r)$ *satisfies (1.4);* $\psi(rq(r)) < \psi(r)$; *for any analytic set* $E \subset R^m$, *if* $h\text{-mes}\, E > 0$ *then* $C_K(E) > 0$.

3. Behaviour of a meromorphic function on the sequence of points

There are more general theorems which state that an analytic or meromorphic function of some class cannot have a very small modulus of it's values on a frequent sequence of points. As an application of estimation of δ-subharmonic functions we obtain theorems of such type. For simplicity we consider meromorphic functions $f(z)$ in the unit disk $\mathcal{D} = \{z : |z| < 1\}$. By $T_f(r)$ we mean the Nevanlinna characteristic function of $f(z)$.

Let $\Phi(t)$ be a positive increasing function on $(-\infty, \infty)$ such that $\Phi(t) \to 0$ as $t \to -\infty$. Let F be some class of meromorphic functions in $\mathcal{D}$. The following problem is considered: *to find sufficient conditions for a sequence of points* $z_n \in \mathcal{D}$, $|z_n| \to 1$ *to satisfy* $\sum_{n=1}^{\infty} \Phi(\ln |f(z_n)|) = \infty$ *for* $\forall f \in F$, $f \not\equiv 0$.

Let $\nu(r)$ be a continuous increasing function on $[0, 1]$ and $\nu(0) = 0$. Let $\eta(r)$

be the inverse function of $\nu(r)$, let $\Psi(t)$ be the inverse function of $\Phi(t)$. Let $W_\nu = \{f(z) : \nu(1-r)T_f(r)/(1-r) \leq A_f < \infty\}$. We assume that

$$\Phi(t/\alpha) \leq \alpha\Phi(t), \quad \nu(\alpha r) \leq \alpha\nu(r), \quad 0 < \alpha \leq 1.$$

Theorem 3.1. *For a sequence of real numbers p_n such that $p_n \to \infty$, $p_n \leq p_{n+1}$, $p_n = o(-\Psi(n^{-1}))$, and a number $L > 1$, if $|z_n| = 1 - L\eta(-p_n/\Psi(n^{-1}))$, then $\sum_{n=1}^{\infty} \Phi(\ln|f(z_n)|) = \infty$ for $\forall f \in W_\nu$, $f \not\equiv 0$.*

Every condition in this theorem is essential. If we put $\Phi(t) = -1/t$, $t \leq t_0$ or $\Phi(t) = \exp t$ or $\Phi(t) = \exp(-\exp(-t))$ then we get the following

Corollary. *If $p_n \to \infty$, $p_n \leq p_{n+1}$, $L > 1$ then for $\forall f(z) \in W_\nu$, $f \not\equiv 0$*

$$\sum_n |\ln|f(z_n)||^{-1} = \infty \quad \textit{where} \quad |z_n| = 1 - L\eta(p_n/n) \to 1, \tag{3.1}$$

$$\sum_n |f(z_n)| = \infty \quad \textit{where} \quad |z_n| = 1 - L\eta(p_n/\ln n) \to 1, \tag{3.2}$$

$$\sum_n \exp(-|f(z_n)|^{-1}) = \infty \quad \textit{where} \quad |z_n| = 1 - L\eta(p_n/\ln\ln n) \to 1. \tag{3.3}$$

For the functions $f(z)$ with a bounded Nevanlinna characteristic (class N) $\nu(r) = \eta(r) = r$. In this case from (3.2) it follows that $\sum_n |f(z_n)| = \infty$ for $\forall f \in N$ when $|z_n| = 1 - p_n/\ln n \to 1$, $p_n \uparrow \infty$. This result essentially improves Theorems 3, 4 in [9] and answers the question in the same paper (see p.184). It is impossible to change the sequence $\{p_n\}$ by any bounded value.

If $W_\nu = N$ then (3.1), (3.2) may be improved.

Theorem 3.2. *Let $t_1 \geq 1$, $t_{n+1} - t_n \geq \delta > 0$, $|z_n| = 1 - t_n^{-1}$. Then $\sum_n |\ln|f(z_n)||^{-1} = \infty$ for $\forall f \in N$ if and only if $\sum_n t_n^{-1} = \infty$.*

Theorem 3.3. *Let $t_1 \geq 1$, $t_{n+1} - t_n \geq \delta > 0$, $\sum_n (t_n \ln t_n)^{-1} = \infty$, $p_n = o(\ln t_n)$, $p_n \to \infty$ and for some $\alpha \in (0,1)$ $p_n/p_{n+1} \geq \max\{(\ln t_n/\ln t_{n+1})^\alpha, 1/2\}$. Let $|z_n| = 1 - p_n/\ln t_n$. Then $\sum_n |f(z_n)| = \infty$ for $\forall f(z) \in N$, $f \not\equiv 0$.*

This result is best possible in the sense that we cannot replace the condition $\sum_n (t_n \ln t_n)^{-1} = \infty$ by $\sum_n (t_n g(t_n))^{-1} = \infty$, where $g(t)$ is an increasing function such that $g(t) = o(\ln t)$, $t \to \infty$.

The results in this section are generalized in two directions. First they are extended to some classes of δ-subharmonic functions in the unit ball of R^m. Secondly more general classes of sequences are considered. But the formulations of these results are more complicated and will be published in other papers. Some of the previous results were published in [10].

References

[1] Azarin, V.S., *On asymptotic behavior of subharmonic functions of finite order*, Mat. Sbornik **108(150)** (1979), 147-167.

[2] Levin, B.Ya., "Distribution of zeros of entire functions," (Trans. Math. Monographs. 5), Amer. Math. Soc., 1964.

[3] Carleson, L., "Selected problems on exceptional sets," Van Nostrand Co., Toronto - London - Melbourne, 1967.

[4] Favorov, S.Yu., *Sets of growth reduction for entire and subharmonic functions*, Mat Zametki **40** (1986), 460-467.

[5] Hayman, W.K., Kennedy, P.B., "Subharmonic functions," Vol.**1**, Acad. Press, London - New York - San Francisco, 1976.

[6] Eiderman, V.Ya., *On exceptional set in asymptotic estimations of subharmonic functions*, Sibirsk. Mat. J. **29** (1988), 185-196.

[7] Essen, M., Jackson, H.L., Rippon, P.J., *On minimally thin and rarefied sets in* R^p, $p \geq 2$, Hiroshima Math. J. **15** (1985), 393-410.

[8] Aikawa, H., *Comparison of* L^p*-capacity and Hausdorff measure*, Complex Variables **13** (1990).

[9] Chow, Y.S., Trent, T., Li-Ming Wang, J., *A separating problem on function spaces*, J. Math. Anal. and Appl. **111** (1985), 177-187.

[10] Eiderman, V.Ya, *On sum of modulus of values of meromorphic function on the sequence of points*, Deposited at VINITI 8139-B86 (1986).

Balayage spaces, standard H-cones and hyperharmonic cones

Sirkka-Liisa A. Eriksson-Bique

Abstract

In this paper we state necessary and sufficient conditions for a balayage space to be a standard H-cone in terms of an infimum condition concerning a weak unit and an increasingly dense subset of universally continuous elements. We also show a relationship between balayage spaces and hyperharmonic cones. This result gives a characterization of balayage spaces without fine topology.

1980 Mathematics Subject Classification (1985 Revision): 31D05, 06A10.

Introduction

In this paper we present an interesting relationship between balayage spaces defined by J. Bliedtner and W. Hansen and standard H-cones defined by N. Boboc, Gh. Bucur and A. Cornea using only basic properties of balayage spaces and standard H-cones.

The concept of fine topology is used in the definition of balayage spaces. We state a characterization of balayage spaces in terms of hyperharmonic cones without using that concept.

Let $(X, \mathcal{W})$ be a balayage space. An element $w \in \mathcal{W}$ is called *cancellable* if the condition $u + w \leq v + w \iff u \leq v$ holds for all u and v in $\mathcal{W}$. It is relatively easy to see that w is cancellable in a balayage space $(X, \mathcal{W})$ if and only if w is finite on a finely dense subset of X.

We shall show that the set of cancellable elements of a balayage space is a standard H-cone possessing a weak unit u such that an increasingly dense subset $\mathcal{F}$ of universally continuous elements satisfies the condition $\inf(\alpha u, f_1, f_2)$ is u-continuous for all $f_1, f_2 \in \mathcal{F}$ and $\alpha \in \mathbf{R}$. Conversely, a standard H-cone satisfying this property is isomorphic with an increasingly dense solid subcone of a balayage space.

1. Review of the basic concepts

We recall some essential definitions and results which will be utilized subsequently.

A partially ordered abelian semigroup $(W, +, \leq)$ with a neutral element 0 is called an *ordered convex cone* if it satisfies the conditions

$$u \geq 0 \tag{1.1}$$

Potential Theory, ed. by M. Kishi

and

$$u \le v \Longrightarrow u + w \le v + w \tag{1.2}$$

for all u, v and w in W and if it admits an operation of multiplication with strictly positive real numbers such that for all α, $\beta \in \mathbf{R}_+ \setminus \{0\}$ and x, $y \in W$,

$$\alpha(x+y) = \alpha x + \alpha y, \qquad (\alpha+\beta)x = \alpha x + \beta y,$$
$$(\alpha\beta)x = \alpha(\beta x), \qquad 1 \cdot x = x, \qquad x \le y \Longrightarrow \alpha x \le \alpha y.$$

Ordered convex cones $(W_1, +, \le)$ and $(W_2, +, \le)$ are called *isomorphic* if there exists a one-to-one and onto mapping $\varphi : W_1 \to W_2$, called an *isomorphism*, satisfying the conditions

$$\varphi(\alpha x + \beta y) = \alpha\varphi(x) + \beta\varphi(y), \qquad x \le y \iff \varphi(x) \le \varphi(y)$$

for all α, $\beta \in \mathbf{R}_+ \setminus \{0\}$ and x, $y \in W_1$.

In an ordered convex cone $(W, +, \le)$ another partial order, called *specific order*, is defined by $u \preceq v$ if $v = u + u'$ for some $u' \in W$.

Definition 1.1. A partially ordered abelian semigroup $(W, +, \le)$ satisfying (1.1) and (1.2) is called a *hyperharmonic structure* if for all u and v in W the set $\{ w \in W \mid u \le w + v \}$ has the least element m and $m \preceq u$.

The theory of hyperharmonic structures is studied by Arsove and Leutwiler in [1].

Definition 1.2. An ordered convex cone $(W, +, \le)$ is called a *hyperharmonic cone* if the following axioms hold:

(H_1) any non-empty upward directed family $F \subset W$ has a least upper bound $\bigvee F$ satisfying $\bigvee(x + F) = x + \bigvee F$ for all $x \in W$,

(H_2) any non-empty family $F \subset W$ has a greatest lower bound $\bigwedge F$ satisfying $\bigwedge(x + F) = x + \bigwedge F$ for all $x \in W$,

(H_3) dominated decomposition property: for any u, v_1 and v_2 in W such that $u \le v_1 + v_2$ there exist u_1 and u_2 in W satisfying $u = u_1 + u_2$, $u_1 \le v_1$ and $u_2 \le v_2$.

An element u in a hyperharmonic cone $(W, +, \le)$ is called *cancellable* if $x + u \le y + u$ implies $x \le y$ for all x and y in W.

The theory of hyperharmonic cones is developed in [6] and [7].

Definition 1.3. An ordered convex cone $(S, +, \le)$ is called an *H-cone*, if S is the set of cancellable elements of a hyperharmonic cone.

The preceding definition is equivalent to one given by Boboc, Bucur and Cornea in [3].

Definition 1.4. A hyperharmonic cone $(W, +, \le)$ is called a *(order) completion* of an H-cone S if W has a solid order dense subset isomorphic with S and W possesses the property

(H_4) $\bigvee_{f\in F} s \wedge f = s \wedge (\bigvee F)$ for all upward directed sets $F \subset W$ and $s \in W$.

Proposition 1.5. *Every H-cone possesses a completion.*

Proof. See Proposition 2.2 of [5] or Theorem 2.7 of [8]. □

An element u of an ordered convex cone $(S, +, \leq)$ is called a *weak unit* if $s = \bigvee_{n\in\mathbf{N}} s \wedge nu$ for all $s \in S$. If u is a weak unit in an ordered convex cone $(S, +, \leq)$ then $s \in S$ is called *u-continuous* if for any $\varepsilon > 0$ and for any upward directed subset $F \subseteq W$ with $s = \bigvee F$ there exists $f \in F$ such that $s \leq f + \varepsilon u$. An element s is *universally continuous* if s is u-continuous for all weak units u.

Definition 1.6 An H-cone $(S, +, \leq)$ is called a *standard H-cone* if the following axioms hold:

(S_1) there exists a weak unit $u \in S$,

(S_2) there exists a countable increasingly dense subset of universally continuous elements (i.e. for any $s \in S$ there exists an increasing sequence (s_n) of universally continuous elements such that $s = \bigvee s_n$).

For the reference to the theory of H-cones and standard H-cones we mention [3]. In this work we need the following result:

Theorem 1.7. *Let S be a standard H-cone and $u \in S$ be a weak unit. Denote by K_u the set of additive mappings $\mu : S \to \mathbf{R}_+ \cup \{\infty\}$ satisfying the conditions*

$\mu(s) = \sup_{f\in F} \mu(f)$ for all upward directed sets $F \subset S$ and $s \in S$, $\bigvee F = s$

$\{ s \in S \mid \mu(s) < \infty \}$ is dense in order,

and

$\mu(u) \leq 1$.

Then K_u is compact with respect to the coarsest topology in which the mappings $\widetilde{s} : K_u \to \mathbf{R}_+ \cup \{\infty\}$, defined by

$$\widetilde{s}(\mu) = \mu(s), \quad s \in S,$$

are continuous for all u-continuous s in S. Moreover, the mappings $\widetilde{s}$ are lower semicontinuous for all $s \in S$ and

$$\widetilde{s} \leq \widetilde{t} \iff s \leq t.$$

Proof. See Proposition 4.1.4, 4.1.5 and Corollary 4.1.6 of [3]. □

2. Balayage spaces

We are going to study relationships between balayage spaces, standard H-cones and hyperharmonic cones. First we recall some concepts needed for the definition of a balayage space given by Bliedtner and Hansen in [2].

Let X be a locally compact space with a countable base. Let $\mathcal{W}$ be a convex cone of positive lower semicontinuous functions on X. The *fine topology* on X is the coarsest topology which is finer than the initial topology and for which all functions of $\mathcal{W}$ are continuous. The fine lower semicontinuous regulation of a function $g : X \to \mathbf{R} \cup \{-\infty, \infty\}$ is denoted by $\widehat{g}^f$.

A convex cone $\mathcal{P} \subset C(X)$ is called a *function cone* if the following axioms hold:

(F_1) there exists $p \in \mathcal{P}$ such that $p > 0$ on X;

(F_2) $\mathcal{P}^+$ is linearly separating (i.e. for all x, $y \in X$ and $\lambda \in \mathbf{R}_+$ there exists $s \in \mathcal{P}^+$ satisfying $s(x) \neq \lambda s(y)$);

(F_3) $\mathcal{P}$ is adapted (i.e. for any $f \in \mathcal{P}$ there exists $g \in \mathcal{P}^+$ such that for every $\varepsilon > 0$ the inequality $|f(x)| \leq \varepsilon g(x)$ holds for every $x \in X \setminus K$ for some compact subset K of X).

A function cone $\mathcal{P}$ is called min-*stable* if $\inf(f, g) \in \mathcal{P}$ for all f and g in $\mathcal{P}$.

A helpful result to form a function cone is given next.

Lemma 2.1. *Let X be a locally compact space with a countable base. If $\mathcal{P} \subset C^+(X)$ is* min-*stable and satisfies the axioms* (F_2),

(F'_1) *for every $x \in X$ there exists $p \in \mathcal{P}$ with $p(x) > 0$*

and

(F'_3) *for every $p \in \mathcal{P}$, $\varepsilon > 0$ and $x \in X$ there exists $q \in \mathcal{P}$ for which $q(x) < \varepsilon$ and $p \leq q$ on $X \setminus K$ for some compact K*

then the convex cone

$$\mathcal{P}_\sigma = \left\{ \sum_{n=1}^{\infty} p_n \in C(X) \mid p_n \in \mathcal{P} \text{ for all } n \in \mathbf{N} \right\}$$

is a function cone.

Proof. See p. 15 of [2]. □

Definition 2.2. Let X be a locally compact space with a countable base and $\mathcal{W}$ be a convex cone of positive lower semicontinuous functions on X. Then the pair $(X, \mathcal{W})$ is called a balayage space if the following axioms hold

(B_1) for any increasing sequence $(f_n)_{n \in \mathbf{N}}$ of $\mathcal{W}$, the function $\sup_{n \in \mathbf{N}} f_n$ belongs to $\mathcal{W}$;

(B_2) $\widehat{\inf \mathcal{V}}^f \in \mathcal{W}$ for every subset of $\mathcal{V}$ of $\mathcal{W}$;

(B_3) if u, v', $v'' \in \mathcal{W}$ such that $u \leq v' + v''$ there exist u', $u'' \in \mathcal{W}$ satisfying the conditions $u = u' + u''$, $u' \leq v'$ and $u'' \leq v''$;

(B_4) there exists a function cone $\mathcal{P} \subset C^+(X)$ such that

$$\mathcal{W} = \{ \sup_{n \in \mathbf{N}} f_n \mid f_n \in \mathcal{P} \text{ for all } n \in \mathbf{N} \text{ and } (f_n)_{n \in \mathbf{N}} \text{ increasing} \}.$$

Balayage spaces are special type of hyperharmonic cones given below.

Proposition 2.3. *Let $(X, \mathcal{W})$ be a balayage space. Then $\mathcal{W}$ is a hyperharmonic cone in which addition, multiplication by strictly positive real numbers and partial ordering are defined pointwise. Moreover, the set of cancellable elements in $\mathcal{W}$ is increasingly dense.*

Proof. Let $F \subset \mathcal{W}$ be directed upwards. Then there exists a countable increasing sequence $(f_n)_{n\in\mathbf{N}}$ such that $\sup F = \sup_{n\in\mathbf{N}} f_n$ and $f_n \in F$ for all $n \in \mathbf{N}$. Hence by (B_1) $\sup F$ belongs to $\mathcal{W}$ and therefore (H_1) holds.

Let $\mathcal{V}$ be a non-empty subset of $\mathcal{W}$. Then we have $\widehat{\inf \mathcal{V}}^f = \widehat{\inf \mathcal{V}}$ since $\widehat{\inf \mathcal{V}}^f \in \mathcal{W}$ and the fine topology is finer than the initial topology. Consequently, we have

$$\bigwedge \mathcal{V} = \widehat{\inf \mathcal{V}} = \widehat{\inf \mathcal{V}}^f \in \mathcal{W}$$

and

$$a + \bigwedge \mathcal{V} = \bigwedge (a + \mathcal{V})$$

for all $a \in \mathcal{W}$. From (B_4) it follows that the set of cancellable elements are increasingly dense. □

Corollary 2.4. *If $(X, \mathcal{W})$ is a balayage space then $(\mathcal{W}, +, \leq)$ is a hyperharmonic structure in which addition and partial ordering are defined pointwise.*

Proof. See Theorem 2.3 of [6] or Proposition 4.4.4 of [2]. □

Necessary and sufficient conditions for a hyperharmonic cone to be a balayage space are given in the following result.

Theorem 2.5. *Let $\mathcal{W}$ be a convex cone of positive lower semicontinuous functions on a locally compact space X with a countable base. Partial order in $\mathcal{W}$ is defined pointwise. Then $(X, \mathcal{W})$ is a balayage space if and only if $\mathcal{W}$ is a* min-*stable hyperharmonic cone satisfying axioms* (B_1) *and* (B_4).

Proof. It is enough to show that a min-stable hyperharmonic cone satisfying (B_1) and (B_4) is a balayage space. Let $\mathcal{P} \subset \mathcal{W}$ be a function cone given in (B_4). If p is a strictly positive function of $\mathcal{P}$ then

$$\frac{\mathcal{W}}{p} = \left\{ \frac{w}{p} \mid w \in \mathcal{W} \right\}$$

is also a min-stable hyperharmonic cone satisfying (B_1) and (B_4). Hence we can assume that $f \equiv 1$ belongs to a function cone $\mathcal{P}$.

In order to verify (B_2) we first prove the assertion

$(*)$ If V is a fine neighborhood of $x \in X$ then there exist v and w in $\mathcal{W}$ such that $v = w$ on $X \setminus V$, $0 \leq v - w \leq 1$ and $v(x) = 1 + w(x)$.

Let V be a fine neighborhood of $x \in X$. Applying Lemma 4.4.1 of [2] we find a compact neighborhood K of x and $h \in \mathcal{W}$ such that $V \supseteq K \cap \{ h \leq \alpha \}$ and $h(x) < \alpha$ for some $\alpha > 0$. Let $f \in \mathcal{K}^+(X)$ satisfy the conditions $f(x) = 1$ and supp f $\subseteq$ K. Choose $\varepsilon < 1$. Then by Proposition 1.1.2 of [2] there exist s and t in $\mathcal{P}$ such that

$$0 \leq s - t \leq f \leq s - t + \varepsilon.$$

Let us denote $h' = \inf(h, \alpha)$. Then we have

$$x \in \{ y \in X \mid t(y) < s(y) \text{ and } h'(y) < \alpha \} \subset V.$$

Putting $v' = t + h'$ and $w' = \inf(s + h', \alpha + t)$ we obtain $v' \leq w'$ and

$$x \in \{ y \in X \mid v'(y) < w'(y) \} \subset V.$$

Set $\beta = 1/(w'(x) - v'(x))$. Choosing $v = \inf(\beta w', 1 + \beta v')$ and $w = \beta v'$ we establish the assertion $(*)$.

Let now $\mathcal{V} \subset \mathcal{W}$ be non-empty. Set $g = \widehat{\inf \mathcal{V}}^f$. Let x belong to X and β be strictly positive with $g(x) > \beta$. Then there exists a fine neighborhood V of X in which $g(y) > \beta$ for all $y \in V$. According to $(*)$ there exist v and w in $\mathcal{W}$ possessing the properties $v = w$ on $X \setminus V$, $v(x) = w(x) + 1$ and $0 \leq v - w \leq 1$. Then $\beta v - \beta w \leq g \leq s$ for every $s \in \mathcal{V}$. This yields $\beta v \leq s + \beta w$ for every $s \in \mathcal{V}$. Since $\mathcal{W}$ is a hyperharmonic cone, we have $\beta v \leq \beta w + \bigwedge \mathcal{V}$. Thus for all $\beta < g(x)$ the inequality $\beta \leq \bigwedge \mathcal{V}(x)$ holds. Hence $g \leq \bigwedge \mathcal{V}$. Consequently $g = \bigwedge \mathcal{V} \in \mathcal{W}$. The proof of (B_2) is completed. □

K. Janssen has proved that a balayage space is associated with a standard H-cone in Proposition 1 of [9]. We will verify a similar result by using just the definitions of a standard H-cone and a balayage space. The key idea is that universally continuous elements are obtained from reduced continuous functions with compact support.

Let $(X, \mathcal{W})$ be a balayage space. If f is a numerical function on X we shall define a reduced function of f by

$$Rf = \inf\{ w \in \mathcal{W} \mid w \geq f \}.$$

From (B_2) it follows that $\widehat{Rf} = \widehat{Rf}^f \in \mathcal{W}$. Moreover, if f is finely lower semicontinuous then $Rf = \widehat{Rf} = \widehat{Rf}^f \in \mathcal{W}$.

Proposition 2.6. *Let $(X, \mathcal{W})$ be a balayage space and $\mathcal{P} \subset \mathcal{W}$ be a function cone. If $f : X \to \mathbf{R}_+$ is continuous and bounded by some function of $\mathcal{P}$ then Rf is continuous and bounded by some function of $\mathcal{P}$.*

Proof. Assume first that $f \in \mathcal{K}^+(X)$. According to Proposition 1.1.2 of [2] if p is a strictly positive function in $\mathcal{P}$ there exist s_n and t_n in $\mathcal{W} \cap C^+(X)$ such that

$$0 \leq s_n - t_n \leq f \leq s_n - t_n + \frac{p_0}{n}.$$

Then we have
$$R(s_n - t_n) \le Rf \le R(s_n - t_n) + \frac{p_0}{n}.$$
Hence $\lim_{n\to\infty} R(s_n - t_n)(x) = Rf(x)$ for all $x \in X$. Moreover, the sequence $\big(R(s_n - t_n)\big)_{n\in\mathbf{N}}$ converges to Rf uniformly on every compact set K. Since $\mathcal{W}$, by Corollary 2.4, is a hyperharmonic structure, $s_n = R(s_n - t_n) + w_n$ for some $w_n \in \mathcal{W}$ and therefore $R(s_n - t_n)$ is continuous. Hence Rf is continuous. Clearly Rf is also bounded by some function of $\mathcal{P}$.

Suppose now that $f : X \to \mathbf{R}_+$ is continuous. Assume that there exists $p \in \mathcal{P}$ such that $f \le p$. Then $f = \sup_{n\in\mathbf{N}} \varphi_n$ for some increasing sequence $(\varphi_n)_{n\in\mathbf{N}}$ in $\mathcal{K}^+(X)$. Since $\mathcal{P}$ is a function cone there exists $p_0 \in \mathcal{P}$ such that $p \in o(p_0)$ and $p_0 > 0$. Hence for any $\varepsilon > 0$ there exists a compact subset K_ε of X such that $f \le p \le \varepsilon p_0$ on $X \setminus K_\varepsilon$. Applying Dini's theorem we find φ_n for $n \in \mathbf{N}$ such that
$$\varphi_n \le f \le \varphi_n + \frac{1}{n} p_0 \quad \text{on } K_{1/n}.$$
Hence $\varphi_n \le f \le \varphi_n + \frac{1}{n} p_0$ on X. Using the same deductions as in the case $f \in \mathcal{K}^+(X)$ we obtain $Rf = \lim_{n\to\infty} R\varphi_n$ and Rf is continuous. From $f \le p$ it follows that $Rf \le p$ completing the proof. □

Using reduced functions we are able to show a relationship between balayage spaces and standard H-cones.

Theorem 2.7. *Let $(X, \mathcal{W})$ be a balayage space. Then the set of cancellable elements of $\mathcal{W}$ is a standard H-cone in which addition, multiplication by $\mathbf{R}_+$ and the partial ordering are defined naturally.*

Proof. By Lemma p. 167 of [4] there exists a countable set $\mathcal{F} \subseteq \mathcal{K}^+(X)$ enjoying the properties $g = \sup\{f \in \mathcal{F} \mid f \le g\}$ for all $g \in \mathcal{K}^+(X)$ and if $f_1, f_2 \in \mathcal{F}$ then $\sup(f_1, f_2) \in \mathcal{F}$. Our aim is to show that the set $D = \{Rf \mid f \in \mathcal{F}\}$ is an increasingly dense subset of universally continuous elements of the set $\widetilde{\mathcal{W}}$ of cancellable elements of $\mathcal{W}$. Let $w \in \mathcal{W}$. Then $w = \sup\{f \in \mathcal{K}^+(X) \mid f \le w\}$, which implies $w = \sup\{Rf \mid f \in \mathcal{F}, f \le w\}$. The set $\{Rf \mid f \in \mathcal{F}, f \le w\}$ is obviously directed upwards. Hence D is increasingly dense in $\mathcal{W}$ and therefore in $\widetilde{\mathcal{W}}$.

Every weak unit is strictly positive on X. Indeed, there exists a strictly positive $p_0 \in \mathcal{W}$ since $(X, \mathcal{W})$ is a balayage space. From $w = \sup_{n\in\mathbf{N}} \inf(np_0, w)$ for all $w \in \mathcal{W}$ it follows that p_0 is a weak unit. Hence all weak units have to be strictly positive and (S_1) holds in $\widetilde{\mathcal{W}}$. Let u be an arbitrary weak unit and f be an element of $\mathcal{F}$. Assume $F \subset \mathcal{W}$ is directed upwards and $\sup F = Rf$. Since $f \le Rf \le \sup F$ and $f \in \mathcal{K}^+(X)$, by Dini's theorem the relation
$$f < \inf(f, g) + \varepsilon\big(\min_{x\in \operatorname{supp} f} u(x)\big) \le g + \varepsilon u$$
holds for some $g \in F$ on supp f. But then $g + \varepsilon u \ge f$ on the whole space X. Hence we have $g + \varepsilon u \ge Rf$. Since Rf is real valued by Proposition 2.6 the function Rf

is cancellable and universally continuous. Therefore the axiom (S_2) holds in $\widetilde{\mathcal{W}}$. Consequently $\widetilde{\mathcal{W}}$ is a standard H-cone, finishing the proof. □

Corollary 2.8. *If $(X, \mathcal{W})$ is a balayage space then $\mathcal{W}$ is a completion of a standard H-cone of cancellable elements of $\mathcal{W}$.*

A standard H-cone is associated with a balayage space if and only if it can be represented on a Bauer-simplex. This result is given by K. Janssen in Theorem p. 201 of [9]. We are interested in finding necessary and sufficient conditions in terms of elements of standard H-cones under which a standard H-cone can be extended to a balayage space.

Theorem 2.9. *Let $(X, \mathcal{W})$ be a balayage space. Then there exists a weak unit u such that the set of u-continuous elements has a countable increasingly dense and* min-*stable subset C satisfying the condition* $\inf(\alpha u, w)$ *is u-continuous for all $w \in C$ and $\alpha \in \mathbf{R}_+$.*

Proof. Let $\mathcal{P} \subset C^+(X)$ be a function cone such that $S(\mathcal{P}) = \mathcal{W}$ and let $p \in \mathcal{P}$ be strictly positive. By Theorem 2.7 there exists a countable increasingly dense subset D in which all elements are universally continuous. We are going to prove that the set $C = \{\inf(w_1, \ldots, w_n) \mid w_i \in D,\ i = 1, \ldots, n,\ n \in \mathbf{N}\}$ has the desired properties.

Let $\alpha > 0$ and w_1, $w_2 \in D$. Since $\mathcal{P}$ is a function cone there exists a strictly positive function $p \in \mathcal{P}$ and $p \leq u$ for some $u \in \mathcal{P}$ with $p \in o(u)$. Then u is a weak unit. Assume $F \subset \mathcal{W}$ is directed upwards and $\sup F = \inf(\alpha u, w_1, w_2)$ for w_1 and w_2 in D. Since universally continuous elements are bounded by all weak units by Proposition 4.1.2 of [3] there exists $\beta > 0$ such that $w_i \leq \beta p$ for $i = 1, 2$. From $p \in o(u)$ it follows that for any $\varepsilon > 0$ the inequality $p \leq \frac{\varepsilon}{\beta} u$ holds on $X \setminus K$ for some compact subset K of X.

Applying Dini's theorem we find $f \in F$ such that $\inf(\alpha u, w_1, w_2) \leq f + \varepsilon u$ on K. The inequality $\inf(\alpha u, w_1, w_2) \leq \beta p \leq \varepsilon u$ holds on $X \setminus K$, whence $\inf(\alpha u, w_1, w_2) \leq f + \varepsilon u$ on the whole space X. Therefore $\inf(\alpha u, w_1, w_2)$ is u-continuous for all $\alpha > 0$ and w_1, $w_2 \in C$. Choosing $\alpha \in \mathbf{R}^+$ enough large we obtain $\inf(w_1, w_2)$ is u-continuous. This completes the proof. □

One part of the next result is stated on p. 151 of [3] by Boboc, Bucur, Cornea in certain types of standard H-cones.

Lemma 2.10. *Let S be a standard H-cone. If $u \in S$ is a weak unit then there exists a u-continuous weak unit.*

Moreover, suppose that for some weak unit $u \in S$ there exists a countable increasingly dense set C of u-continuous elements enjoying the property:

$$\alpha u \wedge w_1 \wedge w_2 \text{ is } u\text{-continuous for all } \alpha \geq 0 \text{ and } w_1,\ w_2 \in C. \tag{2.1}$$

Then the set $S_0(u)$ of u-continuous elements possesses the same property (2.1).

Proof. Let $u \in S$ be a weak unit and $(s_n)_{n\in\mathbf{N}}$ be an increasing sequence of universally continuous elements with $\bigvee s_n = u$. Then by Proposition 4.1.1 of [3] the element $s = \sum_{n=1}^{\infty} \frac{1}{2^n} s_n$ is a weak unit. To show u-continuity of s, let $\varepsilon > 0$ and $F \subset S$ be directed upwards with $\bigvee F = s$.

Pick $n \in \mathbf{N}$ for which $\sum_{k=n+1}^{\infty} 1/2^k < \varepsilon/2$. Then we have

$$\sum_{k=n+1}^{\infty} \frac{1}{2^k} s_k \leq \sum_{k=n+1}^{\infty} \frac{1}{2^k} u < \frac{\varepsilon}{2} u. \tag{2.2}$$

Since

$$\bigvee_{f\in F} f \wedge (\sum_{k=1}^{n} \frac{1}{2^k} s_k) = \sum_{k=1}^{n} \frac{1}{2^k} s_k$$

and $\sum_{k=1}^{n} \frac{1}{2^k} s_k$ is u-continuous by Proposition 4.1.2 of [3] we can find $f \in F$ satisfying the inequality

$$\sum_{k=1}^{n} \frac{1}{2^k} s_k \leq f + \frac{\varepsilon}{2} u.$$

Combining this with (2.2) we conclude

$$s = \sum_{k=1}^{\infty} \frac{1}{2^k} s_k \leq f + \varepsilon u.$$

Thus s is u-continuous.

Let v_1 and v_2 be u-continuous. We shall show that $v_1 \wedge v_2$ is u-continuous provided that (2.1) holds in S. Denote by $(v_{1n})_{n\in\mathbf{N}}$ and $(v_{2n})_{n\in\mathbf{N}}$ increasing sequences in C satisfying the condition $v_1 = \bigvee_{n\in\mathbf{N}} v_{1n}$ and $v_2 = \bigvee_{n\in\mathbf{N}} v_{2n}$, respectively. Then we have

$$v_1 \wedge v_2 = \bigvee_{n\in\mathbf{N}} v_{1n} \wedge v_{2n}.$$

Since v_1 and v_2 are u-continuous there exist for any $\varepsilon > 0$ elements v_{1n} and v_{2n} for some $n \in \mathbf{N}$ satisfying the inequalities $v_1 \leq v_{1n} + \frac{\varepsilon}{2} u$ and $v_2 \leq v_{2n} + \frac{\varepsilon}{2} u$. Hence the inequality

$$v_1 \wedge v_2 \leq v_{1n} \wedge v_{2n} + \frac{\varepsilon}{2} u \tag{2.3}$$

holds. Let $F \subset S$ be directed upwards and satisfy $\bigvee F = v_1 \wedge v_2$. Since $v_{1n} \wedge v_{2n} \leq \bigvee F$ and $v_{1n} \wedge v_{2n}$ is u-continuous there exists $f \in F$ such that

$$v_{1n} \wedge v_{2n} \leq f + \frac{\varepsilon}{2} u.$$

Substituting this inequality into (2.3) we obtain $v_1 \wedge v_2 \leq f + \varepsilon u$. Hence $v_1 \wedge v_2$ is u-continuous for all v_1 and v_2 in $S_0(u)$. Let $\alpha > 0$ and $G \subset S$ be directed upwards with $\bigvee G = \alpha u \wedge v_1$. Using the inequality $v_1 \leq v_{1n} + \frac{\varepsilon}{2} u$ it follows that

$$v_1 \wedge \alpha u \leq v_{1n} \wedge \alpha u + \frac{\varepsilon}{2} u.$$

Since $v_{1n} \wedge \alpha u \leq \bigvee G$ and $v_{1n} \wedge \alpha u$ is u-continuous we obtain $v_1 \wedge \alpha u \leq g + \varepsilon u$ for some $g \in G$. Hence $v_1 \wedge \alpha u$ is u-continuous. This implies $v_1 \wedge v_2 \wedge \alpha u$ is u-continuous completing the proof. □

Theorem 2.11. *Let S be a standard H-cone. The completion of S is isomorphic with a balayage space if there exist a weak unit $u \in S$ and a countable increasingly dense subset T of universally continuous elements enjoying the property $s \wedge t \wedge \alpha u$ is u-continuous for all s and t in T and $\alpha \geq 0$. Conversely, a balayage space has this property.*

Proof. Let u be a weak unit satisfying the assumptions. We use the same notation K_u as in Theorem 1.7. Then, by Theorem 1.7, K_u is compact with respect to the coarsest topology which makes the function $\mu \to \mu(s)$ continuous for all u-continuous s. Moreover the functions $\widetilde{s} : K_u \to \mathbf{R} \cup \{\infty\}$ defined by

$$\widetilde{s}(\mu) = \mu(s), \qquad \mu \in K_u$$

are lower semicontinuous for all $s \in S$ and

$$\widetilde{s} \leq \widetilde{t} \iff s \leq t$$

for all s, t in S. It is easy to see that K_u is a locally compact space with a countable base. Denote by S_u the set of extreme points of K_u. Then we have

$$s \leq t \iff \widetilde{s} \leq \widetilde{t} \iff \widetilde{s}_{|S_u} \leq \widetilde{t}_{|S_u}.$$

We show that S_u is closed. Let $\varphi \not\equiv 0$ be an accumulation point of S_u. Assume that s and t are u-continuous. Then $s \wedge t$ is u-continuous by Lemma 2.10. Applying Corollary 2.3.5 of [3] we see that $\lambda(s \wedge t) = \inf(\lambda(s), \lambda(t))$ for all $\lambda \in S_u$ and therefore $\varphi(s \wedge t) = \inf(\varphi(s), \varphi(t))$. Since the set of u-continuous function is order dense and φ is order continuous, we have $\varphi(s \wedge t) = \inf(\varphi(s), \varphi(t))$ for all s, $t \in S$. From the existence of a countable base of K_u it follows that there exists a sequence $(\lambda_n)_{n \in \mathbf{N}}$ in S_u converging to φ. Using the assumption $\varphi \not\equiv 0$ we find $s \in T$ and $\alpha > 0$ such that $\lambda_n(s) \geq \alpha$ for all $n \in \mathbf{N}$. Assume that $0 < \varphi(u) < 1$. Then we have

$$\widetilde{(\alpha u \wedge s)}(\varphi) = \varphi(\alpha u \wedge s) \leq \alpha \varphi(u) < \alpha. \tag{2.4}$$

Using u-continuity of $\alpha u \wedge s$ we find a neighborhood V of φ satisfying $\mu(\alpha u \wedge s) < \alpha$ for all $\mu \in V$. On the other hand, $\lambda_n(u) = 1$ for all $n \in \mathbf{N}$ since λ_n is a non-zero extreme point of K_u. Hence

$$\lambda_n(\alpha u \wedge s) = \inf(\lambda_n(\alpha u), \lambda_n(s)) = \alpha$$

for all $n \in \mathbf{N}$. This implies $\varphi(\alpha u \wedge s) = \alpha$ which is an contradiction by (2.4). Thus a non-zero function $\varphi : S \to \mathbf{R} \cup \{\infty\}$ satisfying $0 < \varphi(u) < 1$ is not an extreme point of K_u. This establishes the result

$$S_u \subseteq \{0\} \cup \{\varphi \in K_u \mid \varphi(u) = 1, \inf(\varphi(s), \varphi(t)) = \varphi(s \wedge t) \text{ for all } s,\, t \in S\}.$$

By virtue of Corollary 2.3.5 of [3] the functions $\varphi : S \to \mathbf{R} \cup \{\infty\}$ satisfying the conditions $\varphi(u) = 1$ and $\varphi(s \wedge t) = \inf(\varphi(s), \varphi(t))$ for all s, $t \in S$ are extreme points of K_u. Consequently, S_u is closed in K_u and therefore compact. This implies that $X = S_u \setminus \{0\}$ is a locally compact subspace of K_u with a countable base.

Let W be a completion of S to a hyperharmonic cone.This completion indeed exists by Proposition 1.5. In order to simplify notations we identify S and a subset of W isomorphic with S.

Define a mapping $\phi : W \to \{ f : X \to \mathbf{R} \cup \{\infty\} \mid f \text{ lower semicontinuous} \}$ by

$$\phi(w) = \sup\{ \widetilde{s}_{|X} \mid s \leq w,\, s \in S \}.$$

Our aim is to show that $\phi(W)$ is a balayage space isomorphic with W.

We first show that

$$\phi(w_1) \leq \phi(w_2) \iff w_1 \leq w_2 \tag{2.5}$$

for all w_1, $w_2 \in W$. It is clear that $\phi(w_1) \leq \phi(w_2)$ if $w_1 \leq w_2$. Assume that $\phi(w_1) \leq \phi(w_2)$ for some w_1 and w_2 in W. Suppose that $s \in S$ satisfies $s \leq w_1$. Set

$$\mathcal{A}_{w_2} = \{ t \in S \mid t \leq w_2 \}.$$

Then $\mathcal{A}_{w_2}$ is directed upwards, since S is solid in W and from $t_1 \vee t_2 \leq t_1 + t_2$ it follows that $t_1 \vee t_2 \in \mathcal{A}_{w_2}$ for all t_1, $t_2 \in \mathcal{A}_{w_2}$. Moreover the following equalities hold in X

$$\widetilde{s} = \phi(s) = \inf(\phi(s), \phi(w_2)) = \sup_{t \in \mathcal{A}_{w_2}} \inf(\phi(s), \phi(t)) = \sup_{t \in \mathcal{A}_{w_2}} \inf(\widetilde{s}, \widetilde{t})$$

The equality $\inf(\widetilde{s}, \widetilde{t})(\mu) = (\widetilde{s \wedge t})(\mu)$ for all $\mu \in X$ results in

$$\widetilde{s}(\mu) = \sup_{t \in \mathcal{A}_{w_2}} (\widetilde{s \wedge t})(\mu) = \mu\Big(\bigvee_{t \in \mathcal{A}_{w_2}} (s \wedge t) \Big)$$

for all $\mu \in X$. Hence we have

$$s = \bigvee_{t \in \mathcal{A}_{w_2}} (s \wedge t) \leq w_2$$

for all $s \leq w_1$. This implies $w_1 \leq w_2$ which establishes (2.5).

Using the dominated decomposition property in W and the fact that S is solid in W we can deduce that ϕ is also additive. Hence ϕ is an isomorphism between W and $\phi(W)$ which shows that $\phi(W)$ is a hyperharmonic cone.

According to Theorem 2.5 $\phi(W)$ is a balayage space if (B_1) and (B_4) holds in $\phi(W)$. To prove (B_1) we show that ϕ is order continuous from below. Let $F \subseteq W$ be directed upwards and $\bigvee F = w$. Assume that $s \in S$ satisfies $s \leq w$. Then we have

$$s = s \wedge (\bigvee F) = \bigvee_{f \in F} s \wedge f,$$

since W is a completion of S. The property $s \wedge f \in S$ yields

$$\widetilde{s} = \sup_{f \in F} \widetilde{s \wedge f} \leq \sup_{f \in F} \phi(f).$$

Thus we have verified that

$$\phi(w) \leq \sup_{f \in F} \phi(f) \leq \phi(w).$$

which implies the order continuity of ϕ.

To show (B_1) let $F \subseteq \phi(W)$ be directed upwards. Then $\phi^{-1}(F)$ is by (2.5) also directed upwards in W and

$$\phi(\bigvee \phi^{-1}(F)) = \sup \phi(\phi^{-1}(F)) = \sup F.$$

Hence $\sup F \in \phi(W)$ and (B_1) holds.

Denote by $S_0(u)$ the set of all u-continuous elements of S. We show that a convex cone

$$\mathcal{P}_\sigma = \left\{ \sum_{n=1}^{\infty} \phi(p_n) \in C^+(X) \mid p_n \in S_0(u) \right\}$$

is a function cone. The set $\phi(S_0(u))$ is min-stable by Lemma 2.10. According to the same result there exists a weak unit p which is u-continuous. Then $\phi(p) > 0$ since $u = \bigvee_{n \in \mathbf{N}} u \wedge np$ and $\phi(u) \equiv 1$. Hence the condition (F_1') of Lemma 2.1 holds. To show (F_3') of Lemma 2.1, let v be u-continuous, $\varepsilon > 0$ and $\mu \in X$. Choose $n \in \mathbf{N}$ for which $\frac{1}{2^n} < \varepsilon$. Then we have

$$\phi(v \wedge \frac{1}{2^n} u)(\mu) = \widetilde{(v \wedge \frac{1}{2^n} u)}(\mu) = \inf(\mu(v), \frac{1}{2^n}) < \varepsilon.$$

Furthermore,

$$\widetilde{(v \wedge \frac{1}{2^n} u)}(\varphi) < \widetilde{v}(\varphi) \iff \widetilde{v}(\varphi) = \varphi(v) > \frac{1}{2^n}$$

for every $\varphi \in X$. Since $\widetilde{v}$ is continuous on K_u the set $K = \{ \varphi \in X \mid \varphi(v) \geq \frac{1}{2^n} \}$ is closed in K_u and therefore compact. Hence $K \subset X$ is also compact in the space X. This implies

$$\widetilde{v}(\varphi) = \varphi(v) \leq \inf(\widetilde{v}(\varphi), \frac{1}{2^n} \widetilde{u}(\varphi) = \phi(v \wedge \frac{1}{2^n} u) \quad \text{on } X \setminus K.$$

Since $v \wedge \frac{1}{2^n} u$ is u-continuous by assumption the condition (F_3') holds in $\phi(S_0(u))$. We still have to prove that $\phi(S_0(u))$ separates linearly the points of X. Let $\lambda > 0$ and μ, $\varphi \in X$(in case of $\lambda = 0$, the separation property is trivially verified) . If $\lambda = 1$, then there exists $v \in S_0(u)$ satisfying $\varphi(v) \neq \mu(v)$ since u-continuous functions separate the points of X. On the contrary assume that $\varphi(v) = \lambda\mu(v)$ for all $v \in S_0(u)$. If necessary interchanging the roles of φ and μ we may assume that $0 < \lambda < 1$. Then we have

$$\begin{aligned} \phi(p \wedge \lambda\mu(p)u)(\varphi) &= \lambda\phi(p \wedge \lambda\mu(p)u)(\mu) \iff \\ \inf(\varphi(p), \lambda\mu(p)) &= \lambda\Big(\inf(\mu(p), \lambda\mu(p))\Big), \end{aligned}$$

whence

$$\lambda\mu(p) = \varphi(p) = \lambda^2\mu(p).$$

Since $\mu(p) > 0$ we obtain $\lambda = 1$ which is a contradiction. By Lemma 2.1 $\mathcal{P}_\sigma$ is a function cone. It is clear that $\mathcal{P}_\sigma$ satisfies (B_4).

Applying Theorem 2.5 we establish the assertion $\phi(W)$ is a balayage space. □

References

[1] Arsove, M., H. Leutwiler, Algebraic potential theory. Mem. Amer. Math. Soc. No. 226 (1980)

[2] Bliedtner, J., W. Hansen, Potential theory. Springer Verlag, Berlin-Heidelberg 1986

[3] Boboc, N., Gh. Bucur, A. Cornea, Order and the convexity in potential theory: H-cones. Lecture Notes in Math. 853, Springer Verlag, Berlin-Heidelberg 1981

[4] Constantinescu, C., A. Cornea, Potential theory on harmonic spaces. Springer Verlag, Berlin-Heidelberg 1972

[5] Eriksson, S.-L., Hyperharmonic cones and cones of hyperharmonics. An. Stiint. Univ. "Al. I. Cuza" Iasi Sect. Ia Mat., Vol 31 (1985), 109–116

[6] Eriksson, S.-L., Hyperharmonic cones and hyperharmonic morphisms. Ann. Acad. Sci. Fenn. Ser. I A. Dissertationes 49 (1984)

[7] Eriksson, S.-L., Representations of hyperharmonic cones. Trans. Amer. Math. Soc. 305 (1988), 247–262

[8] Eriksson-Bique, S.-L., Completions of H-cones. Ann. Acad. Sci. Fenn. Ser. A. I 15 (1990),185–199

[9] Janssen, K., Standard H-cones and balayage spaces. In: Complex Analysis, 197–203. Lecture Notes in Math. 1014, Springer Verlag, Berlin-Heidelberg 1983

On thermoelastic potential in three-dimensional hyperbolic thermoelasticity theory

Jerzy A. Gawinecki

Abstract

A system of linear hyperbolic partial differential equations describing isotropic, homogeneous thermoelastic medium has been considered. Basing on the matrix of fundamental solutions of this system of equations and applying the method of potential theory we obtained the solution of the Cauchy problem for this system of equations. We use Strauss – Klainerman approach to the $L^p - L^q$ – time decay estimates for the solution of the Cauchy problem for this system. Such $L^p - L^q$ – time decay estimates play an important role in the proof of global (in time) existence of solution to the Cauchy problem for nonlinear hyperbolic system of equations describing thermoelastic medium.

1980 Mathematics Subject Classification (1985 Revision): 35A07, 35A08, 35E05, 35E15, 35L15, 35L40, 35L45, 73B30.

0. Introduction

We consider the Cauchy problem for a hyperbolic system of partial differential equations of linear thermoelasticity (cf. [13]):

$$\varrho\,\partial_t^2 u - \mu\Delta u - (\lambda+\mu)\,\mathrm{grad}\,\mathrm{div}\,u + \beta\,\mathrm{grad}\,\partial_t T = 0 \tag{0.1}$$

$$\beta\,\mathrm{div}\,\partial_t u + \varrho\tau\partial_t^2 T - k\Delta T = 0 \tag{0.2}$$

with initial conditions:

$$\begin{aligned} u(+0,x) &= u^0(x), \qquad (\partial_t u)(+0,x) = u^1(x)\\ T(+0,x) &= T^0(x), \qquad (\partial_t T)(+0,x) = T^1(x) \end{aligned} \tag{0.3}$$

where: $u = (u_1, u_2, u_3)$ – is the displacement vector field of the medium, T – the temperature of the medium, $t \geq 0$, $x \in \mathbb{R}^3$, $\partial_t = \frac{\partial}{\partial t}$, $\partial_j = \frac{\partial}{\partial x_j}$, $\Delta = \sum_{j=1}^{3} \partial_j^2$; ϱ, μ, λ, β, τ, k – are positive physical constants; u^0, u^1, T^0, T^1 – are given functions.

Remark 0.1. The system (0.1) – (0.2) is the principal part of a hyperbolic system of partial differential equations describing the evolution of a thermoelastic medium (cf. [13], p. 199 formulae (2.7.36)). For the sake of simplicity we assume in system (2.7.36) that $\nu = 1$ and $T_0 = 1$.

Under the assumption that Cauchy data u^0, u^1, T^0, T^1 are smooth enough (cf. [4] formulae (5.1) – (5.3)) the solution of the problem (0.1) – (0.3) is given by convolutions (which are called the thermoelastic potential in three-dimensional hyperbolic

Potential Theory, ed. by M. Kishi

thermoelasticity):

$$U(t,x) = (u(t,x),T(t,x)) = H(t,\cdot) * \overline{g}(x) + \partial_t H(t,\cdot) * \widetilde{h}(x) \tag{0.4}$$

where: $\overline{g}(\cdot) = \tilde{g}(\cdot) + D(\partial)\widetilde{h}(\cdot)$; $\tilde{g}(x) = (u^1(x), T^1(x))$, $\widetilde{h}(x) = (u^0(x), T^0(x))$;

$$D(\partial) = \begin{bmatrix} 0 & 0 & 0 & \beta\partial_1 \\ 0 & 0 & 0 & \beta\partial_2 \\ 0 & 0 & 0 & \beta\partial_3 \\ \beta\partial_1 & \beta\partial_2 & \beta\partial_3 & 0 \end{bmatrix}$$

$*$ – denotes the three–dimensional convolution in $\mathbb{R}^3$ and $H(t,x)$ – is the matrix of fundamental solutions of the system (0.1) – (0.2) constructed in [4] of the form:

$$\begin{aligned} H_{jk}(t,x) = (32\pi 7!\, \varrho^4\, \tau)^{-1} \Bigg\{ & \delta_{jk}(1-\delta_{k4})\Big[\frac{A_1}{a_1^2}\frac{1}{|x|}\delta\Big(t-\frac{|x|}{a_1}\Big) - \frac{A_2}{a_2^2}\frac{1}{|x|}\delta\Big(t-\frac{|x|}{a_2}\Big) \\ & - \frac{B}{b^2}\frac{1}{|x|}\delta\Big(t-\frac{|x|}{b}\Big)\Big] + \delta_{jk}\delta_{k4}\Big[\frac{A_3}{a_1^2}\frac{1}{|x|}\delta\Big(t-\frac{|x|}{a_1}\Big) - \frac{A_4}{a_2^2}\frac{1}{|x|}\delta\Big(t-\frac{|x|}{a_2}\Big)\Big] \\ & + \delta_{jj}(1-\delta_{k4})\Big[\frac{A_5}{a_1^2}\frac{x_jx_k}{|x|^3}\delta\Big(t-\frac{|x|}{a_1}\Big) - \frac{A_6}{a_2^2}\frac{x_jx_k}{|x|^3}\delta\Big(t-\frac{|x|}{a_2}\Big) + \frac{B}{b^2}\frac{x_jx_k}{|x|^3}\delta\Big(t-\frac{|x|}{b}\Big)\Big] \\ & - \delta_{jj}(1-\delta_{k4})\Big[A_7 t\Big(\frac{\delta_{jk}}{|x|^3} - \frac{3x_jx_k}{|x|^5}\Big)\Big[\varepsilon\Big(t-\frac{|x|}{a_1}\Big) - \varepsilon\Big(t-\frac{|x|}{a_2}\Big)\Big] \\ & + A_8 t\Big(\frac{\delta_{jk}}{|x|^3} - \frac{3x_jx_k}{|x|^5}\Big)\Big[\varepsilon\Big(t-\frac{|x|}{b}\Big) - \varepsilon\Big(t-\frac{|x|}{a_1}\Big)\Big] \\ & + A_9 t\Big(\frac{\delta_{jk}}{|x|^3} - \frac{3x_jx_k}{|x|^5}\Big)\Big[\varepsilon\Big(t-\frac{|x|}{b}\Big) - \varepsilon\Big(t-\frac{|x|}{a_2}\Big)\Big]\Big] \\ & + A_{10}\frac{x_j}{|x|^3}\Big[\varepsilon\Big(t-\frac{|x|}{a_1}\Big) - \varepsilon\Big(t-\frac{|x|}{a_2}\Big)\Big]\delta_{4j}(1-\delta_{kj}) \\ & + A_{10}\frac{x_j}{|x|^2}\Big[\delta\Big(t-\frac{|x|}{a_1}\Big) - \delta\Big(t-\frac{|x|}{a_2}\Big)\Big]\delta_{4j}(1-\delta_{jk}) \Bigg\} \quad j,k = 1,2,3,4 \end{aligned} \tag{0.5}$$

where:

δ_{jk} – denotes the Kronecker symbol $\delta_{jk} = \begin{cases} 1 & \text{for } j = k \\ 0 & \text{for } j \neq k \end{cases}$

$\delta(\cdot)$ – Dirac's distribution,

$\varepsilon(\cdot)$ – Heaviside's function $\varepsilon(t) = \begin{cases} 1 & \text{for } t > 0 \\ 0 & \text{for } t < 0 \end{cases}$

a_1, a_2, b, $A_1, \dots, A_{10}$, B are constants.

The aim of this paper is to prove the $L^\infty - L^1$ – time decay estimates for the solution of the problem (0.1) – (0.3) using the formula (0.4), then to prove the $L^2 - L^2$ – time decay estimates for this solution applying some theorems of the theory of symmetric hyperbolic system of first order (cf. Yu. V. Egorov [3] p. 320–333) and finally the $L^p - L^q$ – time decay estimates using interpolation inequalities in Sobolev spaces (cf. [5], [10], [8], [14]).

Some notation:

Besides other standard notation we use the symbol $W^{m,p}(\mathbb{R}^3)$ $(1 \leq p \leq \infty$, $m \in \mathbb{N} \cup \{0\})$ for the well–known Sobolev spaces with norm $\|\cdot\|_{W^{m,p}(\mathbb{R}^3)}$ (cf. [1], [11]); $W^{o,p}(\mathbb{R}^3) = L^p(\mathbb{R}^3)$ with norm $\|\cdot\|_{L^p(\mathbb{R}^3)}$. Norm $\|\cdot\|_{L^p_{,m}(\mathbb{R}^3)}$ stands for the Sobolev norm $\|\cdot\|_{W^{m,p}(\mathbb{R}^3)}$. Instead of $W^{m,p}(\mathbb{R}^3)$ we write $L^p_{,m}(\mathbb{R}^3)$.

We use the notation:

$\nabla f = (\partial_t f, \partial_{x_1} f, \partial_{x_2} f, \partial_{x_3} f)$ – the space–time gradient of a function f, while
$Df = (\partial_{x_1} f, \partial_{x_2} f, \partial_{x_3} f)$ – the space gradient of a function f.
$[s]$ – denotes the smallest integer larger than or equal to s for $s \in \mathbb{R}$.

1. The $\mathbf{L}^\infty - \mathbf{L}^1$ – time decay estimates

We shall prove the following theorem:

Theorem 1.1 ($L^\infty - L^1$ – time decay estimates). *Let the Cauchy data u^0, u^1, T^0, T^1 be functions vanishing at infinity. Moreover, let $(u^1, Du^0, T^1, DT^0) \in L^1_{,3}(\mathbb{R}^3)$ then the solution (u,T) of the problem* (0.1) – (0.3) *given by the formula* (0.4) *satisfies the following estimates:*

$$\|(u(t,\cdot), T(t,\cdot))\|_{L^\infty(\mathbb{R}^3)} \leq C(1+t)^{-1} \|(u^1, Du^0, T^1, DT^0)\|_{L^1_{,3}(\mathbb{R}^3)} \tag{1.1}$$

$$\|(\nabla u(t,\cdot), \nabla T(t,\cdot))\|_{L^\infty(\mathbb{R}^3)} \leq C(1+t)^{-1} \|(u^1, Du^0, T^1, DT^0)\|_{L^1_{,3}(\mathbb{R}^3)} \tag{1.2}$$

for $t \geq 0$, where: C is a constant independent of u^0, u^1, T^0, T^1 and t.

Proof. We prove the estimate (1.2). The proof of the estimate (1.1) runs in the same way. Writing the solution $U(t,x) = (u(t,x), T(t,x))$ given by the formula (0.4) in the form:

$$U_j(t,x) = \sum_{k=1}^4 H_{jk}(t,\cdot) * \overline{g}^k(x) + \sum_{k=1}^4 \partial_t H_{jk}(t,\cdot) * \widetilde{h}^k(x) \tag{1.3}$$

$j = 1,2,3,4$; where $U_j(t,x) = u_j(t,x)$, $j = 1,2,3$; $U_4(t,x) = T(t,x)$ and differentiating (1.3) with respect to t and x_l (for $l = 1,2,3$) we get:

$$\begin{aligned}\partial_t U_j(t,x) =& \sum_{k=1}^4 \partial_t H_{jk}(t,\cdot) * \overline{g}^k(x) + \sum_{k=1}^4 \sum_{l=1}^3 \left[\frac{\mu}{\varrho} \partial_l H_{jk}(t,\cdot) + \frac{\lambda+\mu}{\varrho} \partial_j H_{lk}(t,\cdot)\right] * \widetilde{h}^k_l(x) \\ &- \sum_{k=1}^4 \frac{\beta}{\varrho} \partial_t H_{jk}(t,\cdot) * \widetilde{h}^k_j(x), \qquad j = 1,2,3\end{aligned} \tag{1.4}$$

$$\begin{aligned}\partial_t U_4(t,x) =& \sum_{k=1}^4 \partial_t H_{4k}(t,\cdot) * \overline{g}^k(x) + \sum_{k=1}^4 \sum_{l=1}^3 \frac{k}{\varrho} \partial_l H_{4k}(t,\cdot) * \widetilde{h}^k_l(x) \\ &- \sum_{k=1}^4 \sum_{l=1}^3 \frac{\beta}{\varrho} \partial_t H_{lk}(t,\cdot) * \widetilde{h}^k_l(x)\end{aligned} \tag{1.5}$$

$$\partial_l U_j(t,x) = \sum_{k=1}^{4} \partial_l H_{jk}(t,\cdot) * \overline{g}^k(x) + \sum_{k=1}^{4} \partial_t H_{jk}(t,\cdot) * \widetilde{h}_l^k(x) \tag{1.6}$$

$l = 1,2,3$, $j = 1,2,3,4$, $\widetilde{h}_m^k = \partial_m \widetilde{h}^k$, $m = 1,2,3$.

We can write relations (1.4) – (1.6) in the vector form as follows:

$$V(t,x) = R(t,\cdot) * V^0(x) \tag{1.7}$$

where:

$$V(t,x) = (\nabla u, \nabla T), \qquad V^0(x) = (u^1, Du^0, T^1, DT^0) \tag{1.8}$$

and $R(t,x)$ is a 16×16 matrix with elements which are linear combinations of the terms $\partial_l H_{jk}(t,x)$ and $\partial_t H_{jk}(t,x)$ (cf. (1.3) – (1.6)).

From (1.3) – (1.6) and (1.7) it follows that in order to prove the estimate (1.2) it is sufficient to prove the following estimates:

$$||\partial_t H_{jk}(t,\cdot) * f||_{L^\infty(\mathbb{R}^3)} \le C(1+t)^{-1} ||f||_{L^1_{,3}(\mathbb{R}^3)} \tag{1.9}$$

$$||\partial_l H_{jk}(t,\cdot) * f||_{L^\infty(\mathbb{R}^3)} \le C(1+t)^{-1} ||f||_{L^1_{,3}(\mathbb{R}^3)} \tag{1.10}$$

for $j,k = 1,2,3,4$ and any scalar function $f(x)$ satisfying the assumptions of theorem 1.1. Taking into account the form of the matrix $H(t,x)$ (cf. (0.5)) we get:

$$\begin{aligned}
H_{jk}(t,\cdot) * f(x) = (32\pi 7!\, \varrho^4 \tau)^{-1} \Bigg\{ & \delta_{jk}(1-\delta_{k4}) \Bigg[\frac{A_1}{a_1} \int_{|z|=a_1} t\frac{1}{a_1} f(x+tz)dS_z \\
& - \frac{A_2}{a_2} \int_{|z|=a_2} t\frac{1}{a_2} f(x+tz)dS_z - \frac{B}{b} \int_{|z|=b} t\frac{1}{b} f(x+tz)dS_z \Bigg] \\
& + \delta_{jk}\delta_{k4} \Bigg[\frac{A_3}{a_1} \int_{|z|=a_1} t\frac{1}{a_1} f(x+tz)dS_z - \frac{A_4}{a_2} \int_{|z|=a_2} t\frac{1}{a_2} f(x+tz)dS_z \Bigg] \\
& + \delta_{jj}(1-\delta_{k4}) \Bigg[\frac{A_5}{a_1} \int_{|z|=a_1} \frac{t z_j z_k}{a_1^3} f(x+tz)dS_z - \frac{A_6}{a_2} \int_{|z|=a_2} \frac{t z_j z_k}{a_2^3} f(x+tz)dS_z \\
& + \frac{B}{b} \int_{|z|=b} \frac{t z_j z_k}{b^3} f(x+tz)dS_z \Bigg] - \delta_{jj}(1-\delta_{k4}) \Bigg[A_7 \int_{a_1 \le |z| \le a_2} t\Big(\frac{\delta_{jk}}{|z|^3} - \frac{3z_j z_k}{|z|^5}\Big) f(x+tz)dz \\
& + A_8 \int_{b \le |z| \le a_1} t\Big(\frac{\delta_{jk}}{|z|^3} - \frac{3z_j z_k}{|z|^5}\Big) f(x+tz)dz + A_9 \int_{b \le |z| \le a_2} t\Big(\frac{\delta_{jk}}{|z|^3} - \frac{3z_j z_k}{|z|^5}\Big) f(x+tz)dz \Bigg] \\
& + \delta_{4j}(1-\delta_{kj}) A_{10} \int_{a_1 \le |z| \le a_2} \frac{z_j}{|z|^3} f(x+tz)dz \\
& + \delta_{4j}(1-\delta_{kj}) A_{10} \Bigg[\int_{|z|=a_1} t\frac{z_j}{a_1^2} f(x+tz)dS_z - \int_{|z|=a_2} t\frac{z_j}{a_2^2} f(x+tz)dS_z \Bigg] \Bigg\}
\end{aligned} \tag{1.11}$$

where: dS_z – is the area element of the sphere $|z| = a_j$, $j = 1,2$ or $|z| = b$.

For simplicity we consider two typical integrals occurring on the right hand side of (1.11) (other integrals in (1.11) are estimated similarly):

$$I^1 = \int_{|y|=b} tf(x+ty)dS_y \tag{1.12}$$

$$I^2 = \int_{b\le|y|\le a} t\frac{f(x+ty)}{|y|^3}dy \tag{1.13}$$

Differentiating the integrals I^1 and I^2 with respect to t and x_l ($l = 1,2,3$) we obtain:

$$I_t^1 = \int_{|y|=b} f(x+ty)dS_y + \int_{|y|=b} t\partial_t f(x+ty)dS_y \tag{1.14}$$

$$I_t^2 = \int_{b\le|y|\le a} \frac{f(x+ty)}{|y|^3}dy + \int_{b\le|y|\le a} \frac{t}{|y|^3}\partial_t f(x+ty)dy \tag{1.15}$$

and

$$I_{x_l}^2 = \int_{|y|=b} t\partial_l f(x+ty)dS_y \tag{1.16}$$

$$I_{x_l}^2 = \int_{b\le|y|\le a} \frac{t}{|y|^3}\partial_l f(x+ty)dy \tag{1.17}$$

where we use notation: $I_t^j = \partial_t I^j$ and $I_{x_l}^j = \partial_l I^j$ for $l = 1,2,3$, $j = 1,2$.

Following S. Klainerman (cf. [7] p. 53–59 and [5]) we get:

$$|I_t^1| + |I_{x_l}^1| \le C||f||_{L^1_{,3}(\mathbb{R}^3)} \qquad \text{for } t \ge 0 \tag{1.18}$$

$$|I_t^2| + |I_{x_l}^2| \le C||f||_{L^1_{,1}(\mathbb{R}^3)} \qquad \text{for } 0 \le t \le 1 \tag{1.19}$$

$$|I_t^2| + |I_{x_l}^2| \le Ct^{-1}||D_x^1 f||_{L^1(\mathbb{R}^3)} \qquad \text{for } t \ge 1 \tag{1.20}$$

Thus from (1.19) and (1.18) we obtain:

$$||\nabla H_{jk}(t,\cdot) * f(\cdot)||_{L^\infty(\mathbb{R}^3)} \le C||f||_{L^1_{,3}(\mathbb{R}^3)} \qquad \text{for } 0 \le t \le 1 \tag{1.21}$$

and $j,k = 1,2,3,4$.

Now, in view of: $1 \le 2(1+t)^{-1}$ for $0 \le t \le 1$ and $t^{-1} \le 2(1+t)^{-1}$ for $t \ge 1$ we conclude (cf. [5]) that:

$$||\nabla H_{jk}(t,\cdot) * f(\cdot)||_{L^\infty(\mathbb{R}^3)} \le C(1+t)^{-1}||f||_{L^1_{,3}(\mathbb{R}^3)} \qquad \text{for } t \ge 0 \tag{1.22}$$

and $j,k = 1,2,3,4$. □

2. The $L^2 - L^2$ – time decay estimates

We derive the $L^2 - L^2$ – time decay estimates for solution of the Cauchy problem (0.1) – (0.3). More precisely we formulate the following theorem:

Theorem 2.1 ($L^2 - L^2$ – time decay estimates). *Let the Cauchy data u^0, u^1, T^0, T^1 be functions vanishing at infinity. Moreover, let $(u^1, Du^0, T^1, DT^0) \in L^2(\mathbb{R}^3)$ then the solution (u,T) of the problem* (0.1) – (0.3) *given by the formula* (0.4) *satisfies the following estimates:*

$$||(u(t,\cdot),T(t,\cdot))||_{L^2(\mathbb{R}^3)} \le C||(u^1,Du^0,T^1,DT^0)||_{L^2(\mathbb{R}^3)} \qquad \text{for } t \ge 0 \quad (2.1)$$

$$||(\nabla u(t,\cdot),\nabla T(t,\cdot))||_{L^2(\mathbb{R}^3)} \le C||(u^1,Du^0,T^1,DT^0)||_{L^2(\mathbb{R}^3)} \quad \text{for } t \ge 0 \quad (2.2)$$

where C is constant independent from u^0, u^1, T^0, T^1 and t.

Sketch of proof. Following Yu. V. Egorov (cf. [3] p. 320–322, 326–333) we reduce the Cauchy problem (0.1) – (0.3) to an equivalent Cauchy problem for a linear symmetric hyperbolic system of first order. Next, applying the existence and uniqueness theorem (cf. [3] Theorem 3.2 p. 329) of the theory of such hyperbolic first order systems we obtain the estimates (2.1), (2.2).

3. The $L^p - L^q$ – time decay estimates

In this section we express the $L^p - L^q$ – time decay estimates for solutions of the Cauchy problem (0.1) – (0.3) in terms of their gradient.

We consider the operator $\prod_*$ defined as follows:

$$\prod\nolimits_* f(x) = R(t,\cdot) * f(x) \tag{3.1}$$

for any function $f(x)$ satisfying the assumptions of theorems 1.1 and 2.1, where $R(t,x)$ is defined by (1.7).

From theorems 1.1 and 2.1 it follows that:

$$\prod\nolimits_*^0 : L^1_{,3}(\mathbb{R}^3) \longrightarrow L^\infty(\mathbb{R}^3) \quad \text{and} \quad ||\prod\nolimits_*^0|| \le C(1+t)^{-1} \tag{3.2}$$

$$\prod\nolimits_*^1 : L^2(\mathbb{R}^3) \longrightarrow L^2(\mathbb{R}^3) \quad \text{and} \quad ||\prod\nolimits_*^1|| \le C \tag{3.3}$$

By interpolation (cf. [5]) we can prove the following theorem:

Theorem 3.1 ($L^p - L^q$ – time decay estimates). *Let the Cauchy data u^0, u^1, T^0, T^1 be functions vanishing at infinity. Moreover, let $(u^1, Du^0, T^1, DT^0) \in L^p_{,s_0}(\mathbb{R}^3)$ for $p = \frac{2\alpha+2}{2\alpha+1}$, $s_0 = \left[\frac{3\alpha}{\alpha+1}\right]$ and α be an integer greater than or equal to 0, then the solution of the problem* (0.1) – (0.3) *given by the formula* (0.4) *satisfies the following estimates:*

$$||(\nabla u(t,\cdot),\nabla T(t,\cdot))||_{L^{2\alpha+2}(\mathbb{R}^3)} \le C(1+t)^{-\frac{\alpha}{\alpha+1}}||(u^1,Du^0,T^1,DT^0)||_{L^p_{,s_0}(\mathbb{R}^3)} \tag{3.4}$$

for $t \ge 0$, where C is constant independent from u^0, u^1, T^0, T^1 and t.

Remark 3.1. We used theorem 3.1 in the proof of global (in time) existence of solution of the Cauchy problem for nonlinear hyperbolic systems of partial differential equations describing a thermoelastic medium (cf. [6]).

References

[1] Adams, R., Sobolev Spaces. Academic Press, New York 1975

[2] Friedrichs, K.O., Symmetric hyperbolic linear differential equations. Comm. Pure Appl. Math. 7 (1954), 345–392

[3] Egorov, Yu.V., Linear differential equations of principal type [in Russian], Nauka, Moskva 1984

[4] Gawinecki, J., Matrix of fundamental solutions for the system of equations of hyperbolic thermoelasticity with two relaxation times and solution of the Cauchy problem. Bull. Acad. Polon. Sci. Sér. Sci. Tech. 36 (1988), 449–466

[5] Gawinecki, J., $L^p - L^q$ – time decay estimate for solution of the Cauchy problem for hyperbolic partial differential equations of linear thermoelasticity, to appear in: Ann. Polon. Math.

[6] Gawinecki, J., Global solution to the Cauchy problem in nonlinear hyperbolic thermoelasticity, to appear in: Math. Meth. in the Appl. Sci.

[7] Klainerman, S., Global existence for nonlinear wave equations. Comm. Pure Appl. Math. 33 (1980), 43–101

[8] Klainerman, S., Ponce, G., Global, small amplitude solutions to nonlinear evolution equations. Comm. Pure Appl. Math. 36 (1983), 131–141

[9] Klainerman, S., Long – time behaviour of solutions to nonlinear evolution equations. Arch. Rat. Mech. Anal. 78 (1982), 72–98

[10] Shatah, J., Global existence of small solutions to nonlinear evolution equations. J. Differential Equations 46 (1982), 409–425

[11] Sobolev, S.L., Applications of Functional Analysis in Mathematical Physics. Leningrad [in Russian], 1950

[12] Strauss, W.A., Nonlinear scattering theory at love energy. J. Funct. Anal. 41 (1981), 101–113

[13] Suhubi, E.S., Thermoelastic solids in: Continuum Physics (ed. by A.C.Eringen), New York 1975

[14] Triebel, H., Interpolation Theory, Function Spaces, Differential Operators. VEB Deutscher Verlag der Wissenschaften, Berlin 1978

Representations and estimates for inverse operators of the potential theory integral equations in a polyhedron

Nikolai V. Grachev

Abstract

Integral equations of the harmonic potential theory on the boundary of a polyhedron are considered. The point-wise estimates for kernels of inverse operators of these equations are stated. As a corollary the invertibility of the double layer potential operator in the space of continuous functions and in L_p is obtained.

1980 Mathematics Subject Classification(1985 Revision): 31B20, 31B35,35A08.

The lecture is devoted to the exposition of results obtained by the author and V.G. Maz'ya. It is closely related to our papers [3]–[5] in which integral equations of the potential theory on surfaces with conic points were considered. Here, similar to [9] (see also [10]), the study of the integral equations is reduced to the study of some auxiliary boundary value problems. We obtain representations for inverse operators of the integral equations in question in terms of the inverse operators of the Dirichlet and Neumann problems in the interior and exterior of the surface. Using the estimates of the fundamental solutions of these boundary value problems, we arrive at the estimates for kernels of inverse operators of integral equations. Such estimates lead to theorems on the invertibility of the integral equations in various function spaces. In particular the solvability in the space of continuous functions C for the integral equations associated with the Dirichlet problem is established without any assumptions on the shape of the polyhedron. The question of the validity of the last result was known for a long time. The solvability of the boundary integral equations in the space C over surfaces of a fairly wide class was established by Burago, Maz'ya [2] and Kral [7]. This approach requires that the essential norm $|T|$ of the double layer potential T is less than 1. This condition can be formulated in geometrical terms. However, it does not always hold even for sufficiently simple cones. Angell, Kleinman, Kral [1] and Kral, Wendland [8] succeeded in compelling the inequality $|T| < 1$ to hold by replacing the usual norm in C with an equivalent weighted norm. The polyhedral surfaces considered in [8] are constituted by a finite number of rectangles parallel to the coordinate planes.

In papers [3], [5] by Grachev and Maz'ya the solvability in the space C for the integral equation mentioned above on surfaces with a finite number of conical points was proved without any complementary geometrical assupmtions. Hence, it was shown that the use of the essential norm had been unnecessary and dictated only by the method of proof. We (and independently Rathsfeld [12]) extended this result

Potential Theory, ed. by M. Kishi

to arbitrary polyhedra. In [12] a direct approach, based upon the Mellin transform, was used.

Let us suppose that S is a polyhedron in the three-dimensional Euclidean space. We denote by G^+ the interior of the polyhedron and by G^- its exterior. We will consider two problems for the Laplace's operator. One of them is the Dirichlet problem in G^+ and the other is the Neumann problem in G^-:

$$\Delta u = 0 \text{ on } G^+; \quad u = f \text{ on } S; \tag{1}$$

$$\Delta v = 0 \text{ on } G^-; \quad \partial v/\partial n = g \text{ on } S\backslash M. \tag{2}$$

Here M is the singularity set of the polyhedron, i.e. the union of the edges and the vertices, and $\partial/\partial n$ denotes the derivative in the direction of the outer normal to $S\backslash M$.

Let $O_1, O_2, \dots, O_m$ be the vertices of the polyhedron, let $M_1, M_2, \dots, M_k$ be the edges and $r_i(x) = \text{dist}(x, M_i)$, $\rho_i(x) = \text{dist}\,(x, O_i)$, $r(x) = \min\limits_{1\le i\le k}\{r_i(x)\}$, $\rho(x) = \min\limits_{1\le i\le m}\{\rho_i(x)\}$. We denote by ω_i, $i = 1, 2, \dots, k$, the spread of the dihedral angle with the edge M_i coinciding with G^+ near M_i and let $\lambda_i^+ = \pi/\omega_i$, $\lambda_i^- = \pi/(2\pi-\omega_i)$, $\lambda_i = \min\{\lambda_i^+, \lambda_i^-\}$.

Let K_i, $i = 1, 2, \dots, m$, be the cone with vertex O_i which coincides with G^+ near the point O_i. The open set that the cone K_i cuts from the unit sphere S^2 centred at O_i is denoted by Ω^+ and the set $S^2\backslash\bar{\Omega}^+$ by Ω^-. Let δ_i and ν_i be such positive numbers that $\delta_i(\delta_i+1)$ and $\nu_i(\nu_i+1)$ are the first eigenvalues of the Dirichlet problem on Ω^+ and of the Neumann problem on Ω^- for the Beltrami operator. Then we will denote by κ_i the minimum of $\delta_i, \nu_i, 1$.

We denote by $W\psi$ the classical double layer potential with the density ψ

$$(W\psi)(x) = -\frac{1}{4\pi}\int_S \frac{\partial}{\partial n_\xi}\left(\frac{1}{|x-\xi|}\right)\psi(\xi)ds_\xi, \quad x \in G^{\pm}.$$

We will seek the solution of the equation (1) in the form of a double layer potential. It is well known that the density satisfies the integral equation

$$(1+T)\psi = 2f. \tag{3}$$

Here T is the operator on S defined by the equation

$$(T\psi)(x) = 2W_0\psi(x) + (1-d(x))\psi(x),$$

where $d(x) = 1$ for $x \in S\backslash M$, $d(x) = \omega_i/\pi$ for $x \in M_i$, $d(x) = \text{meas}\ \Omega_i^+/2\pi$ for $x \in O_i$ and $W_0\psi$ is the direct value on S of the double layer potential.

Now we will formulate the main result for the integral equation associated with the Dirichlet problem.

Theorem. *Let* $\kappa = \min_i\{\kappa_i\}$, $\lambda = \min_i\{\lambda_i\}$. *If*

$$p > 2/(1+\kappa), \quad p > 1/\lambda \tag{4}$$

then the integral operator

$$1 + T: \; L_p(S) \to L_p(S)$$

as well as the operator

$$1 + T: \; C(S) \to C(S)$$

both realize the isomorphisms. The inverse operator admits the following representation

$$(1+T)^{-1} f = (1 + L + M) f, \tag{5}$$

where L *and* M *are integral operators on* S *whose kernels* $\mathcal{L}(x,y)$ *and* $\mathcal{M}(x,y)$ *admit the following estimates:*

If M_j *is the nearest edge from the point* y *and* O_i *is the nearest vertex from* y, *then*

$$|\mathcal{M}(x,y)| \leq c\rho(y)^{\kappa_i - 1 - \varepsilon} \left(\frac{r(y)}{\rho(y)}\right)^{\lambda_j - 1 - \varepsilon}. \tag{6}$$

The kernel $L(x,y)$ *is different from zero if the point* x *lies near the point* y. *Let us suppose that the points* x *and* y *lie in a neighbourhood of a vertex* O_i, $i = 1, 2, \ldots, m$ *and this neighbourhood contains no vertices of the polyhedron except* O_i. *If* M_j *and* M_l *are the nearest edges from the points* y *and* x, *then*

$$\begin{aligned} |\mathcal{L}(x,y)| &\leq c\rho(y)^{-2}(r(y)/\rho(y))^{\lambda_j - 1 - \varepsilon} \\ &+ c(r(y) + |x-y|)^{-2} \left(\frac{r(x)}{r(x) + |x-y|}\right)^{\lambda_l - \varepsilon} \left(\frac{r(y)}{r(y) + |x-y|}\right)^{\lambda_j - 1 - \varepsilon} \end{aligned} \tag{7}$$

for $\rho(x)/2 < \rho(y) < 2\rho(x)$, *and*

$$|\mathcal{L}(x,y)| \leq c\rho(y)^{-1}(\rho(x)+\rho(y))^{-1} \left(\frac{\min\{\rho(x),\rho(y)\}}{\rho(x)+\rho(y)}\right)^{\kappa_i - \varepsilon} \left(\frac{r(y)}{\rho(y)}\right)^{\lambda_j - 1 - \varepsilon} \tag{8}$$

in the contrary case. Here ε *is any small positive number.*

Remark. A similar theorem is valid for the integral equation

$$(1 + T^*)\psi = -2g, \tag{9}$$

associated with the Neumann problem, where T^* is the operator formally adjoint to T. In that case it is sufficient to replace the inequalities (4) by estimates $1 \leq p < 2/(1-\kappa)$, $p < 1/(1-\lambda)$ and to put x in place of y and y in place of x in the estimates (6)–(8). However, the invertibility in the space $C(S)$ is not valid.

Now we will brief on our method. At first we consider the interior Dirichlet problem and the exterior Neumann problem in some weighted Hölder spaces, where weight is $\rho^\beta r^\gamma$, β and γ are real numbers. It is known (see [11]), that there exists a unique solution satisfying (1) and the representation

$$u(x) = \int_S \mathcal{P}^+(x,\xi) f(\xi) ds_\xi \tag{10}$$

holds, where the derivatives of the kernel $\mathcal{P}^+(x,\xi)$ admit the following estimates:

We suppose that the points x, ξ lie in a neighbourhood of a vertex O_i, $i = 1, 2, \ldots, m$, and we denote by M_j and M_l the nearest edges from point x and ξ, respectively. If $2\rho(\xi) < \rho(x)$ or $\rho(\xi) > 2\rho(x)$, then

$$|\partial_x^\sigma \partial_\xi^\tau P^+(x,\xi)| \leq c_{\sigma\tau} \rho(x)^{-|\sigma|} \rho(\xi)^{-1-|\tau|} (\rho(x)+\rho(\xi))^{-1} \times \left(\frac{\min\{\rho(x),\rho(\xi)\}}{\rho(x)+\rho(\xi)}\right)^{\delta_i-\varepsilon} \left(\frac{r(x)}{\rho(x)}\right)^{\lambda_j-|\sigma|-\varepsilon} \left(\frac{r(\xi)}{\rho(\xi)}\right)^{\lambda_l-|\tau|-1-\varepsilon}.$$

In the intermediate zone $\rho(\xi) < 2\rho(x) < 4\rho(\xi)$ the estimates of the derivatives of the function $\mathcal{P}^+(x,\xi)$ have the form

$$|\partial_x^\sigma \partial_\xi^\tau \mathcal{P}^+(x,\xi)| \leq c_{\sigma\tau} |x-\xi|^{-2-|\sigma|-|\tau|} \times \left(\frac{r(x)}{r(x)+|x-\xi|}\right)^{\lambda_j-|\sigma|-\varepsilon} \left(\frac{r(\xi)}{r(\xi)+|x-\xi|}\right)^{\lambda_l-1-|\tau|-\varepsilon}.$$

In the case with $x \in U_i$, $\xi \in U_q$, where U_i, U_q are small neighbourhoods of vertices O_i and O_q, respectively, $i \neq q$, the estimates have the form

$$|\partial_x^\sigma \partial_\xi^\tau \mathcal{P}^+(x,\xi)| \leq c_{\sigma\tau} \rho(x)^{\delta_i-|\sigma|-\varepsilon} \rho(\xi)^{\delta_q-|\tau|-\varepsilon} \times \left(\frac{r(x)}{\rho(x)}\right)^{\lambda_j-|\sigma|-\varepsilon} \left(\frac{r(\xi)}{\rho(\xi)}\right)^{\lambda_l-|\tau|-1-\varepsilon}.$$

Here σ and τ are arbitrary multy-indices, ε is any sufficiently small positive number.

A similar representation

$$v(x) = \int_S \mathcal{Q}^-(x,\xi) g(\xi)\, ds_\xi \tag{11}$$

holds for the solution of the problem (2) and the kernel $\mathcal{Q}^+(x,\xi)$ admit the following relations [6]:

We suppose that the points x, ξ lie in a neighbourhood of a vertex $O_i, i = 1,2,\dots,m$, and we denote by M_j and M_l the nearest edges from point x and ξ, respectively. If $2\rho(\xi) < \rho(\xi)$ or $\rho(\xi) > 2\rho(\xi)$, then

$$\mathcal{Q}^-(x,\xi) = \mathcal{Q}^-(0,\xi) + \mathcal{R}^-(x,\xi) \text{ for } 2\rho(x) < \rho(\xi),$$
$$\mathcal{Q}^-(x,\xi) = \mathcal{Q}^-(x,0) + \mathcal{R}^-(\xi,x) \text{ for } 2\rho(\xi) < \rho(x),$$

where

$$\mathcal{Q}^-(0,\xi) = \mathcal{Q}^-(\xi,0) = a_i^-/\rho(\xi) + b_i^- + d_i^-(\xi),$$
$$a_i^- = 1/\text{meas}\,(\Omega_i^-).$$

For $\mathcal{R}(x,\xi)$ and $d_i^-(\xi)$ one has the estimates

$$|\partial_\xi^\sigma d_i^-(\xi)| \le c_\sigma \rho(x)^{\nu_i-|\sigma|-\varepsilon}\left(\frac{r(\xi)}{\rho(\xi)}\right)^{\lambda^l_{\sigma\varepsilon}}$$
$$|\partial_x^\sigma \partial_\xi^\tau \mathcal{R}^-(x,\xi)| \le c_{\sigma\tau}\rho(x)^{\nu_i-|\sigma|-\varepsilon}\rho(\xi)^{-1-\nu_i-|\tau|+\varepsilon} \times \left(\frac{r(x)}{\rho(x)}\right)^{\lambda^j_{\sigma\varepsilon}}\left(\frac{r(\xi)}{\rho(\xi)}\right)^{\lambda^l_{\tau\varepsilon}}.$$

In the intermediate zone $\rho(x) < 2\rho(\xi) < 4\rho(x)$, the estimate of the derivatives of the function $\mathcal{Q}^+(x,\xi)$ takes the form

$$|\partial_x^\sigma \partial_\xi^\tau \mathcal{Q}^-(x,\xi) \le \frac{c_{\sigma\tau}}{|x-\xi|^{1+|\sigma|+|\tau|}}\left(\frac{r(x)}{r(x)+|x-\xi|}\right)^{\lambda^j_{\sigma\varepsilon}}\left(\frac{r(\xi)}{r(\xi)+|x-\xi|}\right)^{\lambda^l_{\tau\varepsilon}}.$$

In the case with $x \in U_i$, $\xi \in U_q$, where U_i, U_q are small neighbourhoods of vertices O_i and O_q, respectively, $i \ne q$, the estimates have the form

$$|\partial_x^\sigma \partial_\varepsilon^\tau \mathcal{Q}^-(x,\xi)| \le c_{\sigma\tau}\rho(x)^{\nu^i_{\sigma\varepsilon}}\rho(\xi)^{\nu^q_{\tau\varepsilon}}\left(\frac{r(x)}{\rho(x)}\right)^{\lambda^j_{\sigma\varepsilon}}\left(\frac{r(\xi)}{\rho(\xi)}\right)^{\lambda^l_{\tau\varepsilon}}.$$

Here

$$\lambda^j_{\sigma\varepsilon} = \min\{0, \lambda_j - |\sigma| - \varepsilon\}, \qquad \lambda^l_{\tau\varepsilon} = \min\{0, \lambda_l - |\tau| - \varepsilon\},$$
$$\nu^i_{\sigma\varepsilon} = \min\{0, \nu_i - |\sigma| - \varepsilon\}, \qquad \nu^q_{\tau\varepsilon} = \min\{0, \nu_q - |\tau| - \varepsilon\}.$$

One can show that the representation

$$(1+T)^{-1} = (1 - Q^- \frac{\partial}{\partial n} P^+)/2 \tag{12}$$

for the inverse operator of the integral equation associated with the Dirichlet problem holds in the space of traces on S of functions from the weighted Hölder space mentioned above. Here P^+ and Q^- are the integral operators defined by equalities (10) and (11), respectively.

The estimates of derivatives of the kernels $\mathcal{P}^+(x,\xi)$, $Q^-(x,\xi)$ and the equality (12) allow to establish the representation (5) in the Hölder spaces and to obtain the estimates (6)–(8) of the kernels $\mathcal{L}(x,y)$, $\mathcal{M}(x,y)$. With help of the estimates one can show that the operator $(1+T)^{-1}$ is continuous in the continuous function space as well as in an appropriate L_p-space and can extend the representation (5) on these spaces.

Remark. The representation (12) for the inverse operator of the equation (9) has the form

$$(1+T^*)^{-1} = (1 - \frac{\partial}{\partial n}P^+Q^-)/2.$$

References

[1] T.S.Angell, R.E.Kleinman, J.Kral, Double layer potentials on boundaries with corners and edges, Comment. Math. Univ. Carol. 27 (1986), p.419.

[2] Yu.D.Burago, V.G.Maz'ya, Multidimensional potential theory and the solution of boundary value problems in domains with non-regular boundaries, Zap. Nauchn. Semin. Leningr. Otd. Mat. Inst. Steclova 3 (1967), 1–152. Engl. translation: Sem. Math. V.A.Steklov Math. Inst. Leningr. 3, 1–68 (1969).

[3] N.V.Grachev, V.G.Maz'ya, Representations and estimates for inverse operators of the potential theory integral equations on surfaces with conic points Soobshch. Akad. Nauk Grusin. SSR 1321 (1988), 21–23. [Russian]

[4] N.V.Grachev, V.G.Maz'ya, Estimates for kernels of the inverse operators of the integral equations of elasticity on surfaces with conic points, Preprint N26, Akad. Nauk SSSR, Inst of Engin. Studies, Leningrad, 1989. [Russian]

[5] N.V.Grachev, V.G.Maz'ya, On invertibility of the boundary integral operators of elasticity on surfaces with conic points in the spaces generated by norms C, C^α, L_p, Preprint N30, Akad. Nauk SSSR, Inst. of Engin. Studies, Leningrad, 1990. [Russian]

[6] N.V.Grachev, V.G.Maz'ya, Estimates of the fundamental solution of the Neumann problem in domains with piece-wise smooth boundaries, (to appear).

[7] J.Kral, Integral Operators in Potential Theory, Lect. Notes Math., 823, Springer-Verlag, Berlin, Heidelberg, New York, 1980.

[8] J.Kral, W.Wendland, Some examples concerning applicability of the Fredholm-Radon method in potential theory, Apl. Mat. 31 (1986), 293–318.

[9] V.G.Maz'ya, Boundary integral equations of elasticity in domains with piecewise smooth boundaries, Equadiff 6, Proc. Int. Conf., Brno, 1985 (Lect. Notes Math. 1192, 235–242).

[10] V.G.Maz'ya, Boundary integral equations, (Encyclopaedia of Mathematical Sciences, 27 (1991), 127–223).

[11] V.G.Maz'ya, B.A.Plamenevsky, The first boundary-value problem for the classical equations of the mathematical physics in domain with piece-wise smooth boundary, 1, Z. Anal. Anwend. 24 (1983), 335–339; 2, 26 (1983), 523–551. [Russian]

[12] A.Rathsfeld, The invertibility of the double layer potential operator n the space of continuous functions defined on a polyhedron,in "Preprint – math – 01/90", AdW d. DDR, Karl-Weierstraß-Inst. f. Math., Berlin, 1990.

Use of single layer potential in mixed problems of elasticity

Ashot Hakobyan

Abstract

The plane problem of linear elasticity with mixed boundary conditions is considered on domains with piece-wise smooth boundary, together with corresponding integral equations given rise from single layer elastic potential. We obtain propositions about solvability of the boundary problem, integral equations and asymptotic forms of their solutions in neighborhoods of irregular points of boundary cites.

1980 Mathematics Subject Classification (1985 Revision): 53C30, 22E15.

1. Integral equations of mixed problems

The mixed plane problem of linear elasticity consists of finding displacements u which satisfies Lame's equation

$$\mathsf{A}(\partial_x)u(x) = \mu\Delta u(x) + (\lambda+\mu)\nabla \mathrm{div} u(x) = 0 \tag{1.1}$$

in a domain Ω and conditions on four parts of the boundary $\partial\Omega = \cup_{j=1}^{4}\partial\Omega_j$

$$\begin{aligned} u(y) &= f_1(y), & y \in \partial\Omega_1, \\ \mathsf{T}(\partial_y, n(y))u(y) &= f_2(y), & y \in \partial\Omega_2, \\ \mathsf{H}(\partial_y, n(y))u(y) &= f_3(y), & y \in \partial\Omega_3, \\ \mathsf{R}(\partial_y, n(y))u(y) &= f_4(y), & y \in \partial\Omega_4. \end{aligned} \tag{1.2}$$

Here T is a stress operator given by

$$\mathsf{T}(\partial_y, n(y)) = \lambda n(x)\mathrm{div} + 2\mu\frac{\partial}{\partial n(x)} + \mu[n(x)\times \mathrm{rot}], \tag{1.3}$$

and H and R are special operators defined by

$$\begin{aligned} \mathsf{H}(\partial_y, n(y)) &= \{s(y)\mathsf{T}(\partial_y, n(y)), n(y)\}, \\ \mathsf{R}(\partial_y, n(y)) &= \{s(y), n(y)\mathsf{T}(\partial_y, n(y))\}, \end{aligned}$$

which are represented correspondingly by the tangential stress and normal displacement, and by the tangential displacement and the normal stress. Let us signify this problem $V^{\pm}$ for a bounded or an unbounded domain Ω.

Potential Theory, ed. by M. Kishi

Let us represent a solution u by a modified single layer elastic potential

$$\mathsf{V}(\varphi)(x) = \int_{\partial\Omega} \Gamma(x,y)\varphi(y)ds_y + \overset{\circ}{\Gamma}(x,x_0)\int_{\partial\Omega}\varphi(y)ds_y + A, \tag{1.4}$$

where $\Gamma(x,y) = \| \Gamma_{i,j}(x,y) \|_{2\times 2}$ is the fundamental solution of the equation (1.1) with elements

$$\Gamma_{ij}(x,y) = a\delta_{ij}\ln\frac{1}{r} + b\frac{(y_i - x_i)(y_j - x_j)}{r^2},$$

$r = |x-y|, a = (\lambda+3\mu)/(2\pi\mu(\lambda+2\mu)), b = (\lambda+\mu)/(2\pi\mu(\lambda+2\mu)), \overset{\circ}{\Gamma}(x,x_0)$ is the matrix $\|\overset{\circ}{\Gamma}_{ij}(x,x_0)\|_{2\times 2}$ with elements

$$\overset{\circ}{\Gamma}_{ij}(x,x_0) = a\delta_{ij}\ln\frac{1}{|x-x_0|},$$

x_0 is a point and A is a constant vector. This representation leads to the singular integral equation of mixed type

$$\begin{array}{rcll} \mathsf{V}(\varphi) & = & f_1(x) - A, & x \in \partial\Omega_1, \\ \pm\varphi(x) + \mathsf{T}(\partial_x, n(x))\mathsf{V}(\varphi)(x) & = & f_2(x), & x \in \partial\Omega_2, \\ \pm(s(x)\varphi(x), O) + \mathsf{H}(\partial_x, n(x))\mathsf{V}(\varphi)(x) & = & f_3(x) - (s(x)A, O), & x \in \partial\Omega_3, \\ \pm(O, n(x)\varphi(x)) + \mathsf{R}(\partial_x, n(x))\mathsf{V}(\varphi)(x) & = & f_4(x) - (O, n(x)A), & x \in \partial\Omega_4. \end{array} \tag{1.5}$$

Let us consider together with the source problem (1.1) – (1.2) some auxiliary problems in the supplement of Ω

$$\mathsf{A}(\partial_x)v(x) = 0,\ x \in R^2 \setminus (\bar{\Omega} \cup \{x_0\}),\ v(y) = u(y),\ y \in \partial\Omega. \tag{1.6}$$

Then from properties of single layer potentials it follows that

$$\varphi(x) = \pm\frac{1}{2}([\mathsf{T}(\partial_x, n(x))u(x)]^{\pm} - [\mathsf{T}(\partial_x, n(x))v(x)]^{\pm}). \tag{1.7}$$

This formula allows us to carry out properties of a solution of the integral equations (1.5) from the boundary problem (1.1) – (1.2) and (1.6). First let us formulate the boundary problem $V^{\pm}$.

Proposition 1.1. a) *There is only one solution of the mixed problem* $V^{\pm}$ *if* $\partial\Omega_1 \neq \emptyset$.
b) *If* $\partial\Omega = \partial\Omega_2$, *then any two solutions of the problem* $V^{\pm}$ *differ by an additive vector*

$$\overset{\circ}{u}(x) = [a \times (x - x_0)] + b, \tag{1.8}$$

where $a = (0,0,a_3)$ *and* $b = (b_1, b_2, 0)$ *are arbitrary constants and* x_0 *is an arbitrary point. Any two solutions of problem* V^- *differ by arbitrary constant vector* b.
c) *If* $\partial\Omega_3$ *is a non empty subset of straight segments parallel to a vector* b, *or if* $\partial\Omega_4$ *is a non empty subset of straight segments perpendicular to vector* b, *then any two solutions of problem* $V^{\pm}$ *differ by an additive vector* b.

d) *If $\partial\Omega_3$ is a non empty subset of concentric arcs with center x_0 or if $\partial\Omega_4$ is a non empty set of straight segments converging to point x_0, then any two solutions of problem V^+ differ by an additive vector $[a \times (x - x_0)]$, while the problem V^- allows only one solution.*
e) *If $\partial\Omega_3$ or $\partial\Omega_4$ is not empty and not the sets given in c) and d), then problem $V^\pm$ allows at most one solution.*

Proof. From a known Betty's formula

$$\int_\Omega [u(x)\mathsf{A}(\partial_x)v(x) + E(u,v)(x)]dx = \int_{\partial\Omega} u(y)\mathsf{T}(\partial_y, n(y))v(y)ds_y \tag{1.9}$$

where

$$E(u,v) = \lambda \operatorname{div} u \operatorname{div} v + \mu \sum_{i,j}^{2} \frac{\partial v_i}{\partial x_j}\left(\frac{\partial u_i}{\partial x_j} + \frac{\partial u_j}{\partial x_i}\right).$$

Put $v(x) \equiv u(x)$. From that u is a solution of the equation (1.1) follows

$$\int_\Omega e(u,u)(x)dx = \int_{\partial\Omega} u(y)\mathsf{T}(\partial_y, n(y))u(y)ds_y. \tag{1.10}$$

Taking to account that

$$u\mathsf{T}u = (s\mathsf{T}u)(su) + (n\mathsf{T}u)(nu) = \mathbf{H}u\mathbf{R}u,$$

one can rewrite the integral on the right hand side of (1.10) in the form

$$\int_{\partial\Omega_1\cup\partial\Omega_2} u(y)\mathsf{T}(\partial_y, n(y)ds_y + \int_{\partial\Omega_3\cup\partial\Omega_4} \mathbf{H}(\partial_y, n(y)u(y)\mathbf{R}(\partial_y, n(y))u(y)ds_y. \tag{1.11}$$

If u is a solution of the homogeneous mixed problem (the problem (1.1) – (1.2) with $f_1 \equiv 0$), then (1.11) gives that integral on the right hand side of (1.10) is equal to zero. Furthermore, it coincides almost everywhere with a solution of the equation

$$E(u,u)(x) = 0. \tag{1.12}$$

The same is true for unbounded domain Ω if and only if

$$u(y) = O(1), \quad \mathsf{T}(\partial_y, n(y))u(y) = o(1/|y|), \quad y \to \infty. \tag{1.13}$$

With a substitution one can be convinced that the function (1.8) (that is, the rigid displacements and rotations of an elastic domain) is a solution of the equation (1.12). It is clear that if Ω is unbounded and u satisfies (1.13), a in (1.8) must be zero. A substitution of (1.8) in the boundary conditions (1.2) gives us nontrivial cases of a solution of the homogeneous problem. And finally from the linearity of operators of problem (1.1) – (1.2) follows Proposition 1.1. □

The existence of solutions of the problem $V^{\pm}$ can be proved in weighted Sobolev spaces (cf. [Ma]). Now, notified that the problem (1.6) is simply the problem $V^{\mp}$ with $\partial\Omega = \partial\Omega_1$, by Proposition 1.1 and the formula (1.7) we obtain

Proposition 1.2. *A solution of the integral equation* (1.6) *exists if and only if there exists a solution of the corresponding boundary value problem* $V^{\pm}$*, and this solution is unique if the solution of the problem* $V^{\pm}$ *is unique.*

2. Special solutions

In a well-known paper [Ko] it was proved that the asymptotic form of an elliptic operator in neighborhoods of corner points of a plane domain represents as a sum of a smooth function and series of eigen solutions of the homogeneous problem on a wedge, "freezed" on the top of the wedge. Applied to the mixed boundary value problem of elasticity, we obtain that in neighborhoods of corner points, the solution satisfying the boundary conditions (1.2) changing points of the boundary can be represented in the following asymptotic form

$$u(x) = \sum_{0<\mathrm{Re}\lambda_j<1} \sum_{s=0}^{q_j-1} C_{js}\rho^{\lambda_j}(\ln\rho)^s\psi_{sj}(\theta) + w(x). \tag{2.1}$$

Here the members of series are eigen solutions on the infinite wedge. These solutions in polar coordinates ρ, θ are

$$\begin{aligned} & e_\rho C\rho^k\{(k-\kappa)(C_1\cos(k-1)\theta + C_2\sin(k-1)\theta) \\ & \quad + C_3\cos(k+1)\theta + C_4\sin(k+1)\theta\} \\ + \; & e_\theta C\rho^k\{-(k-\kappa)(C_1\sin(k-1)\theta + C_2\cos(k-1)\theta) \\ & \quad - C_3\sin(k+1)\theta + C_4\cos(k+1)\theta)\}, \end{aligned} \tag{2.2}$$

where $\kappa = (\lambda + 3\mu)/(\lambda + \mu)$ and coefficients C_i are determined by the systems

$$\begin{cases} \mathbf{B}^+ u|_{\theta=\alpha} & = 0 \\ \mathbf{B}^- u|_{\theta=-\alpha} & = 0 \end{cases} \tag{2.3}$$

and $\mathbf{B}^{\pm}$ is a boundary operator on the wedge's sides $\theta = \pm\alpha$. The system (2.3) have nontrivial solutions if the main determinant is equal to zero. For example, if $\mathbf{B}^+ = \mathbf{E}$ is a unitary operator, and $\mathbf{B}^- = \mathbf{T}$ is a stress operator, then the characteristic equation is

$$\sin^2 2k\alpha - (1+\kappa)^2/4\kappa - k^2/\kappa \sin^2 2\alpha = 0. \tag{2.4}$$

Now the members of series in (2.1) are represented in the form (2.2), where an exponent k is determined by (2.4) and the coefficients C_i are determined by the system (2.3).

Finally, we can formulate a proposition about asymptotic form of a solution of the integral equation (1.5).

Proposition 2.1. *If $\varphi(x)$ is a solution of the integral equation* (1.5) *of the mixed problem satisfying the boundary conditions* (1.2) *changing points of boundary,*

$$\begin{aligned}\varphi(x) &= \sum_{0<\mathrm{Re}\lambda_j<1} \sum_{js}^{q_j-1} C_{js}\rho^{\lambda_j-1}(\ln\rho)^s\psi_{sj}(\theta) \\ &+ \sum_{0<\mathrm{Re}\lambda_{1j}<1} \sum_{js}^{q_{1j}-1} C_{1ij}\rho^{\lambda_{1j}-1}(\ln\rho)^s\psi_{1sj}(\theta) + w_1(x),\end{aligned} \tag{2.5}$$

where the members of the first series are solutions of the source problem $V^{\pm}$ on the wedge, while in the second there are solutions of problem (2.1).

Proof. The truth of this proposition follows from the asymptotic form (2.1) and the formula (1.7). □

References

[Ma] Maz'ja, V.G., On potential theory for Lame's system in domains with piece-wise smooth boundary. Proceedings of symposium in Tbilisi, 21-23 april 1982, Mitsnereba, Tbilisi, 1982, 123-129.

[Ko] Kondrat'ev, V.A., Boundary problems for elliptic equations on domains with cone or angle points, Proceedings of Moscow Math. ass. vol.16 (1967) 209-292.

Discretization of bounded harmonic functions on Riemannian manifolds and entropy

Vadim A. Kaimanovich *

Abstract
We give conditions under which the space of bounded harmonic functions on a Riemannian manifold M is naturally isomorphic to the space of bounded harmonic functions of a Markov chain on a discrete net $X \subset M$ arising from a discretization procedure for the pair (M, X). If, further, M is a regular covering manifold and the net is invariant with respect to the deck transformation group, then the entropy of the arising random walk on X equals the entropy of the Brownian motion on M times the average stopping time of the discretization procedure.

1980 Mathematics Subject Classification (1985 Revision): 31C12, 58G32, 60J50.

0. Introduction

During the last few years a lot of papers devoted to the discrete potential theory have appeared. It turns out that this theory is to a large extent parallel to the potential theory on Riemannian manifolds (see, e.g., a survey [1]). Thus one can naturally ask about any direct relationships between the potential theory on Riemannian manifolds and on graphs adapted in some way to the Riemannian structure. In probabilistic terms this is the question about a correspondence between the Brownian motion on a Riemannian manifold and appropriately taken Markov chain on a net in the manifold.

The situation with the connection between the *transience* of the Brownian motion on a Riemannian manifold (i.e. existence of the Green function on this manifold) and the transience of a reversible Markov chain on a net in this manifold (*hyperbolicity,* in terminology of [13]) turned out to be quite simple. Namely, under natural cocompactness conditions the net is transient if and only if the manifold is transient itself [12]. The same is true about the isoperimetric properties of the manifold and its cocompact net [7]. In particular, the spectral radius of the random walk on a cocompact net is strictly less than 1 iff zero doesn't belong to the spectrum of the Laplacian on the manifold.

Another important property of Riemannian manifolds and Markov chains is the *Liouville property* – non-existence of non-constant bounded harmonic functions, or, in

* The author gratefully acknowledges a travel grant of the International Mathematical Union and financial support of the Organizing Committee ICPT-90. This paper was written during the author's stay at the Universities of Sydney and of New South Wales.

Potential Theory, ed. by M. Kishi

probabilistic terms, triviality of the stationary σ-algebra of the corresponding Markov process [11] (i.e., of the Brownian motion for the case of manifolds). More generally, one can ask about a correspondence between the bounded harmonic functions on the manifold and on a net, i.e. about a correspondence between the *Poisson boundaries* [3,6] of the Brownian motion and an approximating Markov chain.

The situation here is more complicated, because the Liouville property (for manifolds as well as for reversible Markov chains) isn't stable with respect to the quasi-isometric changes [8], hence the answer, in principle, depends on the chosen approximation. Even more, when Riemannian manifold is a regular covering space of another manifold, and we look for a discrete approximation invariant with respect to the deck transformation group, it turns out that it is sometimes impossible (and, in fact, it seems to be a general rule) to find an approximation with the same space of harmonic functions (i.e., the same Poisson boundary) within the class of finite range walks [2].

Lyons and Sullivan [9] basing upon an earlier work by Furstenberg [3] proposed a discretization procedure which permits one to obtain a Markov (infinite range) chain on a net $X \subset M$ starting from the Brownian motion on a Riemannian manifold M. The net here is subject to fairly weak recurrence properties. The resulting chain has properties close to those of the initial Brownian motion. Particularly, in the cocompact case its sample paths closely follow the Brownian trajectories on M and transience of the Brownian motion on M is equivalent to transience of the random walk on X. They also showed that the restriction on X of every bounded harmonic function on M is harmonic with respect to the chain on X.

In this paper we prove that in fact for the Furstenberg–Lyons–Sullivan construction the spaces of the bounded harmonic functions (in other words, the Poisson boundaries) on M and on X are naturally isomorphic. In the group invariant case, when the manifold M is a regular covering space of another manifold and the net X is invariant with respect to the deck transformation group G, we show that this construction gives a good approximation of the Brownian motion on the initial manifold in yet another sense: the *entropy* of the arising random walk on X equals the entropy of the Brownian motion on M times the average stopping time of the discretization process. These results were first announced in the author's paper [4].

1. Discretization procedure

We begin with a description of the discretization procedure in a slightly more general situation than in the paper [9] by Lyons and Sullivan. In order to do it, we introduce first the notions of balayage and Harnack constant. We prefer to do this in probabilistic terms (rather than in purely potential theory terms), because it is the probabilistic meaning of these notions that we shall use below.

Let M be a complete connected Riemannian manifold. We shall always assume that

M has *bounded geometry,* i.e. the sectional curvatures of M are uniformly bounded and the injectivity radii are bounded away from zero. The *Brownian motion* on M is a diffusion process whose generating operator is the Laplace–Beltrami operator Δ of the Riemannian metric on M (see, e.g., [10]). Let P_x (resp., P_λ) denote the probability measure in the space of (continuous) Brownian sample paths $\xi = \{\xi(t)\}_{t\geq 0}$ corresponding to the initial distribution δ_x (resp., λ), and let π_x^t (resp., π_λ^t) denote the one-dimensional distribution of the measure P_x (resp., P_λ) at the time t. Let $p^t(x,y)$ denote the density of π_x^t with respect to the Riemannian volume m at the point y (in other words, $p^t(x,y)$ is the fundamental solution of the heat equation on M). A smooth function f on M is called *harmonic* if $\Delta f \equiv 0$, or, equivalently,

$$f(x) = \langle f, \pi_x^t \rangle = \int f(y) p^t(x,y) dm(y)$$

for all $x \in M$ and $t > 0$. Below we shall use the following fact proved in [4] (cf. also [1]).

Proposition 1. *If M has bounded geometry, then*

$$\lim_{t\to\infty} \|\pi_x^{t+\tau} - \pi_x^t\| = 0$$

for all $x \in M$ and $\tau > 0$, where $\|\mu\|$ is the total variance of a measure μ.

Let now V be an open subset of M, and $x \in M$. Denote by $\varepsilon(x,V)$ the measure obtained by allowing the Brownian motion to start at x and run until it leaves V and then setting $\varepsilon(x,V)$ equal to its exit distribution. If $x \in M \setminus V$, then $\varepsilon(x,V) = \delta_x$. If $x \in V$, this measure is concentrated on the boundary ∂V and is called the *harmonic measure* on ∂V with the pole at the point x. For every sample path ξ define the (Markov stopping) time

$$T^{M\setminus V}(\xi) = min\{t \geq 0 : \xi(t) \in M \setminus V\}.$$

Then the measure $\varepsilon(x,V)$ is just the image of the measure P_x by the map $\xi \mapsto \xi(T^{M\setminus V}(\xi))$ (provided that almost every sample path eventually leaves the set V). The measure $\varepsilon(x,V)$ is always a probability measure if V is relatively compact in M. In general, to say that a closed set K is recurrent for the Brownian motion on M is precisely the same as saying that $\varepsilon(x, M\setminus K)$ is a probability measure for each $x \in M$.

If a function f is bounded and harmonic on V and $\|\varepsilon(x,V)\| = 1$, then f has fine boundary values at $\varepsilon(x,V)$ almost every Martin boundary point and $f(x) = \langle f, \varepsilon(x,V)\rangle$. Conversely, if f is a bounded Borel function on ∂V, its extension on V given by $x \mapsto \langle f, \varepsilon(x,V)\rangle$ is harmonic. The *balayage* of a measure μ onto ∂V is defined to be $\mu^V = \int \varepsilon(x,V) d\mu(x)$, so that $\langle f, \mu\rangle = \langle f, \mu^V\rangle$ for every bounded harmonic function on V providing the set $M \setminus V$ is recurrent. Remark that Proposition 1 implies that

$$\lim_{t\to\infty} \|\pi_\mu^t - \pi_{\mu^V}^t\| = 0\,.$$

The *Harnack constant* $C = C(V, E)$ of a pair (V, E), where V is open and $E \subset V$, is defined by

$$C = sup\{\frac{f(x)}{f(y)} : x, y \in E,\ f \text{ is a positive harmonic function on } V\}$$
$$= sup\{\frac{d\varepsilon(x, V)}{d\varepsilon(y, V)}(z) : x, y \in E, z \in \partial V\}$$

Clearly, if E is relatively compact in V, the Harnack constant $C = C(V, E)$ is finite.

Definition [9]. Let X be a discrete subset of a Riemannian manifold M. We shall say that X is *$*$-recurrent* if for each $x \in X$ there exist sets $E_x, V_x \subset M$ such that:

(1) $x \in E_x \subset V_x$ for all $x \in X$;
(2) V_x are relatively compact open;
(3) E_x are Borel pairwise disjoint;
(4) $E = \bigcup_{x \in X} E_x$ is closed and recurrent for the Brownian motion on M;
(5) The supremum of the Harnack constants of the pairs (V_x, E_x) is $C < \infty$.

From now on we fix a $*$-recurrent net $X \subset M$ and sets $V_x \supset E_x \ni x$ satisfying the conditions (1-5).

If $y \in E$, denote by $\chi(y) \in X$ the point $\chi(y) = x$ uniquely determined by the condition $y \in E_x$. For the sake of convenience put also $E_y = E_{\chi(y)}$ and $V_y = V_{\chi(y)}$.

Describe now a discretization procedure for the Brownian motion on M. In order to do it introduce the Markov stopping times $(R_n)_{n \geq 1}$ and $(S_n)_{n \geq 0}$ for the Brownian motion. Let $\xi = \{\xi(t)\}_{t \geq 0}$ be a sample path of the Brownian motion. Put

$$S_0(\xi) = \begin{cases} 0 & \text{, if } \xi(0) \notin X ; \\ min\{t > 0 : \xi(t) \in M \setminus V_{\xi(0)}\} & \text{, otherwise ;} \end{cases}$$

and define by induction

$$R_n(\xi) = min\{t \geq S_{n-1}(\xi) : \xi(t) \in E\} ,$$
$$S_n(\xi) = min\{t > R_n(\xi) : \xi(t) \in M \setminus V_{\xi(R_n)}\} .$$

Put also

$$X_n(\xi) = \chi(\xi(R_n)) \in X$$

Take a sequence of i.i.d. random variables $\alpha = (\alpha_n)_{n \geq 1}$ which are independent on the initial Brownian motion and have the Lebesgue measure on the interval $(0, 1)$ as their common distribution. Define

$$N_0(\xi, \alpha) = 0 ,$$

and for $k \geq 1$ by induction

$$N_k(\xi, \alpha) = min\{n > N_{k-1} : \alpha_n < \frac{1}{C} \frac{d\varepsilon(X_n, V_{X_n})}{d\varepsilon(\xi(R_n), V_{X_n})}(\xi(S_n))\} .$$

Now for every point $y \in M$ let

$$\nu_y(x) = P_y\left[X_{N_1} = x\right]$$

be the distribution on X of the first coordinate X_{N_1} of the random X-valued sequence $(X_{N_k})_{k\geq 1}$. The uniform Harnack condition guarantees that all the times N_k are almost everywhere finite, i.e. all the measures $\nu_y, y \in M$ are probability measures on X.

The measures ν_y can be also described in the following way [9]. Start with the distribution δ_y. If $y \in X$, balayage δ_y onto $M \setminus V_y$ (this corresponds to the stopping time S_0). Now apply to the resulting measure μ_0 the following recurrent procedure (we describe the inductive step for an arbitrary measure μ_n). Balayage μ_n first on E (it corresponds to the stopping time R_n). Let $\mu_n(x), x \in X$ be the restrictions of μ_n onto the sets $E_x, x \in X$ and $\widetilde{\mu}_n(x)$ be the balayage of the measure $\mu_n(x)$ onto the set ∂V_x (it corresponds to the stopping time S_n). Now put

$$\mu_{n+1} = \sum_{x\in X}(\widetilde{\mu}_n(x) - \frac{\|\widetilde{\mu}_n(x)\|}{C}\varepsilon(x, V_x))$$

and

$$\tau_n = \sum_{x\in X} \frac{\|\widetilde{\mu}_n(x)\|}{C}\delta_x .$$

Then the sought for measure ν_y can be written as the sum

$$\nu_y = \sum_{n\geq 1} \tau_n .$$

From Proposition 1 we get

Proposition 2. *The measures $(\nu_y)_{y\in M}$ on X satisfy the following stationarity property:*

$$\lim_{t\to\infty} \|\pi_y^t - \sum_{x\in X} \nu_y(x)\, \pi_x^t\| = 0 .$$

for all points $y \in M$.

Now from the mean value property of harmonic functions we immediately get the following corollary.

Proposition 3. *Every bounded harmonic function on the Riemannian manifold M satisfies the relations*

$$f(y) = \sum_{x\in X} \nu_y(x) f(x) , \qquad y \in M .$$

The sequence $(X_{N_k})_{k\geq 1}$ has also the following remarkable property. It was proved in [9] by Lyons and Sullivan for the *cocompact case* (see below Section 3), but their proof can be *verbatim* reproduced in our slightly more general situation.

Proposition 4. *The sequence* $(X_{N_k})_{k\geq 1}$ *is a time homogeneous Markov chain with the state space* X *and with the transition probabilities* $p(x,z) = \nu_x(z)\,, x, z \in X$.

Propositions 3 and 4 imply the following corollary:

Proposition 5. *The restriction on* X *of every bounded harmonic function on* M *is a harmonic function for the Markov chain on* X *with the transition probabilities* $p(\cdot,\cdot)$ *constructed above.*

2. Bounded harmonic functions

Our aim now is to find out conditions under which the converse statement to Proposition 5 is also true, namely, to find out when the spaces of bounded harmonic functions on M and on X are in fact isomorphic.

Definition 2. We shall say that the system of sets $(E_x, V_x)_{x\in X}$ satisfies the *uniform core* condition if there exist closed sets $U_x \subset V_x, x \in X$ such that

$$U_x \cap V_z = \emptyset \qquad \text{whenever}\, x \neq z \in X\,,$$

and

$$\varepsilon(x, V_x \setminus U_x)\,(\partial U_x) \geq \delta$$

for a certain constant $\delta > 0$. We shall say that the sets U_x are *cores* of the corresponding sets V_x.

In probabilistic terms, the uniform core condition means that starting from any point $y \in E_x$ the Brownian trajectory with probability not less than δ first visits its core U_x before leaving the set V_x, and the cores of different sets V_x are pairwise disjoint. This condition is always satisfied if the sets V_x are pairwise disjoint themselves.

Theorem 1. *Let the system of sets* $(E_x, V_x)_{x\in X}$ *satisfy the conditions of Definitions 1, 2 (*-recurrence and the uniform core condition). Then the spaces of bounded harmonic functions for the Riemannian manifold* M *and for the Markov chain on the net* X *with the transition probabilities* $p(\cdot,\cdot)$ *defined in Proposition 4 are naturally isomorphic. For every bounded harmonic function* f *on* M *its restriction to* X *is harmonic with respect to the transition probabilities* $p(\cdot,\cdot)$ *on* X. *Conversely, for every bounded harmonic function* f *of the chain* $p(\cdot,\cdot)$ *on* X *its extension to* M *given by the formula*

$$f(y) = \langle f, \nu_y\rangle = \sum_{x\in X} f(x)\nu_y(x) \tag{2.1}$$

is a harmonic function on M.

Proof. In view of Proposition 5 what we have to prove is that for every bounded p-harmonic function on the net X its extension to M given by the formula (2.1) is harmonic on M. Recall that the chain on X with the transition probabilities $p(\cdot,\cdot)$ was obtained in Section 1 as a result of a series of subsequent measure preserving mappings of Markov processes starting with the initial Brownian motion on M:

$$(\xi_t) \xrightarrow{1} (\xi(R_n),\xi(S_n)) \xrightarrow{2} (\xi(R_n),\xi(S_n),\alpha_n) \xrightarrow{3} (\xi(R_{N_k}),\xi(S_{N_k}),\alpha_{N_k}) \xrightarrow{4} (X_{N_k}).$$

On the first step we obtain a discrete time process with the state space $M \times M$, then we add the sequence (α_n), pass to the subsequence of times (N_k), and finally take the images $X_{N_k} = \chi(\xi(R_{N_k}))$. The processes obtained on the steps (1) and (2) are evidently Markov, because the stopping times R_n and S_n are Markov and the sequence (α_n) is i.i.d. and independent on the Brownian motion. The process obtained on step (3) can be considered as the process induced by the process $(\xi(R_n),\xi(S_n),\alpha_n)$ on the recurrent set

$$\{\alpha_n < \frac{1}{C}\frac{d\varepsilon(X_n,V_{X_n})}{d\varepsilon(\xi(R_n),V_{X_n})}(\xi(S_n))\}.$$

Finally, the fact that the process (X_{N_k}) is Markov is the statement of Proposition 4.

Now we have to show that every bounded harmonic function of a subsequent process is a bounded harmonic function for the preceding one, or, more precisely, that every stationary event of a subsequent process is a stationary event of a preceding one. Recall that the Poisson boundary of a Markov process $(Z_t)_{t\geq 0}$ can be interpreted as the factor-space of its path space with respect to the *stationary σ-algebra* , which can be obtained as the σ-algebra generated by the *stationary equivalence relation* in the path space: two trajectories (Z_t^1) and (Z_t^2) are equivalent iff there exist $T_1, T_2 > 0$ such that $Z^1_{T_1+t} = Z^2_{T_2+t}$ for all $t \geq 0$ [3,6,11].

It is evident that the mapping (2) doesn't change the space of bounded harmonic functions (α_n are independent, hence the transition probabilities of the obtained process don't depend on α_n). On the step (3) we obtained an induced process on a recurrent set, hence the bounded harmonic functions aren't changed [3,11]. Finally, on the step (4) new Markov process is obtained by factorizing the state space of the preceding one using the map $(\xi(R_{N_k}),\xi(S_{N_k}),\alpha_{N_k}) \mapsto (X_{N_k})$ from $E \times M \times (0,1)$ to X, hence the stationary σ-algebra of the new process isn't larger than the stationary σ-algebra of the old one.

What we have to show now is that the mapping (1) doesn't extend the stationary σ-algebra, i.e. the behaviour at infinity of the process $(\xi(R_n),\xi(S_n))$ isn't "richer" than that of the initial Brownian motion. This is the only place where we use the uniform core condition (Definition 2). The problem here is that in principle stationary equivalence of two Brownian paths (ξ_t^1) and (ξ_t^2) might not imply stationary equivalence of the corresponding sequences $(\xi(R_n^1),\xi(S_n^1))$ and $(\xi(R_n^2),\xi(S_n^2))$. In other words, running the same Brownian path with different "initial conditions" we might get two sequences $(\xi(R_n^i),\xi(S_n^i))\,, i = 1,2$ which aren't stationary equivalent.

Notice that once in a "core", the Brownian particle loses all the information about preceding stopping times R_n and S_n, and the first stopping time S_n after leaving the core is determined uniquely. Hence if a path visits infinitely many cores, running it from any starting time we eventually get the same sequence $(\xi(R_n), \xi(S_n))$. But the uniform core condition guarantees that almost every Brownian path satisfies this condition, which proves the Theorem. □

Remarks 1. In fact we have proved that the *Poisson boundaries* of the Brownian motion on M and of the Markov chain on X with the transition probabilities $p(\cdot,\cdot)$ are naturally isomorphic as measure spaces.
2. The stationarity condition (Proposition 2) in general doesn't imply that every bounded ν-harmonic function is also harmonic with respect to the Laplacian Δ of the Riemannian metric. It may well happen that the manifold M satisfies the Liouville property (hence the stationarity condition is satisfied for *all* families $(\nu_y)_{y\in M}$), but nonetheless there exist chains on X with non-trivial bounded harmonic functions. The simplest example is given by the polycyclic covers of compact manifolds: they always satisfy the Liouville property, whereas the polycyclic groups themselves can have non-trivial Poisson boundary [4]. On the other hand, if M is the universal covering space of a negatively curved manifold and the system $(\nu_y)_{y\in M}$ is invariant with respect to the fundamental group action, then the stationarity condition *does* imply that the spaces of harmonic functions on M and X are naturally isomorphic (cf. [3]).

3. Discretization and entropy

In this Section M is a regular covering space of a Riemannian manifold N with the deck transformation group G (particular case: $M = \widetilde{N}$ is the universal covering space, $G = \pi_1(N)$ is the fundamental group of N). The net X and the system of sets (E_x, V_x) are invariant with respect to the action of the deck transformations group G.

In this case the discretization process described in Section 1 is group invariant, so that we get a Markov chain on X satisfying the condition

$$p(x,z) = p(gx,gz) \qquad \text{for all } x,z \in X\,, g \in G\,.$$

For the sake of simplicity assume that the set X consists of only one G-orbit, i.e. the set X can be identified with the group G by fixing a reference point $x_0 \in X$ and using the map $g \mapsto gx_0$ (this assumption can be easily omitted). In other words, we have a group-invariant Markov chain on G, i.e. a *random walk* on G.

The group invariance permits one to identify all the sets V_x with a single set $V = V_{x_0}$ and to consider the images $\pi(\xi(S_n))$ of random variables $\xi(S_n)$ on ∂V. The uniform Harnack condition (Definition 1) implies that the resulting Markov chain on ∂V has a unique stationary distribution, and the distributions of $\pi(\xi(S_n))$ converge to this limiting distribution exponentially fast in variance. Now the ergodic theorem applied to this Markov chain on ∂V implies that there exist an *average stopping time* T (not necessarily finite, in general) such that $S_{N_k}(\xi)/k \to T$ as $k \to \infty$ almost everywhere and in the space $L^1(P_y)$ for every starting point $y \in M$.

Recall that the *entropy of a random walk* on a discrete group G determined by a probability measure μ is defined as the limit

$$h(G, \mu) = \lim_{n\to\infty} H(\mu_n)/n\,,$$

where

$$H(\mu_n) = -\sum log(\mu_n(g))\mu_n(g)$$

is the entropy of the n-fold convolution of the measure μ, i.e. the one-dimensional distribution of the corresponding random walk at time n. The entropy of random walk has the following remarkable property: it vanishes if and only if the random walk has no non-trivial bounded harmonic functions (i.e., satisfies the Liouville property). Remark that the entropy of random walks was formally defined only for the random walks on groups, but the definition and all the theory [6] can be easily extended to the group-invariant chains which are regular covers of recurrent chains. It can be done in nearly the same way as for the Brownian motion on a covering space M of a Riemannian manifold. In this case the entropy is defined as the limit of the normalized differential entropies of one-dimensional distributions of the Brownian motion

$$h(M) = \lim_{t\to\infty} -\frac{1}{t}\int log(p^t(x,y))p^t(x,y)dm(y)\,,$$

and is independent on the starting point $x \in M$. The entropy criterion holds true for the Brownian motion as well: $h(M) = 0$ iff there exist no non-constant bounded harmonic functions on the manifold M [4].

Theorem 2. *Let M be a regular covering space of a Riemannian manifold N with the deck transformation group G; the $*$-recurrent net X and the coresponding system of sets (E_x, V_x) are invariant with respect to the action of the deck transformations group G. If the average stopping time T of the discretization procedure is finite, then the entropy of the random walk (G, μ) on the deck transformation group is finite and is given by the formula*

$$h(G, \mu) = T\, h(M)\,,$$

where $h(M)$ is the entropy of the Brownian motion on M.

Proof. First remark that $h(M) < \infty$, because the geometry of M is bounded [4]. Consider two families of maps of the path space of the Brownian motion on M onto the net X (identified with the group G):

$$\pi_k^0(\xi) = \chi(\xi(kT))\,,$$
$$\pi_k^1(\xi) = X_{N_k}(\xi) = \chi(\xi(R_{N_k}))\,.$$

From the fact that $S_{N_k}/k \to T$ follows that

$$\frac{1}{k}\rho(\pi_k^0(\xi), \pi_k^1(\xi)) \to 0 \tag{3.1}$$

almost everywhere and in $L^1(P_y)$ for any starting point $y \in M$, where ρ is the Riemannian distance (cf. [4]). The Shannon–McMillan–Breiman theorem for the entropy of random walks and Brownian motions [4,6] combined with the Harnack-type estimates and uniformity of growth of balls in M (which follows from the boundedness of geometry) imply that there exist a constant K such that for all sufficiently small $\varepsilon > 0$

$$\limsup_{k\to\infty} \mid k\,T\,h(M) + \log(\pi_k^0 \circ P_y)\,B(\varepsilon n, \pi_k^0(\xi)) \mid / \,k \leq K\varepsilon$$

and

$$\limsup_{k\to\infty} \mid k\,h(G,\mu) + \log(\pi_k^1 \circ P_y)\,B(\varepsilon n, \pi_k^1(\xi)) \mid / \,k \leq K\varepsilon$$

almost everywhere, where $B(r,x)$ is the Riemannian ball of radius r centered at x. In combination with the inequality (3.1) it gives the desired result. □

Remarks. 1. Theorem 2 also holds true for the Brownian motion on leaves of Riemannian foliations (cf. [5]). In this case one should take for X the intersection of a finite number of transversals with a leaf of foliation.

2. A particular case of this theorem is when M is a regular covering space of a *compact* manifold N and the set E coincides with the whole manifold M, i.e. every set E_x is a fundamental domain with respect to the action of G. This is a group-invariant "cocompact net" in the terminology of Lyons and Sullivan [9]. In this situation the stopping times R_n and S_n introduced in Section 1 are connected with the relation $R_n = S_{n-1}$, so that in fact one can use here only one sequence of stopping times (as it was done in [9]). The arising measure μ on G determining the random walk has exponential decay at infinity, as follows from the estimates given in [9], and hence has finite entropy. The fact that $T < \infty$ follows from the uniform estimates for the exit times of the Brownian motion on Riemannian manifolds [10].

References

[1] Ancona, A., Theorie du potentiel sur les graphes et les variétés. Cours de l'Ecole d'été de Saint-Flour, Août 1988, to appear in Springer Lecture Notes in Math.

[2] Chassaing, P., G.Letac, M.Mora, Brocot sequences and random walks in SL(2,R). Springer Lecture Notes in Math. 1064 (1984), 36–48

[3] Furstenberg, H., Random walks and discrete subgroups of Lie groups. In: Advances in Probability and Related Topics. I (ed. by P.Ney), 1–63. Dekker, New York 1971

[4] Kaimanovich, V.A., Brownian motion and harmonic functions on covering manifolds. An entropy approach. Soviet Math. Dokl. 31 (1985), 193–197

[5] Kaimanovich, V.A., Brownian motion on foliations: entropy, invariant measures, mixing. Funct. Anal. Appl. 22 (1988), 326–328

[6] Kaimanovich, V.A., Vershik, A.M., Random walks on discrete groups: boundary and entropy. Ann. Prob. 11 (1983), 457–490

[7] Kanai, M., Analytic inequalities, and rough isometries between non-compact Riemannian manifolds. Springer Lecture Notes in Math. 1201 (1986), 122–137

[8] Lyons, T., Instability of the Liouville property for quasi-isometric Riemannian manifolds and reversible Markov chains. J. Diff. Geom. 26 (1987), 33–66

[9] Lyons, T., Sullivan D. Function theory, random paths and covering spaces. J. Diff. Geom. 19 (1984), 299–323

[10] Pinsky, M.A., Stochastic Riemannian Geometry. In: Probabilistic Methods in Analysis and Related Topics (ed. by A.T. Bharucha), 199–236. Academic Press, New York 1978

[11] Revuz, D., Markov chains. North-Holland, Amsterdam 1984

[12] Varopoulos, N.T., Brownian motion and random walks on manifolds. Ann. Inst. Fourier 34 (1984), 243–269

[13] Yamasaki, M., Parabolic and hyperbolic infinite networks. Hiroshima Math. J. 7 (1977), 135–146

Capacities and mappings quasiconformal in the mean

Victor I. Kruglikov

The main goal of the present paper is to describe a geometric method for studying properties of mappings quasiconformal in the mean. We use the well-known notion of the α-capacity of condenser for this description.

By a condenser in R^n, $n \geq 2$, we understand a pair (E, G) of an open set G and a compact set E in R^n with $E \subset G$. The α-capacity (for $1 \leq \alpha \leq n$) of this condenser is

$$\operatorname{cap}_\alpha(E, G) = \inf \int_G |\nabla \varphi(x)|^\alpha dx,$$

where the infimum is taken over all continuous ACL-functions $\varphi : G \to [0,1]$ equal to 1 on E and having compact supports in G.

Let $f : D \to \Delta$ be a homeomorphic mapping between bounded domains D and Δ in R^n. For $p, q > n-1$ the inner p-mean and the outer q-mean deviations are defined as

$$I_p(f) = \left[\frac{1}{|\Delta|} \sup \sum_{k=1}^{m} \frac{\operatorname{cap}_{pn/(p+1)}^{p+1}(f(E_k), f(B_k))}{\operatorname{cap}_n^p(E_k, B_k)}\right]^{1/p}$$

and

$$O_q(f) = \left[\frac{1}{|D|} \sup \sum_{k=1}^{m} \frac{\operatorname{cap}_{qn/(q+1)}^{q+1}(E_k, B_k)}{\operatorname{cap}_n^q(f(E_k), f(B_k))}\right]^{1/q}$$

correspondingly. Here the suprema are taken over all possible disjoint finite systems $\{(E_k, B_k)\}_{k=1}^m$, $m = 1, 2, \ldots$, of condensers (E_k, B_k) lying in D such that the fractions in the formulas are meaningful.

Let us observe the following fundamental properties of the mean deviations:

1. $I_p(f) = O_p(f^{-1})$ and $O_q(f) = I_q(f^{-1})$;

2. $I_p(f) \geq 1$ and $O_q(f) \geq 1$;

3. $I_p(f) \leq \liminf_{j\to\infty} I_p(f_j)$ and $O_q(f) \leq \liminf_{j\to\infty} O_q(f_j)$ for any sequence $(f_j)_{j=1}^\infty$ of homeomorphisms $f_j : D \to \Delta$ converging uniformly to a homeomorphism $f : D \to \Delta$ on any compact set in D.

These three properties are similar to the corresponding ones for the geometric deviations $K_I(f)$ and $K_O(f)$ for the homeomorphism $f : D \to \Delta$ used in the theory of quasiconformal mappings. In that theory these deviations are defined as

$$K_I(f) = \sup \frac{\operatorname{cap}_n(f(E), f(B))}{\operatorname{cap}_n(E, B)}$$

Potential Theory, ed. by M. Kishi

and

$$K_O(f) = \sup \frac{\operatorname{cap}_n(E,B)}{\operatorname{cap}_n(f(E), f(B))},$$

where the suprema are taken over all condensers (E, B) lying in D such that the fractions are meaningful.

This analogy does not continue to hold in the next property.

4. For any $p, q > n-1$ there exists a homeomorphism $f : D \to \Delta$ (in particular from a ball onto a ball) such that $I_p(f) < \infty$ but $O_q(f) = \infty$.

The next two properties complement the specification of our object and give a connection between the geometric deviations.

5. $I_p(f)$ and $O_q(f)$ are nondecreasing in parameters.

6. $\lim_{p\to\infty} I_p(f) = K_I(f)$ and $\lim_{q\to\infty} O_q(f) = K_O(f)$.

The following assertion gives an analytic characterization of mean deviation.

Theorem 1. *For a homeomorphism $f : D \to \Delta$ such that f and f^{-1} are ACL-mappings nondegenerately differentiable a.e. in their domains, the equalities*

$$I_p(f) = \left[\frac{1}{|\Delta|}\int_D H_I^p(x,f)|J(x,f)|dx\right]^{1/p}$$

and

$$O_q(f) = \left[\frac{1}{|D|}\int_D H_O^q(x,f)dx\right]^{1/q}$$

hold.

Here $J(x, f)$ is the Jacobian of f at a point x; $H_I(x, f)$ and $H_O(x, f)$ are the inner and outer analytic deviations defined by formulas

$$H_I(x,f) = \frac{|J(x,f)|}{l^n(x,f)} \quad \text{and} \quad H_O(x,f) = \frac{|f'(x)|^n}{|J(x,f)|},$$

in which

$$|f'(x)| = \sup_{|h|=1} |f'(x)h| \quad \text{and} \quad l(x,f) = \inf_{|h|=1} |f'(x)h|.$$

The next theorem gives the complete description of mappings quasiconformal in the mean.

Theorem 2. *Suppose that $f : D \to \Delta$ is a homeomorphism and $p, q > n-1, K \geq 1$ are constants. The following two conditions are equivalent:*

(1) $I_p(f) \leq K$ *and* $O_q(f) \leq K$;

(2) $f \in ACL^n(D)$,

$$\left[\frac{1}{|\Delta|}\int_D H_I^p(x,f)|J(x,f)|dx\right]^{1/p} \le K \quad and \quad \left[\frac{1}{|D|}\int_D H_O^q(x,f)dx\right]^{1/q} \le K.$$

Accordingly in the terminology of quasiconformal mappings, (1) is a geometric definition and (2) is an analytic one of the mapping quasiconformal in the (p, q, K)-mean.

The characteristic properties of the distorsion of the capacities of condensers under mappings quasiconformal in the mean, given by formula (1), are the main tools in the investigation of various properties of such mappings. Their role is the same as the well-known property of the quasi-invariance of the n-capacity for quasiconformal mappings.

The fundamental metric and geometric properties of mappings quasiconformal in the mean have been established. In particular, such mappings and their converses have the $\mathcal{N}$-property. We have also proved an estimate of the distortion of Euclidean distances under mappings quasiconformal in the mean which is exact in the order, the compactness of this class, the property of semicontinuity for integral means of the analytic deviations, etc.

The problem of the boundary correspondance under mappings quasiconformal in the mean for domains with fixed and variable boundaries is solved. For this we have constructed a new theory of prime α-ends by using the notion of α-capacity of condenser.

We are also investigating not necessarily univalent mappings quasiconformal in the mean. The distorsion of the capacities of condensers and some important results for these mappings are established.

The problems under consideration were investigated in more detail in the following articles:

1. Kruglikov, V.I.: Capacity of condensers and spatial mappings quasiconformal in the mean, Math. USSR Sbornik 58 (1987), 185-205.

2. Kruglikov, V.I.: Mappings quasiconformal in the mean, Soviet Math. Dokl. 32 (1985), 332-334.

3. Kruglikov, V.I.: Prime ends of spatial domains with variable boundaries, Soviet Math. Dokl. 36 (1988), 565-567.

4. **Кругликов, В.И.: Кольцевое определение средних отклонений и отображений, квазиконформных в среднем, Докл. АН УССР, Сер. А,** 1987, No.9, 21-24.

5. **Кругликов, В.И.: Структура носителя простого конца пространственной области, Докл. АН УССР, Сер. А,** 1987, No.3, 19-22.

6. **Кругликов, В.И., Пайков, В.И.: Емкости и простые концы пространственной области, Докл. АН УССР, Сер. А,** 1987, No.5, 10-13.

7. **Борчук С.М., Кругликов В.И.: Отображения с искажением, ограниченным в среднем, Докл. АН УССР, Сер. А,** 1988, No.11, 3-5.

Beppo Levi spaces and Riesz potential spaces

Takahide Kurokawa

1. Introduction

The relation between the Sobolev spaces and the Bessel potential spaces is well known ([2],[4]). Let $0 < r < \infty$ and G_r be the Bessel kernel of order r on n-dimensional Euclidean space R^n. For a function f on R^n the convolution

$$G_r * f(x) = \int G_r(x-y)f(y)dy$$

is called the Bessel potential of order r of f. The Bessel potential space B_r^p is defined by

$$B_r^p = \{G_r * f; f \in L^p\}$$

and the norm of $u = G_r * f$ is defined to be the L^p-norm of f, i.e.,

$$(1.1) \qquad ||u||_{B_r^p} = ||f||_p.$$

For a positive integer m and $p > 1$, the Sobolev space W_m^p is defined by

$$W_m^p = \{u \in \mathcal{D}'; ||u||_{m,p} = \sum_{|\gamma| \le m} ||D^\gamma u||_p < \infty\}$$

where $\mathcal{D}'$ is the class of distributions, γ is a multi-index $(\gamma_1, \cdots, \gamma_n)$, $|\gamma| = \gamma_1 + \cdots + \gamma_n$ and $D^\gamma = D_1^{\gamma_1} \cdots D_n^{\gamma_n}$. A.P.Caldéron showed

$$(1.2) \qquad W_m^p = B_m^p$$

$$(1.3) \qquad ||u||_{m,p} \approx ||f||_p \ \text{ if } \ u = G_m * f$$

where '≈' means that $C^{-1}||f||_p \le ||u||_{m,p} \le C||f||_p, C > 0$. We use the letter C for a generic positive constant whose value may be different at each occurrence.

Let $\mathcal{D}$ be the class of infinitely differentiable functions with compact support. It is also known [1] that the closure of $\mathcal{D}$ in W_m^p is W_m^p, i.e.,

$$(1.4) \qquad \overline{\mathcal{D}} = W_m^p.$$

In this note we investigate the facts corresponding to (1.2), (1.3) and (1.4) for Beppo Levi spaces and Riesz potential spaces.

Potential Theory, ed. by M. Kishi

2. Riesz potential spaces

For a positive integer m, the Riesz kernel of order m is given by

$$\kappa_m(x) = \begin{cases} |x|^{m-n}, & m < n \text{ or } m \geq n, m-n \text{ odd}, \\ (\delta_{m,n} - \log|x|)|x|^{m-n}, & m \geq n, m-n \text{ even} \end{cases}$$

(see [3]). For an L^p-function f,

$$U_m^f(x) = \int \kappa_m(x-y)f(y)dy$$

does not necessarily exist. In order to define the Riesz potential of order m of an L^p-function, we introduce the following kernel: For a positive integer $k \leq m-1$

$$\kappa_{m,k}(x,y) = \begin{cases} \kappa_m(x-y) - \sum_{|\gamma|\leq k}(x^\gamma/\gamma!)D^\gamma \kappa_m(-y), & 0 \leq k \leq m-1, \\ \kappa_m(x-y), & k \leq -1. \end{cases}$$

Further we write

$$U_{m,k}^f(x) = \int \kappa_{m,k}(x,y)f(y)dy.$$

Concerning the existence of $U_{m,k}^f$, we have

Proposition 2.1 ([3:Corollaries 3.22 and 3.23]). *Let f be an L^p-function and $k = [m-(n/p)]$ be the integral part of $m-(n/p)$. If $m-(n/p)$ is not a nonnegative integer, then $U_{m,k}^f$ exists, and if $m-(n/p)$ is a nonnegative integer, then $U_{m,k-1}^g$ and $U_{m,k}^h$ exist where g is the restriction of f to the unit ball $B_1 = \{|x| < 1\}$ and $h = f - g$.*

Taking Proposition 2.1 into account, we define the Riesz potential of order m of an L^p-function f as follows:

$$K_{m,p}^f = \begin{cases} U_{m,k}^f, & \text{if } m-(n/p) \text{ is not a nonnegative integer}, \\ U_{m,k-1}^g + U_{m,k}^h & \text{if } m-(n/p) \text{ is a nonnegative integer} \end{cases}$$

where g, h and k are as in Proposition 2.1. Further, the Riesz potential space R_m^p is defined by

$$R_m^p = \{K_{m,p}^f; f \in L^p\}.$$

If $f \in L^p \cap L^r$, then the Riesz potentials $K_{m,p}^f$ and $K_{m,r}^f$ can be defined. We have the following relation between $K_{m,p}^f$ and $K_{m,r}^f$.

Proposition 2.2. *Let $p > r, k = [m-(n/p)], j = [m-(n/r)]$ and $f \in L^p \cap L^r$. Then*

$$K_{m,p}^f - K_{m,r}^f = \begin{cases} \sum_{j+1\leq|\gamma|\leq k} a_\gamma x^\gamma & \text{if } m-(n/r) \text{ is not a nonnegative integer} \\ \sum_{j\leq|\gamma|\leq k} a_\gamma x^\gamma & \text{if } m-(n/r) \text{ is a nonnegative integer.} \end{cases}$$

Proof. By [3;Theorem 4.4] we have

$$D^\gamma K^f_{m,p} = D^\gamma K^f_{m,r} \quad \text{for } |\gamma| = m.$$

Hence $P = K^f_{m,p} - K^f_{m,r}$ is a polynomial of degree $m-1$. In case $m-(n/p)<0$, the conclusion is clear. Hence let $m-(n/p)\geq 0$. First we assume that $m-(n/r)\geq 0$. Since $p>r$ implies $-q(m-(n/p))-n < -q(m-(n/r))-n$ and $-(q/p')-1 < -(q/r')-1$, by [3;Corollaries 5.9 and 5.15] we have

$$\int_{|x|\geq 1}(1+|\log|x||)^{(-q/p')-1}|x|^{-q(m-(n/p))-n}|P(x)|^q dx < \infty, \quad q \geq p,$$

where $(1/p)+(1/p')=1$. This implies that the degree of P is equal to or less than k. In case $m-(n/r)$ is not a nonnegative integer, by [3;Corollaries 5.9 and 5.15] we have

$$\int_{|x|<1}|x|^{-q(m-(n/r))-n}|P(x)|^q dx < \infty, \quad q \geq p.$$

This implies that $P(x) = \sum_{j+1\leq|\gamma|\leq k} a_\gamma x^\gamma$. In case $m-(n/r)$ is a nonnegative integer, we see that $P(x) = \sum_{j\leq|\gamma|\leq k} a_\gamma x^\gamma$ by the same corollaries. Next we assume that $m-(n/r)<0$. In case $m-(n/p)$ is not a nonegative integer, we have $K^f_{m,p} = U^f_{m,k}$ and $K^f_{m,r} = U^f_m$. The conditions $m-(n/p)>0$, $m-(n/r)<0$ and $|\gamma|\leq k$ imply that

$$\int|\kappa_m(x-y)f(y)|dy < \infty, \quad \sum_{|\gamma|\leq k}\int|D^\gamma\kappa_m(-y)f(y)|dy < \infty.$$

Hence we have

$$\begin{aligned} K^f_{m,p} &= \int\Big(\kappa_m(x-y) - \sum_{|\gamma|\leq k}(x^\gamma/\gamma!)D^\gamma\kappa_m(-y)\Big)f(y)dy \\ &= \int\kappa_m(x-y)f(y)dy - \sum_{|\gamma|\leq k}(x^\gamma/\gamma!)\int D^\gamma\kappa_m(-y)f(y)dy \\ &= K^f_{m,r} + \sum_{|\gamma|\leq k} a_\gamma x^\gamma. \end{aligned}$$

In case $m-(n/p)$ is a nonnegative integer, the proof is similar. Thus we obtain the proposition. □

Corresponding to (1.1) the norm of $u \in R^p_m$ is defined by

$$||u||_{R^p_m} = ||f|| \quad \text{if } u = K^f_{m,p}.$$

To see that this gives a consistent definition of $||u||_{R^p_m}$ we must observe that

$$f = g \quad \text{if } K^f_{m,p} = K^g_{m,p}. \tag{2.1}$$

For a multi-index γ with $|\gamma| = m$, by [3;Theorem 4.4] we have

$$D^\gamma K^f_{m,p} = C\, R^\gamma f$$

where $R^\gamma = R_1^{\gamma_1} \cdots R_n^{\gamma_n}$ and R_j $(j = 1, \dots, n)$ are the Riesz transforms. It follows from $K^f_{m,p} = K^g_{m,p}$ that $R^\gamma f = R^\gamma g$ for $|\gamma| = m$, and hence $R^{2\gamma} f = R^{2\gamma} g$ for $|\gamma| = m$. By [3; Lemma 4.1] we have

$$\begin{aligned} f &= (-1)^m \sum_{|\gamma|=m} (m!/\gamma!) R^{2\gamma} f \\ &= (-1)^m \sum_{|\gamma|=m} (m!/\gamma!) R^{2\gamma} g = g. \end{aligned}$$

It is easy to show

Proposition 2.3. *The space R^p_m is a Banach space.*

3. Beppo Levi spaces

For a positive integer m and $p > 1$, we define the Beppo Levi space L^p_m by

$$L^p_m = \{u \in \mathcal{D}'; |u|_{m,p} = \sum_{|\gamma|=m} ||D^\gamma u||_p < \infty\}.$$

We note that functions in L^p_m are locally integrable. The norm of $u \in L^p_m$ is defined by

$$||u||_{L^p_m} = \int_{|x|<1} |u(x)| dx + |u|_{m,p}.$$

It is not so difficult to see that

$$(3.1) \qquad ||u||_{L^p_m} \approx \left(\int_{|x|<1} |u(x)|^p dx \right)^{1/p} + |u|_{m,p}.$$

We denote by P_d the space of all polynomials of degree smaller than or equal to d. For a subset E of R^n, we set $L^p(E) = \{f \in L^p; f(x) = 0 \text{ on } E^c\}$ where E^c is the complement of E. Corresponding to the facts (1.2) (1.3), we have the following.

Theorem 3.1. (i) *$L^p_m = R^p_m \oplus P_{m-1}$, where $\oplus$ means the direct sum.*
(ii) *If $L^p_m \ni u = K^f_{m,p} + \sum_{|\delta| \le m-1} a_\delta x^\delta$, then*

$$(3.2) \qquad |u|_{m,p} \approx ||f||_p,$$

$$(3.3) \qquad ||u||_{L^p_m} \approx ||f||_p + \sum_{|\delta| \le m-1} |a_\delta|.$$

Proof. (i) By [3;Theorems 6.1 and 6.3] we see that $L^p_m = R^p_m + P_{m-1}$. We shall show that $R^p_m \cap P_{m-1} = \{0\}$. Let $u \in R^p_m \cap P_{m-1}$. In case $m - (n/p)$ is not a nonnegative integer, we have

$$u = U^f_{m,k} = \sum_{|\delta| \le m-1} a_\delta x^\delta, \quad f \in L^p$$

with $k = [m - (n/p)]$. For $|\gamma| = m$, by [3;Theorem 4.4(i)] we see that

$$CR^{\gamma} f = D^{\gamma} U^{f}_{m,k} = D^{\gamma}\left(\sum_{|\delta|\le m-1} a_{\delta} x^{\delta}\right) = 0,$$

and hence $f = 0$. In case $m - (n/p)$ is a nonnegative integer, we have

$$u = U^{g}_{m,k-1} + U^{h}_{m,k} = \sum_{|\delta|\le m-1} a_{\delta} x^{\delta},$$

where $g \in L^p(B_1)$ and $h \in L^p(B_1^c)$. By [3;Theorem 4.4(ii)] for $|\gamma| = m$ we have

$$CR^{\gamma} g + CR^{\gamma} h = D^{\gamma} U^{g}_{m,k-1} + D^{\gamma} U^{h}_{m,k} = D^{\gamma}\left(\sum_{|\delta|\le m-1} a_{\delta} x^{\delta}\right) = 0.$$

Hence $R^{\gamma} g = -R^{\gamma} h$ for $|\gamma| = m$, and so $g = -h$. Since $g \in L^p(B_1)$ and $h \in L^p(B_1^c)$, we see that $g = h = 0$.

(ii) we note that if $u = K^{f}_{m,p} + \sum_{|\delta|\le m-1} a_{\delta} x^{\delta}$, then such f is unique by (i) and (2.1), and f is given by

$$f = C \sum_{|\gamma|=m} R^{\gamma} D^{\gamma} u$$

(see [3;Theorems 6.1 and 6.3]). Hence we have

$$|u|_{m,p} = \sum_{|\gamma|=m} ||D^{\gamma} u||_p = C \sum_{|\gamma|=m} ||R^{\gamma} f||_p \le C||f||_p$$

and

$$||f||_p \le C \sum_{|\gamma|=m} ||R^{\gamma} D^{\gamma} u||_p \le C \sum_{|\gamma|=m} ||D^{\gamma} u||_p = C|u|_{m,p}.$$

Therefore we obtain (3.2). Further, by [3;Corollaries 5.9 and 5.15] we see that

$$\begin{aligned} &\int_{|x|<1} |u(x)|dx \\ &\le \int_{|x|<1} |K^{f}_{m,p}(x)|dx + \sum_{|\delta|\le m-1} |a_{\delta}| \int_{|x|<1} |x|^{|\delta|} dx \\ &\le C\left(||f||_p + \sum_{|\delta|\le m-1} |a_{\delta}|\right). \end{aligned}$$

On the other hand, by (3.1), [3;(6.3)] and [3;(6.8)] we have

$$\sum_{|\delta|\le m-1} |a_{\delta}| \le \int_{|x|<1} |u(x)|dx + |u|_{m,p}.$$

Hence we have (3.3). The theorem is proved. □

Remark 3.2. In case $m - (m/p)$ is a nonnegative integer k, it follows from the definition that

$$R_m^p = S_m^p + T_m^p$$

where $S_m^p = \{U_{m,k-1}^g; g \in L^p(B_1)\}$ and $T_m^p = \{U_{m,k}^h; h \in L^p(B_1^c)\}$. The sum is a direct sum. In fact, let $u \in S_m^p \cap T_m^p$. Then

$$u = U_{m,k-1}^g = U_{m,k}^h, \quad g \in L^p(B_1), h \in L^p(B_1^c).$$

For $|\gamma| = m$, by [3;Theorem 4.4(ii)] we have

$$CR^\gamma g = D^\gamma U_{m,k-1}^g = D^\gamma U_{m,k}^h = CR^\gamma h.$$

Thus we have $g = h$, and hence $g = h = 0$.

Corresponding to (1.4) we have

Theorem 3.3. *Let* $\overline{\mathcal{D}}$ *be the closure of* $\mathcal{D}$ *in* L_m^p *and* $k = [m - (n/p)]$. *Then*

$$\overline{\mathcal{D}} = R_m^p \oplus P_k.$$

Proof. This theorem follows from [3.Theorem 6.7] and Theorem 3.1. □

Remark 3.4. In particular, if $m - (n/p) < 0$ and $u \in \overline{\mathcal{D}}$, then $u = U_m^f$, $f \in L^p$ and

$$||u||_{L_m^p} \approx ||f||_p \approx |u|_{m,p}.$$

References

1. R.A.Adams, "Sobolev spaces," Academic Press, New York, 1975.
2. A.P.Calderon, *Lebesgue spaces of differentiable functions and distributions*, Proc. Symp. in Pure Math. **5** (1961), 33-49.
3. T.Kurokawa, *Riesz potentials, higher Riesz transforms and Beppo Levi spaces*, Hiroshima Math. J. **18** (1988), 541-597.
4. E.M.Stein, "Singular integrals and differentiability properties of functions," Princeton Univ. Press, Princeton, N.J., 1970.

Boundary limits of harmonic functions in Sobolev-Orlicz classes

Yoshihiro Mizuta

1980 Mathematics Subject Classification (1985 Revision): 31B25.

1. Introduction

Let G be a bounded Lipschitz domain in R^n, $n \geq 2$. The aim in this paper is to study boundary limits of harmonic functions u on G satisfying a condition of the form:

$$\int_G \Psi_p(|\text{grad } u(x)|)\lambda(\rho(x))dx < \infty. \tag{1.1}$$

Here $\rho(x) = \rho_G(x)$ denotes the distance of x from the boundary ∂G, and, throughout this paper, the following conditions are assumed to hold:

(i) $\Psi_p(r)$ is of the form $r^p\psi(r)$, where $p > 1$ and ψ is a positive nondecreasing function on the interval $(0, \infty)$ satisfying

$$M^{-1}\psi(r) \leq \psi(r^2) \leq M\psi(r) \qquad \text{for any } r > 0. \tag{1.2}$$

(ii) λ is a positive monotone function on $(0, \infty)$ satisfying the so-called (Δ_2) condition:

$$M^{-1}\lambda(r) \leq \lambda(2r) \leq M\lambda(r) \qquad \text{for any } r > 0.$$

Here M denotes a positive constant.

For a positive nondecreasing function φ on $(0, \infty)$, $\xi \in \partial G$ and $a > 0$, define

$$T_\varphi(\xi, a) = \{x \in G; \varphi(|x - \xi|) < a\rho(x)\}.$$

Our first aim is to study the existence of (tangential) boundary limits

$$\lim_{x \to \xi, x \in T_\varphi(\xi, a)} u(x)$$

at $\xi \in \partial G$ except those in a suitable exceptional set. The exceptional set appearing in this discussion will be evaluated by Bessel-type capacities as well as Hausdorff measures.

Potential Theory, ed. by M. Kishi

In case $\Psi_p(r) = r^p$, that is, $\psi \equiv 1$, $\lambda(r) = r^\beta$ and $\varphi(r) = r^\alpha$ with $\alpha \geq 1$, many authors studied the existence of (non)tangential boundary limits; see, for example, Carleson [2], Cruzeiro [3], Mizuta [7], [8], [11], Murai [14] and Wallin [17].

By using the same methods as in these works, we discuss weighted boundary limits of the function u. Our second aim in this paper is to find a positive function k such that

$$\lim_{x \to \xi, x \in T_\varphi(\xi,a)} k(x)u(x) = 0,$$

depending on the choice of $\xi \in \partial G$; or globally,

$$\lim_{x \to \partial G} k(x)u(x) = 0.$$

In this paper, let M, M_1, M_2,..., denote various constants independent of the variables in question. Further, we denote by $B(x,r)$ the open ball with radius r and center at x.

2. T_φ-limits

In this section we are concerned with the global existence of tangential boundary limits of harmonic functions u on G satisfying (1.1).

Let φ be a positive nondecreasing function on the interval $(0,\infty)$ satisfying (Δ_2) condition such that $\varphi(r)/r$ is nondecreasing on $(0,\infty)$. We say that u has a T_φ-limit ℓ at $\xi \in \partial G$ if

$$\lim_{x \to \xi, x \in T_\varphi(\xi,a)} u(x) = \ell$$

for any $a > 0$; in case $\varphi(r) = r^\alpha$, we use the terminology T_α-limit, instead of T_φ-limit, for simplicity.

As in [11], we consider the sets

$$E_0 = \left\{\xi \in \partial G; \int_{G \cap B(\xi,1)} |\xi - y|^{1-n} |\text{grad } u(y)| dy = \infty\right\}$$

and

$$E_h = \left\{\xi \in \partial G; \limsup_{r \to 0} \, [h(r)]^{-1} \int_{G \cap B(\xi,r)} \Psi_p(|\text{grad } u(y)|)\lambda(\rho(y)) dy > 0\right\},$$

where h is a positive nondecreasing function on the interval $(0,\infty)$ satisfying (Δ_2) condition.

First we quote the following result from [11], which is our fundamental tool in this section.

Theorem 2.1. *Set $\eta(r) = \psi(r^{-1})\lambda(r)$ and define*

$$\kappa(r) = \left(\int_r^1 [t^{n-p}\eta(t)]^{1/(1-p)}\frac{dt}{t}\right)^{(p-1)/p}$$

for $0 < r \leq 2^{-1}$; define $\kappa(r) = \kappa(2^{-1})$ for $r > 2^{-1}$. Let u be a harmonic function on G satisfying condition (1.1). *If $\xi \in \partial G - (E_0 \cup E_h)$ and $\kappa(\rho(x)) \leq M(a)[h(|\xi - x|)]^{-1/p}$ on $T_\varphi(\xi, a)$, with a positive constant $M(a)$, then $u(x)$ has a finite limit as $x \to \xi$, $x \in T_\varphi(\xi, a)$.*

To study the size of E_0 and E_h, we consider the capacity

$$C_{\alpha,\Psi_p,\lambda}(E; G') = \inf \int \Psi_p(f(y))\lambda(\rho_G(y))dy,$$

where G' is a bounded open set in R^n and the infimum is taken over all nonnegative measurable functions f such that

$$\int_{G'} |x-y|^{\alpha-n} f(y)dy \geq 1 \qquad \text{for every } x \in E.$$

If $C_{\alpha,\Psi_p,\lambda}(E \cap G'; G') = 0$ for any bounded open set $G' \subset R^n$, then we write simply

$$C_{\alpha,\Psi_p,\lambda}(E) = 0.$$

From the definition of $C_{\alpha,\Psi_p,\lambda}$ we see easily that if u satisfies condition (1.1), then

$$C_{1,\Psi_p,\lambda}(E_0) = 0;$$

we see further that

$$H_h(E_h) = 0,$$

where H_h denotes the Hausdorff measure with the measure function h.

Thus Theorem 2.1 gives the following result.

Theorem 2.2. *Let u be as in Theorem 2.1. If $\kappa(0) = \infty$, then there exist E_1 and E_2 such that*

$$C_{1,\Psi_p,\lambda}(E_1) = 0,$$

$$H_h(E_2) = 0 \qquad \text{with } h(r) = [\kappa(\varphi(r))]^{-p}$$

and $u(x)$ has a finite T_φ-limit at any $\xi \in \partial G - (E_1 \cup E_2)$.

To see the relationship between capacities and Hausdorff measures, we prepare the following result, which can be proved in the same manner as in Meyers [6] (cf. [12, Lemma 7.3]).

Lemma 2.1. *If $\xi \in \partial G$, then*

$$M^{-1}[\kappa(r;a)]^{-p} \leq C_{\alpha,\Psi_p,\lambda}(B(\xi,r); B(\xi,a)) \leq M[\kappa(r;a)]^{-p}$$

for any r with $0 < r < a/2$, where

$$\kappa(r;a) = \left(\int_r^a [t^{n-\alpha p}\eta(t)]^{1/(1-p)} \frac{dt}{t} \right)^{(p-1)/p}$$

with $\eta(r) = \psi(r^{-1})\lambda(r)$.

By Lemma 2.1, we can prove the following result.

Lemma 2.2. *Suppose $\kappa(0) = \infty$, and set $h(r) = [\kappa(r)]^{-p}$. If $E \subset \partial G$ and $H_h(E) = 0$, then $C_{1,\Psi_p,\lambda}(E) = 0$.*

Remark 2.1. Let $\lambda(r) = r^\beta$. If $n - p + \beta > 0$, then

$$\kappa(r) \sim [r^{n-p+\beta}\psi(r^{-1})]^{-1/p} \qquad \text{as } r \to 0.$$

If $n - p + \beta = 0$, then

$$\kappa(r^\alpha) \sim \kappa(r) = \left(\int_r^1 [\psi(t^{-1})]^{1/(1-p)} \frac{dt}{t} \right)^{1/p'}$$

as $r \to 0$, for any $\alpha > 1$.

To obtain an estimate for the size of E_0 in the special case where $\lambda(r) = r^\beta$, we need the following lemma (see [9]).

Lemma 2.3. *Let f be a measurable function on R^n such that*

$$\int_{R^n} (1 + |y|)^{1-n}|f(y)|dy < \infty$$

and

$$\int_{R^n} \Psi_p(|f(y)|\ |y_n|^{\beta/p})dy < \infty,$$

where $-1 < \beta < p - 1$. If we set $u(x) = \int |x - y|^{1-n}f(y)dy$, then

$$\int \Psi_p(|\text{grad } u(x)|\ |x_n|^{\beta/p})dx \le M \int \Psi_p(|f(y)|\ |y_n|^{\beta/p})dy$$

with a positive constant M independent of f.

Lemma 2.4. *Let f be a nonnegative measurable function on G such that*

$$\int_G \Psi_p(f(y))\rho(y)^\beta dy < \infty,$$

and set

$$E = \left\{ \xi \in \partial G; \int_{G \cap B(\xi,1)} |\xi - y|^{1-n}f(y)dy = \infty \right\}.$$

If $-1 < \beta < p-1$, *then* $C_{1-\beta/p,\Psi_p}(E) = 0$, *where* $C_{\alpha,\Psi_p} = C_{\alpha,\Psi_p,\lambda}$ *with* $\lambda \equiv 1$.

Proof. By a change of variables, we may assume that G is the half space $D = \{x_n > 0\}$ and f vanishes outside some ball $B(0,N)$; that is,

$$\int_D \Psi_p(f(y))y_n^\beta dy < \infty.$$

Here note that for $\gamma \geq 1$, $\psi(r^\gamma) \leq M_\gamma \psi(r)$ whenever $r > 0$, by (1.2), and

$$\begin{aligned}
&\int_D \Psi_p(f(y)y_n^{\beta/p})dy \\
&= \int_{\{y\in D; f(y)^\varepsilon > y_n^{\beta/p}\}} \Psi_p(f(y)y_n^{\beta/p})dy + \int_{\{y\in D; f(y)^\varepsilon < y_n^{\beta/p}\}} \Psi_p(f(y)y_n^{\beta/p})dy \\
&\leq \int_D y_n^\beta f(y)^p \psi(f(y)^{1+\varepsilon})dy + \int_{\{y\in D; f(y)>0\}} \Psi_p(y_n^{(1+\varepsilon^{-1})\beta/p})dy < \infty
\end{aligned}$$

if $\varepsilon > 0$ and $\beta(1+\varepsilon^{-1}) > -1$. Consider the function

$$u(x) = \int_D |x-y|^{1-n} f(y)dy.$$

For simplicity, set $U = |\text{grad } u|$. From Lemma 2.3 it follows that

$$\int_{R^n} \Psi_p(U(x)|x_n|^{\beta/p})dx < \infty.$$

Since $U(x) = O(|x|^{-n})$ as $|x| \to \infty$, we see that

$$\int_{R^n} \Psi_p(U(x))|x_n|^\beta dx < \infty.$$

Define

$$E^* = \left\{x \in \partial D; \int_D |x-y|^{1-\beta/p-n}[U(y)y_n^{\beta/p}]dy = \infty\right\}.$$

Then, by definition, $C_{1-\beta/p,\Psi_p}(E^*) = 0$. If $\xi \in \partial D - E^*$ and $a > 0$, then

$$\int_{\Gamma(\xi,a)} |\xi - y|^{1-n} U(y)dy < \infty,$$

where $\Gamma(\xi,a) = \{x \in D; |x-\xi| < ax_n\}$. We see readily that

$$\int_0^{r_0} |\text{grad } u(\xi + r\Theta)|dr < \infty \quad \text{for almost every } \Theta \in D \cap \partial B(0,1),$$

which implies that $u(\xi + r\Theta)$ has a finite limit for almost every $\Theta \in D \cap \partial B(0,1)$. If $\xi \in E$, then

$$\liminf_{r\to 0} u(\xi + rx) \geq u(\xi) = \infty$$

for any $x \in D$, by the lower semicontinuity of potentials. Thus if $\xi \in \partial D - E^*$, then $\xi \in \partial D - E$, so that $E \subset E^*$. Consequently, $C_{1-\beta/p,\Psi_p}(E) = 0$. □

Corollary 2.1. *If $\lambda(r) = r^\beta$, $\beta \geq p - n$, then E_0 has Hausdorff dimension at most $n - p + \beta$.*

By Theorem 2.2 together with Corollary 2.1, we have

Theorem 2.3. *Let u be a function which is harmonic in G and satisfies*

$$\int_G \Psi_p(|\text{grad } u(x)|)\rho(x)^\beta dx < \infty.$$

If $\beta > p - n$, then for any $\alpha > 1$ there exists a set $E_\alpha \subset \partial G$ such that

(i) $H_{h_\alpha}(E_\alpha) = 0$ *for* $h_\alpha(r) = \inf_{t \geq r} t^{\alpha(n-p+\beta)}\psi(t^{-1})$;

(ii) $u(x)$ *has a finite T_α-limit at any $\xi \in \partial G - E_\alpha$.*

Using Theorem 2.2, Lemmas 2.2 and 2.4, we establish the following result.

Theorem 2.4. *Let u be a function which is harmonic in G and satisfies*

$$\int_G \Psi_p(|\text{grad } u(x)|)[\rho(x)]^{p-n} dx < \infty.$$

If $p > n - 1$, then there exists a set $E \subset \partial G$ such that $C_{n/p,\Psi_p}(E) = 0$ and u has a finite T_α-limit for any $\alpha \geq 1$, at any $\xi \in \partial G - E$.

Remark 2.2. In case $p = n$ and $\Psi_n(r) = r^n$, Theorem 2.4 was proved by Cruzeiro [3]. Nagel, Rudin and Shapiro [15] proved the existence of T_α-limits of harmonic functions represented as Poisson integrals in a half space.

3. Weighted T_φ-limits

Let $\eta(r) = \psi(r^{-1})\lambda(r)$, as in Theorem 2.1. We consider the following condition:

$$\int_0^r [\eta(t)]^{-1/(p-1)} dt \leq Mr[\eta(r)]^{-1/(p-1)}. \tag{3.1}$$

Further, let $\omega(r)$ and $\omega^*(r)$ be positive nondecreasing functions on $(0, \infty)$ satisfying (Δ_2) condition,

$$\lim_{r \to 0} \omega^*(r) = 0 \tag{3.2}$$

and

$$\int_r^1 \frac{1}{\omega^*(t)} \frac{dt}{t} \leq \frac{M}{\omega(r)} \quad \text{for } 0 < r < 2^{-1}. \tag{3.3}$$

Theorem 3.1. *Let u be a harmonic function on G satisfying condition* (1.1), *and suppose $\kappa(0) = \infty$. Set*

$$\tau_1(r) = \inf_{r \le t \le 1} [\omega^*(t)]^{-p} t^{n-p} \eta(t), \qquad \tau_2(r) = \inf_{r \le t \le 1} [\omega(t)\kappa(t)]^{-p},$$

$$\tau(r) = \min\{\tau_1(r), \tau_2(r)\} \qquad \text{and} \qquad h(r) = \tau(\varphi(r)).$$

Then there exists a set $E \subset \partial D$ such that $H_h(E) = 0$ and

$$\lim_{x \to \xi, x \in T_\varphi(\xi, a)} \omega(\varphi(|x - \xi|))u(x) = 0$$

for any $\xi \in \partial D - E$ and any $a > 0$.

To prove Theorem 3.1, we need several lemmas.

Lemma 3.1. *Let f be a nonnegative measurable function on G satisfying*

$$\int_G \Psi_p(f(y))\lambda(\rho(x))dx < \infty. \tag{3.4}$$

If we set

$$E = \left\{\xi \in \partial G; \limsup_{r \to 0} \omega^*(r) r^{1-n} \int_{G \cap B(\xi, r)} f(y)dy > 0\right\},$$

then $H_{\tau_1}(E) = 0$.

Proof. Without loss of generality we may assume that $f = 0$ outside G. Let $0 < \delta < 1$. By Hölder's inequality and (3.1), we have

$$\begin{aligned}
\int_{B(\xi,r)} f(y)dy &\le \left(\int_{B(\xi,r)} f(y)^p \psi(f(y))\lambda(\rho(y))dy\right)^{1/p} \\
&\quad \times \left(\int_{\{y \in B(\xi,r); f(y) > \rho(y)^{-\delta}\}} [\psi(f(y))]^{-p'/p}[\lambda(\rho(y))]^{-p'/p}dy\right)^{1/p'} + \int_{G \cap B(\xi,r)} \rho(y)^{-\delta}dy \\
&\le M_1 \left(\int_{B(\xi,r)} \Psi_p(f(y))\lambda(\rho(y))dy\right)^{1/p} \left(\int_{G \cap B(\xi,r)} [\eta(\rho(y))]^{-p'/p}dy\right)^{1/p'} + M_1 r^{n-\delta} \\
&\le M_2 r^{n/p'} [\eta(r)]^{-1/p} \left(\int_{B(\xi,r)} \Psi_p(f(y))\lambda(\rho(y))dy\right)^{1/p} + M_1 r^{n-\delta}.
\end{aligned}$$

Hence it follows that

$$\omega^*(r) r^{1-n} \int_{B(\xi,r)} f(y)dy \le M_2 \left([\tau_1(r)]^{-1} \int_{B(\xi,r)} \Psi_p(f(y))\lambda(\rho(y))dy\right)^{1/p} + M_1 \omega^*(r) r^{1-\delta}.$$

Thus Lemma 3.1 is obtained. □

Lemma 3.2. *Let f satisfy* (3.4), *and define E as in Lemma 3.1. If $\xi \in \partial G - E$, then*

$$\lim_{r\to 0} \omega(r) \int_{G\cap B(\xi,1)-B(\xi,r)} f(y)|\xi - y|^{1-n} dy = 0.$$

Proof. As above, assume that $f = 0$ outside G. Define

$$\varepsilon(r) = \sup_{0<t\leq r} \omega^*(t) t^{1-n} \int_{B(\xi,t)} f(y) dy.$$

Then note that $\lim_{r\to 0} \varepsilon(r) = 0$ for $\xi \in \partial G - E$. From (3.3) it follows that $\omega(r) \leq M_1 \omega^*(r)$ for $0 < r < 2^{-1}$, so that ω satisfies (3.2), too. If r_0 is fixed so that $0 < r_0 < 1$, then

$$\lim_{r\to 0} \omega(r) \int_{B(\xi,1)-B(\xi,r_0)} f(y)|\xi - y|^{1-n} dy = 0.$$

On the other hand, using (3.3), we have

$$\begin{aligned}
&\omega(r) \int_{B(\xi,r_0)-B(\xi,r)} f(y)|\xi - y|^{1-n} dy \\
&\leq (n-1)\omega(r) \int_r^{r_0} \left(\int_{B(\xi,t)} f(y) dy \right) t^{-n} dt + \omega(r) r_0^{1-n} \int_{B(\xi,r_0)} f(y) dy \\
&\leq (n-1)\omega(r)\varepsilon(r_0) \int_r^{r_0} \frac{1}{\omega^*(t)} \frac{dt}{t} + M_2 \varepsilon(r_0) \\
&\leq M_3 \varepsilon(r_0).
\end{aligned}$$

Hence the required equality holds for $\xi \in \partial G - E$. □

Proof of Theorem 3.1. For $\xi \in \partial G$, we can find $r_0 > 0$ and $x_0 \in G$ such that for any $x \in G \cap B(\xi, r_0)$, there corresponds an open set $E(x, x_0) \subset G$ with the following properties (see Lemma 1 in [8]):

(i) $E(x, x_0) \subset G \cap B(\xi, 2r_0)$.

(ii) $\rho(x) + |x - y| \leq M_1 \rho(y)$ whenever $y \in E(x, x_0)$.

(iii) $|u(x) - u(x_0)| \leq M_1 \int_{E(x,x_0)} |\text{grad } u(y)| [\rho(y)]^{1-n} dy.$

For simplicity, set $f(y) = |\text{grad } u(y)|$. As in the proof of Lemma 3.1, using Hölder's inequality, we obtain

$$\begin{aligned}
|u(x) - u(x_0)| &\leq M_1 \int_{E(x,x_0)-B(\xi,2|x-\xi|)} f(y)[\rho(y)]^{1-n} dy \\
&\quad + M_1 \int_{E(x,x_0)} [\rho(y)]^{1-\delta-n} dy
\end{aligned}$$

$$+M_1 \left(\int_{G\cap B(\xi,2|x-\xi|)} \Psi_p(f(y))\lambda(\rho(y))dy \right)^{1/p}$$
$$\times \left(\int_{E(x,x_0)} [\rho(y)]^{p'(1-n)} [\psi(\rho(y)^{-\delta})\lambda(\rho(y))]^{-p'/p} dy \right)^{1/p'}$$
$$= M_1[I_1(x;\xi) + I_2(x;\xi) + I_3(x;\xi)],$$

where $0 < \delta < 1$. If $y \in E(x, x_0)$ and $|y - \xi| \geq 2|x - \xi|$, then

$$\rho(y) \geq M_1^{-1}|x-y| \geq M_1^{-1}(|y-\xi| - |x-\xi|) \geq (2M_1)^{-1}|y-\xi|,$$

so that

$$I_1(x;\xi) \leq M_2 \int_{E(x,x_0)-B(\xi,2|x-\xi|)} f(y)|\xi - y|^{1-n}dy.$$

Define

$$E' = \left\{ \xi \in \partial G; \limsup_{r\to 0}\ \omega^*(r) r^{1-n} \int_{G\cap B(\xi,r)} f(y)dy > 0 \right\}.$$

Then Lemma 3.1 implies that $H_{\tau_1}(E') = 0$, and hence $H_h(E') = 0$. Further Lemma 3.2 implies that

$$\lim_{x\to\xi} \omega(|x-\xi|) I_1(x;\xi) = 0$$

for any $\xi \in \partial D - E'$. On the other hand, we have by (ii)

$$I_2(x;\xi) \leq M_3|x-\xi|^{1-\delta}.$$

Finally we find

$$I_3(x;\xi) \leq M_4\kappa(\rho(x)) \left(\int_{G\cap B(\xi,2|\xi-x|)} \Psi_p(f(y))\lambda(\rho(y))dy \right)^{1/p}$$

and

$$\omega(\varphi(|x-\xi|))\kappa(\rho(x)) \leq M(a)[h(|x-\xi|)]^{-1/p} \quad \text{for } x \in T_\varphi(\xi, a).$$

Consequently, if $\xi \in \partial G - (E' \cup E_h)$, then

$$\lim_{x\to\xi, x\in T_\varphi(\xi,a)} \omega(\varphi(|x-\xi|))u(x) = 0.$$

Thus Theorem 3.1 is obtained. □

Suppose η is continuous on $(0, \infty)$ and $\kappa(0) = \infty$. If $\omega(r) = [\kappa(r)]^{-\varepsilon}$ for $0 < \varepsilon < 1$, then we can take

$$\omega^*(r) = [\kappa(r)]^{p'-\varepsilon}[r^{n-p}\eta(r)]^{p'/p}.$$

In this case,

$$\begin{aligned} [\omega^*(r)]^{-p}[r^{n-p}\eta(r)] &= [\kappa(r)]^{-p(p'-\varepsilon)}[r^{n-p}\eta(r)]^{-p'/p} \\ &\leq M[\kappa(r)]^{-p(1-\varepsilon)} = M\tau_2(r), \end{aligned}$$

so that, in Theorem 3.1, we may take h as

$$h(r) = \tau(\varphi(r)) \quad \text{with} \quad \tau(r) = [\kappa(r)]^{-p(p'-\varepsilon)}[r^{n-p}\eta(r)]^{-p'/p}.$$

Therefore, as a consequence of Theorem 3.1, we have the following result.

Theorem 3.2. *Suppose η is continuous on $(0, \infty)$ and $\kappa(0) = \infty$. If u is a harmonic function on G satisfying (1.1) and $0 < \varepsilon < 1$, then*

$$u(x) = o([\kappa(\varphi(|x-\xi|))]^{\varepsilon}) \qquad \text{as } x \to \xi, x \in T_{\varphi}(\xi, a),$$

for any $\xi \in \partial G - E_{\varepsilon,\varphi}$ and any $a > 0$, where $H_{h_{\varepsilon,\varphi}}(E_{\varepsilon,\varphi}) = 0$ with

$$h_{\varepsilon,\varphi}(r) = \tau_{\varepsilon}(\varphi(r)), \quad \tau_{\varepsilon}(r) = [\kappa(r)]^{-p(p'-\varepsilon)}[r^{n-p}\eta(r)]^{-p'/p}.$$

Theorem 3.3. *Let u be a function which is harmonic in G and satisfies*

$$\int_G \Psi_p(|\text{grad } u(x)|)\rho(x)^{\beta}dx < \infty.$$

(i) *If $\alpha \geq 1$, $\beta < p-1$ and $n-p+\beta > 0$ and $0 < \varepsilon < 1$, then*

$$u(x) = o([\omega_{\varepsilon}(|x-\xi|^{\alpha})]^{-1}) \qquad \text{as } x \to \xi,\ x \in T_{\alpha}(\xi, a),$$

for any $\xi \in \partial G - E_{\varepsilon,\alpha}$ and any $a > 0$, where

$$\omega_{\varepsilon}(r) = \inf_{t \geq r} [t^{n-p+\beta}\psi(t^{-1})]^{\varepsilon/p}$$

and $H_{h_{\varepsilon,\alpha}}(E_{\varepsilon,\alpha}) = 0$ with

$$h_{\varepsilon,\alpha}(r) = \inf_{t \geq r} [t^{\alpha(n-p+\beta)}\psi(t^{-1})]^{1-\varepsilon}.$$

(ii) *If $\beta = p - n$ and $0 < \varepsilon < 1$, then*

$$u(x) = o([\omega_{\varepsilon}(|x-\xi|)]^{-1}) \qquad \text{as } x \to \xi, x \in T_{\alpha}(\xi, a),$$

for any $\xi \in \partial G - E_{\varepsilon}$, any $\alpha \geq 1$ and any $a > 0$, where $H_{h_{\varepsilon}}(E_{\varepsilon}) = 0$,

$$\omega_{\varepsilon}(r) = \left(\int_r^1 [\psi(t^{-1})]^{-p'/p}\frac{dt}{t}\right)^{-\varepsilon/p'}$$

and

$$h_{\varepsilon}(r) = \left(\int_r^1 [\psi(t^{-1})]^{-p'/p}\frac{dt}{t}\right)^{-p(1-\varepsilon/p')} [\psi(r^{-1})]^{-p'/p}.$$

In [11], we proved the following result concerning global weighted boundary limits.

Theorem 3.4. *Let u be a harmonic function on G satisfying condition* (1.1), *and suppose $\kappa(0) = \infty$. Then*

$$\lim_{x \to \partial G} [\kappa(\rho(x))]^{-1} u(x) = 0.$$

Remark 3.1. If $\lambda(r) = r^{p-n}$ and ψ satisfies the condition:

$$\int_0^1 [\psi(r^{-1})]^{1/(1-p)} \frac{dr}{r} < \infty,$$

then $\kappa(0) < \infty$; in this case it is shown that the function u has a finite limit at any boundary point of G. In particular, in case $p = n$, the same conclusion follows without assuming harmonicity; see Maz'ya [5, Theorem 5.4] and Mizuta [10].

4. Boundary limits of higher derivatives

Let ζ be a nonnegative function in $C_0^\infty(R^1)$ such that $\zeta(r) = 0$ for $|r| > 1$. For $\delta > 0$ and $x \in R^n$, define

$$\zeta_\delta(x) = M\delta^{-n}\zeta(|x|/\delta),$$

where M is chosen so that

$$\int_{R^n} \zeta_\delta(x)dx = 1.$$

Then, by the mean value theorem for harmonic functions, we find that

$$u(x) = u * \zeta_\delta(x) \tag{4.1}$$

for a harmonic function u on G and $\delta < \rho(x)$. Hence it follows that

$$|\text{grad } u(x)| \le M_1 \delta^{-n-1} \int_{B(x,\delta)} |u(y)|dy.$$

Using Hölder's inequality, we have

$$|\text{grad } u(x)| \le M_2 \delta^{-n/q-1} \left(\int_{B(x,\delta)} |u(y)|^q dy \right)^{1/q}$$

for $1 < q < \infty$. Thus, by interpolation, we establish

$$\int_{B(x_0,r_0)} \Psi_p(r_0 |\text{grad } u(x)|)dx \le M_3 \int_{B(x_0,2r_0)} \Psi_p(|u(y)|)dy$$

whenever $B(x_0, 2r_0) \subset G$ (see [13]). Applying a covering lemma with $\{B(x, \rho(x)/4); x \in G\}$ (see [16, Lemma 1.6, Chapter 1]), we have

Lemma 4.1. *If u is a harmonic function on G, then*

$$\int_G \Psi_p(|\text{grad } u(x)|\rho(x))\lambda(\rho(x))dx \le M \int_G \Psi_p(|u(x)|)\lambda(\rho(x))dx.$$

Corollary 4.1. *If u is a harmonic function on G satisfying*

$$\int_G \Psi_p(|\nabla_1 u(x)|)[\rho(x)]^\beta dx < \infty, \qquad \beta > -1, \tag{4.2}$$

then

$$\int_G \Psi_p(|\nabla_m u(x)|)[\rho(x)]^{p(m-1)+\beta} dx < \infty$$

for any positive integer m, where $|\nabla_m u(x)| = \left(\sum_{|\mu|=m} |(\partial/\partial x)^\mu u(x)|^2\right)^{1/2}$.

By using our theorems and Corollary 4.1, we can derive various boundary limits for higher derivatives of harmonic functions. Among them, by Theorem 3.3, we give the following theorem.

Theorem 4.1. *Let u be a harmonic function on G satisfying (4.2). If $\alpha \ge 1$ and $0 < \varepsilon < 1$, then*

$$\rho(x)^{m-1}|\nabla_m u(x)| = o([\omega_\varepsilon(|x-\xi|^\alpha)]^{-1}) \qquad \text{as } x \to \xi,\ x \in T_\alpha(\xi, a),$$

for any $\xi \in \partial G - E_{\varepsilon,\alpha}$ and any $a > 0$, where $\omega_\varepsilon(r) = \inf_{t \ge r} [t^{n+\beta}\psi(t^{-1})]^{\varepsilon/p}$ and

$$H_{h_{\varepsilon,\alpha}}(E_{\varepsilon,\alpha}) = 0$$

with $h_{\varepsilon,\alpha}(r) = \inf_{t \ge r} [t^{\alpha(n+\beta)}\psi(t^{-1})]^{1-\varepsilon}$.

By (4.1), we find

$$|\nabla_m u(x)| \le M(m)\, r^{-n-(m-1)} \int_{B(x,r)} |\nabla_1 u(y)| dy$$

whenever $B(x, 2r) \subset G$, for any positive integer m. Hence Hölder's inequality yields

$$r^{m-1}|\nabla_m u(x)| \le M \left([r^n\psi(r^{-1})]^{-1} \int_{B(x,r)} \Psi_p(|\nabla_1 u(y)|)dy \right)^{1/p} + Mr^{-\delta},$$

where $0 < \delta < n$.

Theorem 4.2. *Let u be a harmonic function on G satisfying condition (1.1). If $\lim_{r \to 0} r^{n-\delta}\omega(r) = 0$ for some $\delta > 0$, then*

$$\lim_{x \to \partial G} [\rho(x)^{n+(m-1)p}\eta(\rho(x))]^{1/p}|\nabla_m u(x)| = 0$$

for any positive integer m.

Remark 4.1. If u is a harmonic function on G with finite Dirichlet integral, that is,

$$\int_G |\nabla\, u(x)|^n dx < \infty,$$

then :

(i) $|\nabla_m\, u(x)| = o([\rho(x)]^{-m})$ as $x \to \partial G$;

(ii) $|\nabla_m\, u(x)| = o(|x-\xi|^{-\alpha m\varepsilon})$ as $x \to \xi, x \in T_\alpha(\xi, a)$, for any $\xi \in \partial G - E_{\varepsilon,\alpha,m}$, where $0 < \varepsilon < 1$ and $H_{\alpha mn(1-\varepsilon)}(E_{\varepsilon,\alpha,m}) = 0$.

These results give generalizations of the results of Hallenbeck-Samotij [4] in the plane case.

5. Best possibility

Here we discuss the best possibility of our theorems, in case G is the upper half space $D = \{x = (x', x_n) \in R^{n-1} \times R^1; x_n > 0\}$.

The following results are given in [11].

Proposition 5.1. *Let λ, ψ and η be as in Theorem 2.1. Suppose $\kappa(0) = \infty$ and $r^\delta \eta(r)^{-1}$ is bounded above on $(0,1]$ for some $\delta > 1-n$. If $a(r)$ is a nonincreasing positive function on the interval $(0,\infty)$ such that $\lim_{r\to 0} a(r) = \infty$, then there exists a nonnegative measurable function f such that $f = 0$ outside $B(0,1)$, $\int_{R^n} \Psi_p(f(y))\lambda(|y_n|)dy < \infty$ and*

$$\limsup_{r\to 0}\; a(r)[\kappa(r)]^{-1} Uf(r\xi) = \infty \quad \textit{for any } \xi \in D,$$

where $Uf(x) = \int_{R^n - D} \frac{x_n - y_n}{|x-y|^n} f(y)dy.$

Proposition 5.2. *Let ψ, λ and η be as in Theorem 2.1. Let φ be a nonnegative nondecreasing function on $(0,\infty)$ such that $\varphi(r) \le Mr$ for any $r > 0$, with a positive constant M, and set*

$$\varphi^*(r) = \int_{\varphi(r)}^{2Mr} [t^{n-p}\eta(t)]^{-p'/p} t^{-1} dt.$$

Suppose further that the following assertions hold.

(i) *$r^{\delta_1}\lambda(r)^{-1}$ is nondecreasing on $(0,\infty)$ for some $\delta_1 > 1/p - n$.*

(ii) *$r^{\delta_2}\lambda(r)$ is nondecreasing on $(0,\infty)$ for some $\delta_2 < 1$.*

(iii) $\varphi^*(r) \to \infty$ *as* $r \to 0$.

(iv) $\varphi^*(r) \le M^* \varphi^*(s)$ *whenever* $0 < s < r$, *with a positive constant* M^*.

Define $h(r) = \inf_{r \le s \le 1} [\varphi^*(s)]^{-p/p'}$. *Then, for a compact set* $K \subset \partial D$ *with* $H_h(K) = 0$, *there exists a nonnegative measurable function* f *on* R^n *such that*

$$\int_{R^n} \Psi_p(f(y))\lambda(|y_n|)dy < \infty$$

and $Uf(x)$ *does not have a finite limit as* $x \to \xi$, $x \in T_\varphi(\xi, 1)$, *for any* $\xi \in K$.

Remark 5.1. As in Lemma 2.3, if $\lambda(r) = r^\beta$ with $-1 < \beta < p-1$, then

$$\int_{R^n} \Psi_p(|\nabla\, Uf(x)|)|x_n|^\beta dx < \infty.$$

Remark 5.2. Proposition 5.1 shows the best possibility of Theorem 3.4 as to the order of infinity.

Remark 5.3. Proposition 5.2 shows the best possiblity of Theorem 2.3 as to the size of the exceptional sets.

References

[1] M. Brelot, Élément de la théorie classique du potentiel, 4 édition, Centre de Documentation Universitaire, Paris, 1969.

[2] L. Carleson, Selected problems on exceptional sets, Van Nostrand, Princeton, 1967.

[3] A. B. Cruzeiro, Convergence au bord pour les fonctions harmoniques dans R^d de la classe de Sobolev W_1^d, C. R. A. S., Paris, 294 (1982), 71-74.

[4] D. J. Hallenbeck and K. Samotij, Radial growth and variation of Dirichlet finite holomorphic functions in the disk, Colloq. Math. 58 (1990), 317-325.

[5] V. G. Maz'ya, Sobolev Spaces, Springer-Verlag, Berlin, 1985.

[6] N. G. Meyers, A theory of capacities for potentials in Lebesgue classes, Math. Scand. 26 (1970), 255-292.

[7] Y. Mizuta, On the boundary limits of harmonic functions with gradient in L^p, Ann. Inst. Fourier 34 (1984), 99-109.

[8] ———, On the boundary limits of harmonic functions, Hiroshima Math. J. 18 (1988), 207-217.

[9] ———, Continuity properties of Riesz potentials and boundary limits of Beppo Levi functions, Math. Scand. 63 (1988), 238-260.

[10] ———, Boundary limits of locally n-precise functions, Hiroshima Math. J. 20 (1990), 109-126.

[11] ———, On the existence of weighted boundary limits of harmonic functions, to appear in Ann. Inst. Fourier 40 (1990).

[12] ———, Continuity properties of potentials and Beppo-Levi-Deny functions, preprint.

[13] ———, Boundary limits of polyharmonic functions in Sobolev-Orlicz classes, preprint.

[14] T. Murai, On the behavior of functions with finite weighted Dirichlet integral near the boundary, Nagoya Math. J. 53 (1974), 83-101.

[15] A. Nagel, W. Rudin and J. H. Shapiro, Tangential boundary behavior of functions in Dirichlet-type spaces, Ann. of Math. 116 (1982), 331-360.

[16] E. M. Stein, Singular integrals and differentiability properties of functions, Princeton Univ. Press, Princeton, 1970.

[17] H. Wallin, on the existence of boundary values of a class of Beppo Levi functions, Trans. Amer. Math. Soc. 120 (1965), 510-525.

Positive harmonic functions on rotationary symmetric Riemannian manifolds

Minoru Murata

Abstract

Milnor's criterion on hyperbolicity of a two dimensional rotationary symmetric Riemannian manifold together with the Riesz-Herglotz theorem and Liouville theorem determines the structure of positive harmonic functions on the manifold. The author gives a higher dimensional version of it, and also gives observations concerning relation between Schrödinger equations and Riemannian geometry.

1980 Mathematics Subject Classification (1985 Revision): 53C20, 58G99, 31B10, 35J10.

0. Introduction

Recently much attention has been paid to close relation between the structure of positive (or bounded) harmonic functions on a Riemannian manifold and the geometry of the underlying manifold (cf. [AS], [C], [GW], [K], [LT], [Marc], [Mi], [S], [T]). In this paper we determine completely and explicitly the structure of positive (or bounded) harmonic functions on a rotationay symmetric Riemannian manifold by applying results and methods in [Mu1&2], and also give observations concerning relation between Schrödinger equations and Riemannian geometry. (As for positive solutions of a Schrödinger equation, see [KN], [Mu3] and references in [Mu1&2].)

In order to state our main results we first recall a theorem of Milnor (cf. [Mi] and [GW]). Let M be a Riemannian manifold with a fixed pole p, rotationary symmetric at p, with $\dim M = n \geq 2$ (cf. [C] and [GW]). Let the Riemannian metric in terms of geodesic polar coordinates at p be of the form

$$ds^2 = dr^2 + f(r)^2\, d\Theta^2, \tag{0.1}$$

where $d\Theta^2$ is the metric of the standard sphere S^{n-1} and f is a nonnegative function on $[0,\infty)$ satisfying

$$f''(r) + K(r)f(r) = 0 \quad \text{on} \quad (0,\infty), \quad f(0) = 0, \quad f'(0) = 1, \tag{0.2}$$

and $f > 0$ on $(0,\infty)$. Here $K(r)$ denotes the radial curvature function.

Theorem (Milnor). *Let $n = 2$. Then one has:*

(a) If $K(r) \geq -1/(r^2 \log r)$ for large r, then M is conformally diffeomorphic to the complex plane $\mathbf{C}$.

(b) If $K(r) \leq -(1+\varepsilon)/(r^2 \log r)$ for large r, where ε is some positive constant, and

Potential Theory, ed. by M. Kishi

$f(r)$ is unbounded, then M is conformally diffeomorphic to an open disc.
(c) M is conformally diffeomorphic to an open disc iff $\int_1^\infty f(r)^{-1}\,dr < \infty$.

This theorem together with the Riesz-Herglots theorem, Louville theorem, etc. determines completely the structure of positive (or bounded) harmonic functions on M with $\dim M = 2$; for example every positive harmonic function is constant iff the integral in (c) diverges. Our main results are the following higher dimensional version of it. Note that the critical constant is $1/2$ for $n \geq 3$, while it is 1 for $n = 2$.

Theorem A. *Let $n \geq 3$. Suppose that*

$$K(r) \geq -(1/2)/(r^2 \log r) \tag{0.3}$$

for sufficiently large r. Then every positive (or bounded) harmonic function is constant.

Theorem B. *Let $n \geq 3$. Suppose that*

$$K(r) \leq -(1/2+\varepsilon)/(r^2 \log r) \tag{0.4}$$

for sufficiently large r, where ε is some positive constant, and that f is unbounded. Then there exists a family $\{P(\cdot;\omega)\}_{\omega\in S^{n-1}}$ of positive harmonic functions continuous on $M \times S^{n-1}$ such that to each positive harmonic function there corresponds a unique finite positive Borel measure μ on S^{n-1} with

$$u(\cdot) = \int_{S^{n-1}} P(\cdot;\omega)d\mu(\omega), \tag{0.5}$$

and for any such a measure the right hand side of (0.5) represents a positive harmonic function. Furthermore, a bounded harmonic function corresponds to a signed measure such that it is absolutely continuous with respect to the standard surface measure dS of S^{n-1} and its Radon-Nikodym derivative is bounded; and for any continuous function φ on S^{n-1} there exists a unique harmonic function u on M satisfying

$$\lim_{|x|\to\infty,\, x/|x|\to\omega} u(x) = \varphi(\omega) \quad \textit{for any} \quad \omega \in S^{n-1}$$

which is represented by (0.5) with $d\mu = (\varphi/|S^{n-1}|)dS$, where $|S^{n-1}|$ is the area of S^{n-1}.

Remark. The positive harmonic function $P(\cdot;\omega)$ is given by the formula (1.7) in Section 1 below.

Theorem C. *Let $n \geq 3$. Put*

$$J(f) = \int_1^\infty \left(f(t)^{n-3}\int_t^\infty f(s)^{1-n}\,ds\right)dt. \tag{0.6}$$

Then one has: (i) If $J(f) < \infty$, then the conclusions of Theorem B hold; and (ii) If there exists a nonconstant positive (or bounded) harmonic function, then $J(f) < \infty$.

In the rest of this paper we assume that $n \geq 3$. The proof of Theorems A, B and C is given in Section 1. We emphasize the naturality of the decay rate $-1/(r^2 \log r)$ of radial curvature, for it is associated with a class of Schrödinger equations on R^n which have infinitely many linearly independent positive solutions (see Section 2, in particular, Example 2.5 and Remark 2.6).

Acknowledgment. After completing a preliminary version of this paper the author has learned that combinig results in [P] and [Marc] one can give a probabilistic proof of Theorems A, B and C except for the latter half of Theorem B, the explicit formula (1.7) of $P(\cdot;\omega)$, and a related part of Theorem C. He thanks Ross G. Pinsky for sending a preprint [P], from which results of [Marc] concerning bounded harmonic functions has come to his notice. He wishes to cordially thank Seiki Nishikawa for interest in the work [Mu1,2,3] and valuable advice concernig Riemannian geometry.

1. Martin boundary, Dirichlet problem at infinity, and a curvature comparison theorem

This section is devoted to a proof of Theorems A, B and C. Let Δ_M be the Laplace-Beltrami operator for M:

$$\Delta_M = f^{1-n}\, \partial/\partial r(f^{n-1}\, \partial/\partial r) + f^{-2}\Lambda, \tag{1.1}$$

where Λ is the Laplace-Beltrami operator of the sphere. We shall say that M is *subcritical* (resp. *critical*) if there exists a (resp. no) minimal positive Green function for $-\Delta_M$ on M. When M is subcritical, we denote by G, $\bar{M}$, Σ and P the Green function, Martin compactification of M, Martin boundary and Martin kernel for M with reference point p, respectively (cf. [Mart] and [Mu1,2,3]).

For $j = 0, 1, \cdots$, let $\lambda_j = j(j+n-2)$ and g_j be a solution of the initial value problem

$$f^{1-n}(f^{n-1}g_j')' - \lambda_j f^{-2} g_j = 0 \quad \text{on} \quad (0,\infty), \tag{1.2}$$

$$g_j(r) = r^j(1+o(1)) \quad \text{as} \quad r \to 0. \tag{1.3}$$

Clearly, $g_0 \equiv 1$, $g_j > 0$ on $(0,\infty)$ and $\inf\{g_{j+1}(t)/g_j(t);\, t \geq \delta\} > 0$ for any j and $\delta > 0$.

Theorems 3.2 and 1.6 of [Mu2] yield the following theorem.

Theorem 1.1. *M is subcritical if and only if*

$$I(f) \equiv \int_1^\infty f(r)^{1-n}\, dr \quad < \infty. \tag{1.4}$$

Furthermore, if (1.4) *does not hold, then every positive (or bounded) harmonic function is constant.*

Suppose that (1.4) holds. Put

$$f_j(r) = g_j(r)\int_r^\infty g_j(t)^{-2}f(t)^{1-n}dt,$$

and $H_j(r,s) = g_j(r)f_j(s)$ for $r \le s$, and $H_j(r,s) = f_j(r)g_j(s)$ for $r \ge s$. We see that H_j is equal to the Green function for (1.2) on $(0,\infty)$ with respect to the measure $f^{n-1}dr$. Theorem 3.4 in [Mu2] then yields the following theorem (see also results in Section 3 of [Mu1]).

Theorem 1.2. *If* (1.4) *holds, then the Green function G is represented by*

$$G(r,\theta;s,\omega) = \sum_{j=0}^\infty g_j(r)f_j(s)\nu_j P_j(\theta\cdot\omega) \quad for \quad r \le s, \tag{1.5}$$

where $\nu_j = (2j+n-2)\Gamma(n/2)/[2(n-2)\pi^{n/2}]$ with Γ being the gamma function and P_j are Gegenbauer's polynomials defined by

$$\sum_{j=0}^\infty P_j(t)z^j = (1-2tz+z^2)^{-(n-2)/2}, \quad |t| \le 1, \quad |z| < 1. \tag{1.6}$$

Theorem 3.5 of [Mu2], the proof of Theorem 3.4 of [Mu1], and a lemma analogous to Lemma 3.7 of [Mu1] (for the lemma, see also Theorem 9.1 and Exercise 9.4 in [H]) yield the following theorem.

Theorem 1.3. *Suppose that* (1.4) *holds. Then one has:*
(i) Suppose that $J(f) = \infty$, where $J(f)$ is the integral defined by (0.6). *Then the Martin boundary Σ (resp. the Martin compactification M) is homeomorphic to one point (resp. one point compactification of M); and every positive (or bounded) harmonic function is constant.*
(ii) Suppose that $J(f) < \infty$. Then Σ and M are homeomorphic to S^{n-1} and the disjoint union $R^n \cup S^{n-1}$ of R^n and S^{n-1} equipped with the standard topology, respectively; for any j, $a_j = \lim_{t\to\infty} g_j(t)^{-1}$ exists and is a positive number; the Martin kernel P is represented by

$$P(r,\theta;\omega) = \sum_{j=0}^\infty [(2j+n-2)/(n-2)]a_j\, g_j(r)P_j(\theta\cdot\omega) \tag{1.7}$$

and the conclusions of Theorem B holds with $P(\cdot;\omega)$ being equal to the right hand side of (1.7).

Corollary 1.4. *Every positive (or bounded) harmonic function is constant if and only if $J(f) = \infty$.*

Remark. We assumed that $n \ge 3$. But note that if $n = 2$, the following three conditions are equivalent: (i) $I(f) < \infty$; (ii) $J(f) < \infty$; and (iii) M is hyperbolic (cf. [Mi]).

In proving Theorems A and B the following comparison theorem plays a crucial role.

Proposition 1.5. *For $j = 1,2$, let f_j be a positive C^2-function on $[0,\infty)$ satisfying the equation $f_j''(r) + K_j(r)f_j(r) = 0$. Suppose that $K_1 \leq K_2$ and $K_1 \leq 0$ outside some compact set, and f_1 is unbounded. Then one has:*
(*i*) $J(f_1) = \infty$ *implies* $J(f_2) = \infty$.
(*ii*) $J(f_2) < \infty$ *implies* $J(f_1) < \infty$.

We omit the proof of this lemma, since the original proof in the preliminary version of this paper is similar to that in Appendix of [Marc].

We are now ready to prove Theorems A and B. Note that Theorem C has already been shown by Theorems 1.1 and 1.3.

Proof of Theorem A. With $h_\alpha(r) = r(\log r)^{1/2}(\log\log r)^\alpha$, we have that

$$h_\alpha'(r)/h_\alpha(r) = \{1 + 1/(2 \log r) + \alpha/[(\log r)(\log \log r)]\}/r, \tag{1.8}$$

$$h_\alpha''(r)/h_\alpha(r) = \{1/2 + \alpha/(\log\log r) + O(1/\log r)\}/(r^2 \log r) \tag{1.9}$$

as $r \to \infty$.

Let f_1 be a smooth function on $[0,\infty)$ such that $f_1(0) = 0, f_1'(0) = 1$, $f_1(r) > 0$ for $r > 0$, and $f_1(r) = h_{1/2}(r)$ for $r > 3$. Put $K_1 = -f_1''/f_1$. Then we see that $K_1 < 0$ and $K \geq K_1$ near infinity and $J(f_1) = \infty$. Hence Proposition 1.5 and Corollary 1.4 imply the theorem. Q.E.D.

Proof of Theorem B. Let f_2 be a smooth function on $[0,\infty)$ such that $f_2(0) = 0$, $f_2'(0) = 1$, $f_2(r) > 0$ for $r > 0$, and $f_2(r) = h_{1/2+\varepsilon}(r)$ for $r > 3$ with h_α being the function given in the proof of Theorem A. Put $K_2 = -f_2''/f_2$. Then we see that $K \leq K_2 < 0$ outside a compact set, and $J(f_2) < \infty$. Hence Proposition 1.5 and Theorem 1.3 imply the theorem. Q.E.D.

2. Schrödinger equations and decay rate of curvature

In this section we give several observations concerning relation between Schrödinger equations and Riemannian geometry. Consider a stationary Schrödinger equation

$$Lu \equiv (-\Delta + V)u = 0 \quad \text{on} \quad R^n, \tag{2.1}$$

where V, which is called a potential of the Schrödinger operator L, is a real-valued smooth function on R^n. Suppose that (2.1) has a positive solution g, and put $v =$

u/g. Then we have

$$gLu = (gLg)v = -\sum_{j=1}^{n} \partial/\partial x_j(g^2\, \partial v/\partial x_j). \tag{2.2}$$

On the other hand, consider the Riemannian metric

$$ds^2 = g^{4/(n-2)} \sum_{j=1}^{n} (dx_j)^2 \tag{2.3}$$

on R^n. Then the Laplace-Beltrami operator Δ_M of the Riemannian manifold $M = (R^n, ds^2)$ is represented by

$$\Delta_M = g^{-2n/(n-2)} \sum_{j=1}^{n} (\partial/\partial x_j)(g^2\, \partial/\partial x_j). \tag{2.4}$$

Thus, $Lu = 0$ if and only if $\Delta_M v = 0$ with $v = u/g$. We are interested in connection between the potential V and the curvature of M. Now suppose that g is a radial function (i.e. it depends only on $r = |x|$) such that $g(0) = 1$. Then V must be radial, and g satisfies the equation

$$-r^{1-n}(r^{n-1}g'(r))' + V(r)g(r) = 0 \quad \text{on} \quad (0, \infty). \tag{2.5}$$

Introduce a function $\rho(r)$ by

$$\rho(r) = \int_0^r g(t)^{2/(n-2)}\, dt. \tag{2.6}$$

With ρ being a new variable for r, we define a function f of ρ by

$$f(\rho) = rg(r)^{2/(n-2)}. \tag{2.7}$$

Then the metric (2.3) and the Laplace-Beltrami operator (2.4) become (0.1) and (1.1) with r replaced by ρ, respectively. Since the radial curvature $K(\rho)$ is equal to $-[(d/d\rho)^2 f(\rho)]/f(\rho)$, we obtain from (2.5) $\sim$ (2.7) that

$$K(\rho) = -\frac{2}{n-2}\, g(r)^{-4/(n-2)} \left\{ V(r) - \frac{n-2}{r}\frac{g'(r)}{g(r)} - \left(\frac{g'(r)}{g(r)}\right)^2 \right\}. \tag{2.8}$$

We shall give several examples which suggest a close connection between V and K. Before giving them we should recall results in [Mu1] (see Theorems 3.1 and 3.3 there).

Fact. (i) (2.1) has a positive Green function on R^n if and only if

$$I^{\sim}(g) \equiv \int_1^{\infty} s^{1-n}\, g(s)^{-2}\, ds < \infty. \tag{2.9}$$

(ii) The Martin boundary for (2.1) is S^{n-1} if and only if

$$J^{\sim}(g) \equiv \int_1^{\infty} t^{n-3}\, g(t)^2 \int_t^{\infty} s^{1-n}\, g(s)^{-2}\, ds \;<\; \infty. \tag{2.10}$$

Putting $h(t) = t^{(n-1)/2}\, g(t)$, we have that

$$-h'' + [V(r) + (n-1)(n-3)/4r^2]h = 0 \qquad \text{on} \qquad (0,\infty), \tag{2.11}$$

$$J^{\sim}(g) = \int_1^{\infty} dt\; t^{-2} \int_t^{\infty} [h(t)/h(s)]^2\, ds. \tag{2.12}$$

Since $h(t)/h(s) = \exp\{-\int_t^s [h'(x)/h(x)]dx\}$, elementary calculations yield the following proposition.

Proposition 2.1. *For $j = 1$ or 2, let g_j be a positive C^2-function on $[0,\infty)$ satisfying the equation*

$$-r^{1-n}(r^{n-1}g_j'(r))' + V_j(r)g_j(r) = 0 \qquad on \qquad (0,\infty).$$

Suppose that $V_1 \geq V_2$ and $V_1 \geq 0$ outside some compact set, and g_1 is unbounded near infinity. Then one has:
(i) $J^{\sim}(g_1) = \infty$ *implies* $J^{\sim}(g_2) = \infty$.
(ii) $J^{\sim}(g_2) < \infty$ *implies* $J^{\sim}(g_1) < \infty$.

Remark. This kind of monotonicity was first observed by Kawamura and Nakai [KN].

We are now ready to give examples, where we assume (2.9). In what follows, when $\lim_{t\to\infty} f(t)/g(t)$ exists and is a nonzero number, we write $f \sim g$; and r and ρ are supposed to be sufficiently large.

Example 2.2. Let $V(r) = r^{-2-\alpha}$ for some $\alpha > 0$ with $\alpha \neq n-2$. Since we assumed (2.9), we obtain by (2.5) that $g(r) \sim 1$ and

$$g'(r)/g(r) = (r^{-1-\alpha}/|n-2-\alpha|)(1 + o(1)) \qquad \text{as} \qquad r \to \infty.$$

Clearly, $J^{\sim}(g) = \infty$. By (2.6), $\rho(r) \sim r$, which together with (2.8) shows that

$$K(\rho) \sim [(n-2)/|n-2-\alpha| - 1]\rho^{-2-\alpha}.$$

Note that $K(\rho) > 0$ when $0 < \alpha < n-2$ or $n-2 < \alpha < 2(n-2)$, and $K(\rho) < 0$ when $\alpha \geq 2(n-2)$. Compare this with the fact that $K(\rho) = 0$ if $V(r) = 0$ or r^{-2} (see the following example).

Example 2.3. Let $V(r) = ar^{-2}$ for some $a > 0$. Then $g(r) \sim r^{\gamma}$ with γ being the positive root of the equation: $\gamma^2 + (n-2)\gamma - a = 0$. We see that $J^{\sim}(g) = \infty$, $g'(r)/g(r) = \gamma + o(1)$ and $K(\rho) = 0$.

Example 2.4. Let $V(r) = (\log r)^{2\beta} r^{-2}$ for some $\beta > 1$. By applying Theorem (ε) of [HW] to (2.11) and using (2.9) we obtain that

$$g(r) \sim r^{-(n-1)/2}\, q(r)^{-1/4} \exp\{ \int^r q^{1/2}[1 + q'^2/16q^3]^{1/2}\, dr\}, \tag{2.13}$$

where $q(t) = V(t) + (n-1)(n-3)/4t^2$. We have that

$$\int^r q^{1/2}(1 + q'^2/16q^3)^{1/2} dr = (\log r)^{1+\beta}/(1+\beta) + O(1),$$

$$r^{-(n-1)/2}\, q^{-1/4}(r) = r^{-(n-2)/2}\, (\log r)^{-\beta/2}\, (1 + o(1)).$$

We then obtain that

$$g'(r)/g(r) = r^{-1}[(\log r)^{\beta} - (n-2)/2 - (\beta/2)(\log r)^{-1} + O((\log r)^{-\beta})].$$

Thus

$$V(r) - [(n-2)/r][g'/g] - [g'/g]^2 = r^{-2}[\beta(\log r)^{\beta-1} + O(1)].$$

By (2.6),

$$\rho(r) = [(n-2)/2] r (\log r)^{-\beta} g(r)^{2/(n-2)}[1 + o(1)].$$

Hence we obtain by (2.8) that

$$K(\rho) = -[\beta/(\beta+1) + o(1)]/(\rho^2 \log \rho).$$

Note that $\beta/(\beta+1) > 1/2$ and $J^{\sim}(g) < \infty$, since $\beta > 1$; but $J^{\sim}(g) = \infty$ if $\beta \leq 1$.

Example 2.5. Let $V(r) = r^{-2+\alpha}$ for some $\alpha > 0$. Then we obtain as in the Example 2.4 that $J^{\sim}(g) < \infty$ and

$$K(\rho) = -[1 + o(1)]/(\rho^2 \log \rho) \quad \text{as} \quad \rho \to \infty.$$

Remark 2.6. Let M be a rotationary symmetric Riemannian manifold with metric (0.1) and radial curvature K such that f is unbounded and $K(\rho) \leq -(1+\varepsilon)/(\rho^2 \log \rho)$ for sufficiently large ρ with $\varepsilon > 0$. Then M cannot be realized by any Schrödinger equation on R^n. If it could, then we have by (2.6) and (2.7) that $f(\rho)^{-1} d\rho = r^{-1} dr$. But $f(\rho) \geq C\, \rho(\log \rho)^{1+\varepsilon}$ for some positive constant C, and $\int^{\infty} f(\rho)^{-1} d\rho < \infty$. Thus, when $0 \leq \rho < \infty$, the variable r runs in the interval $[0, R)$ for some positive constant R. This is a contradiction. Hence, M must be realized, if possible, by a Schrödinger equation on a bounded ball.

References

[AS] Anderson, M.T., Shoen, R., Positive harmonic functions on complete manifolds of negative curvature, Ann. of Math. 121(1985), 429-461

[C] Choi,H.I., Asymptotic Dirichlet problems for harmonic functions on Riemannian manifolds, Trans. Amer. Math. Soc., 281(1984), 691-716

[GW] Green,R.E., Wu,H., Function Theory on Manifolds Which Possess a Pole, Lecture Notes in Math. No. 699, Springer Verlag, Berlin-Heidelberg-New York, 1979

[H] Hartman, P., Ordinary Differential Equations, John Wiley & Sons, New York-London-Sydney, 1964

[HW] Hartman P., Wintner,A., Asymptotic integrations of linear differential equations, Amer. J. Math., 77(1955), 45-86

[K] Kasue, A., Harmonic functions with growth conditions on a manifold of asymptotically non-negative curvature I , Lecture Notes in Math. No. 1339, Springer-Verlag, Berlin-Heidelberg-New York, 1988, 158-181

[KN] Kawamura, M., Nakai, M., A test of Picard principle for rotation free densities, II , J. Math. Soc. Japan, 28(1976), 323-342

[LT] Li, P., Tam, L.F., Positive harmonic functions on complete manifolds with non-negative curvature outside a compact set, Ann. of Math., 125(1987), 171-207

[Marc] March, P., Brownian motion and harmonic functions on rotationary symmetric manifolds, Ann. Prob., 14(1986), 793-801

[Mart] Martin, R.S., Minimal positive harmonic functions, Trans. Amer. Math. Soc., 49(1941), 137-172

[Mi] Milnor, J., On deciding whether a surface is parabolic or hyperbolic, Amer. Math. Monthly, 84(1977), 43-46

[Mu1] Murata, M., Structure of positive solutions to $(-\Delta + V)u = 0$ in R^n, Duke Math. J., 53(1986), 869-943

[Mu2] Murata, M., On construction of Martin boundaries for second order elliptic equations, Publ. RIMS, Kyoto Univ., 26(1990), 585-627

[Mu3] Murata, M., Martin compactification and asymptotics of Green functions for Schrödinger operators with anisotropic potentials, to appear in Math. Ann.,(1990)

[P] Pinsky, R.G., A new approach to the Martin boundary via conditioned diffusions and h-transforms, Preprint, Aug. 1990

[S] Shiga, K., A geometric characterization of C^n and open balls, Nagoya Math. J., 75(1979), 145-150

[T] Takegoshi, K., Energy estimates and Liouville theorems for harmonic maps, Preprint, Sep. 1988

Examples of stable Martin boundaries of Markov chains

Massimo Picardello and Wolfgang Woess

Abstract

Let P be the transition operator of a countable Markov chain, and denote by ρ its spectral radius. A Martin boundary $\mathcal{M}(P,t)$ is defined for every $t \geq \rho$. In all the examples known to the authors, $\mathcal{M}(P,t)$ has a remarkable stability property: it does not depend on t if $t > \rho$ (say, $\mathcal{M}(P,t)$ is homeomorphic to the same compact space $\mathcal{M}(P)$ for $t > \rho$). On the other hand, in the available examples $\mathcal{M}(P,\rho)$ is either homeomorphic to $\mathcal{M}(P)$ or smaller (i.e., it is a homeomorphic image of $\mathcal{M}(P)$), and finally, the Poisson-Martin kernels $K(x,\xi|t)$ are jointly continuous on $\mathcal{M}(P) \times [\rho,\infty)$ for every state x. The latter property may turn out crucial in determining the Martin topology of $\mathcal{M}(P,t)$: see [PW2]). In this paper we give a precise definition of stability of the Martin boundary and discuss several examples in detail. In particular, we deal with transition operators on a (not necessarily homogeneous) tree T which are not group-invariant but satisfy natural boundedness properties. For these operators, $\mathcal{M}(P,t)$ is known to be homeomorphic to the space of ends of T. We show that their Martin boundaries are stable: namely, the Poisson-Martin kernels are jointly continuous. In particular, group-invariant bounded-range transition operators on trees have stable Martin boundaries.

1980 Mathematics Subject Classification(1985 Revision): 60J50, 05C05, 60J15.

1. Introduction, preliminaries

Let X be a countable, discrete *state space*. A *stochastic transition operator* on X is given by a matrix $P = (p(x,x'))_{x,x' \in X}$ with nonnegative entries and row sums one. It gives rise to a time-homogeneous Markov chain Z_n, $n = 0,1,2,\cdots$, on X, such that the entries of P are the one-step transition probabilities. If $x,x' \in X$ and $n \geq 0$, then the (x,x')-entry of the matrix power P^n is

$$p^{(n)}(x,x') = \Pr[Z_n = x' \mid Z_0 = x] \, .$$

($P^0 = I$, the identity matrix). We shall always assume that P is *irreducible:* for every $x,x' \in X$, we have $p^{(n)}(x,x') > 0$ for some $n \geq 0$. On real valued functions f on X, P acts by

$$Pf(x) = \sum_{x'} p(x,x')f(x') \, ,$$

whenever this sum converges for every $x \in X$. A t-*harmonic function* is an eigenfunction h of P with respect to eigenvalue t: $Ph = t \cdot h$ ($t \in \mathbf{R}$). The linear space

Potential Theory, ed. by M. Kishi

of t-harmonic functions is denoted by $\mathcal{H}(P,t)$. In particular, we shall be interested in the cone $\mathcal{H}^+(P,t)$ of positive t-harmonic functions.

Consider the *convergence norm* (sometimes also called spectral radius)

$$\rho(P) = \limsup_{n\to\infty} p^{(n)}(x,x')^{1/n} .$$

By irreducibility, $\rho(P)$ is independent of x and x'. We say that P has *finite range* if $\{x' \mid p(x,x') > 0\}$ is finite for every $x \in X$.

Theorem 1.1 [Pr]. *$\mathcal{H}^+(P,t)$ is nonvoid only if $t \geq \rho(P)$. If X is infinite and P has finite range then $\mathcal{H}^+(P,t)$ is nonvoid if and only if $t \geq \rho(P)$.*

Let $e \in X$ be a reference point. A function $h \in \mathcal{H}^+(P,t)$ is called *minimal* or *extremal* if $h(e) = 1$ and, whenever $0 \leq h_1 \leq h$ and $h_1 \in \mathcal{H}(P,t)$, then h_1/h is constant. The set of minimal t-harmonic functions is denoted by $\mathcal{E}(P,t)$. The positive t-harmonic functions are determined by the minimal ones.

In addition to the n-step transition probabilities $p^{(n)}(x,x')$, define

$$f^{(n)}(x,x') = Pr[\mathcal{X}_n = x' \; ; \; \mathcal{X}_i \neq x' \text{ for } 0 \leq i < n \mid \mathcal{X}_0 = x] ,$$
$$f^{(0)}(x,x') = 0$$

and

$$F(x,x'|t) = \sum_{n=0}^{\infty} f^{(n)}(x,x')\frac{1}{t^n} , \quad t \geq \rho(P) .$$

The *Green kernel* is

$$G(x,x'|t) = \sum_{n=0}^{\infty} p^{(n)}(x,x')\frac{1}{t^{n+1}} .$$

It converges for real $t > \rho(P)$. Furthermore, by [V-J], for $t = \rho(P)$

$$\text{either} \quad G(x,x'|\rho(P)) < \infty \quad \forall x,x' \in X$$
$$\text{or} \quad G(x,x'|\rho(P)) = \infty \quad \forall x,x' \in X .$$

P is called *$\rho(P)$-transient* in the first case and *$\rho(P)$-recurrent* in the second. The following is essentially due to [Pr].

Theorem 1.2. *Assume that P is $\rho(P)$-recurrent.*

(a) *There is precisely one function h in $\mathcal{E}(P,\rho(P))$, and every function in $\mathcal{H}^+(P,\rho(P))$ is a constant multiple of h.*

(b) *$F(x,x'|\rho(P))$ is finite for all $x,x' \in X$, and*

$$h(x) = F(x,x'|\rho(P))h(x') .$$

Proof. (a) is proved in [Pr]. (b) can be deduced in several ways. Consider the h-process [Do]:

$$p_h(x,x') = \frac{p(x,x')h(x')}{\rho(P)h(x)} .$$

It is recurrent in the usual sense, i.e., $\rho(P_h) = 1$ and $G_h(x,x'|1) = \infty$ for all $x, x' \in X$. In particular,

$$1 = F_h(x,x'|1) = \frac{F(x,x'|\rho(P))h(x')}{h(x)} \quad \forall\, x,x' \in X .$$

This yields (b). □

The *Martin kernel* is

$$K(x,x'|t) = \frac{F(x,x'|t)}{F(e,x'|t)} , \quad t \geq \rho(P) .$$

Recall the formula $G(x,x'|t) = F(x,x'|t)G(x',x'|t)$. Hence, if $t > \rho(P)$ or if $t = \rho(P)$ and P is $\rho(P)$-transient, then $K(x,x'|t) = G(x,x'|t)/G(e,x'|t)$.

The *Martin compactification* $\widehat{X}(P,t)$ of X with respect to P and $t \geq \rho(P)$ is uniquely determined up to homomorphism by

(I) $\widehat{X}(P,t)$ is compact, and X is discrete and open,
(II) $K(\cdot,\cdot|t)$ extends continuously to $X \times \widehat{X}(P,t)$, and
(III) the extended kernels (also denoted by $K(\cdot,\cdot|t)$) separate the points of the *Martin boundary* $\mathcal{M}(P,t) = \widehat{X}(P,t) \setminus X$.

The restrictions of the Martin kernels to $X(P,t) \times \mathcal{M}(P,t)$ are sometimes called *Poisson-Martin kernels.*

It is known [Do], [Hu], [KSK] that every minimal t-harmonic function is of the form $K(\cdot,\xi|t)$, where $\xi \in \mathcal{M}(P,t)$. We identify $\mathcal{E}(P,t)$ with the corresponding (Borel) set of points in the Martin boundary. The importance of the Martin boundary relies on the following fact.

Theorem 1.3 [Do], [Hu]. *Every $h \in \mathcal{H}^+(P,t)$ has a unique integral representation*

$$h = \int_{\mathcal{M}(P,t)} K(\cdot,\xi|t)\,\nu^h(d\xi) ,$$

where ν^h is a nonnegative Borel measure on $\mathcal{M}(P,t)$ satisfying

$$\nu^h(\mathcal{M}(P,t) \setminus \mathcal{E}(P,t)) = 0 .$$

We remark that in our definition of the Martin compactification, $\widehat{X}(P,\rho(P))$ is the one point-compactification of X in the $\rho(P)$-recurrent case. This is not the recurrent Martin boundary, as defined in [KSK,Ch. 11]. In the $\rho(P)$-recurrent case we have in particular by Theorem 1.2(b)

$$K(x,x'|\rho(P)) = F(x,e|\rho(P)) = h(x) \quad \forall x,x' \in X ,$$

where h is the unique function in $\mathcal{E}(P,\rho(P))$.

2. Stability of the boundary

Let (X,P) be as in the preceding section, X infinite.

Definition 2.1. We say that (X,P) has *stable boundary* if the following conditions are satisfied

(1) For $t_1,t_2 > \rho(P)$, id_X extends to a homeomorphism $\widehat{X}(P,t_1) \to \widehat{X}(P,t_2)$. (That is, $K(\cdot,x_n|t_1)$ converges pointwise on X as $n \to \infty$ if and only if $K(\cdot,x_n|t_2)$ converges). In this case, we write $\mathcal{M}(P) = \mathcal{M}(P,t)$, $t > \rho(P)$.

(2) For $t > \rho(P)$, id_X extends to a continuous surjection $\tau : \widehat{X}(P,t) \to \widehat{X}(P,\rho(P))$. We write $K(\cdot,\xi|\rho(P)) = K(\cdot,\tau(\xi)|\rho(P))$, if $\xi \in \mathcal{M}(P)$.

(3) The map $(\xi,t) \mapsto K(x,\xi|t)$ is jointly continuous on $\mathcal{M}(P) \times [\rho(P),\infty)$.

In addition, we say that (X,P) has *strictly stable boundary* if (1) holds for all $t_1,t_2 \geq \rho(P)$ (or, in other words, τ in (2) is a homeomorphism).

For all irreducible Markov chains with infinite state space where the boundary is known, there is at least strong evidence that it is stable. However, the authors do not know any general theorem proving this fact. Furthermore, in many cases $\mathcal{M}(P,t)$ contains no nonminimal elements. i.e., $\mathcal{M}(P,t) = \mathcal{E}(P,t)$ (but there are examples where this does not hold [PW2]). Boundary stability is useful, for example, in the study of the Martin boundary of a Cartesian product: see [PW2].

We list some examples. Given a countable group Γ and a probability measure μ on Γ, the *random walk* on Γ with *law* μ is the Markov chain with state space $X = \Gamma$ and transition probabilities $p(x,y) = \mu(x^{-1}y)$, $x,y \in \Gamma$. "Irreducible" then means that the support of μ generates Γ as a semigroup.

A. Random walks on $\mathbf{Z}^d$ [He],[NS]

On the d-dimensional integer lattice, consider a probability measure μ giving rise to an irreducible random walk. Assume that μ has finite support. Consider the function

$$\Phi(c) = \sum_x e^{c\cdot x}\mu(x) , \quad c \in \mathbf{R}^d ,$$

where $c \cdot x$ is scalar product. Then by [He], [NS]

$$\rho(\mu) = \min \Phi$$

and

$$\mathcal{E}(\mu,t) = \{h(x) = e^{c\cdot x} \mid \Phi(c) = t\} .$$

Furthermore $\mathcal{M}(\mu,t) = \mathcal{E}(\mu,t)$ carries the Euclidean topology of $\{c \in \mathbf{R}^{\mathrm{d}} \mid \Phi(c) = t\}$, which is homeomorphic with the unit sphere $\mathbf{S}^{d-1}$ for $t > \rho(\mu)$ and collapses to a point for $t = \rho(\mu)$ [NS]. For a sequence (x_n) in $\mathbf{Z}^d$ and $t > \rho(\mu)$, the kernels $K(\cdot, x_n|t)$ converge pointwise to a boudary kernel if and only if

$$|x_n| \to \infty \quad \text{and} \quad \frac{x_n}{|x_n|} \to \xi \in \mathbf{S}^{d-1} .$$

In this case $K(x, x_n|t) \to e^{c\cdot x}$, where c is such that $\nabla\Phi(c)$ is proportional with ξ. In particular, the Martin boundary is (nonstrictly) stable.

B. Random walks on free products [W2]

Let Γ_1 and Γ_2 be discrete groups with identity e carrying irreducible probability measures μ_1 and μ_2 with finite support, respectively. On the free product $\Gamma = \Gamma_1 * \Gamma_2$, consider $\mu = a\cdot\mu_1 + (1-a)\cdot\mu_2$, where $0 < a < 1$. Recall that every $x \in \Gamma$ can be written uniquely as a word

$$x = x_1 \cdots x_k , \quad \text{where } k \geq 0 , \ x_j \in \Gamma_{i_j} \setminus \{e\} , \ i_j \neq i_{j+1}$$

(e is represented by the empty word, with $k = 0$). By [W1] and [W2], for $t \geq \rho(\mu)$ there are $t_i = t_i(t) \geq \rho(\mu_i)$, $i = 1, 2$, such that one can decompose $\mathcal{M}(\mu,t)$ into three parts:

$$\mathcal{M}(\mu,t) = \Omega^{(0)} \cup \mathcal{M}^{(1)}(t_1) \cup \mathcal{M}^{(2)}(t_2) ,$$

where

$$\Omega^{(0)} = \{x_1x_2x_3\cdots \mid x_j \in \Gamma_{i_j} \setminus \{e\} , \ i_j \neq i_{j+1}\}$$

is the set of infinite words, and

$$\mathcal{M}^{(i)}(t_i) = \bigcup x_1 \cdots x_k \mathcal{M}(\mu_i, t_i) ,$$

a disjoint union with respect to all $x_1 \cdots x_k$ as above such that $x_k \in \Gamma_{3-i} \setminus \{e\}$ if $k > 0$. The topology is governed by convergence of words and such that the sets $x_1 \cdots x_k \mathcal{M}(\mu_i, t_i)$ are closed and carry the topology of $\mathcal{M}(\mu_i, t_i)$, see [W2] for the precise description.

In particular, suppose that μ_1 and μ_2 have stable boundaries. Then also μ has stable boundary. Furthermore, μ has strictly stable boundary in many cases: this happens

(a) if μ_1 and μ_2 have strictly stable boundaries, or
(b) if for $i = 1, 2$, Γ_i is finite or μ_i has stable boundary and $t_i(\rho(\mu)) > \rho(\mu_i)$.

Furthermore, if all boundary functions for μ_i ($i = 1, 2$) are minimal, then also $\mathcal{M}(\mu, t) = \mathcal{E}(\mu, t)$. We remark that (b) is satisfied in many cases, for example if the groups Γ_i have polynomial growth of degree not exceeding 4 (e.g., Γ_i virtually abelian of rank ≤ 4) [W1], [W2].

Let Γ_0 be any finitely generated group with at least three elements. [C2] shows that there is a finitely supported, symmetric, irreducible probability measure μ_0 on Γ_0 with the following property: if in the above setting, we choose Γ_1, Γ_2 equal to Γ_0 and μ_1, μ_2 identical copies of μ_0, then for $\mu = \frac{1}{2}(\mu_1 + \mu_2)$ one has $t_i(\rho(\mu)) > \rho(\mu_i)$. Therefore, if μ_0 has stable boundary, then μ has strictly stable boundary. On the other hand, if we choose $\Gamma_0 = \mathbf{Z}^d$, $d \geq 5$, and μ_0' to be the equidistribution on the unit vectors and their inverses, then the same construction of Γ and $\mu' = \frac{1}{2}(\mu_1' + \mu_2')$ as above yields the simple random walk on $\mathbf{Z}^d * \mathbf{Z}^d$ which satisfies $t_i(\rho(\mu)) = \rho(\mu_0')$ [C1]. Hence, for μ' the boundary is stable but not strictly stable. We remark that the measure μ described above can be chosen to have the same support as μ'. Therefore on $\Gamma = \mathbf{Z}^d * \mathbf{Z}^d$ there are two symmetric probability measures μ and μ' with the same finite support such that the boundary is strictly stable for μ, while it is nonstrictly stable for μ'. In both cases, all boundary points are minimal.

3. Stability of boundaries of random walks on trees

In this section we complete the characterization of the Martin boundary of random walks on trees given in [PW1] by proving its stability. We consider $X = T$ to be the vertex set of an infinite, locally finite tree and require that P satisfies the assumptions of [PW1]: there are $\delta > 0$ and integers $M, s > 0$, such that

(i) if x and x' are neighbours then $p^{(k)}(x, x') > 0$ for some $k \leq M$, and
(ii) if $p(x, x') > 0$ then $p(x, x') \geq \delta$ and $d(x, x') \leq s$.

Here, d denotes the discrete metric of T.

The space of *ends* Ω of T consists of equivalence classes of one-sided infinite paths without repeated vertices: two such paths are equivalent if they differ only by finitely many vertices. If $x, x' \in T$, then there is a unique shortest path in T connecting the two, called the *geodesic segment* $\overline{xx'}$. Analogously, if $x \in T$, $\omega \in \Omega$, there is a unique infinite path which represents ω and starts at x, called the *geodesic ray* $\overline{x\omega}$. If $x \in T$ and $\xi, \eta \in T \cup \Omega$, then the *confluent* $c(x, \xi, \eta)$ is the last common vertex of $\overline{x\xi}$ and $\overline{x\eta}$. The sets

$$B_{x,x'} = \{\xi \in T \cup \Omega \mid x' \in \overline{x\xi}\}\ , \quad x, x' \in T\ , \quad x \neq x'$$

constitute a subbasis of the *end topology*, and $T \cup \Omega$ becomes a compact, totally disconnected Hausdorff space with T discrete and Ω compact. See [Ca], [PW1] for all details.

Theorem 3.1. *Under conditions (i) and (ii), (T, P) has stable boundary $\mathcal{M}(P) = \Omega$. In the $\rho(P)$-transient case, the boundary is strictly stable.*

By [PW1], (T, P) satisfies requirements (1) and (2) of Definition 2.1. The Martin compactification coincides with the end compactification unless P is $\rho(P)$-recurrent and $t = \rho(P)$. The proof of joint continuity of $(\omega, t) \mapsto K(x, \omega|t)$ needs some preparation. We recall the technical prerequisites of [PW1], which have to be slightly adapted. It is enough to restrict attention to eigenvalues t bounded by some (arbitrary) constant $t_0 > \rho(P)$. From now on, we fix t_0 and let $t \in [\rho(P), t_0]$. If $W \subset T$ then S_W denotes the first hitting time of $\mathcal{X}_n$ in W, and

$$\alpha^W(x, x'|t) = \sum_{n=0}^{\infty} \Pr[S_W = n\,,\ \mathcal{X}_n = x'|\mathcal{X}_0 = x] \cdot t^{-n}\ .$$

This series converges for $t \geq \rho(P)$ by Theorem 1.2(b). If $V, W \subset T$ are finite then we define the matrix

$$A(V, W|t) = (\alpha^W(x, x'|t))_{x \in V, x' \in W}\ .$$

It depends continuously on $t \geq \rho(P)$.

As in [PW1], we consider the nonnegative cone $\mathcal{C} = \mathcal{C}_m$ and the standard simplex $\mathcal{S} = \mathcal{S}_m$ spanned by the unit vectors in m-dimensional space, the central projection $\mathcal{P} = \mathcal{P}_m$ from the nonzero elements of $\mathcal{C}_m$ onto $\mathcal{S}_m$ and the hyperbolic (projective) metric $\theta = \theta_m$ on $\mathcal{S}^o = \mathcal{S}^o_m = \mathcal{S}_m \cap \mathcal{C}^o_m$. For $\gamma > 0$, set $\mathcal{S}^\gamma = \mathcal{S}^\gamma_m = \{(\alpha_1, \cdots, \alpha_m) \in \mathcal{S}^0_m \mid \alpha_i \geq \gamma\, \alpha_j \,\forall\, i, j \leq m\}$ and $\Delta_m(\gamma) = \operatorname{diam}_{\theta_m}(\mathcal{S}^\gamma_m)$. Observe that θ induces the Euclidean topology on $\mathcal{S}^o$. We shall usually omit the specific dimension index m.

Consider two finite sets $V, W \subset T$ such that

$$d(V, W) > Ms + \operatorname{diam}(V)\ .$$

(We need the term $\operatorname{diam}(V)$ only when V is not connected in T.) Then, in analogy with [PW1, Lemma 5],

$$\alpha^W(x_1, x'|t) \geq (\delta/t)^{M\, d(x_1,x_2)} \alpha^W(x_2, x'|t)$$

$$\text{for every } x_1, x_2 \in V\,,\ x' \in W\,,\ t \in [t_0, \rho(P)]\ .$$

It follows that the (Birkhoff) contraction coefficient of $A(V, W|t)$ as an operator on $(\mathcal{S}^o_{|W|}, \theta_{|W|}) \to (\mathcal{S}^0_{|V|}, \theta_{|V|})$ is bounded by

$$\tau(V) = \tanh(\Delta_{|V|}(\gamma)/4) < 1\,, \quad \text{where} \quad \gamma = \gamma(V) = (\delta/t_0)^{M \operatorname{diam}(V)}\,,$$

see [Bi], [Se], [PW1].

If $x_1, x_2 \in T$, $d(x_1, x_2) = s - 1$, then the set $V = V(x_1, x_2)$ of all $x \in T$ with $d(x, x_i) \leq s - 1$ $(i = 1, 2)$ is called an *intermediate set* with base points x_1, x_2. As

T has bounded vertex degrees by assumptions (i) and (ii) (see [PW1, Lemma 1]), there is a finite upper bound m_0 on the size of all intermediate sets in T. Two intermediate sets V, W with $d(V,W) > Ms$ give rise to a matrix $A(V,W|t)$ with contraction coefficient bounded by

$$\tau(V) \le \tau_0 = \tanh(\Delta/4) \ , \text{ where}$$

$$\Delta = \max\{\Delta_\ell(\gamma_0) | 1 \le \ell \le m_0\} \text{ and } \gamma_0 = (\delta/\rho(P))^{M(s-1)} \ .$$

If $V_0 \subset T$ is finite and $x' \in T$, then we define the column vector

$$\mathbf{f}(V_0, x'|t) = (F(x,x'|t))_{x \in V_0} \ , \quad t \ge \rho(P) \ .$$

It follows from Proposition 2 and its proof in [PW1] that for every $\omega \in \Omega$ there is a vector

$$\mathbf{f}(V_0, \omega|t) \in \mathcal{S}^{\gamma(V_0)}$$

with the following properties:

(a) If $x_n \in T$ and $x_n \to \omega$ then

$$\lim_{n\to\infty} \mathcal{P}\,\mathbf{f}(V_0, x_n|t) = \mathbf{f}(V_0, \omega|t) \ .$$

(b) If $v = c(V_0, \omega)$ is the confluence point of all the rays $\overline{x\omega}$ starting at some $x \in V_0$, and if $V_1, \dots, V_k$ are consecutive intermediate sets with base points on $\overline{v\omega}$ (V_1 closest to v) then

$$\mathbf{f}(V_0, \omega|t) = \mathcal{P}A(V_0, V_1|t) \cdots A(V_{k-1}, V_k|t)\,\mathbf{f}(V_k, \omega|t) \ .$$

Proof of Theorem 3.1. We have to verify point (3) of Definition 2.1. Let $x \in T$, $V_0 = \{e, x\}$, and write $f(e,\omega|t)$ and $f(x,\omega|t)$ for the respective entries of $\mathbf{f}(V_0,\omega|t)$. Then $K(x,\omega|t) = f(x,\omega|t)/f(e,\omega|t)$, see [PW1, §4]. Therfore we must prove that for finite $V_0 \subset T$ the mapping

$$(\omega, t) \mapsto \mathbf{f}(V_0, \omega|t)$$

is jointly continuous on $\Omega \times [\rho(P), \infty)$. Let $\omega \in \Omega$. Fix an infinite sequence of consecutive intermediate sets $V_1, V_2, \dots$ as in (b) above, such that $d(V_0, V_1) > Ms + \operatorname{diam}(V_0)$ and $d(V_k, V_{k+1}) > Ms$ for $k \ge 1$. Write $A_k(t) = A(V_k, V_{k+1}|t)$ and $\mathbf{f}_k(t) = \mathbf{f}(V_k, \omega|t)$ for $k \ge 0$.

Step 1. We show that $\mathbf{f}(V_0, \omega'|t) \to \mathbf{f}(V_0, \omega|t)$ uniformly in $t \in [\rho(P), t_0]$ as $\omega' \to \omega$.

Let k be the last of the V_j whose base points lie on the segment $\overline{vc}$, where $v = c(V_0, \omega)$ as above and $c = c(v, \omega, \omega')$. Then by (b)

$$\mathbf{f}(V_0, \omega'|t) = \mathcal{P} A_0(t) \cdots A_{k-1}(t) \mathbf{f}(V_k, \omega'|t)$$

and

$$\theta(\mathbf{f}(V_0, \omega|t)\ ,\ \mathbf{f}(V_0, \omega'|t)) \leq \tau(V_0)\, \tau_0^{k-2}\, \Delta\ .$$

If $\omega' \to \omega$ then $k = k(\omega') \to \infty$, and the last term tends to zero independently of $t \in [\rho(P), t_0]$.

Step 2. We prove that for our fixed ω, the map $t \mapsto f(V_0, \omega|t)$ is continuous on $[\rho(P), t_0]$. Choose $k \geq 3$ in (b). Then, for $\rho(P) \leq t, t' \leq t_0$ we have

$$\begin{aligned}
&\theta(\mathbf{f}(V_0, \omega|t)\ ,\ \mathbf{f}(V_0, \omega|t')) \\
&\quad \leq \sum_{j=0}^{k-1} \theta(\mathcal{P} A_0(t') \cdots A_{j-1}(t') A_j(t) \mathbf{f}_{j+1}(t)\ ,\ \mathcal{P} A_0(t') \cdots A_j(t') \mathbf{f}_{j+1}(t)) \\
&\qquad + \theta(\mathcal{P} A_0(t') \cdots A_{k-1}(t') \mathbf{f}_k(t)\ ,\ \mathcal{P} A_0(t') \cdots A_{k-1}(t') \mathbf{f}_k(t')) \\
&\quad \leq \theta(\mathcal{P} A_0(t) \mathbf{f}_1(t)\ ,\ \mathcal{P} A_0(t') \mathbf{f}_1(t)) + \tau(V_0) \cdot \theta(\mathcal{P} A_1(t) \mathbf{f}_2(t)\ ,\ \mathcal{P} A_1(t') \mathbf{f}_2(t)) \\
&\qquad + \sum_{j=2}^{k-1} \tau(V_0)\, \tau_0^{j-2} \cdot \theta(\mathcal{P} A_j(t) \mathbf{f}_{j+1}(t)\ ,\ \mathcal{P} A_j(t') \mathbf{f}_{j+1}(t)) + \tau(V_0)\, \tau_0^{k-1}\, \Delta\ .
\end{aligned}$$

Given $\varepsilon > 0$, fix k such that $\tau(V_0)\tau_0^{k-1}\Delta < \varepsilon$. Now, for fixed $t < t_0$,

$$\lim_{t' \to t} A_j(t') \mathbf{f}_{j+1}(t) = A_j(t) \mathbf{f}_{j+1}(t)\,, \quad j = 0, \ldots, n-1$$

in the Euclidean topology. As θ induces the same topology, the first three terms in the last expression of the inequality tend to zero, and continuity in t is proved.

Combining the two steps yields joint continuity of $\mathbf{f}(V_0, \cdot|\cdot)$ as proposed. □

Corollary 3.2. *Let Γ be a finitely generated group having a free subgroup with finite index on two or more generators. Let μ be a finitely supported probability measure on Γ giving rise to an irreducible random walk. Then (Γ, μ) has strictly stable boundary, and the Martin compactifications for $t \geq \rho(\mu)$ coincide with the end compactification of Γ.*

Proof. By [W3], the Cayley graph of Γ with respect to some finite set of generators has a uniformly spanning tree T. By [PW1], the end compactifications of T and Γ coincide, and the transition operator of T given by μ has properties (i) and (ii). Finally, the random walk is $\rho(\mu)$-transient by [Gu, p. 85]. Therefore Theorem 3.1 applies. □

Remarks. (a) Consider a *nearest neighbour* random walk on a locally finite tree T: $p(x, x') > 0$ if and only if x and x' are neighbours in T. In this case it is not necessary to assume that the nonzero transition probabilities are bounded below by $\delta > 0$. It is easy to deduce boundary stability (which is strong in the $\rho(P)$-transient case) from [Ca].

(b) Observe also that trees are particular instances of hyperbolic graphs in the sense of [Gr]. The Martin boundary of a large class of transition operators P on hyperbolic graphs has been worked out in [An] for $t > \rho(P)$: it coincides with the hyperbolic boundary. In particular, the property of Definition 2.1(1) is satisfied for these operators, while we do not know if (2) and (3) are also true.

References

[An] A. Ancona, *Positive harmonic functions and hyperbolicity*, in "Potential Theory - Surveys and Problems," (ed. by J. Král et al.) Lect. Notes in Math. **1344**, Springer, Berlin - Heidelberg - New York, 1988, pp. 1–23.

[Bi] G. Birkhoff, *Extensions of Jentzsch's theorem*, Trans. Amer. Math. Soc. **85** (1957), 219–227.

[Ca] P. Cartier, *Fonctions harmoniques sur les arbres*, Symposia Math. **9** (1972), 203–270.

[C1] D. I. Cartwright, *Some examples of random walks on free products of discrete groups*, Annali Mat. Pura Appl. **151** (1988), 1–15.

[C2] ———, *On the asymptotic behaviour of convolution powers of probabilities on discrete groups*, Monatsh. Math. **107** (1989), 287–290.

[Do] J. L. Doob, *Discrete potential thoery and boundaries*, J. Math. Mech. **8** (1959), 433-458.

[Gr] M. Gromov, *Hyperbolic groups*, in "Essays in Group Theory," (ed. by S. M. Gersten) MSRI Publ. **8**, Springer, New York - Berlin - Heidelberg, 1987, pp. 75–263.

[Gu] Y. Guivarc'h, *Sur la loi des grands nombres et le rayon spectral d'une marche aléatoire*, Astérisque **74** (1980), 47–98.

[He] P. L. Hennequin, *Processus de Markoff en cascade*, Ann. Inst. H. Poincaré **18** (1963), 109–196.

[Hu] G. A. Hunt, *Markoff chains and Martin boundaries*, Illinois J. Math. **4** (1960), 313–340.

[KSK] J. G. Kemeny, J. L. Snell and A. W. Knapp, "Denumerable Markov Chains," 2nd ed., Springer, New York - Heidelberg - Berlin, 1976.

[NS] P. Ney and F. Spitzer, *The Martin boundary for random walk*, Trans. Amer. Math. Soc. **121** (1966), 116-132.

[PW1] M. A. Picardello and W. Woess, *Martin boundaries of random walks: ends of trees and groups*, Trans. Amer. Math. Soc. **302** (1987), 185–205.

[PW2] ———, *Martin boundaries of Cartesian products of Markov chains*, Preprint, Univ. Milano.

[Pr] W. E. Pruitt, *Eigenvalues of nonnegative matrices*, Ann. Math. Statistics **35** (1964), 1797–1800.

[Se] E. Seneta, "Nonnegative Matrices and Markov chains," (2nd ed.) Springer, Berlin - Heidelberg - New York, 1981.

[V-J] D. Vere-Jones, *Geometric ergodicity in denumerable Markov chains*, Quart. J. Math. Oxford **13** (1962), 7–28.

[W1] W. Woess, *Nearest neighbour random waks on free products of discrete groups*, Boll. Un. Math. Ital. **B-5** (1986), 961–982.

[W2] ———, *A description of the Martin boundary for nearest neighbour random walks on free products*, in "Probability Measures on Groups," (ed. by H. Heyer) Lect. Notes in Math. **1210**, Springer, Berlin - Heidelberg - New York, 1986, pp. 203–215.

[W3] ———, *Graphs and groups with tree-like properties*, J. Combin. Th. B. **47** (1989), 361–371.

Comparison of liminf and fine liminf of positive superharmonic functions

Sundararaja Ramaswamy

Abstract

Let B_n be the open unit ball in $\mathbf{R}^n$ $(n \geq 2)$. Let ∂B_n be its boundary. One knows that ∂B_n is also the fine boundary of B_n and also the Martin boundary of B_n. Prof. M. Brelot in a personal communication suggested to me to find out whether for any positive superharmonic function v in the ball, for almost all $X \in \partial B_n$. (1): $\operatorname{fine} \liminf_{\substack{y \to X \\ y \in B_n}} v(y) = \liminf_{\substack{y \to X \\ y \in B_n}} v(y)$ and (2): $\operatorname{Minfine} \liminf_{\substack{y \to X \\ y \in B_n}} v(y) = \liminf_{\substack{y \to X \\ y \in B_n}} v(y)$.

The object here is to prove that answer to (2) is negative through a counter example and exhibit a method of attack to prove or disprove (1).

Introduction and results

Let B_n be the unit ball in $\mathbf{R}^n$ $(n \geq 2)$. Let ∂B_n be its boundary. ∂B_n is also its fine and Martin boundaries as is well known. Prof. M. Brelot in a personal communication raised the question whether it is true for any positive superharmonic function v that, for almost all $X \in \partial B_n$

(1)
$$\operatorname{fine} \liminf_{\substack{y \to X \\ y \in B_n}} v(y) = \liminf_{\substack{y \to X \\ y \in B_n}} v(y);$$

(2)
$$\operatorname{Minfine} \liminf_{\substack{y \to X \\ y \in B_n}} f(y) = \liminf_{\substack{y \to X \\ y \in B_n}} v(y).$$

We prove here that the answer to (2) is negative in general. We also give a necessary and sufficient condition for (1) to be true.

Theorem 1. *There exists a positive superharmonic function v in B_n for which the answer to* (2) *is negative.*

Proof. Suppose (2) is true for any positive superharmonic function v. Then, for every upperbounded subharmonic function w

$$\operatorname{Minfine} \limsup_{\substack{y \to X \\ y \in B_n}} w(y) = \limsup_{\substack{y \to X \\ y \in B_n}} w(y)$$

for almost all X.

Potential Theory, ed. by M. Kishi

Thus, for any positive and bounded harmonic function h

$$\text{Minfine} \liminf_{\substack{y \to X \\ y \in B_n}} h(y) = \liminf_{\substack{y \to X \\ y \in B_n}} h(y)$$

and

$$\text{Minfine} \limsup_{\substack{y \to X \\ y \in B_n}} h(y) = \limsup_{\substack{y \to X \\ y \in B_n}} h(y)$$

for almost all X.

But we know that for a bounded harmonic function h in B_n $\text{Minfine} \lim_{\substack{y \to X \\ y \in B_n}} h(y)$ exists for almost all X.

Therefore, if (2) is true for any positive superharmonic function v, we get that for any positive bounded harmonic function h,

$$\liminf_{\substack{y \to X \\ y \in B_n}} h(y) = \limsup_{\substack{y \to X \\ y \in B_n}} h(y)$$

for almost all X, i.e. $\lim_{\substack{y \to X \\ y \in B_n}} h(y)$ exists for almost all X. □

But this is not true as the following example due to Littlewood shows (For the proof of this theorem, see [1], page 137.).

Theorem (Littlewood). *Let C be any Jordan arc in $|z| < 1$ which touches $|z| = 1$ at $z = 1$ internal and let C_θ be the curve obtained from C by rotating C about $z = 0$ through an angle θ, which touches $|z| = 1$ at $e^{i\theta}$. Then there exists a bounded harmonic function $u(z)$ and a bounded regular function $f(z)$ in $|z| < 1$, such that for almost all $e^{i\theta}$, if z tends to $e^{i\theta}$ along C_θ, then* $\lim u(z)$ *and* $\lim f(z)$ *do not exist.*

Next, we exhibit a method of attack to prove or disprove (1).

Theorem 2. (1) *is true for any positive superharmonic function v, if and only if for any two positive superharmonic functions v_1 and v_2,*

$$\liminf_{\substack{y \to X \\ y \in B_n}} v_1(y) \geq \liminf_{\substack{y \to X \\ y \in B_n}} v_2(y)$$

for almost all X implies that

$$\text{fine} \liminf_{\substack{y \to X \\ y \in B_n}} v_1(y) \geq \text{fine} \liminf_{\substack{y \to X \\ y \in B_n}} v_2(y)$$

for almost all X.

Proof. The necessity of this condition is obvious.

Suppose the condition is true. Let v be a positive superharmonic function. Let $f(X) = \liminf_{\substack{y \to X \\ y \in B_n}} v(y)$. Let $g(X) = \text{fine}\liminf_{\substack{y \to X \\ y \in B_n}} v(y)$. Then $f(X) \leq g(X)$ for all $X \in \partial B_n$.

To prove that $f(X) = g(X)$ almost everywhere it is sufficient to prove that $\overline{H}_f = \overline{H}_g$ when $\overline{H}_f$ (resp. $\overline{H}_g$) is the upper Perron Wiener-Brelot function associated with f (resp. g). Let w be a superharmonic function such that

$$\liminf_{\substack{y \to X \\ y \in B_n}} w(y) \geq f(X) \quad \text{for all } X \in \partial B_n$$

i.e. $\liminf_{\substack{y \to X \\ y \in B_n}} w(y) \geq \liminf_{\substack{y \to X \\ y \in B_n}} v(y)$ for all $X \in \partial B_n$. Our condition implies that

$$\text{fine}\liminf_{\substack{y \to X \\ y \in B_n}} w(y) \geq \text{fine}\liminf_{\substack{y \to X \\ y \in B_n}} v(y)$$

for almost all X, i.e. $\text{fine}\liminf_{\substack{y \to X \\ y \in B_n}} w(y) \geq g(X)$ for almost all X.

It follows from a result of M. Brelot which we state below that $\overline{H}_g \leq w$. Thus, $\overline{H}_g \leq \overline{H}_f$.

Since $f \leq g$, $\overline{H}_f \leq \overline{H}_g$. Hence, $\overline{H}_f = \overline{H}_g$. This implies that $f = g$ a.e. □

Result of M. Brelot used in the above proof:

Let Ω be an open set such that the fine boundary of Ω is the same as the usual boundary of Ω. Let f be an extended real-valued function defined on $\partial\Omega$. Let ψ_f be the family of all superharmonic functions v such that

$$\text{fine}\liminf_{\substack{y \to X \\ y \in B_n}} v(y) \geq f(X) \quad \text{for all } X \in \partial\Omega$$

Then, $\overline{H}_f = \inf_{v \in \psi_f} v$.

References

[1] M.Tsuji, Potential theory in modern function theory, Maruzen, Tokyo 1959

Some problems of the potential theory for Schrödinger operator with singular potential

Valeri S. Serov

Abstract

This article deals with the problem of the smoothness of solutions of Schrödinger's equation with singular potential. The basic idea of these considerations is connected with Green-Faddeev's function.

1. Introduction

We propose to discuss here some problems of the potential theory concerning the Schrödinger operator $-\Delta + q(x)$ in $\mathbb{R}^N$ for $N = 2, 3$, which have applications to scattering theory. Assume that the support of $q(x)$ is in a bounded domain Ω of $\mathbb{R}^N$. Besides, assume that the potential $q(x)$ belongs to $L^p(\Omega)$ for some p, $N/2 < p \leq 2$, i.e., local singularities are permitted for the potential q.

2. Green-Faddeev's function

Let $x \in \mathbb{R}^N$, $N \geq 2$ and $\zeta \in \mathbb{C}^N$ be a complex vector with $\zeta \cdot \zeta = k^2$, where $k \geq 0$. Define in the sense of distribution's theory

(2.1) $$g_\zeta(x) = \frac{1}{(2\pi)^N} \int_{\mathbb{R}^N} \frac{e^{ix\cdot\xi} d\xi}{\xi^2 + 2\zeta \cdot \xi} .$$

It is easy to check that $g_\zeta(x)$ is a tempered distribution satisfying

(2.2) $$(-\Delta_x - 2i\zeta \cdot \nabla_x) g_\zeta(x) = \delta(x),$$

where Δ_x is the Laplacian, $\nabla_x = (\frac{\partial}{\partial x_1}, \dots, \frac{\partial}{\partial x_N})$ and $\delta(x)$ is the δ-function. The distribution $G_\zeta(x) = e^{ix\cdot\zeta} g_\zeta(x)$ is then a fundamental solution for the operator $-\Delta - k^2$ in $\mathbb{R}^N$, i.e.

(2.3) $$-(\Delta + k^2) G_\zeta(x) = \delta(x)$$

In his work on quantum scattering, Faddeev introduced this function $G_\zeta(x)$ in 1965 (see [1]).

Potential Theory, ed. by M. Kishi

It follows that $G_\zeta(x)$ differs from the Green's function of classical potential theory by a global harmonic function on $\mathbb{R}^N$:

$$G_\zeta(x) = G_k^+(x) + H_\zeta(x), \quad \zeta\cdot\zeta = k^2, \quad \Delta H_\zeta(x) = 0, \tag{2.4}$$

where

$$G_k^+(x) = \frac{1}{(2\pi)^N}\lim_{\varepsilon\to 0}\int_{\mathbb{R}^N}\frac{e^{ix\cdot\xi}d\xi}{\xi^2-k^2-i\varepsilon} = \frac{i}{4}\left(\frac{k}{2\pi|x|}\right)^{\frac{N-2}{2}} H^{(1)}_{\frac{N-2}{2}}(k|x|) \tag{2.5}$$

is the outgoing Green's function.

Recall that L^p_δ denotes the weighted L^p spaces, $\delta\in\mathbb{R}$, $p\geq 1$, defined as

$$L^p_\delta = \{f\in L^p_{loc} : \|f\|_{p,\delta} = \left(\int_{\mathbb{R}^N}(1+|x|)^{\delta p}|f(x)|^p dx\right)^{1/p} < \infty\}.$$

Proposition 1. *Let p and δ satisfy the following conditions:*

1) $\frac{2N}{N+1} < p \leq 2$ $(N\geq 3)$, $1 < p\leq 2$ $(N=2)$;

2) $-1<\delta<1-\frac{2}{p}$ $(N\geq 3)$, $-\frac{2}{p}<\delta<1-\frac{2}{p}$ $(N=2)$;

3) $|\delta| > 2N\left(\frac{1}{p}-\frac{1}{2}\right)$.

Then for any $\zeta\in\mathbb{C}^N$, $\zeta\cdot\zeta=0$ (or $\zeta\cdot\zeta=k^2$) with $|\zeta|$ sufficiently large, the operator of convolution with g_ζ satisfies

$$\|g_\zeta * f\|_{p,\delta} \leq \frac{C(\varepsilon)}{|\zeta|^{1-N(\frac{1}{p}-\frac{1}{2})-\varepsilon}}\|f\|_{p,\delta+1}, \tag{2.6}$$

where $\varepsilon>0$ is any small number.

Remark 1. For $p=2$ we may put $\varepsilon=0$.

Sylvester and Uhlmann were the first to prove in 1987 a kind of weighted L^2 estimate for $g_\zeta(x)$ (the case $p=2$ and $\varepsilon=0$) (see [2]).

Remark 2. This estimate (2.6) has the applications in the scattering theory for the Schrödinger operator with only L^∞-potential.

Proposition 2. *Let $\Omega\subset\mathbb{R}^N$, $N=2,3$, be a bounded domain, $f\in L^p(\Omega)$ for some p, $N/2<p\leq 2$, $k\geq 0$ be fixed, $\zeta\in\mathbb{C}^N$, $\zeta\cdot\zeta=k^2$ and $\mathrm{Im}\,\zeta\neq 0$. Then for $|\zeta|$ sufficiently large, the convolution map $f\mapsto g_\zeta * f$ is bounded from $L^p(\Omega)$ to $L^\infty(\mathbb{R}^N)$ and satisfies the estimate*

$$\|g_\zeta * f\|_{L^\infty(\mathbb{R}^N)} \leq \frac{c}{|\zeta|^{\frac{\alpha}{\alpha+1}}}\cdot\|f\|_{L^p(\Omega)}, \tag{2.7}$$

where $0 < \alpha < \frac{N}{2}\left(\frac{2}{N} - \frac{1}{p}\right)$.

Remark 3. This estimate (2.7) has applications in the scattering theory for the Schrödinger operator with singular potential.

Let $W_\delta^{2,p}(\mathbb{R}^N) = \{f : D^\alpha f \in L_\delta^p(\mathbb{R}^N),\ 0 \le |\alpha| \le 2\}$ be the weighted Sobolev spaces with the norm

$$\|f\|_{W_\delta^{2,p}} = \left(\sum_{|\alpha|\le 2} \|D^\alpha f\|_{p,\delta}^p\right)^{1/p}. \tag{2.8}$$

Proposition 3. *Let Ω be a bounded domain in $\mathbb{R}^N$, $N \ge 2$. Define the integral operator $\hat{G}_\zeta$ by*

$$\hat{G}_\zeta w(x) = \int_\Omega G_\zeta(x-y)w(y)dy, \tag{2.9}$$

where $G_\zeta(x)$ is Green-Faddeev's function given in (2.3)-(2.5). Then the following inequality holds for any function f in $L^p(\Omega)$

$$\|\hat{G}_\zeta f\|_{W^{2,p}(\Omega')} \le c\|f\|_{L^p(\Omega)}, \tag{2.10}$$

where Ω' is an arbitrary bounded domain in $\mathbb{R}^N$, p satisfies conditions 1) in Proposition 1 and the constant c depends on Ω' and Ω.

Since Ω is a bounded domain, Proposition 3 is an easy consequence of the inequality

$$\|g_\zeta * f\|_{W_\delta^{2,p}} \le C(\delta,\zeta)\|f\|_{p,\delta+1} \tag{2.11}$$

which follows from Proposition 1 and well-known techniques (see, for example, [3]).

Representation (2.4) and Proposition 3 allow us to get the same results (2.10) for the integral operator $\hat{G}_k^+$ with the kernel $G_k^+(|x-y|)$ which is defined in (2.5). That is,

$$\|\hat{G}_k^+ f\|_{W^{2,p}(\Omega')} \le c\|f\|_{L^p(\Omega)} \tag{2.12}$$

where p satisfies condition 1) in Proposition 1, and c depends on Ω' and Ω.

This function $G_k^+(|x-y|)$ is the well-known Green's function in scattering theory.

3. Applications in scattering theory

Let us consider the operator $-\Delta + q(x)$ in $\mathbb{R}^N$ for $N = 2$ and 3 with real-valued potential q. We assume that the support of q is in a bounded domain Ω in $\mathbb{R}^N$ ($N = 2$ or 3) and the potential $q(x)$ belongs $L^p(\Omega)$ for some p, $N/2 < p \leq 2$. It is not so hard to show that $-\Delta + q$ admits a unique self-adjoint realization in $L^2(\mathbb{R}^N)$ in the sense of quadratic forms. Moreover, it is known (see, for example, [4]) that for any $k \in \mathbb{R}$, $k \neq 0$, the operator $(-\Delta + q - k^2)^{-1}$ exists and this operator is an integral operator with the kernel $G_q(x, y)$ satisfying the following integral equation ("kernel" equation)

$$G_q(x, y) = G_k^+(|x - y|) - \int_\Omega G_k^+(|x - u|)q(u)G_q(u, y)du, \tag{3.1}$$

where $G_k^+(|x - y|)$ is given by (2.5), i.e., Green's function of the operator $(-\Delta - k^2)$ satisfying the classical Sommerfeld radiation condition at the infinity. This equation (3.1) has a unique solution for any $k \in \mathbb{R}$, $k \neq 0$. The following theorem is true.

Theorem. *The integral operator $\hat{G}_q$ with the kernel $G_q(x, y)$ maps $L^p(\Omega)$ into $W^{2,p}_{loc}(\mathbb{R}^N)$ ($N = 2$ or 3) and the following estimate holds:*

$$\|\hat{G}_q f\|_{W^{2,p}(\Omega')} \leq c\|f\|_{L^p(\Omega)}, \tag{3.2}$$

where Ω' is an arbitrary bounded domain in $\mathbb{R}^N$, the constant c depends on Ω and Ω', and the number p is taken from the condition for the potential q.

As an example one can consider a potential q with compact support $\bar{\Omega}$ and satisfying the condition

$$|q(x)| \leq c|x - x_0|^{-2+\varepsilon}, \quad \varepsilon > 0, \tag{3.3}$$

where x_0 is an arbitrarily fixed point in the domain Ω, $c > 0$ is a constant and $\varepsilon > 0$ is any small fixed number.

In particular, it follows from this theorem that the bounded solutions $u(x, k)$ of the equation

$$(-\Delta + q - k^2)u(x, k) = 0 \tag{3.4}$$

i.e., the solutions satisfying Sommerfeld's radiation condition at infinity, are the unique solutions of the Lippmann-Schwinger equations

$$u(x, k, \theta) = e^{ik(x,\theta)} - \int_\Omega G_k^+(|x - y|)q(y)u(y, k, \theta)dy, \tag{3.5}$$

where $x \in \mathbb{R}^N$, $k \in \mathbb{R}$, $k \neq 0$ and $\theta \in S^{N-1}$ (the unit sphere in $\mathbb{R}^N$). Moreover, these functions $u(x,k,\theta)$ belong to the space $W^{2,p}_{loc}(\mathbb{R}^N)$ for p, $\frac{N}{2} < p < \frac{N}{2-\epsilon}$.

Using this fact and equation (3.4), we can get an exact formula for the reconstruction of the potential $q(x)$ on scattering data

$$q(x) = C_N \theta \cdot \nabla_x \int_{-\infty}^{\infty} \operatorname{sgn} k \cdot |k|^{N-2} dk \int_{S^{N-1}} d\theta' A(k,\theta',\theta) e^{-ik(\theta',x)} u(x,k,\theta'), \tag{3.6}$$

where $\theta, \theta' \in S^{N-1}$ and $A(k,\theta',\theta)$ is the scattering amplitude which is defined as

$$A(k,\theta',\theta) = -\int_{\Omega} e^{-ik(\theta',y)} q(y) u(y,k,\theta) dy. \tag{3.7}$$

The formula (3.6) is R. Newton's formula (see [5]).

References

[1] Faddeev, L.D., Growing solutions of the Schrödinger equation. Dokl. Akad. Nauk SSSR. 165, 514-517 (1965) (transl. Sov. Phys. Dokl. 10, 1033).

[2] Sylvester, J. and Uhlemann, G., A global uniqueness theorem for an inverse boundary value problem. Annals of Math. 125, 153-169 (1987).

[3] Lavine, R., Classical limit of the number of quantum states. Quantum Mechanics in mathematics, chemistry and physics, edited by K. Gustafson and W. Reinhardt, Plenum Press.

[4] Read, M. and Simon, B., Methods of modern mathematical physics. Vols. III, IV, Academic Press, New York (1975-1978).

[5] Newton, R.G., Inverse scattering. II-IV. Three dimensions. J. Math. Phys. - 1980-1982 - Vols. 21-23. - N 7,10,4. - p. 1698-1715, p. 2191-2200, p. 594-604.

Potential theory for Helmholtz operator and nonlinear eigenvalue problems

Yuri V. Shestopalov

Abstract

Trace properties of generalized potentials and Green's potentials in two-dimensional domains of a certain class are considered. Applications for the transmission boundary eigenvalue problems for the system of Helmholtz equations are presented.

1980 MOS Subject Classification (1985 Revision): 35B40, 35J05, 35L05, 35P25.

1. Introduction

In the theory of boundary value problems the integrals

$$u(x) = \int_C E(x,y)\xi(y)dl_y; \quad v(x) = \int_C \frac{\partial}{\partial n_y} E(x,y)\eta(y)dl_y$$

are usually called generalized potentials. Here $x = (x_1, x_2)$, $y = (y_1, y_2) \in R^2$; $E(x,y)$ is the fundamental solution of the second order elliptic differential operator, $\frac{\partial}{\partial n_y}$ is the normal derivative in the point y of the closed piece-wise smooth boundary C of the domain in R^2; $\xi(y)$, $\eta(y)$ are sufficiently smooth functions defined on C. In case of the Helmholtz operator $\mathcal{L}(k) = \Delta + k^2$ one can choose $E(x,y)$ in the form

$$E(x,y) = \mathcal{E}(x-y) = \frac{i}{4} H_0^{(1)}(k|x-y|),$$

where $\mathcal{E}(x-y)$ satisfies the Sommerfeld radiaton condition [1] in the domain $\Omega^+ = \{x \,|\, x_2 > 0\}$ for real values of k; $H_0^{(1)}(z) = -2ig(z) + h(z)$ is the Hankel function, $\frac{1}{2}g(x-y) = \frac{1}{2\pi} ln \frac{1}{|x-y|}$ is the kernel of the two-dimensional single layer potential, and the second derivative of $h(z)$ has logarithmic singularity.

The study of potentials and the method of boundary integral equations with strongly elliptic Fredholm operators for 'transmission' boundary value problems for the Helmholtz equation in Sobolev spaces is carried out in [4] and in some other works (see References in [4]).

Potential Theory, ed. by M. Kishi

We will consider here some properties of generalized potentials and Green's potentials together with an application of the method of generalized potentials for transmission-type boundary eigenvalue problems arising from the electrodynamics of guided waves and resonators. Detailed statements of the corresponding boundary value problems for Maxwell equations are given in [2], [3].

2. Generalized potentials

We will give the definitions and some assertions about the most important properties of the potentials in two-dimensional domains of a certain class.

Let us consider finite connected domains $S_\Pi = S_\Pi(L) \subset R^2$, bounded by a closed piece-wise smooth contour $\Gamma = \partial S_\Pi$, such that

$$L \subset L_0 \subset \Gamma; \ \ L_0 \subset \{x_2 = 0\},$$

i.e., $L_0 = [c, d]$; $L = (a, b)$; $c < a < b < d$.

Let us say that the functions $U_j(x)$ have *a behavoir like the potential of a single layer* (SL) ($j = 1$) *or a double layer*(DL) ($j = 2$) if

$$U_j(x) = \int_L K_j(x,t)\phi_j(t)dt; \ \ x = (x_1, x_2) \in S_\Pi(L), \tag{1}$$

where

$$K_j(x,t) = g_j(x,t) + F_j(x,t); \ \ j = 1, 2,$$

with

$$g_1(x,t) = g(x, y^0) = \frac{1}{\pi}\ln\frac{1}{|x - y^0|}; \ \ g_2(x,t) = \frac{\partial}{\partial y_2}g(x, y^0); \ \ y^0 = (t, 0).$$

F_1 and F_2 are smooth functions, and we shall assume that in every closed domain $S_{0\Pi}(L) \subset S_\Pi(L)$:

1) $F_1(x,t)$ is once continuously differentiable with respect to the variables of x and continuous in t;

2) $F_2(x,t)$ and $F_2^1(x,t) = \frac{\partial}{\partial x_2}\int_q^t F_2(x,s)ds$, $q \in R^1$ are continuous.

We shall also assume that for *the densities*, $\phi_1 \in X_a(L)$, $\phi_2 \in X_0(L)$, where the functional spaces X_a, X_0 are defined in (6) and (10).

The following lemmas contain known assertions about the properties of logarithmic potentials. They can be proved straightforwardly for the S_Π-domains.

Lemma 1.

$$\frac{\partial}{\partial x_2}\int_L g_2(x,t)\phi_2(t)dt = \int_L \frac{\partial}{\partial x_1}g_1(x,t)\frac{d\phi_2}{dt}(t)dt;$$

$$x \in S_{\Pi}; \quad \phi_2(t) \in X_0(L).$$

Lemma 2.

$$\lim_{x \to x^0 \in L} \frac{\partial}{\partial x_2} \int_L g_2(x,t)\phi_2(t)dt = \int_L \lim_{x \to x^0 \in L} \frac{\partial}{\partial x_1} g_1(x,t) \frac{d\phi_2}{dt} dt =$$

$$= S_L \frac{d\phi_2}{dt}; \; \phi_2(t) \in X_0(L); \; S_L f \equiv \frac{1}{\pi} \int_L \frac{f(t)}{t-x} dt.$$

Lemma 3. *The properties of the normal and tangential derivatives of* SL *and of the trace of* DL *on* L *coincide with the ones of logarithmic potentials.*

Lemma 4. *There exists the limiting value (trace) of the normal derivative of* DL *in the* S_{Π}*-domain, and it admits the representation as a singular integral:*

$$\lim_{x \to x^0 \in L} \frac{\partial U_2}{\partial x_2} = \lim_{x \to x^0 \in L} \int_L \{\frac{\partial g_1}{\partial x_1} + F_2^1\}(x,t) \frac{d\phi_2}{dt} dt =$$

$$= (S_L + K_2^1)\frac{d\phi_2}{dt}; \quad \phi_2(t) \in X_0(L),$$

where

$$K_2^1 f = \int_L F_2^1(x,t) f(t) dt.$$

Proof follows from the Lemmas 1,2 and the definition of DL, under the condition that the density of DL is from $X_0(L)$.

3. Transmission boundary eigenvalue problems

For the sake of simplicity we will formulate boundary value problems in two S_{Π}-type rectangular domains $S_1 = \{x \,|\, 0 < x_1 < a;\, 0 < x_2 < b_1\} \subset \Omega^+ = \{x \,|\, x_2 > 0\}$ and $S_2 = \{x \,|\, 0 < x_1 < a;\, -b_2 < x_2 < 0\} \subset \Omega^- = \{x \,|\, x_2 < 0\}$ with common part $\Gamma = \partial S_1 \cap \partial S_2 = \{x \,|\, x_2 = 0,\, 0 \le x_1 \le a\}$ of the boundary, where there is an interval $L = \{x \,|\, x_2 = 0;\, \frac{a}{2} - w = d_1 < x_1 < d_2 = \frac{a}{2} + w\}$ with endpoints $\partial L = \{(d_1, 0), (d_2, 0)\}$; $\epsilon = \epsilon(x) = \epsilon_j;\, x \in S_j;\, j = 1,2$; (classical) statement of *the problem* H w.r.t. (complex) spectral parameter λ

$$\Delta u + \lambda \epsilon u = 0; \qquad x \in S_1 \cup S_2 \tag{2}$$

$$[u]|_L = \left[\epsilon^{-1}\frac{\partial u}{\partial x_2}\right]\bigg|_L = 0 \tag{3}$$

$$\frac{\partial u}{\partial n}\bigg|_{\Gamma_1} = \frac{\partial u}{\partial n}\bigg|_{\Gamma_2} = 0; \qquad \Gamma_j = \partial S_j \backslash L \tag{4}$$

where $\frac{\partial}{\partial n}$ is the normal derivative on ∂S_1 and ∂S_2; $[.]$ means the jump of the limiting values of the function from the different sides of the common part of the boundaries of the domains S_1 and S_2; $u(x)$ is continuously differentiable in the closed domains $\overline{S_1 \backslash S_\rho}$, and $\overline{S_2 \backslash S_\rho}$ where S_ρ are the ρ-neighborhoods of the endpoints of ∂L, continuous in $\overline{S_1}$, $\overline{S_2}$ and with 'edge' conditions

$$\iint_{S_\rho} [|u|^2 + |\nabla u|^2]\, ds < \infty. \tag{5}$$

We shall also demand that for the traces

$$u(x)|_L = u(x^0) = \phi(t) \in Y_a = \widetilde{H^1_\mu}(L);$$

$$\widetilde{H^1_\mu}(L) = \{\phi | \phi \in C(\overline{L}) \cap C^1(\overline{L_\delta}); \phi' R \in H_\mu(L)\}; \tag{6}$$

$$\overline{L_\delta} = [d_1 + \delta, d_2 - \delta];\ \overline{L_\delta} \subset L;$$

$$R(t) = \sqrt{(d_2 - t)(t - d_1)}; \qquad x^0 = (t, 0) \in L;$$

$$\frac{\partial u}{\partial x_2}(x)\bigg|_L = \frac{\partial u}{\partial x_2}(x^0) = \psi(t) \in X_a = \widetilde{H_\mu}(L) = \{\psi\, |\psi R(t) \in H_\mu(L)\},$$

where $H_\mu(L)$ is a class (Banach space) of Hölder functions on the interval $L = (d_1, d_2)$ with constant $\mu \in (0, 1)$, so that X_a, Y_a are the Hölder classes with a weight $R(t)$.

We will also consider the following general transmission boundary eigenvalue *problem V* for the system of Helmholtz equations w.r.t (complex) spectral parameters λ or η:

$$\Delta u + gu = 0; \qquad \Delta v + gv = 0; \qquad x \in S_1 \cup S_2; \tag{7}$$

$$g = g(x) = \lambda \epsilon(x) - \eta;\ \eta = \gamma^2$$

$$v = 0; \qquad \frac{\partial u}{\partial n} = 0, \qquad x \in (\partial S_1 \backslash L) \cup (\partial S_2 \backslash L);$$

$$[u] = 0; \qquad [v] = 0; \tag{8}$$

$$\gamma\left[\frac{\partial v}{\partial x}\right] - \left[\frac{\partial u}{\partial y}\right] = 0;$$

$$\gamma \left[\frac{\partial u}{\partial x}\right] + \left[g\frac{\partial v}{\partial y}\right] = 0 \qquad on\, L, \tag{9}$$

with 'edge' conditions (5) and (6) for $u(x)$; for the function $v(x)$ we assume that the trace

$$v|_L = \psi(t) \in X_0(L) = \{\psi|\psi \in C(\overline{L}); \psi \in \widetilde{H}^1_\mu; v(d_1, 0) = v(d_2, 0) = 0; \}. \tag{10}$$

If we formally put here $\gamma = 0$ we shall get two independent eigenvalue problems w.r.t. the spectral parameter λ, in particular, the problem H for the function u.

4. Green's potentials

Let us introduce Green's functions of the second boundary value problem for the Helmholtz equation (2) in rectangular domains S_j:

$$G_j(\lambda; x, y) = \frac{4}{\epsilon_j a b_j} \sum_{n=0}^{\infty} \sum_{m=0}^{\infty} \epsilon_n \epsilon_m \frac{\psi_n(x_1, a)\psi_n(y_1, a)\psi_m(x_2, b_j)\psi_m(y_2, b_j)}{(\lambda - \lambda^{(j)}_{nm})}, \tag{11}$$

where

$$x = (x_1, x_2);\; y = (y_1, y_2); \qquad \epsilon_0 = 0.5; \epsilon_n = 1,\, n \geq 1;$$

$$\psi_n(t, d) = \cos\frac{\pi n t}{d}; \qquad \lambda^{(j)}_{nm} = \frac{\pi^2}{\epsilon_j}\left(\frac{n^2}{a^2} + \frac{m^2}{b_j^2}\right).$$

As it is shown in [5], if $|x - y| \geq \delta, \delta > 0$, the series (11) converges uniformly with respect to the points $x, y \in S_j$ if one sums with respect to increasing values of $\lambda^{(j)}_{nm}$ in every inner subdomain $S' \subset S_j$ and has a logarithmic singularity:

$$G_j(\lambda; x, y) = c_j \ln|x - y| + N_j(\lambda; x, y); \qquad x, y \in S_j,$$

where c_j =const, $N_j(\lambda; x, y)$ are once continuously differentiable with respect to x, y for $\lambda \notin \Lambda_j = \{\lambda^{(j)}_{nm}\}^{\infty}_{m,n=0}$ and, together with $G_j(\lambda; x, y)$, are meromorphic functions with respect to λ with the same simple (if all of them are different) real poles $\lambda^{(j)}_{nm}; m, n = 0, 1, \ldots; j = 1, 2$. The following representations are true for the Green's functions:

$$G_j|_{y=y^0\in L} = G_j(\lambda; x, y^0) = f(x, y^0) + g_j(\lambda; x, y^0); \tag{12}$$

$$f(x, y^0) = \frac{1}{2\pi}\ln\Phi(x, y^0) = \frac{1}{\pi}\ln|x - y^0| - \frac{\ln 2}{2\pi} + f_1(x, y^0);\; y^0 \in L,$$

where $g_j(\lambda; x, y^0)$ and $f_1(x, y^0)$ are once continuously differentiable with respect to the coordinates of x in every closed domain $\overline{S_0} \subset S_j$; $L \subset \partial S_0; j = 1,2$. It means that when one of the points (y) lies on a part (L) of the boundary of S_j, the Green's function has the same singularity as the kernel of a two-dimensional single layer potential.

It follows from Green's identity that the solutions of the problems H,V admit integral representations as the (Green's) potentials of a single layer and double layer

$$u(x) = u_j(x) = (-1)^{j+1} \int_L G_j(\lambda; x, y^0)\phi_j(t)dt; \; x \in S_j; y^0 = (t, 0) \in L; \tag{13}$$

$$\lambda \notin \Lambda_j; \; \phi_j(t) = \frac{\partial u_j}{\partial x_2}(y^0) \in \widetilde{H_\mu}(L); \; j = 1,2. \tag{14}$$

$$v(x) = v_j(x) = (-1)^{j+1} \int_L \mathcal{K}_j(\mathbf{z}; x, y^0)\psi_j(t)dt; \; y^0 = (t, 0) \in L; \; x \in S_j; \tag{15}$$

$$\mathcal{K}_j(\mathbf{z}; x, y^0) = \frac{\partial}{\partial y_2} G_j^1(\mathbf{z}; x, y^0);$$

$$\psi_j(t) = v_j|_L \in X_0(L); \mathbf{z} = (\lambda, \eta),$$

where G_j^1 is the Green's function of the Dirichlet problem of the Helmholtz equation (7) of the domain S_j, $j = 1,2$.

Theorem 1. *The solutions of problems* H *and* V *in the* S_Π*- domains* S_1, S_2 *admit integral representations as* SL *and* DL *given in formulas* (13) *and* (15). *The functions*

$$G_j(\lambda, x, y^0); \; \frac{\partial}{\partial y_2} G_j^1(z; x, y^0), \; j = 1,2$$

admit representations as the kernels of SL *and* DL.

Proof is based on the study of the integral representation of two auxiliary problems

$$(\Delta + k^2)u = 0; \qquad (\Delta + k^2)v = 0; \; x \in S_\Pi(L) \subset \{x_2 > 0\};$$

$$\left.\frac{\partial u}{\partial n}\right|_L = \phi \in X_a(L); \qquad v|_L = \psi \in X_0(L).$$

To get the required representations it is sufficient to apply Green's identity to the functions $u(x)$, $V_2(x, y) = h_2(x, y) - G_2(x, y)$ and $v(x)$, $V_1(x, y) = h_1(x, y) - G_1(x, y)$, where

$$h_j(x, y) = i/4[H_0^{(1)}(k|x - y|) + (-1)^j H_0^{(1)}(k|x - y^*|)];$$

$$y = (y_1, y_2); \; y^* = (y_1, -y_2).$$

h_j and G_j are Green's functions of the first ($j = 1$) or the second ($j = 2$) boundary-value problem of the Helmholtz equation corresponding to the half-plane $\{x_2 > 0\}$ and the domain S_Π.

The assumed existence of the Green's functions G_1 and G_2 (for only regular values of the parameter k) yields the existence of the solutions of the boundary-value problems for u and v.

Direct application of the results of Lemmas 1-3 gives the following

Theorem 2. *There exists a limiting value (trace) on L of the first (tangential) derivative of u in (13) and it admits the representation as a singular integral*

$$\lim_{x \to x^0 \in L} \frac{\partial u}{\partial x_1} = (S_L + g(\lambda))\,\phi = \frac{1}{\pi} \int\limits_L \frac{\phi(t)}{t - x_1} dt + \int\limits_L g(\lambda; x_1, t)\phi(t)dt, \qquad (16)$$

where $g(\lambda; x_1, t)$ is once continuously differentiable with respect to x_1 on $L; t \in L; x^0 = (x_1, 0)$.

One can prove the same assertion assuming that S_j are arbitrary S_Π-type domains

5. Boundary integral equations and properties of eigenvalues

Now it is possible to reduce problems H,V to the boundary integral equations with logarithmic singularity of the kernels.

Let us start with the problem H. One can take into account that as it follows from Green's identity, theorem 1, representation (12) and the theorem about the normal derivative of the logarithmic potential of a simple layer, (14) may be treated as an identity on L. So we may put $\phi_2(t) = \epsilon_2/\epsilon_1 \phi_1(t)$ and it fulfils the second condition in (3). To satisfy the first condition in (3) we must apply (13) letting x tend to L respectively from domains S_1, S_2, finally obtaining the integral equation

$$K(\lambda)\phi_1 \equiv \int\limits_L \mathcal{K}(\lambda; t, s)\phi_1(s)ds = 0; \qquad t \in L \qquad (17)$$

$$\mathcal{K}(\lambda; t, s) = \epsilon_1 G_1(\lambda; x^0, y^0) + \epsilon_2 G_2(\lambda; x^0, y^0),$$

where $\phi_1 \in \widetilde{H_\mu}(L)$ according to (6).

The inverse assertion is true: if we take any nontrivial solution $\phi_1 \in \widetilde{H_\mu}(L)$ of (17), put $\phi_2 = \epsilon_2/\epsilon_1 \phi_1$ then $u_j(x)$ from the formula (13) satisfy Helmholtz equation in S_j, boundary conditions (4) and (3) (because $\phi_j(t) \equiv \frac{\partial u_j}{\partial x_2}|_L$). It is also easy to show using the properties of singular integrals [6] that $u_j(x)$ in (13) satisfy the 'edge' conditions (5). Hence we have proved

Theorem 3. *If $\lambda \notin \Lambda = \Lambda_1 \cup \Lambda_2$ then the problem* H *is equal to the integral equation* (17).

We treat (17) as a nonlinear eigenvalue problem for *the integral operator-valued function* $K(\lambda)$ of the complex variable λ, which is also the spectral parameter of the problem H. A detailed study of the properties of the point spectrum of (17) is carried out in [2], [3].

In case of the general transmission problem V the corresponding operator-valued function $K_V(\mathbf{z})$ can be constructed in the same way as the one for the problem H given in (17). Applying the transmission conditions (8), (9) to the integral representations (13), (15) and using the theorems about the traces of the derivatives of the potentials, we finally obtain the system of two boundary integral equations which is similar to (17):

$$K_V(\mathbf{z})\phi \equiv \int_L \mathcal{K}_V(\mathbf{z};t,s)\phi(s)ds = 0;\ t \in L \tag{18}$$

$$\mathcal{K}_V(\mathbf{z};t,s) = \sum_{j=1}^{2} \begin{pmatrix} (\frac{\epsilon_j}{\eta} - 1)\overset{0}{G}_j & \overset{0}{G}_j \\ \overset{0}{G}_j & \frac{\eta}{\epsilon_j - \eta}\overset{0}{G}_j + \frac{\epsilon_j}{\epsilon_j - \eta}\overset{0}{F}_j \end{pmatrix}$$

$$\overset{0}{G}_j = G_j(\mathbf{z}; x^0, y^0); \qquad x^0 = (t,0); y^0 = (s,0);$$

$$\overset{0}{F}_j = F_j(\mathbf{z};t,s) = \int_{c_1}^{t} \frac{\partial}{\partial x_2} \int_{c_2}^{s} \frac{\partial}{\partial y_2} G_j^1(\mathbf{z}; x_1^0, y_1^0)|_{x_2=y_2=0} ds_1 dt_1;$$

$$x_1^0 = (t_1, x_2);\ y_1^0 = (s_1, y_2); \qquad \phi(s) = (\phi_1(s), \phi_2(s))^T.$$

(18) yields the problem of characteristic numbers of the operator-valued function $K_V(\mathbf{z})$, of the problem V. We shall consider $K_V(\mathbf{z})$ on the pair of weighted Hölder-type Banach spaces **X**,**Y** of the two-component vector-functions ϕ. This problem is equivalent to the problem V if

$$\eta \notin E = \cup_{i,j=1}^{2} E_j^i;\ E_j^i = \{\eta_{nj}^i = \epsilon_j - \mu_{nj}^i\}_{n=0}^{\infty},$$

where $\{\mu_{nj}^i\}_{n=0}^{\infty}$ are the eigenvalues of the first ($i = 1$) or the second ($i = 2$) boundary value problem for Laplace's equation in the domain S_j. E is the set of poles of the finite-meromorphic operator-valued function $K_V(\mathbf{z})$ with respect to η which coincides with the set of poles of the Green's functions.

In case of the two rectangular domains S_j we may represent the kernel of the operator-valued function in the following explicit form:

$$\mathcal{K}_V(\mathbf{z};t,s) = \sum_{j=1}^{2} \begin{pmatrix} (\frac{\epsilon_j}{\eta} - 1)\overset{0}{G}_j & \overset{0}{G}_j \\ \overset{0}{G}_j & \epsilon_j R_j - \overset{0}{G}_j \end{pmatrix};$$

$$\overset{0}{G}_j = \overset{0}{l}(t,s) - \frac{2}{a}\sum_{n=0}^{\infty}(\alpha_n^j - \frac{a}{\pi n})\psi_n;$$

$$R_j = R_j(\mathbf{z};t,s) = -\frac{2a}{\pi^2}\sum_{n=1}^{\infty}\frac{\alpha_n^j}{n^2}\psi_n;$$

$$\overset{0}{l}(t,s) = -\frac{2}{\pi}\sum_{n=1}^{\infty}\frac{\psi}{n} =$$

$$= \frac{1}{\pi}\left[\ln\left|2\sin\frac{\pi|t-s|}{2a}\right| + \ln\left|2\sin\frac{\pi|t+s|}{2a}\right|\right] \equiv \frac{1}{\pi}\ln\frac{1}{|t-s|} + l_1(t,s),$$

$$\alpha_n^j = \frac{\coth b_j\sqrt{\eta-(\epsilon_j - r_n)}}{\sqrt{\eta-(\epsilon_j - r_n)}}; r_n = n^2 r_1 = \frac{\pi^2 n^2}{a^2},$$

where the series for $\overset{0}{G}_j$ and R_j converge absolutely and uniformly with respect to $s,t \in [0,a]$, and $l_1(t,s)$ is once continuously differentiable with respect to its variables.

It is shown in [2] that $K_V(\mathbf{z})$ admits separation of a singular (invertible) part and pole pencils at every simple pole η_{nj}^i:

$$K_V(\mathbf{z}) = G(\mathbf{z})l(w) + N(\mathbf{z},w) \equiv G(\mathbf{z})l(w) + \frac{A_{-nj}^i}{\eta - \eta_{nj}^i} + U_{nj}^i(\mathbf{z},w) \equiv$$

$$\equiv P_{nj}^i(\mathbf{z},w) + U_{nj}^i(\mathbf{z},w); \tag{19}$$

where

$$G(\mathbf{z}) = \begin{pmatrix} \dfrac{\epsilon_1+\epsilon_2}{\eta} - 2 & 2 \\ 2 & -2 \end{pmatrix}$$

$$l(w)\phi = \frac{1}{\pi}\int_L \ln\frac{1}{|t-s|}\phi(s)ds.$$

$T(\mathbf{z},w) = G(\mathbf{z})l(w)$ is invertible and $N(\mathbf{z},w)$ is completely continuous on the pair of Banach spaces $\mathbf{X},\mathbf{Y}$; A_{-nj}^i is a two-dimensional operator of the pole pencil P_{nj}^i corresponding to the simple pole η_{nj}^i; $n = 0,1,2\ldots; i,j = 1,2$.

The representations (18) – (19) allow us to prove that $K_V(\mathbf{z})$ is a Fredholm (Φ)-operator. Under the assumptions of the absence of point spectrum in a deleted neighborhood of the origin in the complex η–plane (see [3]) we can prove now that problem V has not more than a finite number of elements of the point spectrum in every finite circle $\{\eta | \, |\eta| < r\}$.

To prove *the existence* of points in the spectrum we can use, similarly to problem H, the generalized theorems [2] about the pole pencils.

References

[1] Tikhonov, A. N. and Samarskij, A.A., *Equations of Mathematical Physics*, Pergamon Press, Oxford-New York, 1963.
[2] Illinskij, A.S. and Shestopalov, Yu.V., *Applications of the Methods of Spectral Theory in the Problems of Wave Propagation (in Russian)*, Moscow State Univ., Moscow, 1989.
[3] Shestopalov, Yu.V., *On the Theory of Cylindrical Resonators*, Preprint TRITA-TET No 1, Royal Institute of Technology, Stockholm, 1989.
[4] T.von Petersdorff, *Boundary Integral Equations for Mixed Dirichlet, Neumann and Transmission Problems*, Math. Meth. in the Appl. Sci. **11** (1989), 185-213.
[5] Il'in, V.A., *Kernels of Fractional Order*, Dokl. Akad. Nauk SSSR **109** (1956), 341–395.
[6] Muskhelishvili, N.I., *Singular Integral Equations (in Russian)*, Fizmatgiz, Moscow, 1962.

On stationary points of the energy integral

Keiichi Shibata

Abstract
Every sufficiently smooth non-singular surface is known classically to admit a conformal structure at least in the small. In this short note we will give a characterization for such conformal metric in terms of Dirichlet's energy integral, complementing our previous paper [4]. To be precise, the conformal metric provides a stationary point to the Dirichlet functional in a neighbourhood consisting of all its small quasiconformal deformations (Theorem 5 in §3), and vice versa (cf. Shibata [4]).

1980 Mathematics Subject Classification (1985 Revision): 31A05, 30C60, 58E20.

1. Preliminaries

Let S be an arbitrary closed Riemann surface. With a non-negative continuous function $\rho(w)$ in a local coordinate w on S, the conformal metric of S is defined to be an invariant form

$$ds^2 = \rho(w)|dw|^2,$$

which remains unaltered under any conformal change of the coordinate w into $\tilde{w}$, i.e., the relation $\tilde{\rho}(\tilde{w})|d\tilde{w}|^2 = \rho(w)|dw|^2$ holds good with $\tilde{\rho}(\tilde{w}) = \rho(w)|dw/d\tilde{w}|^2$. We assume further that the metric ds^2 is strictly positive with at most finite number of exceptional points, existence of which is guaranteed, e.g., by Abelian differentials of the first kind.

We intend here to change these metrics, connected intimately with the conformal structure of S, by means of a kind of non-singular local deformations, which are not necessarily conformal.

The following notations will be employed repeatedly. Let $\mathcal{U} = \{U_j\}_{j=1}^N$ denote a finite system of coordinate neighbourhoods covering S, upon an arbitrary one of which we fix our eye, say U_1, once and for all. Let ϕ denote a coordinate map of U_1 onto a plane open disk D. The notation $\mathbf{C}$ is always a complex plane, which may be sometimes specified by the preciser ones $\mathbf{C}_z$, $\mathbf{C}_w$ etc. at needs. In accordance with them, the open disks in $\mathbf{C}_z$ will be denoted by D_z : in a similar fashion D_w in $\mathbf{C}_w$, and so on. Let κ denote an arbitrary compact set comprised in U_1, which is disjoint with all other coordinate neighbourhoods than U_1 belonging to $\mathcal{U}$.

Note that there exists a coordinate transformation of the Riemann surface S which is quasiconformal in every coordinate neighbourhood of S and is conformal outside κ. In fact, in the presence of a quasiconformal homeomorphism of $\phi(U_1)$

Potential Theory, ed. by M. Kishi

which is conformal off $\phi(\kappa)$, all the conformal neighbourhood-relations with U_1 are kept invariant. So it will be conceivable to recall the notion of quasiconformal transform of the conformal structure on the Riemann surface S, the existence of which is based upon the following fact:

Let $\kappa \subset U_1 \subset S$ be as above. Suppose we are given on U_1 a measurable Beltrami coefficient ν (i.e., the component of an invariant $(-1,1)$-form $\nu d\overline{w}/dw$ with $\|\nu\|_\infty < 1$) supported by κ. It is known to us (cf. Ahlfors [2]) that there exists a homeomorphism of $\mathbf{C}_w$ onto $\mathbf{C}_z$ which is totally differentiable and satisfies the differential equation

$$\frac{\partial z}{\partial \overline{w}} = \nu(w)\frac{\partial z}{\partial w} \quad \text{a.e. on } \mathbf{C}_w.$$

We want to define the locally quasiconformal transform of the conformal structure of S as the solution of this differential equation restricted to $D_w = \phi(U_1)$, because it keeps the conformal structure entirely invariant everywhere on S except on κ and beside this, $\nu = 0$ implies nothing but a conformal exchange of local coordinate on U_1, much less elsewhere.

As prepared above, our present situation is such that the original conformal coordinate w of the Riemann surface S shall remain maintained in $U_2, \ldots, U_N$, while it should be replaced by the new coordinate z occurring from w as the locally quasiconformal transform that has just offered itself.

Now the new coordinate z decomposes the line element on U_1 into the form

$$ds^2 = \eta = \frac{1}{2}\left[(E+G)\,|dz|^2 + \mathrm{Re}\,\left\{(E - G - 2\sqrt{-1}\,F)\,dz^2\right\}\right] \tag{1}$$

at every point, where the mapping $w \mapsto z$ is totally differentiable: here E, F, and G stand, as customary, for the covariant componments of the first fundamental form in the real local coordinates $(x, y) = (\mathrm{Re}\, z, \mathrm{Im}\, z)$. The first and the second term of the right-hand side in (1) correspond respectively to the direction-independent and direction-dependent part of the infinitesimal arc-length ds relative to the coordinate difference dz.

2. Admissible quasiconformal neighbourhoods to the variational problem

Proposition 1. *Let B be an arbitrary closed subregion of U_1 with analytic Jordan boundary curve, which is disjoint with all U_j $(j = 2, 3, \ldots, N)$. If a Beltrami differential $\mu d\overline{z}/dz$ supported on B belongs, in particular, to the class $Lip\,\alpha$ $(0 < \alpha \leq 1)$ in U_1, then*

$$\frac{\partial \zeta}{\partial \overline{z}} = \mu(z)\frac{\partial \zeta}{\partial z} \tag{2}$$

is solved by some coordinate transformation $z \mapsto \zeta$ in U_1, which is of class $C^{1,\alpha}$.

Proof. Supposing a coordinate map ϕ takes U_1 onto the open disk $D_z \subset \mathbf{C}_z$, we set $\Delta_z = \phi(B)$. It is possible to extend the Beltrami coefficient $\mu(z) = \mu \circ \phi^{-1}(z)$ over the whole plane $\mathbf{C}_z$ by setting $\mu(z) = 0$ outside of Δ_z. Let $\zeta = \phi(z)$ denote the homeomorphism of $\mathbf{C}_z$ onto $\mathbf{C}_\zeta$ solving the Beltrami differential equation (2) under the normalization $f(\infty) = \infty$. The mapping $\zeta = f(z)$ restricted to D_z will be the desired coordinate change (cf. Ahlfors [1]). □

Definition 1. The coordinate transformation introduced in the above Proposition 1 is called a locally μ-conformal deformation of the local coordinate in U_1. If further the Beltrami coefficient in the above local μ-conformality satisfies the condition $|\mu| < |\varepsilon|$ for any complex parameter ε, we say that the local μ-conformality is *small*, or the μ-conformal deformation lies in the ε-neighbourhood of the conformality.

Remark. We need only the case in which $\|\mu\|_\infty$ is small enough to ensure the validity of the main theorem in [1].

Under this additional hypothesis we can state and prove the

Theorem 1 (Ahlfors [1]). *For any locally μ-conformal deformation of coordinate $z \mapsto \zeta$ there exists an $a(z) \in C^{1,\alpha}(D_z, \mathrm{C})$, such that*

$$\zeta = z + a(z). \tag{3}$$

Conversely, suppose an $a(z) \in C^{1,\alpha}(D_z, \mathrm{C})$ admitting the expression (3) *satisfies the condition $|1 + a_z| > |a_{\bar{z}}|$ with a compactly supported $a_{\bar{z}}$. Then* (3) *is a locally μ-conformal deformation of coordinate with $\mu = a_{\bar{z}}/(1 + a_z)$.*

Proof. Let us first recall that in the course of defining the locally μ-conformal deformation $z \mapsto \zeta$ of coordinates we have made full use of the global quasiconformal homeomorphism $\zeta = f(z)$ of $\mathbf{C}_z$ onto $\mathbf{C}_\zeta$, putting $\mu = 0$ outside its support and normalizing by the condition $f(\infty) = \infty$.

In accordance with Ahlfors [1] we introduce the modified pre-Hilbert transform

$$P_0\, q(z) = \frac{1}{2\sqrt{-1}\pi} \iint \left(\frac{1}{\omega - z} - \frac{1}{\omega} \right) q(\omega)\, d\omega \wedge d\bar{\omega} \tag{4}$$

acting on the Pompeiu's derivative $q = f_{\bar{z}}$ of f, where the integration should be taken over the whole finite plane $\mathbf{C}_\omega$. Since $\partial/\partial\bar{z}\,[f(z) - P_0\, q(z)] = q(z) - q(z) = 0$, we see that $f(z) - P_0\, q(z)$ must be holomorphic in the whole finite plane $\mathbf{C}_z$, which has a simple pole at $z = \infty$ by definition. We conclude that $f(z) - P_0\, q(z)$ reduces to an affine linear function. On account of conformal invariance of the μ-conformality, an appropriate parallel displacement together with one more normalization $f'(\infty) = 1$ is admitted, hence we have successively

$$f(z) = z + (P_0\, q)(z), \tag{5}$$

$$p = f_z = 1 + (Tq)(z), \tag{6}$$

$$q = \mu + \mu(Tq)(z), \tag{7}$$

where $z \in \mathbf{C}_z$, and T stands for the integral operator of Hilbert transform. The integral equation (7) is solved by the convergent Neumann series

$$q = \mu + \mu T\mu + \mu T\mu T\mu + \ldots\ldots,$$

which falls again in the space $\mathrm{Lip}\,\alpha$. Then (6) also holds and $p \in \mathrm{Lip}\,\alpha$, since $\mu \neq 0$. Integrating (6), we see that the right-hand side of (5) is written in terms of μ. Therefore, (4) with $a(z) = (P_0\, q)(z)$ is conformally equivalent to the given μ-conformal deformation of coordinate.

The converse is simpler to prove. For such an $a(z)$, the mapping given by (3) is quasiconformal everywhere with the Beltrami coefficient $\mu = a_{\bar{z}}/(1 + a_z)$ which vanishes off some compact subset of $\mathbf{C}_z$. In other words, the coordinate change is locally quasiconformal, and is a μ-conformal deformation, by definition. □

Definition 2. The quasiconformal change of local coordinates described in (3) is called of *quasi-affine type.* When further $a(z)$ in the expression (3) is smaller than $|\varepsilon|$ in modulus with the same parameter ε as in Definition 1, we say that it is *small.*

Proposition 2. *In the quasiconformal change of local coordinates, the smallness in the locally μ-conformal deformation is equivalent to the one in quasi-affine type.*

Proof. If $\mu = O(\varepsilon)$, the sum q of the Neumann series is $O(\varepsilon)$, too. Hence $a(z) = (P_0\, q)(z)$ is so. Conversely if $a(z) = O(\varepsilon)$, then $a_{\bar{z}} = O(\varepsilon)$. Therefore $\mu = a_{\bar{z}}/(1 + a_z) = O(\varepsilon)$. □

Proposition 3. *In order that a μ-conformal deformation ζ of z lies in a neighbourhood of the local coordinate z, it is necessary and sufficient that for any $h \in C^{1,\alpha}(D_z, \mathbf{C}) \cap Hol(D_z \setminus \Delta_z, \mathbf{C})$, there exists an $\varepsilon_0 > 0$ such that ζ admits the expression*

$$\zeta = z + \varepsilon\, h(z) \tag{8}$$

so far as the complex parmeter ε satisfies the condition $|\varepsilon| < \varepsilon_0$.

Now suppose that some μ-conformal deformation $\zeta = \zeta(z)$ is affected on the local coordinate z in (1). Then the metric η will be somehow changed into a new form

$$ds^2 = \tilde{\eta} = \frac{1}{2}\left[\left(\tilde{E} + \tilde{G}\right)|d\zeta|^2 + 2\mathrm{Re}\left\{\left(\tilde{E} - \tilde{G} - 2\sqrt{-1}\tilde{F}\right)d\zeta^2\right\}\right]. \tag{9}$$

Definition 3. In view of the relationship (8) between the coordinates z and ζ in the forms describing η and $\tilde{\eta}$, we say that $\tilde{\eta}$ lies in the ε-neighbourhood of η, so far as ζ lies in a ε-neighbourhood of z.

3. The energy integral associated with Riemannian metrics

M. Gerstenhaber and H. E. Rauch were the first who proposed to assign the double integral

$$I[\eta] = \frac{1}{2} \iint (E + G)\, dx \wedge dy$$

to the direction-independent term $(\tilde{E} + \tilde{G}) d\zeta d\overline{\zeta}$ and investigated an effect of its extremality on the direction-dependent $(2,0)$-form of the second term on the right-hand side of (9). Improving their ideas, we will arrive in quite natural way at the following

Definition 4. If the functional $I[\tilde{\eta}]$ defined in an ε-neighbourhood of η takes a stationary value at $\tilde{\eta} = \eta$, then we say simply that the specific Riemannian metric η appearing in the expression (1) is stationary.

In the argument above the integration should first be taken with respect to the local coordinate which is valid in each member U_j $(j = 1, 2, \ldots, N)$ of the neighbourhood system $\mathcal{U}$ covering S and then the integral over such neighbourhoods should be added after multiplication by the partition of unity.

From Theorem 1 and Definition 4 follows

Theorem 2. *In order that the specific Riemannian metric η is a stationary point of Dirichlet's energy integral, it is necessary and sufficient that the first variation of I vanishes at η compared with all the small deformations quasi-affine type of the local coordinates.*

Now we can have

Theorem 3 (Gerstenhaber-Rauch [3]). *The first variation of the energy integral $I[\eta]$ in the small changes of local coordinate of quasi-affine type* (8) *reads*

$$\delta I[\eta] = 4\varepsilon \iint (E - G - 2\sqrt{-1}\, F)\, h_{\overline{z}}\, dx \wedge dy.$$

Proof. Let the coordinate $z \in D_z = \phi(U_1)$ be changed into ζ by means of a local quasiconformal change of quasi-affine type (8), whose Jacobian is J. Then we have

$$I[\tilde{\eta}] - I[\eta] = \iint \rho(w(\zeta))\,(|\tilde{p}|^2 + |\tilde{q}|^2)\, J\, dx \wedge dy - \iint \rho(w(z))\,(|p|^2 + |q|^2)\, dx \wedge dy, \tag{10}$$

where we have set $p = \partial w/\partial z$, $q = \partial w/\partial \overline{z}$, $\tilde{p} = \partial w/\partial \zeta$, and $\tilde{q} = \partial w/\partial \overline{\zeta}$ for brevity. Differentiation of (8) together with the chain rule provides

$$|\tilde{p}|^2 = \left[\, |p|^2\, |1 + \varepsilon\, h_z|^2 + |q|^2\, |\varepsilon h_{\overline{z}}|^2 - 2\mathrm{Re}\, \{p\overline{q}\,(1 + \overline{\varepsilon h_z}\,)\,\varepsilon h_{\overline{z}}\}\, \right] / J^2,$$

$$|\tilde{q}|^2 = \left[\, |q|^2 \, |1 + \varepsilon\, h_z|^2 + |p|^2 \, |\varepsilon h_{\overline{z}}|^2 - 2\mathrm{Re}\left\{ p\overline{q}\, (1 + \overline{\varepsilon h_z}\,)\, \varepsilon h_{\overline{z}} \right\} \right] / J^2,$$

where

$$J = 1 + 2\mathrm{Re}\left\{\varepsilon h_z\right\} + |\varepsilon|^2 \left(|h_z|^2 - |h_{\overline{z}}|^2\right).$$

Hence

$$I[\tilde{\eta}] - I[\eta] = -4 \iint \mathrm{Re}\, \left\{ \rho(w(z))\, \varepsilon p\overline{q}\, (1 + \overline{\varepsilon h_z})\, h_{\overline{z}} \right\}\, dx \wedge dy + O(\varepsilon^2).$$

Therefore in view of arbitrariness of $\arg \varepsilon$, one obtaines the desired conclusion. □

From Theorems 2 and 3 follows at once

Theorem 4. *In order that the energy value $I[\eta]$ is tationary, it is necessary and sufficient that*

$$\iint (E - G - 2\sqrt{-1}\, F)\, h_{\overline{z}}\, dx \wedge dy = 0$$

for all the same changes of local coordinate of quasi-affine type (8).

Corollary. *If $I[\eta]$ is a stationary value, then the mapping $z \mapsto w$ is harmonic relative to the metric $\rho(w)|dw|^2$.*

Proof. In reference to Theorem 4 one is allowed to adopt the particular choice of $h(z)$ which vanishes outside of Δ_z. The contour integral

$$\int (E - G - 2\sqrt{-1}\, F)\, h(z)\, dz = 0$$

along an arbitrary path enclosing Δ_z together with Green's theorem yields

$$\iint (E - G - 2\sqrt{-1}\, F)_{\overline{z}}\, h(z)\, dx \wedge dy = 0,$$

which guarantees that $(E - G - 2\sqrt{-1}\, F)_{\overline{z}} = 0$ almost everywhere, accordingly everywhere on Δ_z. Since Δ_z was arbitrary in D_z, the quadratic differential $(E - G - 2\sqrt{-1}\, F)\, dz^2$ must be holomorphic everywhere in D_z, which was to be proved. □

Theorem 5. *The conformal metric is stationary compared with all the Riemannian metrics that arise from small quasiconformal deformations of its local coordinate.*

Proof. Suppose that the metric η is, in particular, conformal. Then (1) gives that $E - G = F = 0$. Therefore Theorem 4 together with Definition 4 provides the desired result. □

References

[1] Ahlfors, L. V., Conformality with respect to Riemannian metrics. Ann. Acad. Sci. Fenn. A I 206 (1955), 1–22

[2] Ahlfors, L. V., Lectures on Quasiconformal Mappings. D. van Nostrand Company Inc., Princeton 1966

[3] Gerstenhaber, M., Ruach, H. E. On extremal quasi-conformal mappings I. Proc. Nat. Acad. Sci. US. 40 (1954), 808–812

[4] Shibata, K., Dirichlet Principle and the analytic prolongation theorem onto analytic boundary. In 'Analytic Function Theory of One Complex Variable' (ed. by Y. Komatu et al.), pp. 299–324, Longman Scientific & Technical, Essex 1989

Application of potential theory in biharmonic analysis

Abdullah Shidfar and M.A. Hosseini

Abstract

A method for the solution of some two-dimensional biharmonic problems is described. In this method, we use the Almansi representation of a biharmonic function in terms of two harmonic functions in the plane and conclude that some biharmonic problems may be solved by using fundamental solutions of Laplace's equation.

1. Introduction

Almansi showed that every solution of the biharmonic equation

$$\nabla^4 w = \nabla^2(\nabla^2 w) = 0 \quad \text{in D} \tag{1}$$

may be expressed in the form

$$w = r^2\phi + \psi; \qquad r^2 = x^2 + y^2, \tag{2}$$

where ϕ and ψ are harmonic functions in D [1]. For any given w , we note that [2,3]

$$\phi + r\frac{d\phi}{dr} = 1/4\nabla^2 w, \tag{3}$$

which is a differential equation for ϕ in terms of the harmonic function $\nabla^2 w$. This equation has a general solution

$$\phi = \frac{r^{-1}}{4}\int \nabla^2 w dr + r^{-1}g(\theta). \tag{4}$$

However, ϕ is not necessarily a harmonic function. The appropriate condition is $\nabla^2\phi = 0$ which yields the equation

$$\nabla^2\{r^{-1}g\} = -1/4\nabla^2\{r^{-1}\int \nabla^2 w dr\}. \tag{5}$$

This is a poisson's equation for g which always has a solution, and accordingly there always exists a harmonic function of the form (4) which satisfies equation (3).

Here, we consider two biharmonic problems and by using the Almansi representation (1) we investigate the solutions to these problems.

Potential Theory, ed. by M. Kishi

2. The deflection of a thin plate

We consider the deflection of unloaded elliptic plate with a clambed edge. If w is the deflection of this plate then w satisfies the following problem

$$\begin{aligned} \nabla^4 w &= 0 \quad \text{in D} && (6)\\ w(x,y) &= f(x,y) \quad \text{on } \partial \text{ D} && (7)\\ \frac{\partial w}{\partial n}(x,y) &= g(x,y) \quad \text{on } \partial \text{ D}. && (8) \end{aligned}$$

If $h(x,y)=0$ is the equation of ∂ D then we have

$$\frac{\partial w}{\partial n}|_{\partial D} = \nabla w \frac{\nabla h}{|\nabla h|} = (w_x h_x + w_y h_y)(h_x^2 + h_y^2)^{-1/2}. \tag{9}$$

From (1) and (9) the equations (7) and (8) for any fixed point (x_0, y_0) of the boundary ∂ D become

$$r_0^2\phi(x_0,y_0) + \psi(x_0,y_0) = f(x_0,y_0); \quad r_0^2 = x_0^2 + y_0^2, \tag{10}$$

$$(w_x h_x + w_y h_y)|_{(x_0,y_0)} = [(h_x^2 + h_y^2)^{1/2} g]|_{(x_0,y_0)}. \tag{11}$$

From (10) and (11) we might calculate ϕ and ψ at point (x_0, y_0). For this purpose, we reconsider the equations (10) and (11) for the harmonic functions ϕ and ψ defined at an arbitrary point (x,y) in D and we find

$$\psi(x,y) = f(x_0,y_0) - r_0^2\phi(x,y), \tag{12}$$

$$\begin{aligned} &h_x(2x_0\phi(x,y) + r_0^2\phi_x(x+y) + \psi_x(x,y)) \\ &+ h_y(2y_0\phi(x,y) + r_0^2\phi_y(x,y) - \psi_y(x,y)) = g(h_x^2 + h_y^2)^{1/2}. \end{aligned} \tag{13}$$

By substituting from (12) into (13) we find

$$h_x(x_0,y_0)(2x_0\phi(x,y) + h_y(x_0,y_0)(2y_0\phi(x,y)) = g(x_0,y_0)(h_x^2 + h_y^2)^{1/2}, \tag{14}$$

whence

$$\phi(x_0,y_0) = \frac{g(x_0,y_0)(h_x^2 + h_y^2)^{1/2}}{2(x_0 h_x + y_0 h_y)}, \tag{15}$$

and $\phi(x_0, y_0)$ may be obtained from (12).

Consequently, for any given point (x,y) on the boundary, we have

$$\phi = \frac{g(h_x^2 + h_y^2)^{1/2}}{2(x h_x + y h_y)}, \tag{16}$$

and

$$\psi = f - r^2\phi. \tag{17}$$

As a second example, we consider the deflection of a uniform unloaded equilateral triangular plate, the deflection of this plate satisfies the following problems

$$\nabla^4 w = 0 \quad \text{in D} \tag{18}$$

$$w = f(x,y) \quad \text{on } \partial \text{ D} \tag{19}$$

$$\nabla^2 w = g(x,y) \quad \text{on } \partial \text{ D} \tag{20}$$

Using the Almansi r^2-form representation, $w = r^2\phi + \psi$, we find

$$\nabla^2 w = 4(x\frac{\partial\phi}{\partial x} + y\frac{\partial\phi}{\partial y}) + 4\phi(x,y). \tag{21}$$

By a similar situation we may conclude that

$$\phi = 1/4g \tag{22}$$

$$\psi = f - r^2\phi, \tag{23}$$

on the boundary.

Now let us deal some problems which described by Karageorghis and Fairweather [4].

As a first example we consider the following problem

$$\nabla^4 w = 0 \qquad -0.5 \le x, y \le 0.5,$$

$$w(x, 0.5) = w(x, -0.5) = x^2,$$

$$w(0.5, y) = w(-0.5, y) = 0.25,$$

$$\nabla^2 w = 2 \qquad \text{on the boundary .}$$

This problem was given by Bogomolny [5] and also used in [4]. The above problem has a solution $w = x^2$.

From (22) and (23) we obtain $\phi = 1/2$ and

$\psi = 1/2(0.25 - y^2); \quad x = \pm 0.5, \quad -0.5 \le y \le 0.5,$

$\psi = 1/2(x^2 - 0.25); \quad y = \pm 0.5, \quad -0.5 \le x \le 0.5,$

respectively, on the boundary of D.

A solution for ϕ and ψ are $\phi = 1/2$ and $\psi = 1/2(x^2 - y^2)$ in domain D. These solutions may be obtained by a direct calculation from (4) and the Almansi r^2-form representation, respectively.

A problem possesses symmetry about the line $y = x$ also considered by Bogomolny [5] takes the form

$$\nabla^4 w = 0 \qquad \text{on a unit disk D}$$
$$w = 1 + \sqrt{2}\cos\theta \qquad \text{on } \partial \text{ D}$$
$$\frac{\partial w}{\partial n} = 2 + \sqrt{2}\cos\theta \qquad \text{on } \partial \text{ D} .$$

This problem has an exact solution $w = x^2 + y^2 + x\sqrt{2}$ in D .
From $h(x,y) = x^2 + y^2 - 1 = 0$, (16) and (17), we find $\phi = 1 + \frac{\sqrt{2}}{2}\cos\theta$ and $\psi = \frac{\sqrt{2}}{2}\cos\theta$ on the boundary of D .
Now, the solution for ϕ on the unit disk may be obtained from (4)

$$\phi = 1 + r^{-1}g(\theta).$$

Choosing $g(\theta) = \frac{\sqrt{2}}{2}\cos\theta$ we find

$$\phi = 1 + \frac{\sqrt{2}}{2}r^{-1}\cos\theta$$

and

$$\psi = w - r^2\phi = \frac{\sqrt{2}}{2}\cos\theta.$$

3. Conclusion

It has been shown that the biharmonic problems can be solved by solving the harmonic problems. For this purpose, we used the Almansi r^2-form representation and transformed our biharmonic problem into two harmonic problems. We applied our method for some special problems introduced by Karageorghis and Fairweather [4]. However, this method can be used for the other problems as well as these problems.

References

[1] E. Almansi, Sull' integrazione dell' equazione differentiable $\Delta^{2n} = 0$, Annali di mathematica pure et applicata, series III, 2:1 (1897).

[2] M.A. Jaswon and A. Shidfar, On a differential equation of Bergman and Schiffer, J. Maths. and Phys., 14 (1980).

[3] A. Shidfar, On the representation of x, y, z, and r^2-forms of a biharmonic function in terms of two harmonic functions and its applications to the Papkovich representation, the Mathematics Student, I.M.S., 51 (1987).

[4] A. Karageorghis and G. Fairweather, The Almansi method of fundamental solutions for solving biharmonic problems, International Journal for Numerical Methods in Engineering, 26 (1988).

[5] A. Bogomolny, Fundamental solutions method for elliptic boundary value problems, SIAM J. Numer. Anal., 22 (1985).

Sobolev imbeddings and integrability of harmonic functions on Hölder domains

Wayne Smith[1] *and David A. Stegenga*

Abstract

A proper subdomain D of R^n is called a Hölder domain if, for fixed $x_0 \in D$, the quasi-hyperbolic distance $k_D(x_0, x)$ is less than a constant plus a constant multiple of the logarithm of the Euclidean distance from x to the boundary of D. These domains are characterized by the exponential integrability of $k_D(x_0, x)$ over D. Integrability of positive harmonic functions and some classical Sobolev imbedding inequalities are shown to be consequences of this exponential integrability.

1980 Mathematics Subject Classification (1985 Revision): 46E35, 31B05.

1. Introduction

Let D be an open, connected and proper subdomain of n-dimensional Euclidean space R^n, $n \geq 2$. Following [GO] we define the quasi-hyperbolic metric k_D in D by

$$k_D(x_1, x_2) = \inf_{\gamma} \int_{\gamma} \frac{ds}{\delta_D(x)} \tag{1.1}$$

where the infimum is taken over all rectifiable arcs γ joining x_1 to x_2 in D. Here we denote by $\delta_D(x)$ the Euclidean distance between x and the complement of D.

If D is a simply connected planar domain, then the quasi-hyperbolic metric is comparable to the usual hyperbolic or Poincaré metric on D. See, for example, [BP]. For domains in R^n, the quasi-hyperbolic metric provides a useful substitute for the hyperbolic metric. Applications can be found, for example, in [GM], [GP], [GO], [H], [S], [SS2] and [SS3].

Definition. Let D be a proper subdomain of R^n and let $x_0 \in D$. The domain D is said to be a *Hölder domain* provided there are constants c and C such that

$$k_D(x_0, x) \leq c \log \frac{\delta_D(x_0)}{\delta_D(x)} + C, \quad x \in D. \tag{1.2}$$

The terminology is derived from the fact that in the plane, a simply connected domain D is a Hölder domain if and only if a Riemann mapping function from the

[1]The first author is supported in part by a grant from the National Science Foundation.

Potential Theory, ed. by M. Kishi

unit disk onto D is Hölder continuous. (See [BP].) In general, Hölder domains are bounded and multiply connected domains.

A closely related type of domain is the John domain. Fix a point $x_0 \in D$; we say that D is a John domain provided that for each $x_1 \in D$ there is an arc γ joining x_0 to x_1 in D along which

$$\delta_D(x) \geq \alpha|\gamma(x, x_1)|, \qquad x \in \gamma. \tag{1.3}$$

Here α is a positive constant, $\gamma(x, x_1)$ is the portion of γ joining x to x_1 and $|\gamma(x, x_1)|$ is its arc length. The thickness condition (1.3) can be visualized as a "twisted cone condition". Clearly domains satisfying a uniform inner cone condition form a proper subclass of John domains. John domains are easily seen to be Hölder domains, since the infimum in (1.1) can be estimated by integrating along the curve in (1.3) to show that (1.2) holds. Hölder domains are more general than John domains; see [SS2].

We recently have established a characterization of Hölder domains in terms of an integrability property of the quasi-hyperbolic metric. The following theorem is contained in Theorem A of [SS3].

Theorem 1. *Let D be a proper subdomain of R^n and let $x_0 \in D$. The following are equivalent:*

(a) *D is a Hölder domain;*
(b) *There is a $\tau > 0$ such that*

$$\int_D \exp(\tau k_D(x_0, x))\, dx < \infty .$$

The proof of this theorem in [SS3] is based on the extension of some recent results of Carleson and Jones [CJ] and Jones and Makarov [M] concerning simply connected planar domains to multiply connected domains in R^n by using the quasi-hyperbolic metric. The exponential integrability of the quasi-hyperbolic metric was previously known for John domains [SS2]. In [SS3] we used Theorem 1 to derive results about the geometry of Hölder domains and the Poincaré inequality. In this paper we present further applications of the exponential integrability of the quasi-hyperbolic metric given in Theorem 1.

In section 2 we consider the following problem: For which proper subdomains D of R^n does there exist $p = p(D) > 0$ such that every positive harmonic function u on D is contained in $L^p(D)$, the space of Lebesgue measurable functions on D which are p-integrable with respect to n-dimensional Lebesgue measure? We prove that if $D \subset R^n$ is a Hölder domain, then such a $p > 0$ exists, and for the converse problem, if $D \subset R^2$ is finitely connected and such a $p > 0$ exists, then D is a Hölder domain.

Closely related problems concerning integrability properties of positive superharmonic and positive subharmonic functions have been the subject of recent work of

F-Y. Maeda and N. Suzuki [MS], M. Masumoto [M1], [M2] and Suzuki [Su]. In particular, it is proved in [MS] that a positive superharmonic function on a Lipschitz domain $D \subset R^n$ belongs to $L^p(D)$, where $p = p(D) > 0$. In [M2] this conclusion is obtained for $D \subset R^2$ bounded by finitely many closed Jordan curves and satisfying an interior wedge condition, and also for $D \subset R^2$ bounded by finitely many quasicircles. These domains are all John domains, and thus are Hölder domains. However, our result in section 2 does not provide as good an upper bound for $p(D)$ as is given in [MS] and [M2].

Section 3 is concerned with the classical Sobolev imbedding inequalities on a domain $D \subset R^n$. In [SS2] we developed techniques to establish that some geometric conditions on a domain are sufficient for Poincaré inequalities (which are special cases of the Sobolev imbedding inequalities) to hold on the domain. In section 3 we use these techniques to prove certain geometric conditions are sufficient for Sobolev imbedding inequalities. This along with Theorem 1 is then used to show the existence of these imbeddings on Hölder domains. A further application of Theorem 1 is that Trudinger's inequality holds on Hölder domains. Trudinger's inequality improves on the classical Sobolev imbedding of the Sobolev space $W^{1,n}(D) \to L^q(D)$, $q < \infty$, by establishing the existence of the imbedding into a certain Orlicz space. Definitions and details are given in section 3. We thank Professor S.-Y. Chang for bringing this problem to our attention.

2. Global integrability of positive harmonic functions

Let D be a domain in R^n ($n \geq 2$) and $x_0 \in D$. Denote by $S^+(D)$ ($H^+(D)$) the class of all positive superharmonic (harmonic) functions on D. We say that $S^+(D)$ ($H^+(D)$) imbeds into $L^p(D)$, for some $p > 0$, if there is a finite constant $c(D, x_0, p)$ such that the inequality

$$\int_D u^p(x)\, dx \leq c(D, x_0, p) \cdot u^p(x_0) \tag{2.1}$$

holds for all $u \in S^+(D)$ ($H^+(D)$).

Armitage [Ar] has shown that $S^+(D)$ imbeds into $L^p(D)$ for certain smooth domains (including balls) and all $0 < p < n/(n-1)$. Recently, Maeda-Suzuki [MS] and Masumoto [M1], [M2] have extended this result to certain domains satisfying Lipschitz boundary conditions and various values for p. We consider this problem for the smaller class $H^+(D)$ but more general domains D.

For the remainder of this paper $W = \{Q\}$ will be a Whitney decomposition of D into closed dyadic cubes with disjoint interiors. This means that the vertices of each cube have coordinates that are dyadic rational numbers and that the diameter of each cube $Q \in W$, which we denote by $d(Q)$, is comparable to its Euclidean distance to the complement of D. The volume of a measurable subset E of R^n will be denoted by $m(E)$. The notation $a \approx b$ and $a \lesssim b$ will be used to mean that a and b are either comparable or satisfy an inequality with a constant depending only on the dimension.

Theorem 2. *If D is a Hölder domain, then there is a $p_0 > 0$ such that $H^+(D)$ imbeds into $L^p(D)$ for all $0 < p < p_0$.*

Proof. Let D be a Hölder domain with base point $x_0 \in D$ and let W be a Whitney decomposition of D into cubes. Let $u \in H^+(D)$. Suppose that Q_1, Q_2 are adjacent Whitney cubes, then it follows from Harnack's inequality that

$$\sup_{x \in Q_1} u(x) \leq c \inf_{x \in Q_2} u(x)$$

where c is a constant independent of the cubes. It follows immediately that for $x \in D$, $u(x)$ is dominated by $c^n u(x_0)$ where n is the minimum number of Whitney cubes required to connect x to x_0 in D.

But n is comparable to the quasi-hyperbolic distance $k_D(x, x_0)$. Hence, by increasing c we obtain that

$$\int_D u^p \, dx \leq \int_D e^{cpk_D(x,x_0)} \, dx \cdot u^p(x_0). \tag{2.2}$$

By Theorem 1, $p_0 > 0$ can chosen so that the integral on the right is finite for $0 < p < p_0$. □

Theorem 3. *Suppose that D is a finitely connected domain in R^2 and that $H^+(D)$ imbeds into $L^p(D)$ for some $p > 0$. Then D is a Hölder domain.*

Proof. Suppose that $H^+(D)$ imbeds into $L^p(D)$ for some $p > 0$. Then taking $u = 1$ in (2.1) shows that D has finite area and hence that ∂D is nonempty. Suppose that D is not a Hölder domain . Then there exists a sequence $\{w_n\}$ in D satisfying

$$\lim_{n \to \infty} \frac{k_D(w_0, w_n)}{-\log \delta(w_n)} = \infty. \tag{2.3}$$

We may assume that either $|w_n| \to \infty$ or else that $\lim w_n = a \in \partial D$. If $w = a$ is an isolated singularity for a positive harmonic function, then it can be written as the sum of a harmonic function plus $c \log |w - a|$ and hence is integrable to all orders near $w = a$. Also, D is a Hölder domain if and only if $D \setminus \{a\}$ is a Hölder domain . Hence we may assume that D has no isolated boundary points (including, of course, the point at ∞).

By the Riemann Mapping Theorem we map D onto a finitely connected domain $\tilde{\Delta}$ which is contained in the unit disk Δ in such a way that the outer boundary of $\tilde{\Delta}$ is $\partial\Delta$. Since ∂D has no isolated points, we may also assume that the points z_n corresponding to w_n under the mapping satisfy $\lim |z_n| = 1$.

As in [M2], we need some basic facts concerning the Poincaré metric associated with hyperbolic domains in $\mathbb{C}$. Corresponding to a domain D (with at least two boundary points) there is a metric ρ_D with density $\lambda_D(z)|dz|$ which is invariant under conformal mappings and satisfies the following, see Proposition 1.1 in Chapter II in [Kra].

Lemma 2.1. *If D is a finitely connected domain such that every component of $\hat{\mathbb{C}} \setminus D$ contains at least two points and $\infty \notin D$, then there exists $M_D > 0$ for which*

$$M_D \le \lambda_D(w)\delta_D(w) \le 1, \qquad w \in D. \tag{2.4}$$

By the lemma and our assumptions on D we see that the Poincaré metric is comparable to the quasi-hyperbolic metric on both D and $\tilde{\Delta}$. Since $\tilde{\Delta}$ is finitely connected, there is a $0 < r_0 < 1$ such that $r_0 \le |z| < 1$ implies that $z \in \tilde{\Delta}$. We may assume that $r_1 = (r_0 + 1)/2 < |z_n| < 1$ for all n and define $a_n = r_1 z_n/|z_n|$. Now it is obvious that there is a finite constant A such that

$$\rho_{\tilde{\Delta}}(a_n, z_n) \ge \rho_{\tilde{\Delta}}(z_0, z_n) - A, \qquad n = 1, 2, \ldots. \tag{2.5}$$

Now for each n, we rotate $\tilde{\Delta}$ and then map Δ onto the right half plane Π in such a way that a_n maps to $b_1 = (1 + r_1)/(1 - r_1)$, z_n maps to x_n, $b_1 < x_n < \infty$, and $\tilde{\Delta}$ onto $\tilde{\Pi}$. Put $b_0 = (1 + r_0)/(1 - r_0)$. For $b_1 \le x \le x_n$, we easily see that $\delta_{\tilde{\Pi}}(x) \ge x - b_0 \ge b_1 - b_0 > 0$. Using Lemma 2.1 and (2.5), we estimate x_n from below as follows:

$$\begin{aligned}\log(x_n - b_0) - \log(b_1 - b_0) &\ge \int_{b_1}^{x_n} \frac{dx}{\delta_{\tilde{\Pi}}(x)} \ge k_{\tilde{\Pi}}(b_1, x_n)\\ &\ge \rho_{\tilde{\Pi}}(b_1, x_n) = \rho_{\tilde{\Delta}}(a_n, z_n)\\ &\ge \rho_{\tilde{\Delta}}(z_0, z_n) - A = \rho_D(w_0, w_n) - A\\ &\ge M_D k_D(w_0, w_n) - A.\end{aligned}$$

The above estimate says that

$$\log x_n \ge M_D k_D(w_0, w_n) + B, \qquad n = 1, 2, \ldots \tag{2.6}$$

where B is a constant.

We now obtain an upper bound for x_n using (2.1). For each n, we have constructed a conformal mapping f_n from D into Π whose real part u_n is positive and satisfies $u_n(w_n) = x_n$ and $u_n(w_0) \le b_0$.

Let D_n be the disk centered at w_n of radius $\frac{1}{2}\delta_D(w_n)$. By Harnack's inequality, we have that $x_n = u_n(w_n) \lesssim \inf_{D_n} u$. It follows from (2.1) that

$$x_n^p \delta_D^2(w_n) \lesssim \int_{D_n} u_n^p |dw| \lesssim u_n^p(w_0) \le b_0^p.$$

Taking logarithms we get

$$\log x_n \le \frac{2}{p} \log \frac{1}{\delta_D(w_n)} + C, \qquad n = 1, 2, \ldots.$$

Combining the above with (2.6) we obtain the required contradiction to (2.3). □

Corollary 4. *Suppose that D is a finitely connected domain in $\mathbb{C}$. $H^+(D)$ imbeds into $L^p(D)$ for some $p > 0$ if and only if D is a Hölder domain.*

Corollary 5. *Suppose that D is a finitely connected domain in $\mathbb{C}$ for which $H^+(D)$ imbeds into $L^p(D)$ for some $p > 0$, then D can not have a cusp.*

Proof. This property is proved for simply connected Hölder domains in [SS1] where a general definition of cusps is given. One easily extends that result to finitely connected domains; thus, the proof is immediate from Theorem 3. See also Lemma 3 in [SS2]. □

Remark. Theorem 3 is probably true without any restrictions on D, but we have been unable to prove this.

3. Sobolev imbedding inequalities

In this section we give some geometric conditions that are sufficient for a classical Sobolev inequality to hold for a domain. Let $D \subset R^d$ be a domain with finite volume, $m(D) < \infty$, and $1 \leq p < \infty$. We denote by $W^{1,p}(D)$ the usual Sobolev space of functions on D that together with their first order weak partial derivatives are in $L^p(D)$. The norm for $W^{1,p}(D)$ is given by

$$\|u\|_{W^{1,p}(D)} = (\int_D |u|^p \, dx + \int_D |\nabla u|^p \, dx)^{1/p}.$$

Recall that $C^1(D)$ is dense in $W^{1,p}(D)$.

For $p \leq q$, define

$$M_{p,q}(D) = \sup_u \frac{(\int_D |u - u_D|^q \, dx)^{1/q}}{(\int_D |\nabla u|^p \, dx)^{1/p}}, \tag{3.1}$$

where the supremum is taken over non-constant $u \in W^{1,p}(D)$ and $u_D = \frac{1}{m(D)} \int_D u \, dx$ has been used for the average of u over D. Let $p^* = np/(n-p)$ if $p < n$, and $p^* = \infty$ if $p > n$. A classical result is that if D satisfies a uniform inner cone condition, then $M_{p,q}(D) < \infty$ for $p \neq n$ and $p \leq q \leq p^*$, and $M_{n,q}(D) < \infty$ for $n \leq q < \infty$. See for example [Ma, Theorem 1.4.5]. It is easy to see that $M_{p,q}(D) < \infty$ implies that the classical Sobolev imbedding

$$W^{1,p}(D) \to L^q(D) \tag{3.2}$$

exists.

In [SS2] we studied various geometric conditions on D that implied $M_{p,p}(D) < \infty$. We used that $M_{p,p}(\lambda D) = \lambda M_{p,p}(D)$, and we developed techniques involving chains of Whitney cubes and the quasi-hyperbolic metric to estimate $M_{p,p}(D)$, the Poincaré constant for D. Here λD is the dilation of D by $\lambda > 0$. A simple observation concerning behavior of $M_{p,q}(D)$ under dilations allows us to use these same techniques to get results on the Sobolev imbeddings (3.2). A change of variables argument shows that, for $\lambda > 0$,

$$M_{p,q}(\lambda D) = \lambda^{n(\frac{1}{q}-\frac{1}{p}+\frac{1}{n})} M_{p,q}(D). \tag{3.3}$$

For $p \leq n$ and $q = p^*$, this is often referred to as conformal invariance, since $M_{p,p^*}(\lambda D) = M_{p,p^*}(D)$. Observe that as a consequence of (3.3) and the case of the Sobolev imbedding theorem mentioned above, if D is a domain of finite volume with Whitney decomposition W, then there is a constant c depending only on the size of the largest cube $Q \in W$ such that

$$M_{p,q}(Q) \leq c, \qquad p \leq q \leq p^*, \, Q \in W. \tag{3.4}$$

Theorem 6 below extends Theorem 8 in [SS2], where the case $p = q$ was considered. The proof of Theorem 6 uses the same techniques as were employed in [SS2]. We will concentrate on the changes required when $p \neq q$. Our basic estimate for $M_{p,q}(D)$ is given in the following lemma. It is an extension of Lemma 7 in [SS2], in which the case $p = q$ is considered. Since the proof is a simple modification of that in [SS2], it will not be given here. Fix a cube $Q_0 \in W$ and let x_0 be its center.

Lemma 3.1. *Let $1 \leq p \leq q \leq p^*$ if $p \neq n$, and $n \leq q < \infty$ if $p = n$. Then, for c as in (3.4), we have*

$$M_{p,q}(D) \lesssim c + \sup_{u \in C^1(D)} \frac{(\sum_{Q \in W} |u_Q - u_{Q_0}|^q m(Q))^{1/q}}{\|\nabla u\|_{L^p(D)}}.$$

We now must review some definitions from [SS2]. We continue to assume that Q_0 is a fixed cube in W. We assume that, for each $Q \in W$, there is a set $P(Q) \subset D$ containing a chain $Q_0, \ldots, Q_N$ of cubes in W, starting with Q_0 and ending with $Q_N = Q$. This means that Q_i is adjacent to Q_{i+1} in W, so that a face of the smaller cube is contained in a face of the larger cube, and $Q_i \subset P(Q)$ for $0 \leq i \leq N$. Now define

$$S(Q) = \cup \{Q_1 \in W \mid Q \subset P(Q_1)\}. \tag{3.5}$$

As an example of how to construct the sets $P(Q)$, we could simply take the cubes in W which intersect a quasi-hyperbolic geodesic joining x_0 to the center of Q. A geometric interpretation of $S(Q)$, in this case, is that it is the points in the *shadow* of Q, assuming that light travels from x_0 along the quasi-hyperbolic geodesics in D.

The next lemma is Lemma 8 from [SS2].

Lemma 3.2. *If $u \in C^1(D)$ and $Q \in W$, then*

$$|u_Q - u_{Q_0}| \lesssim \int_{P(Q)} \frac{|\nabla u|}{\delta_D(x)^{n-1}}\, dx.$$

For $x \in D$, let Q_x denote a cube in W containing x.

Theorem 6. *Let $p \le q \le p^*$ if $p \ne n$ and $n \le q < \infty$ if $p = n$. If D is as above and λ is a real number, then*

$$M_{p,q}(D) \lesssim \{\sup_{Q\in W} \frac{1}{d(Q)^{\lambda q(n-1)}} \int_{S(Q)} (\int_{P(Q_x)} \frac{dy}{\delta_D(y)^{p'(1-\lambda)(n-1)}})^{q/p'}\, dx\}^{1/q} \tag{3.5}$$

for $1 < p < \infty$, $p' = \frac{p}{p-1}$ and

$$M_{1,q}(D) \lesssim \sup_{Q\in W} \frac{m(S(Q))^{1/q}}{d(Q)^{n-1}}$$

when $p = 1$.

Proof. The proof is a modification of the proof of Theorem 8 in [SS2]. Let

$$F_p(x_Q, \lambda) = (\int_{P(Q)} \frac{dx}{\delta_D(x)^{(1-\lambda)p'(n-1)}})^{p/p'},$$

for $Q \in W$, λ real, and $1 < p < \infty$. For $p = 1$, let $F_1 = 1$ and set $\lambda = 1$. Then, for $u \in C^1(D)$, by using Lemma 3.2 and Hölder's inequality we see that

$$\begin{aligned}
&\{\sum_{Q\in W} |u_Q - u_{Q_0}|^q m(Q)\}^{1/q} \\
&\lesssim \{\sum_{Q\in W} (\int_{P(Q)} \frac{|\nabla u|}{\delta_D(x)^{n-1}}\, dx)^q m(Q)\}^{1/q} \\
&\le \{\sum_{Q\in W} (\int_D \frac{|\nabla u|^p}{\delta_D(x)^{\lambda p(n-1)}} \chi_{P(Q)}(x) F_p(x_Q,\lambda)\, dx)^{q/p} m(Q)\}^{1/q}
\end{aligned}$$

and hence by the generalized Minkowski inequality

$$\begin{aligned}
&\le \{\int_D |\nabla u|^p \frac{1}{\delta_D(x)^{\lambda p(n-1)}} (\sum_{Q\in W} (\chi_{P(Q)}(x) F_p(x_Q,\lambda))^{q/p} m(Q))^{p/q}\, dx\}^{1/p} \\
&\le \sup_{x\in D} \frac{1}{\delta_D(x)^{\lambda(n-1)}} (\sum_{Q\in W} (\chi_{P(Q)}(x) F_p(x_Q,\lambda))^{q/p} m(Q))^{1/q} \cdot \|\nabla u\|_{L^p(D)} \\
&\lesssim \{\sup_{Q\in W} \frac{1}{d(Q)^{\lambda q(n-1)}} \sum_{Q_1 \subset S(Q)} (F_p(x_{Q_1},\lambda))^{q/p} m(Q_1)\}^{1/q} \cdot \|\nabla u\|_{L^p(D)} \\
&\lesssim \{\sup_{Q\in W} \frac{1}{d(Q)^{\lambda q(n-1)}} \int_{S(Q)} (\int_{P(Q_x)} \frac{dy}{\delta_D(y)^{p'(1-\lambda)(n-1)}})^{q/p'}\, dx\}^{1/q} \cdot \|\nabla u\|_{L^p(D)}.
\end{aligned}$$

We also have, as in the proof of Theorem 8 in [SS2], that the right hand side of (3.5) provides an upper bound for $d(Q)$, $Q \in W$. Thus Lemma 3.1, combined with the above estimates, gives the required bound for $M_{p,q}(D)$ and the proof is complete. □

Corollary 7. *Let $p \geq n$, and suppose $p \leq q \leq p^*$ if $p \neq n$ and $n \leq q < \infty$ if $p = n$. If D is as above, then*

$$M_{p,q}(D) \lesssim m(D)^{\frac{p-n}{np}} \Big(\int_D k_D^{q(1-1/n)}(x_0, x)\, dx \Big)^{1/q}. \tag{3.6}$$

Proof. For a cube Q in the Whitney decomposition W of D, let $P(Q)$ be the union of the cubes in W which intersect a quasi-hyperbolic geodesic joining x_0 to the center of Q. An application of Hölder's inequality shows that, for $p \geq n$,

$$\int_{P(Q)} \frac{dy}{\delta_D(y)^{p'(n-1)}} \lesssim k_D(x_0, x)^{p'(1-1/n)} m(D)^{\frac{p-n}{n(p-1)}} \qquad x \in Q.$$

(See the remark following Theorem 9 in [SS2].) The proof is completed by using Theorem 6 with $\lambda = 0$ and this estimate. □

An immediate consequence of Theorem 1 and Corollary 7 is the existence of Sobolev imbedding inequalities on Hölder domains.

Corollary 8. *Let $p \geq n$, and suppose $p \leq q \leq p^*$ if $p \neq n$ and $n \leq q < \infty$ if $p = n$. If D is a Hölder domain, then*

$$M_{p,q}(D) < \infty.$$

The classical Sobolev imbedding result that $M_{n,q}(D) < \infty$ for $q < \infty$ if D satisfies a uniform inner cone condition doesn't provide a "best" target space for an imbedding of $W^{1,n}(D)$. This gap has been filled by Trudinger, whose result involves Orlicz spaces. A reference for the definition of and basic facts about Orlicz spaces is Chapter 8 of Adams' book [A] . We remind the reader that, for a suitable real valued function $A(t)$, the Orlicz space $L_A(D)$ is a Banach space with norm

$$\|u\|_A = \inf\{s > 0 \mid \int_D A\Big(\frac{|u(x)|}{s}\Big)\, dx \leq 1\}.$$

The following is a generalization of a theorem of Trudinger [T], who established this result for a domain satisfying a uniform inner cone condition. Trudinger's theorem is an improvement of the classical Sobolev imbedding result for $M_{n,q}(D)$, and is in a sense best possible. See [A, 8.25].

Theorem 9. *Let $D \subset R^n$ be a Hölder domain and set $A(t) = \exp(t^{\frac{n}{n-1}}) - 1$. Then there exists the imbedding*

$$W^{1,n}(D) \to L_A(D).$$

Proof. Since a Hölder domain is bounded (see Corollary 1 in [SS2]), we may assume without loss of generality that $m(D) = 1$. Let $A_0(t) = \exp(t^{\frac{n}{n-1}}) - \sum_{j=0}^{n-1} \frac{1}{j!} t^{\frac{jn}{n-1}}$, and observe that $A_0(t) \approx A(t)$ for $t \geq 1$. Consequently, it suffices to establish the existence of the embedding $W^{1,n}(D) \to L_{A_0}(D)$. This follows from Theorem 8.12(b) in [A].

By Theorem 1 , there exists a constant $\tau > 0$ such that

$$\int_D (\exp((a\tau)^{\frac{n}{n-1}} k_D(x_0, x)) - 1)\, dx \leq 1\,,$$

where a is the dimensional constant implicit in (3.6). Thus, by the application of Corollary 7 with $p = n$ and $q = jn/(n-1)$, we have that

$$\begin{aligned}\int_D A_0\Big(\frac{\tau|u - u_D|}{\|\nabla u\|_n}\Big)\, dx &= \sum_{j=n}^{\infty} \frac{1}{j!}\Big(\frac{\tau}{\|\nabla u\|_n}\Big)^{\frac{jn}{n-1}} \int_D |u - u_D|^{\frac{jn}{n-1}}\, dx \\ &\leq \sum_{j=n}^{\infty} \frac{(a\tau)^{\frac{jn}{n-1}}}{j!} \int_D k_D^j(x_0, x)\, dx \\ &\leq \int_D (\exp((a\tau)^{\frac{n}{n-1}} k_D(x_0, x) - 1)\, dx \leq 1\,.\end{aligned}$$

Hence $\|u - u_D\|_{A_0} \leq \frac{1}{\tau}\|\nabla u\|_n \leq \frac{1}{\tau}\|u\|_{W^{1,n}(D)}$. Also, $|u_D| \leq \|u\|_{W^{1,n}(D)}$ (recall that we have assumed $m(D) = 1$), and so $\|u_D\|_{A_0} \leq \frac{1}{\epsilon}\|u\|_{W^{1,n}(D)}$, where ϵ satisfies $A_0(\epsilon) = 1$. This establishes the existence of the imbedding $W^{1,n}(D) \to L_{A_0}(D)$, and the proof is complete. □

References

[A] Adams, Robert A., Sobolev Spaces. Academic Press, New York 1975.

[Ar] Armitage, D. H., Further results on the global integrability of superharmonic functions. J. London Math. Soc.(2) 6 (1972), 109 - 121.

[BP] Becker, J. and Pommerenke, C., Hölder continuity of conformal mappings and non-quasiconformal Jordan curves. Comment. Math. Helv. 57 (1982), 221–225.

[CJ] Carleson, L. and Jones, P., On coefficient problems for univalent functions and conformal dimension. Preprint.

[GM] Gehring, F. W. and Martio O., Lipschitz classes and quasiconformal mappings. Ann. Acad. Sci. Fenn. Ser. A. I. Math. 10 (1985), 203–219.

[GO] Gehring, F. W. and Osgood, B. G., Uniform domains and the quasi-hyperbolic metric. J. Analyse Math. 36 (1979), 50–74.

[GP] Gehring, F. W. and Palka, B. P., Quasiconformally homogeneous domains. J. Analyse Math. 30 (1976), 172–199.

[H] Hurri, R., Poincaré domains in R^n. Annales Academiae Scientiarum Fennicae, Series A. I. Mathematica Dissertationes 71 (1988), 1–42.

[Kra] Kra, I., Automorphic Forms and Kleinian Groups. Springer-Verlag, Berlin-Heidelberg-New York, 1973.

[M] Makarov, N. G., private communication.

[MS] Maeda, F-Y. and Suzuki, N., The integrability of superharmonic functions on Lipschitz domains. Bull. London Math. Soc. 21 (1989), 270–278.

[M1] Masumoto, M., A distortion theorem for conformal mappings with an application to subharmonic functions. Hiroshima Math. J. 20 (1990), 341–350.
[M2] Masumoto, M., Integrability of superharmonic functions on plane domains. Preprint.
[Ma] Maz'ja, V. G., Sobolev Spaces. Springer-Verlag, Berlin and Heidelberg, 1985.
[SS1] Smith, W. and Stegenga, D. A., A geometric characterization of Hölder domains. J. London Math. Soc.35 (1987), 471 – 480.
[SS2] Smith, W. and Stegenga, D. A., Hölder domains and Poincaré domains. Trans. Amer. Math. Soc. 319 (1990), 67–100.
[SS3] Smith, W. and Stegenga, D. A., Exponential integrability of the quasi-hyperbolic metric on Hölder domains. Preprint.
[S] Staples, S. G., L^p averaging domains and the Poincaré inequality. Ann. Acad. Sci. Fenn. Ser. A. I. Math. 14 (1989), 103–127.
[Su] Suzuki, N., Nonintegrability of positive subharmonic functions on Lipschitz domains. Preprint.

An estimate of harmonic measure with an application to subharmonic functions

Noriaki Suzuki *

Let D be a bounded Lipschitz domain in the n-dimensional euclidean space $\boldsymbol{R}^n$, $n \geq 2$. We consider a function σ on D defined by

$$\sigma(x) = \left(\int_{D^C} \frac{1}{|x-y|^{n+1}} dy \right)^{-1} \qquad \text{for } x \in D$$

and set

$$D(t) = \{x \in D; \sigma(x) > t\} \qquad \text{for } t > 0.$$

Then for sufficiently small t, $D(t)$ is a nice domain. We can get the following estimate of harmonic measure, which is a main result in this article.

Theorem. *Let D be a bounded Lipschitz domain. For any compact set K in D, there are positive constants t_0 and c_0 such that $K \subset D(t_0)$ and*

$$\omega^x_{D(t)} \leq c_0 \frac{S_t}{t}$$

for every $x \in K$ and $0 < t < t_0$, where $\omega^x_{D(t)}$ is the harmonic measure for $D(t)$ relative to x and S_t is the surface measure on $\partial D(t)$.

Denoting by $\delta_D(x)$ the distance between $x \in D$ and ∂D, we see that two functions σ and δ_D are comparable (see Lemma 2). Hence the following result is obtained from the above estimate.

Corollary. *For a bounded Lipschitz domain D, we have*

$$\int_D s(x)\delta_D(x)^{-2}(1 + |\log \delta_D(x)|)^{-1} dx = +\infty$$

for every nonnegative nonzero subharmonic function s on D.

Some related remarks will be given in the last section.

*This research was partially supported by Grand-in-Aid for Encouragement of Young Scientist (No. 02740081) of Ministry of Education of Japan.

Potential Theory, ed. by M. Kishi

1. Preliminary

Let $k > 0$. We say that that a domain D in $\boldsymbol{R}^n$ is a k-Lipschitz domain if ∂D is given by a k-Lipschitz function, that is, there is a constant r_0 such that; to each point $X_0 \in \partial D$ there correspond a coordinate system $x = (x', x_n) \in \boldsymbol{R}^{n-1} \times \boldsymbol{R} = \boldsymbol{R}^n$ with $X_0 = (0,0)$ and a function $f : \boldsymbol{R}^{n-1} \mapsto \boldsymbol{R}$ which satisfy the Lipschitz condition $|f(x') - f(y')| \leq k|x' - y'|$, $f(0) = 0$ and

$$D \cap \Psi(X_0, r_0) = \{(x', x_n); |x'| < r_0, f(x') < x_n < 2kr_0\}, \tag{1}$$

where $\Psi(y,r) = \{(x', x_n); |x' - y'| < r, |x_n - y_n| < 2kr\}$. For fixed $X_0 \in \partial D, x = (x', x_n)$ and f, consider the cones

$$\begin{aligned} \Gamma_+(y) &= \{(x', x_n); x_n - y_n > 4k|y' - x'|\} \cap \Psi(y, \frac{r_0}{4}); \\ \Gamma_-(y) &= \{(x', x_n); x_n - y_n < -4k|y' - x'|\} \cap \Psi(y, \frac{r_0}{4}). \end{aligned}$$

We use the following facts (cf. [2, §11]); if $y = (y', y_n) \in D \cap \Psi(X_0, r_0/4)$ then

$$\Gamma_+(\tilde{y}) \subset D \text{ and } \Gamma_-(\tilde{y}) \subset \overline{D}^C, \tag{2}$$

where $\tilde{y} = (y', f(y'))$; and if $y \in \overline{D}^C \cap \Psi(X_0, r_0/4)$ then

$$\Gamma_-(y) \subset \overline{D}^C \cap \Psi(X_0, r_0). \tag{3}$$

A Lipschitz domain means a k-Lipschitz domain for some $k > 0$. We say that a domain D has an r-sphere property ($r > 0$) if for each $X \in \partial D$ there are two balls B_1 and B_2 of radius r such that $B_1 \subset D, B_2 \subset \overline{D}^C$ and $X \in \partial B_1 \cap \partial B_2$. Every bounded $C^{1,1}$-domain has an r-sphere property.

In the sequel the following notations are used; $B(x,r)$: the open ball of radius r centered at x; $G_D(\cdot,\cdot)$: the Green function of a domain D; ω_D^x: the harmonic measure for D relative to x; $c = c(p, q, \cdots)$: a positive constant depending only on $p, q, \cdots$ not necessary the same on any two occurrences.

2. Proof of Theorem

Let D be a bounded k-Lipschitz domain in $\boldsymbol{R}^n$. We use the description (1) freely. The following lemma holds.

Lemma 1. *There are positive constants t_1, k_1 and c_1 such that for each t with $0 < t < t_1, D(t)$ is a k_1-Lipschitz domain with a $c_1 t$-sphere property.*

For the proof of Lemma 1, we use the following

Lemma 2. *For $p > 0$ and $x \in D$, we define*

$$\rho_p(x) = \int_{D^C} |x-y|^{-n-p} dy.$$

Then $c(D,p)^{-1}\delta_D(x)^{-p} \le \rho_p(x) \le c(D,p)\delta_D(x)^{-p}$.

Proof. Since $B = B(x,\delta_D(x)) \subset D$, we have $\rho_p(x) \le \int_{B^C} |x-y|^{-n-p}dy = c(n,p)\delta_D(x)^{-p}$. For the lower estimate, we may consider on $D \cap \Psi(X_0, r_0/4)$ for each $X_0 \in \partial D$. Let $x = (x', x_n) \in D \cap \Psi(X_0, r_0/4)$. From (2) it follows that $\rho_p(x) \ge c(k, r_0) \int_{\Gamma_-(\tilde{x})} (\delta_D(x) + |\tilde{x} - y|)^{-n-p} dy \ge c(D,p)\delta_D(x)^{-p}$, which proves the lemma. □

Proof of Lemma 1. By the above, $c(D)^{-1}\sigma \le \delta_D \le c(D)\sigma$. Further we have

$$|\text{grad } \sigma(x)| \le c(D), \tag{4}$$

$$\left|\frac{\partial^2}{\partial x_i \partial x_j}\sigma(x)\right| \le c(D)\delta_D^{-1} \quad i,j = 1,2,\cdots,n. \tag{5}$$

In fact, for $1 \le i \le n$,

$$\begin{aligned} \left|\frac{\partial}{\partial x_i}\sigma(x)\right| &= \left|\rho_1(x)^{-2} \int_{D^C} \frac{\partial}{\partial x_i}|x-y|^{-n-1}dy\right| \\ &\le (n+1)\rho_1(x)^{-2} \int_{D^C} |x-y|^{-n-2}dy \le c(D), \end{aligned}$$

here the last inequality follows from Lemma 2. Similarly we get (5). Note that these estimates are independent of the choice of a coordinate system $(x_1, x_2, \cdots x_n)$.

Next we show

$$|\text{grad } \sigma(x)| \ge c(D) > 0 \tag{6}$$

provided that $\delta_D(x) < r_1$ for some positive constant r_1, so we consider (6) on $D \cap \Psi(X_0, r_0)$. Let $x = (x', x_n) \in D \cap \Psi(X_0, r_0/8)$ be fixed. Now denote

$$\frac{\partial}{\partial x_n}\sigma(x) = (n+1)\rho_1(x)^{-2} \int_{D^C} (x_n - y_n)|x-y|^{-n-3}dy = \sum_{k=0}^{4} I_k,$$

where $I_k = (n+1)\rho_1(x)^{-2} \int_{D_k} (x_n - y_n)|x-y|^{-n-3}dy$ and where

$$\begin{aligned} D_0 &= \Gamma_-((x', f(x'))), \\ D_1 &= \{(y', y_n); x_n - y_n < 0\} \cap D^C \cap D_0^C \cap \Psi(X_0, r_0/4), \\ D_2 &= \{(y', y_n); x_n - y_n \ge 0\} \cap D^C \cap D_0^C \cap \Psi(x_0, r_0), \\ D_3 &= D^C \cap D_1^C \cap D_2^C \cap \Psi(X_0, r_0/4)^C, \\ D_4 &= D^C \cap \Psi(X_0, r_0)^C. \end{aligned}$$

Then, since $x_n - y_n \geq c(k, r_0, n)|x - y|$ for $y \in D_0$, we have $I_0 \geq c(k, r_0, n)$ as in the proof of (3). Next we shall estimate $I_1 + I_2$. For this we put $y^* = (y', y_n^*) := (y', 2x_n - y_n)$ for $y = (y', y_n) \in D_1$. Then $y^* \in \Gamma_-(y) \cap D_0^C \subset D^C \cap \Psi(X_0, r_0)$ by (2) and hence $y^* \in D_2$. Since the mapping $D_1 \ni y \mapsto y^* \in D_2$ is injective and $(x_n - y_n)|x - y|^{-n-3} = -(x_n - y_n^*)|x - y^*|^{-n-3}$, $I_1 + I_2 \geq 0$ follows. Finally, since $|x - y| \geq c(k, r_0)$ for $y \in D_3 \cup D_4$, we have $|I_3 + I_4| \leq c(k, r_0, n)\rho_1(x)^{-2}$, which tends to zero as $\delta_D(x) \to 0$. From these estimates we deduce $|\text{grad}\ \sigma(x)| \geq \partial\sigma/\partial x_n(x) \geq c(D)$ provided that $\delta_D(x)$ is sufficiently small, and hence (6) is obtained.

By the implicit function theorem, there are positive constants k_1 and c_1' such that $\partial D(t)$ is given locally by a C^2-function whose C^1-norm is less than k_1 (by (4)) and whose C^2-norm is dominated by c_1'/t (by (5)). Lemma 1 follows from this immediately. □

Now we are through with the preparation of the proof of Theorem. Put $t_0 = \min\{t_1, \text{dist}(\partial D, K)\}$, and let t with $0 < t < t_0$ and $x \in K$. Then since $D(t)$ is a smooth domain, we have

$$d\omega_{D(t)}^x = -c(n)\frac{\partial}{\partial n.}G_{D(t)}(x, \cdot)dS_t \tag{7}$$

(cf. [8]. See also [1] and [3]), where $\partial/\partial n$ denotes the exterior normal derivative. Let $X \in \partial D(t)$. With the aid of Lemma 1, we will find a ball $B = B(x_0, c_1 t)$ satisfying $B \subset \overline{D(t)}^C$ and $X \in \partial B$. Then it is well-known that

$$G_{\overline{B}^C}(x, y) = \begin{cases} -\log(c_1 t|x - y|) + \log(|x - y^*||y - x_0|) & (n = 2) \\ |x - y|^{2-n} - (c_1 t)^{n-2}(|x - y^*||y - x_0)^{2-n} & (n \geq 3) \end{cases}$$

where y^* is the inversion image of y with respect to ∂B. Hence a direct computation gives

$$-\frac{\partial}{\partial n_X}G_{\overline{B}^C}(x, X) \leq c(D, K, c_1)t^{-1}$$

On the other hand, since $D(t) \subset \overline{B}^C$ it follows that

$$-\frac{\partial}{\partial n_X}G_{D(t)}(x, X) \leq -\frac{\partial}{\partial n_X}G_{\overline{B}^C}(x, X).$$

Combining these inequalities and (7), we obtain the assertion of Theorem.

3. Proof of Corollary

Let s be a nonnegative subharmonic function on D. We assume that

$$A := \int_D s(x)\delta_D(x)^{-2}(1+|\log\delta_D(x)|)^{-1}dx < +\infty.$$

Let $x_0 \in D$. Then for sufficiently small $r_1 > 0$, Lemmas 1, 2 and Theorem give

$$\begin{aligned} A &\geq \int_{D\setminus D(r_1)} s(x)\delta_D(x)^{-2}(1+|\log\delta_D(x)|)^{-1}dx \\ &= \int_0^{r_1}\left(\int_{\partial D(t)} s(x)\delta_D(x)^{-2}(1+|\log\delta_D(x)|)^{-1}|\mathrm{grad}\sigma(x)|^{-1}dS_t(x)\right)dt \\ &\geq c(D,x_0)\int_0^{r_1} t\left(\int_{\partial D(t)} s(x)d\omega_{D(t)}^{x_0}(x)\right)t^{-2}(1+|\log t|)^{-1}dt \\ &\geq c(D,x_0)s(x_0)\int_0^{r_1}(t(1+|\log t|))^{-1}dt, \end{aligned}$$

so $s(x_0) = 0$. Since x_0 is arbitrary, s must vanish identically.

4. Remarks on nonintegrability

For a subdomain D in $\boldsymbol{R}^n$, we denote by $S^+(D)$ the set of all nonnegative nonzero subharmonic functions on D.

We consider the following problem: Determine the constant $m = m(D)$ satisfying

$$\int_D s(x)\delta_D(x)^{-m}dx = +\infty$$

for every $s \in S^+(D)$. It is known that for any domain $D \neq \boldsymbol{R}^n$ we may take $m = 2$ ([7]) and if D is a bounded $C^{1,1}$-domain then we can take $m = 1$ ([6]). For a bounded k-Lipschitz domain D, it might be true that m(D) is determined only by k and that $m(D) \to 1$ as $k \to 0$ and $m(D) \to 2$ as $k \to \infty$. In fact, in case $n = 2$, Masumoto [5] proved more general results which show that this is true. Unfortunately we can not show this assertion in case $n \geq 3$ as yet, but if we only consider positive harmonic functions then this is true; more precisely, for $n \geq 3$ and $k > 0$, we set

$$m_k = 2 - \alpha_n(\pi - \arctan 1/k),$$

where $\alpha_n(\theta)$ is the maximal order of barriers of the cone in $\boldsymbol{R}^n$ with aperture θ. It should be noted that $1 < m_k < 2, \lim_{k\to 0} m_k = 1$ and $\lim_{k\to\infty} m_k = 2$ (cf. [4, p.271]). Then we have

Proposition. *Let D be a bounded k-Lipschitz domain in $\boldsymbol{R}^n$. Then*

$$\int_D h(x)\delta_D(x)^{-m_k}dx = +\infty$$

for every positive harmonic function h on D.

Proof. By a famed theorem of Hunt and Wheeden [2], there is a positive Borel measure μ on ∂D such that

$$h(\cdot) = \int_{\partial D} P_D(\cdot,\xi)d\mu(\xi),$$

where $P_D(x,\xi)$ is the Martin(Poisson) kernel on $D \times \partial D$ (normalized at some point $x_0 \in D$). Let $\xi \in \partial D$. We choose two cones $\Gamma = \Gamma_\xi$ and $\Gamma' = \Gamma'_\xi$ with vertex at ξ satisfying $\bar{\Gamma} - \{\xi\} \subset \Gamma' \subset D$. Then by [6, Lemma 1], we have

$$P_D(x,\xi) \geq c(\Gamma)|x-\xi|^{2-n-\alpha_n(\pi - \arctan 1/k)}$$

for $x \in \Gamma$. Thus

$$\begin{aligned}\int_D h(x)\delta_D(x)^{-m_k}dx &= \int_{\partial D}\left(\int_D P_D(x,\xi)\delta_D(x)^{-m_k}dx\right)d\mu(\xi)\\ &\geq \int_{\partial D}\left(c(\Gamma_\xi)\int_{\Gamma_\xi}|x-\xi|^{-n}dx\right)d\mu(\xi) = +\infty.\end{aligned}$$

This completes the proof. □

References

[1] B.E.J. Dahlberg, Estimate of harmonic measure, Arch. Rat. Mech. Anal. **65** (1977), 275-288.

[2] R.A. Hunt and R.L. Wheeden, Positive harmonic functions on Lipschitz domains, Trans. Amer. Math. Soc. **147** (1970), 507-527.

[3] R.A. Hunt, Harmonic measure and estimates of Green's function, Bull. Amer. Math. Soc. **84** (1987), 194-200.

[4] F-Y. Maeda and N. Suzuki, The integrability of superharmonic functions on Lipschitz domains, Bull. London Math. Soc. **21** (1989), 270-278.

[5] M. Masumoto, A distortion theorem for conformal mappings with an application to subharmonic functions, Hiroshima Math. J. **20** (1990), 341-350.

[6] N. Suzuki, Nonintegrability of harmonic functions in a domain, Japanese J. Math. **16** (1990), 269-278.

[7] N. Suzuki, Nonintegrability of superharmonic functions, to appear in Proc. Amer. Math. Soc.

[8] K.-O. Widman, On the boundary value of harmonic functions in $\boldsymbol{R}^3$, Ark. Mat. **5** (1963), 221-230.

Kuramochi boundaries of Riemannian manifolds

Hiroshi Tanaka

Abstract

For a non-compact n-dimensional ($n \geq 2$) Riemannian manifold M and a real number p with $1 < p \leq n$, we define the Kuramochi p-compactification $M_p^\#$ of M. We show that $M_p^\#$ is metrizable and resolutive in the sense of Perron-Wiener-Brelot.

1980 Mathmatics Subject Classification(1985 Revision): 31B, 31C.

1. Introduction

In 1956, Z. Kuramochi ([K1],[K2]) introduced an ideal boundary of a Riemann surface which is called the Kuramochi boundary. The theory of Kuramochi boundary is discussed in higher dimensional case, e.g. in [M]. In these discussions the corresponding harmonic space is a linear space.

In this paper we define the Kuramochi p-boundary of a Riemannian manifold in the nonlinear case and show that the Kuramochi p-compactification $M_p^\#$ of M is metrizable and resolutive in the sense of Perron-Wiener-Brelot.

2. Preliminaries and terminologies

We use the same notations as in [T]. Let $\mathbf{N}$ and $\mathbf{R}$ be the set of all natural numbers and the field of real numbers respectively. The n-dimensional ($n \geq 1$) Euclidean space is denoted by $\mathbf{R}^n$ whose points x are n-tuples $x = (x^1, x^2, \cdots, x^n)$ of real numbers.

By a Riemannian manifold M we shall mean a connected separable and orientable n-dimensional ($n \geq 2$) manifold of C^∞-class with a fundamental metric tensor (g_{ij}) where g_{ij} is of C^∞-class in each local coordinate neighborhood. We denoted by

$$dt_M = \sqrt{g}\, dx^1 dx^2 \cdots dx^n$$

the volume element on M with respect to (g_{ij}) where $x = (x^1, x^2, \cdots, x^n)$ is a local coordinate on M and $g = \det(g_{ij})$. Furthermore we denote by $d(x, y)$ the distance of x and $y \in M$ with respect to (g_{ij}).

All functions considered in this paper are real-valued. Let U be an open set in M. Denote by $C_0^\infty(U)$ the family of all C^∞-functions with compact supports in U. For $p \geq 1$ let $L^p(U)$ (resp. $L^p_{loc}(U)$) be the space of all p-integrable functions (resp. locally p-integrable functions) on U with respect to dt_U which is the restriction of dt_M

Potential Theory, ed. by M. Kishi

to U. We denote by $\Lambda^r(U)$ the space of all measurable r-forms ($0 \leq r \leq n$) on U. For $1 \leq p < \infty$, the p-norm of $\alpha \in \Lambda^r(U)$ is defined by

$$\|\alpha\|_p = \|\alpha\|_{p,U} = \left\{ \int_U |\alpha|^p dt_U \right\}^{1/p},$$

where $|\alpha|$ is the length of α defined by the Riemannian metric. We denote by $\Lambda_p^r(U)$ the subclass of $\Lambda^r(U)$ whose r-forms have finite p-norms. We set $\mathcal{L}^p(U)$ (resp. $\mathcal{L}_{loc}^p(U)$) to be the space of all measurable 1-forms α with $|\alpha| \in L^p(U)$ (resp. $L_{loc}^p(U)$). By definition, we have $\mathcal{L}^p(U) = \Lambda_p^1(U)$. We note that $\mathcal{L}^p(M)$ is a Banach space for $1 \leq p < \infty$.

Let f be any function in $L_{loc}^1(U)$. Suppose there is an $\alpha \in \mathcal{L}_{loc}^1(U)$ such that

$$\int_U f d\omega = -\int_U \alpha \wedge \omega$$

for all $(n-1)$-forms ω of C^∞-class with compact supports in U. Since such an α is uniquely determined up to a set of measure zero, we denote it by df and say that df is the weak exterior derivative of f.

3. Royden p-compactifications

For an open set U in M, we denote by $ACL(U)$ the family of all ACL (absolutely continuous on lines) functions on U (cf. [T]). It is known that a finite continuous function f on U has a weak exterior derivative if and only if $f \in ACL(U)$ and $\partial f/\partial x^i \in L_{loc}^1(U)$ for each i.

Let p be fixed with $1 \leq p < \infty$. Denote by $ACL^p(U)$ the subfamily of $ACL(U)$ whose function f has a finite p-Dirichlet integral $D_p(f)$ over U:

$$D_p(f) = D_{p,U}(f) = \int_U |df|^p dt_U.$$

Obviously $\|df\|_p = \{D_p(f)\}^{1/p}$ for any $f \in ACL^p(U)$.

For a relatively compact domain D in M, we denote by $W_p^1(D)$ the Sobolev space on D (cf. [A]). Since every function $f \in W_p^1(D)$ has a weak exterior derivative df, we see that a finite continuous function f on U belongs to $ACL^p(U)$ if and only if $f \in W_p^1(D)$ for all relatively compact domains D in U and $D_{p,U}(f) < \infty$.

Denote by $A^p(M)$ the family of all bounded continuous functions in $ACL^p(M)$. The function space $A^p(M)$ is said to be the Royden p-algebra of M. It is known (cf. [L],[T]) that $A^p(M)$ is a vector lattice with respect to the maximum and minimum operations and is a commutative Banach algebra with the norm

$$\nu_p(f) = \sup_M |f| + \{D_p(f)\}^{1/p}.$$

We say that a sequence $\{f_j\}_{j=1}^\infty$ in $A^p(M)$ converges to $f \in A^p(M)$ in BD_p-topology as $j \to \infty$ if

(i) $\{f_j\}_{j=1}^{\infty}$ is uniformly bounded,

(ii) $f_j \to f$ locally uniformly on M as $j \to \infty$,

(iii) $\|df_j - df\|_p \to 0$ as $j \to \infty$.

Then we see that $A^p(M)$ is complete in the BD_p-topology. Denote by $A_0^p(M)$ the closure of the class

$$\{f \in A^p;\ \text{the support of } f \text{ is compact in } M\}$$

in BD_p-topology. We say that $A_0^p(M)$ is the p-Dirichlet potential subalgebra of $A^p(M)$. By a discussion similar to that in the proof of Theorem 2E in [GN], we can prove that both $A^p(M) \cap C^\infty(M)$ and $C_0^\infty(M)$ are dense in $A^p(M)$ and $A_0^p(M)$ respectively in BD_p-topology.

By the aid of the well-known Clarkson's inequality (cf. [A]), we have the following lemma.

Lemma. *Let $1 < p < \infty$ and let $\{\alpha_k\}_{k=1}^{\infty}$ be a sequence in $\mathcal{L}^p(M)$ with a finite limit $\lim_{k\to\infty} \|\alpha_k\|_p$. If $\|(\alpha_j + \alpha_k)/2\|_p \geq \min\{\|\alpha_j\|_p, \|\alpha_k\|_p\}$ for each j and k, then α_k strongly converges to some $\alpha \in \mathcal{L}^p(M)$.*

Corollary. *Let $\{f_k\}_{k=1}^{\infty}$ be a uniformly bounded sequence in $A^p(M)$ such that f_k converges to a function f locally uniformly on M. If $\{\alpha_k\}_{k=1}^{\infty}$ with $\alpha_k = df_k$ satisfies the assumptions in the Lemma, then $f \in A^p(M)$ and f_k converges to f in BD_p-topology as $k \to \infty$.*

Proof. By théorème 2.3 in [LF], we see that $f \in A^p(M)$. Hence the Corollary follows immediately from the Lemma. □

We refer to [GN, LF, L] for the definition of the Royden p-compactification. Let M_p^* be the Royden p-compactification of M and $\Gamma_p = M_p^* - M$ be the Royden p-boundary of M_p^*. We set

$$\Delta_p = \{x \in \Gamma_p\ ;\ \lim_{y \to x} f(y) = 0 \text{ for all } f \in A_0^p(M)\}$$

and say that Δ_p is the Royden p-harmonic boundary of M_p^*. For the definition of p-parabolicity, we refer to [T] and denote by O_Q^p the class of all p-parabolic open Riemannian manifolds. Then we have the following theorem.

Theorem 1 (cf. [GN], [L]). *The following conditions are mutually equivalent:*

(a) *M is p-parabolic, i.e $M \in O_Q^p$,*

(b) *$1 \in A_0^p(M)$,*

(c) *$A_0^p(M) = A^p(M)$,*

(d) *$\Delta_p = \emptyset$.*

4. Variational problem with an obstacle

Let p be fixed with $1 < p \leq n$ in what follows. We consider a variational integral as follows:

$$I(f) = I(f, G) = \int_G |df|^p dt_G,$$

where G is a domain in M. A finite continuous function u on G is said to be p-harmonic if

(i) for any relatively compact domain D in G, the restriction of u to D belongs to $ACL^p(D)$,

(ii) u is a solution of Euler-Lagrange's equation:

$$\int_D |du|^{p-2}(du, dg)\, dt_M = 0 \quad \text{for all} \quad g \in C_0^\infty(M),$$

which corresponds to the variational integral $I(f, D)$.

The following important results of p-harmonic functions are known ([S]):

(a) local Hölder's continuity,
(b) Harnack's inequality.

Let G be a relatively compact domain in M with smooth boundary and f be any function in $A^p(M)$. Then, by a result of Maz'ya [MA], we see that there exists a uniquely determined p-harmonic function $T_G f$ in G with boundary values f on ∂G (cf. [T]). Furthermore, for any $f \in A^p(M)$, we have the Royden decomposition of f such that $f = T_M f + g$ with a p-harmonic function $T_M f$ and a p-Dirichlet potential g ([T]).

An extended real-valued function s on G is called p-superharmonic if

(1) s is lower semicontinuous,
(2) $-\infty < s(x) \leq \infty$ for all $x \in G$ and $s \not\equiv \infty$,
(3) for each domain $D \subset\subset G$ and each p-harmonic $h \in C(\bar{D})$, $s \geq h$ on ∂D implies $s \geq h$ in D.

Properties of p-superharmonic functions:

(a) if s is p-superharmonic, then so is $as + b$ for any constants a and b with $a \geq 0$,
(b) if s_1 and s_2 are p-superharmonic, then so is $\min(s_1, s_2)$.

The following theorem has been proved by J. H. Michael and W. P. Ziemer (Theorem 3.7 in [MZ]) for a bounded domain D in $\mathbf{R}^n$ (cf. [HK]). Then, by using a method of partition of unity, we can prove the theorem for a bounded domain on a Riemannian manifold.

Theorem 2 (Obstacle problem). *Let φ be any function in $A^p(M)$ and D be any relatively compact domain in M with smooth boundary. If φ is locally Hölder continuous, then there exists a unique continuous function s on $\overline{D}$ such that*

(i) *$s \in A^p(D)$, s is locally Hölder continuous and p-superharmonic in D,*

(ii) *$s \geq \varphi$ in D and $s = \varphi$ on ∂D,*

(iii) *if $g \in A^p(M)$ satisfies $g \geq \varphi$ on D and $g = \varphi$ on ∂D, then $I(s, D) \leq I(g, D)$.*

As for the local Hölder continuity of s, we mean that, for any compact subset K of D, there exist two positive constants c and α $(0 < \alpha < 1)$ such that

$$|s(x) - s(y)| \leq c\,(d(x, y))^{\alpha} \quad \text{for all } x, y \in K,$$

where c and α depend on φ, K, D, p, n.

Theorem 3. *Let φ be any function in $A^p(M) \cap C^\infty(M)$. Then there exists a unique function s in $A^p(M)$ which satisfies the following conditions:*

(i) *s is p-superharmonic in M,*

(ii) *$s \geq \varphi$ on M and $s - \varphi \in A^p_0(M)$,*

(iii) *if $g \in A^p(M)$ satisfies $g \geq \varphi$ on M and $g - \varphi \in A^p_0(M)$, then $D_p(s) \leq D_p(g)$.*

Proof. Let $\{M_k\}_{k=1}^\infty$ be an exhaustion of M. Let s_k be the function determined in Theorem 2 by taking $D = M_k$ and let f_k the extension of s_k to M by setting $f_k = \varphi$ on $M - M_k$. Then we see that $f_k \in A^p(M)$ for each k and $\{f_k\}_{k=1}^\infty$ is a sequence of uniformly bounded locally Hölder continuous functions. Hence we can find a convergent subsequence $\{f_{k_j}\}_{j=1}^\infty$ and denote by s the limiting function. Since $\|(df_j + df_k)/2\|_p \geq \|df_j\|_p$ for $j > k$ by (iii) in Theorem 2, it follows from the Corollary to the Lemma that f_{k_j} converges to $s \in A^p(M)$ in BD_p-topology. Thus we see that s satisfies (i) and (ii).

Let g be a function in $A^p(M)$ such that $g \geq \varphi$ in M and $g - \varphi \in A^p_0(M)$. Then there exists a sequence $\{g_k\}_{k=1}^\infty$ in $A^p(M)$ such that $g_k = 0$ on $M - M_k$ and $g_k \to g - \varphi$ in BD_p-topology. Since $d(g_k \vee 0) \to d(g - \varphi)$ a.e. on M, by Fatou's lemma, we have

$$D_p(g - \varphi) \leq \liminf_{k\to\infty} D_p(g_k \vee 0).$$

On the other hand, it follows that

$$\liminf_{k\to\infty} D_p(g_k \vee 0) \leq \liminf_{k\to\infty} D_p(g_k) = D_p(g - \varphi).$$

Hence we have $\lim_{k\to\infty} D_p(g_k \vee 0) = D_p(g - \varphi)$. Thus we may assume that $g_k \geq 0$ for each k from the beginning. Then $g_k + \varphi \geq \varphi$ in M and $(g_k + \varphi) - \varphi \in A^p_0(M)$ for each k. This implies that $D_p(f_k) \leq D_p(g_k + \varphi)$ and $D_p(s) \leq D_p(g)$ by letting $k \to \infty$. Thus s satisfies (iii). The uniqueness of s folllows from Clarkson's inequality. □

5. Dirichlet problem

Let M^* be an arbitary compactification of M and Γ be the ideal boundary $M^* - M$. We denote by $C(\Gamma)$ the family of all finite continuous functions on Γ. Let us consider the Dirichlet problem for M^* in the sense of Perron-Wiener-Brelot by a standard method (cf. [CC]; [GLM] for the nonlinear case). For any function $\varphi \in C(\Gamma)$, we consider the following two classes of functions:

$$\overline{\mathcal{S}}_\varphi = \left\{ s\,;\ \begin{matrix} s \text{ is } p\text{-superharmonic, bounded below in } M \\ \text{and } \lim_{x\to y} s(x) \geq \varphi(x) \text{ for all } y \in \Gamma \end{matrix} \right\}$$

and

$$\underline{\mathcal{S}}_\varphi = \{-s\,;\ s \in \overline{\mathcal{S}}_{-\varphi}\}.$$

It is known (cf. [GLM]) that both $\overline{H}_\varphi$ and $\underline{H}_\varphi$ are p-harmonic in M. If $\overline{H}_\varphi = \underline{H}_\varphi$, then φ is called p-resolutive and the common value is denoted by H_φ. A compactification M^* is called p-resolutive if any $\varphi \in C(\Gamma)$ is p-resolutive. For a p-resolutive compactification M^*, a boundary point $y \in \Gamma$ is said to be p-regular if $\lim_{x\to y} H_\varphi(x) = \varphi(y)$ for any $\varphi \in C(\Gamma)$.

Theorem 4. *The Royden p-compactification of a p-hyperbolic Riemannian manifold M is p-resolutive.*

Proof. Let φ be a function in $C(\Gamma_p)$ such that $\varphi = f$ on Γ_p for some function f in $A^p(M) \cap C^\infty(M)$. Let s be the function s in Theorem 3 which corresponds to this f. Then we have $s \in \overline{\mathcal{S}}_\varphi$. Let $s_k = T_{M_k}s$ in M_k and $= s$ on $M - M_k$. Then, by a discussion similar to that in the proof of Lemma 2.9 in [GLM], we see that s_k is p-superharmonic in M. Hence $s_k \in \overline{\mathcal{S}}_\varphi$ and $s_k \geq \overline{H}_\varphi$. Since $T_{M_k}s \to T_M s$ as $k \to \infty$ ([T]), we have $T_M s \geq \overline{H}_\varphi$. Since $T_M s = \varphi$ on Δ_p, it follows from the maximum and minimum principle ([T]) that $T_M f = T_M s \geq \overline{H}_\varphi$. Similarly we have $T_M f \leq \underline{H}_\varphi$. Thus we have $\overline{H}_\varphi = \underline{H}_\varphi = T_M f$.

Since $A^p(M) \cap C^\infty(M)$ is dense in $A^p(M)$ with respect to DB_p-topology, for any $\varphi \in C(\Gamma_p)$ and any $\varepsilon > 0$, there exists $g \in A^p(M) \cap C^\infty(M)$ such that $|\varphi - g| < \varepsilon$ on Γ_p (in the extended sense). Then we have

$$T_g - \varepsilon \leq \underline{H}_\varphi \leq \overline{H}_\varphi \leq T_g + \varepsilon \quad \text{in } M.$$

Hence $0 \leq \overline{H}_\varphi - \underline{H}_\varphi \leq 2\varepsilon$. Since ε is arbitary, we have $\overline{H}_\varphi = \underline{H}_\varphi$ in M. Thus φ is p-resolutive and we have the theorem. □

Corollary 1. *Let f be any function in $A^p(M)$. If we denote by φ the continuous extension of f to Γ_p, then $H_\varphi = T_M f$ in M, i.e. $f = H_\varphi + g$ with $g \in A^p_0(M)$.*

Corollary 2. *Any point of Δ_p is p-regular.*

Proof. Since $A^p(M)$ is dense in the space of all finite continuous functions on M_p^*, we have the corollary. □

6. Dirichlet principle

Let us consider the following variational problem. Let K be a non-empty compact subset of M with smooth boundary. For any function f in $A^p(M)$, we set

$$\mathcal{F}_f = \{g \in A^p(M);\ g = f \text{ on } K\}.$$

We shall find a function in the class $\mathcal{F}_f$ which minimizes the variational integrals $D_p(f) = \int_M |dg|^p dt_M$.

Theorem 5. *Let K be a non-empty compact subset in M with smooth boundary and f be any function in $A^p(M)$. Then there exists a uniquely determined $u \in \mathcal{F}_f$ such that*

(a) *u is p-harmonic in $M - K$,*

(b) *$D_p(u) \leq D_p(g)$ for all $g \in \mathcal{F}_f$,*

(c) *$\int_{M-K} (|du|^{p-2}du, dg)\, dt_M = 0$ for all $g \in A^p(M)$ such that $g = 0$ on ∂K.*

Proof. Let $\{M_k\}_{k=1}^{\infty}$ be an exhaustion of M with $M_1 \supset K$. Let $c_1 = \inf_M f$ and $c_2 = \sup_M f$. Let h be a function in $A^p(M)$ such that h is p-harmonic in $M_1 - K$, $h = f$ on K and $= 1$ on $M - M_1$. Let $\{g_k\}_{k=1}^{\infty}$ be a minimizing sequence in $\mathcal{F}_f$, that is, $\|dg_k\|_p \to m = \inf\{\|dg\|_p;\ g \in \mathcal{F}_f\}$. For each k, let $f_k = \min\{c_2, \max\{g_k, c_1\}\}$ and let $u_k = T_{M_k - K} f_k$. Extend u_k by f on K. Since each u_k is p-harmonic in $M_k - K$ and $\nu_p(u_k)$ is bounded, by choosing a subsequence if necessary, it follows from the Corollary to the Lemma that $\{u_k\}$ converges to a function u on M in BD_p-topology as $k \to \infty$. Since $c_1 h \leq u_k \leq c_2 h$ in M, $\lim_{x \to y} u(x) = f(y)$ for all $y \in \partial K$. Hence $u \in \mathcal{F}_f$. Thus (a), (b) and (c) are proved. The uniqueness of u follows by the same method as in the proof of the maximum principle in [MA]. □

Since the function u in Theorem 5 is uniquely determined by f and K, we denote it by f^K (cf. [CC]).

Properties of f^K:

(i) $(af + b)^K = af^K + b$ for constants a, b,

(ii) if $f \leq g$ on ∂K, then $f^K \leq g^K$ in $M - K$,

(iii) $\int_{M-K} |df^K|^{p-2}(df^K, dg)\, dt_M = 0$ for all $g \in A^p(M)$ such that $g = 0$ on ∂K,

(iv) if $K_1 \subset K_2$, then $(f^{K_1})^{K_2} = f^{K_1}$.

7. Kuramochi p-compactification

We denote by Q the subfamily of $A^p(M)$ such that:

$$Q = \{f \in A^p(M);\ f = f^K \text{ for some } K\}.$$

We define the Q-compactification M_Q^* of M by following the method in [CC] and call it the Kuramochi p-compactification of M. In the following we denote by $M_p^\#$ the Kuramochi p-compactification of M. $\Gamma_p^\# = M_p^\# - M$ is called the Kuramochi p-boundary of M. By definition we see that $M_p^\#$ is a quotient space of M_p^*, that is, there exists a canonical continuous surjection of M_p^* onto $M_p^\#$ whose restriction to M is the identity.

By a slight modification of the proof of Theorem 5 in [M] we have the following theorem.

Theorem 6. *The Kuramochi p-compactification $M_p^\#$ is metrizable.*

Proof. Let $\{K_j\}_{j=1}^\infty$ be a sequence of compact sets with smooth boundaries in M such that $K_j \subset K_{j+1}^i$ $(j = 1, 2, \cdots)$ and $\bigcup_{j=1}^\infty K_j = M$. For each j, we set $Q_j = \{f \in A^p(M);\ f = f^{K_j}\}$. Then we choose a countable subfamily $Q_{j,o}$ of Q_j such that $Q_{j,o}|_{\partial K_j}$ is dense in $Q_j|_{\partial K}$ with respect to the uniformly convergence topology on ∂K_j. We set $Q_o = \bigcup_{j=1}^\infty Q_{j,o}$. Since $Q \supset Q_o$, there exists a continuous surjection π from M_Q^* onto $M_{Q_o}^*$. Since Q_o separates points of $\Gamma_p^\#$, we see that π is injective. Hence $M_p^\# = M_Q^*$ is homeomorphic to $M_{Q_o}^*$. The countability of Q_o implies that $M_p^\#$ is metrizable. □

Theorem 7. *The Kuramochi p-compactification is p-resolutive.*

Proof. Let φ be any function in $C(\Gamma_p^\#)$. Since $M_p^\#$ is a quotient space of M_p^*, there exists a continuous surjection π from M_p^* onto $M_p^\#$. Then we have $\overline{H}_\varphi = \overline{H}_{\varphi\circ\pi}$ and $\underline{H}_\varphi = \underline{H}_{\varphi\circ\pi}$. Since $\varphi \circ \pi$ is p-resolutive on M_p^* by Theorem 4, we see that φ is p-resolutive. This completes the proof. □

References

[A] Aubin, T., Nonlinear Analysis on Manifolds. Monge-Ampère Equations. Springer-Verlag, 1982.

[CC] Constantinescu, C. and A. Cornea, Ideale Ränder der Riemannscher Flächen. Springer-Verlag, 1963.

[GLM] Grunlund, S., P. Lindqvist and O. Martio, Note on the PWB-method in the nonlinear case. Pacific J. Math. 125 (1986), 318-395.

[GN] Glasner, M. and M. Nakai, Riemannian manifolds with discontinuous metrics and the Dirichlet integrals. Nagoya Math. J. 46 (1972), 1-48.

[HK] Heinonen, J. and T. Kilpeläinen, A-superharmonic functions and supersolutions of degenerate elliptic equations. Ark. Mat. 26 (1988), 87-105.

[K1] Kuramochi, Z., Mass distributions on the ideal boundaries of abstract Riemann surfaces, II. Osaka Math. J. 8 (1956), 145-186.

[K2] —, Potentials on Riemann surfaces. J. Fac. Sci. Hokkaido Univ. Ser. I, 16 (1962), 5-79.

[L] Lewis, L. G., Quasiconformal mappings and Royden algebras in the space. Trans. Amer. Math. Soc. 158 (1971), 481-492.

[LF] Lelong-Ferrand, F., Etude d'une classe d'applications liées à des homomorphismes d'algebres de fonctions, et generalisant les quasi conformes. Duke Math. J. 40 (1973), 163-185.

[M] Maeda, F-Y., Notes on Green lines and Kuramochi boundary of a Green space. J. Sci. Hiroshima Univ. Ser A-I Math. 28 (1964), 59-66.

[MA] Maz'ya, V. G., On the continuity at a boundary point of solutions of quasi-linear elliptic equations, Vestnik Leningrad Univ. Math. 3 (1976), 225-242.

[MZ] Michael, J. H. and W. P. Ziemer, Interior regularity for solutions to obstacle problems, Nonlinear Analysis, Theory, Methods & Applications, 10 (1986), 1427-1448.

[S] Serrin, J., Local behavior of solutions of quasi-linear equations. Acta Math. 111 (1964), 247-302.

[T] Tanaka, H., Harmonic boundaries of Riemannian manifolds. Nonlinear Analysis, Theory, Methods & Applications, 14 (1990), 55-67.

On axiomatic non-linear potential theory

Frans van Gool

Abstract

An axiomatic non-linear potential theory is presented that has no reference to a linear structure.

1980 Mathematics Subject Classification (1985 Revision): 31D05.

1. Introduction

In recent years there has been an attempt from several people to use the methods of standard potential theory to non-linear (and non-real) problems. The usual approach is to start with a specific example that has some 'potential theoretic properties' and use this example to develop a theory that contains both that standard theory and the example. This approach has led to several, non-compatible, axiomatic systems, each of which assumes some weak form of linearity.

In this text an different approach is tried: An attempt is made to characterize those properties in potential theory that depend only on the topological and lattice properties in linear axiomatic potential theory. This is done in an axiomatic context, i.e. we use a set of axioms, comparable to those in [CC], but without any reference to a linear structure. Hence, for the results stated in this text, the 'valuespace' doesn't need to have a linear structure, also it doesn't need to be totally ordered. For simplicity however, we will suppose in this text that the values are in the set of extended real numbers. For those who are interested in the exact conditions the 'valuespace' should satisfy, and for those who would like to see some more results, we refer to [Go].

Obviously the proposed approach is very general and it might be expected that the resulting axiomatic system is too weak to be interesting. However, the opposite is true: it appears that a substantial part of linear potential theory can be proved in this general axiomatic system. Furthermore almost every result obtained in other non-linear axiomatic potential theories, that can be stated in topological and order-theoretic properties only, can also be proved in this axiomatic theory.

The proposed axiomatic system is applicable to:

- a DB-harmonic space as defined in [Be];
- a quasi-linear harmonic space as defined in [La];
- a non-linear harmonic space with axiom Ac as defined in [Le];
- the solutions of the generalised Monge-Ampere equation $\mathrm{Hess}(u) = f \geq 0$;
- the solutions of the equation $\Delta u = f(x, u)$ with f increasing in u;

Potential Theory, ed. by M. Kishi

- the solutions of the n-Laplace equation $\mathrm{div}(|\nabla u|^{n-2}\nabla u) = 0$;
- and, if we take the values in the extended plane, also to a biharmonic space as defined in [Sm].

2. Basic definitions

In this text, X will be a locally compact Hausdorff space and R will be the set of extended real numbers. We will denote by C(X) (resp. LSC(X), resp. USC(X), resp. NUM(X)) the finite continuous (resp. lower finite lower semicontinuous, resp. upper finite upper semicontinuous, resp. all) numerical functions on X. For any $a, b \in \mathrm{R}$ we will say that $a \ll b$ (resp. $a \gg b$) if $a < b$ or $a = -\infty$ (resp. $a > b$ or $a = \infty$). Note that $\ll$ is the 'way below' relation when we see R as a continuous lattice as defined in [Gi]. Note also that $a \ll b$ is not the same as $b \gg a$.

The first thing we need is an analogue of the upper and lower integrals for non-linear maps. So let us consider a compact subset K of X and an isotone map I from $\mathrm{C}(K)$ to R. We define:

- $\mathrm{I}^*(f) = \sup\{\mathrm{I}(g) | f \geq g \in \mathrm{C}(K)\}$ for $f \in \mathrm{LSC}(K)$;
- $\mathrm{I}^{**}(f) = \inf\{\mathrm{I}^*(g) | f \leq g \in \mathrm{LSC}(K)\}$ for $f \in \mathrm{NUM}(K)$;
- $\mathrm{I}_*(f) = \inf\{\mathrm{I}(g) | f \leq g \in \mathrm{C}(K)\}$ for $f \in \mathrm{USC}(K)$;
- $\mathrm{I}_{**}(f) = \sup\{\mathrm{I}_*(g) | f \geq g \in \mathrm{USC}(K)\}$ for $f \in \mathrm{NUM}(K)$.

It is evident that both I^* and I_* are extensions of I and that I^{**} (resp. I_{**}) is an extension of I^* (resp. I_*). Furthermore it is evident that al the extensions are isotone. We will say that a function $f \in \mathrm{NUM}(K)$ is (I-)integrable if $\mathrm{I}^{**}(f) = \mathrm{I}_{**}(f)$. The following lemma is essential for this notion of integrability.

Lemma 2.1. *All* $f \in \mathrm{LSC}(K) \cup \mathrm{USC}(K)$ *are integrable and for all* $f \in \mathrm{NUM}(K)$ *we have* $\mathrm{I}^{**}(f) \geq \mathrm{I}_{**}(f)$.

Proof. Let $f \in \mathrm{LSC}(K)$ and take $g \in \mathrm{USC}(K)$ such that $f \geq g$. Then there is an $h \in \mathrm{C}(K)$ such that $f \geq h \geq g$. Now it follows from the isotony of I^* and I_* that $\mathrm{I}^*(f) \geq \mathrm{I}(h) \geq \mathrm{I}_*(g)$. But then $\mathrm{I}^*(f) \geq \sup\{\mathrm{I}_*(g) | f \geq g \in \mathrm{USC}(K)\} = \mathrm{I}_{**}(f)$. On the other hand $\mathrm{I}^*(f) = \sup\{\mathrm{I}(g) | f \geq g \in \mathrm{C}(K)\} \leq \sup\{\mathrm{I}_*(g) | f \geq g \in \mathrm{USC}(K)\} = \mathrm{I}_{**}(f))$. So every $f \in \mathrm{LSC}(K)$ is I-integrable and dually also every $f \in \mathrm{USC}(K)$ is I-integrable. Now let $f \in \mathrm{NUM}(K)$ and take $g \in \mathrm{LSC}(K)$ with $f \leq g$. Then we have $\mathrm{I}_{**}(f) \leq \mathrm{I}_{**}(g) = \mathrm{I}^{**}(g) = \mathrm{I}^*(g)$ and hence $\mathrm{I}_{**}(f) \leq \inf\{\mathrm{I}^*(g) | f \geq g \in \mathrm{LSC}(K)\} = \mathrm{I}^{**}(f)$. □

A very important property of the extension I^* of I is that it preserves some continuity of the map I. We will say that I is lower semicontinuous if for every upper directed subset F in $\mathrm{C}(K)$ such that $\sup F \in \mathrm{C}(K)$ we have $\sup \mathrm{I}(F) = \mathrm{I}(\sup F)$. Furthermore we will say that I^* is lower semicontinuous if for every upper directed subset F in $\mathrm{LSC}(K)$ we have $\sup \mathrm{I}^*(F) = \mathrm{I}^*(\sup F)$. It is important to note that the above definitions of semicontinuity are special cases of a general theory of non-real-valued semicontinuous maps. A study on such maps can be found in [Go]. Now we have the lemma below. Of course there are dual definitions of upper semicontinuity for I and $\mathrm{I}_\star$ with a corresponding result.

Lemma 2.2. *If* I *is lower semicontinuous, then* $\mathrm{I}^\star$ *is lower semicontinuous.*

Proof. Take $F \subset \mathrm{LSC}(K)$ upper directed and define $F^c = \{g | \exists h \in F : h \geq g \in \mathrm{C}(K)\}$. Note that $\sup \mathrm{I}^\star(F) = \sup \mathrm{I}(F^c)$. Now take $f \in \mathrm{C}(K)$ with $f \leq \sup F$ and define $G = \{g \wedge f | g \in F^c\}$. Then $G \subset \mathrm{C}(K)$ is upper directed and $\sup G = f$. So $\mathrm{I}(f) = \sup \mathrm{I}(G) \leq \sup \mathrm{I}(F^c) = \sup \mathrm{I}^\star(F)$ Since f was arbitrary we get $\mathrm{I}^\star(\sup F) = \sup \mathrm{I}^\star(F)$. □

In this text, $\mathcal{H}^\star$ will be a map that assigns to every open set U of X a collection $\mathcal{H}^\star(U) \subset \mathrm{LSC}(U)$. Dually $\mathcal{H}_\star$ assigns to every open U a collection $\mathcal{H}_\star(U) \subset \mathrm{USC}(U)$ and $\mathcal{H}$ assigns to every open U the collection $\mathcal{H}(U) = \mathcal{H}^\star(U) \cap \mathcal{H}_\star(U) \subset \mathrm{C}(U)$. If $f \in \mathcal{H}^\star(U)$ (resp. $f \in \mathcal{H}_\star(U)$, resp. $f \in \mathcal{H}(U)$), then f is called hyperharmonic (resp. hypoharmonic, resp. harmonic) on U.

Let U be a relatively compact set. For any $f \in \mathrm{NUM}(\partial U)$ we will define the upper functions of f by $\mathcal{H}^\star(U, f) = \{u \in \mathcal{H}^\star(U) | u \text{ lower bounded, } \hat{u} \geq f \text{ on } \partial U\}$, and the upper solution of f by

$$\overline{\mathrm{H}}(U, f) = \begin{cases} \inf \mathcal{H}^\star(U, f) & \text{on } U \\ f & \text{on } \partial U \end{cases}.$$

The upper harmonic map $\overline{\mathrm{M}}(U, .)$ is the restriction of $\overline{\mathrm{H}}(U, .)$ to $\mathrm{C}(\partial U)$. The lower functions $\mathcal{H}_\star(U, f)$, lower solution $\underline{\mathrm{H}}(U, f)$ and lower harmonic map $\underline{\mathrm{M}}(U, .)$ are defined dually.

Lemma 2.3. *Let U be a relatively compact set. Then:*

1. $\overline{\mathrm{H}}(U, .)$ *is an isotone map;*
2. *if* $f \in \mathrm{LSC}(\overline{U}) \cap \mathcal{H}^\star(U)$, *then* $\overline{\mathrm{H}}(U, f) \leq f$;
3. *for all* $f \in \mathrm{NUM}(\partial U)$ *we have* $\overline{\mathrm{M}}^{\star\star}(U, f) \leq \overline{\mathrm{H}}(U, f)$;
4. *for all* $f \in \mathrm{USC}(\partial U)$ *we have* $\overline{\mathrm{M}}_\star(U, f) = \overline{\mathrm{H}}(U, f) = \overline{\mathrm{M}}^{\star\star}(U, f)$;
5. *if for all* $f \in \mathrm{LSC}(\partial U)$ *we have* $\overline{\mathrm{M}}^\star(U, f) = \overline{\mathrm{H}}(U, f)$, *then* $\overline{\mathrm{M}}^{\star\star}(U, .) = \overline{\mathrm{H}}(U, .)$.

Proof. (3) First take $f \in \mathrm{LSC}(\partial U)$. Then $\overline{\mathrm{M}}^\star(U, f) = \sup\{\overline{\mathrm{H}}(U, g) | f \leq g \in \mathrm{C}(\partial U)\} \leq \overline{\mathrm{H}}(U, f)$. Now take $f \in \mathrm{NUM}(\partial U)$ and $u \in \mathcal{H}^\star(U, f)$. Then $\hat{u} \geq f$ on ∂U and thus $\overline{\mathrm{M}}^{\star\star}(U, f) \leq \overline{\mathrm{M}}^\star(U, \hat{u}) \leq \overline{\mathrm{H}}(U, \hat{u}) \leq \hat{u} = u$ on U. But then also $\overline{\mathrm{M}}^{\star\star}(U, f) \leq \inf \mathcal{H}^\star(U, f) = \overline{\mathrm{H}}(U, f)$.
(4) $\overline{\mathrm{M}}_\star(U, f) = \inf\{\overline{\mathrm{H}}(U, g) | f \leq g \in \mathrm{C}(\partial U)\} \geq \overline{\mathrm{H}}(U, f) \geq \overline{\mathrm{M}}^{\star\star}(U, f) = \overline{\mathrm{M}}_\star(U, f)$.
(5) We have $\overline{\mathrm{M}}^{\star\star}(U, f) = \inf\{\overline{\mathrm{H}}(U, g) | f \leq g \in \mathrm{LSC}(\partial U)\} \geq \overline{\mathrm{H}}(U, f)$. □

Let U be a relatively compact set. If the upper and lower solution are the same for some $f \in \mathrm{NUM}(\partial U)$, then f is called semiresolutive and the common value is called the solution of f and is denoted by $\mathrm{H}(U,f)$ (or by $\mathrm{M}(U,f)$ if $f \in \mathrm{C}(\partial U)$). If also $\mathrm{H}(U,f) \in \mathcal{H}(U)$, then f is called resolutive. We define the following properties of U:

- U is called MP if for all $f \in \mathrm{C}(\partial U)$ we have $\overline{\mathrm{M}}(U,f) \geq \underline{\mathrm{M}}(U,f)$;
- U is called resolutive if all $f \in \mathrm{C}(\partial U)$ are resolutive;
- U is called strong resolutive if U is resolutive, every open subset is MP and $\mathrm{M}(U,.)$ is a continuous map.

Furthermore U is said to have one of the above properties 'in' an open set V if U has the property and $\overline{U} \subset V$.

Note that the following three properties are equivalent:

1. U is an MP-set;
2. for all lower bounded $f \in \mathcal{H}^*(U)$ and all upper bounded $g \in \mathcal{H}_*(U)$ with $\hat{f} \geq \check{g}$ on ∂U we have $f \geq g$;
3. for all $f \in \mathrm{NUM}(\partial U)$ we have $\overline{\mathrm{H}}(U,f) \geq \underline{\mathrm{H}}(U,f)$.

Note also that in linear potential theory, every resolutive set is strong resolutive. In non-linear theory, the extra assumption that $\mathrm{M}(U,.)$ is continuous is not very restrictive: if $\mathcal{H}^*(U) = \mathcal{H}^*(U) + \lambda f$ for some $\infty > f \geq 1$ and all $\lambda \geq 0$, then $\overline{\mathrm{M}}(U,.)$ is lower semicontinuous. The assumption that every subset is MP can present some problems: in the examples usually every relatively compact set is MP, but in the axiomatic theory there is no general way to prove that a set is MP. This is a very serious problem.

3. The axioms of strong resolutivity and completeness

In this section we will assume that the following properties hold:

- the strong resolutive sets form a base of X;
- for all open $U \subset \mathrm{X}$: $\mathcal{H}^*(U) = \{u \in \mathrm{LSC}(U) | \forall V$ strong resolutive in U : $\underline{\mathrm{H}}(V,u) \leq u\}$;
- for all open $U \subset \mathrm{X}$: $\mathcal{H}_*(U) = \{u \in \mathrm{USC}(U) | \forall V$ strong resolutive in U : $\overline{\mathrm{H}}(V,u) \geq u\}$.

Note that $\mathcal{H}^*$ and $\mathcal{H}_*$ are not supposed to be sheaves, but that they are presheaves. Note also that $\mathcal{H}^*(U)$ (resp. $\mathcal{H}_*(U)$) is a lower (resp. upper) sublattice of $\mathrm{NUM}(U)$ for all open U.

Lemma 3.1. *Let $U \subset \mathrm{X}$ be relatively compact, then we have:*

1. *if $F \subset \mathcal{H}^*(U)$ upper directed, then $\sup F \in \mathcal{H}^*(U)$;*
2. *if U resolutive and $f \in \mathrm{LSC}(\partial U)$, then $\underline{\mathrm{H}}(U,f) \in \mathcal{H}^*(U)$;*
3. *if $F \subset \mathcal{H}(U)$ is upper directed and $\sup F \in \mathrm{C}(U)$, then $\sup F \in \mathcal{H}(U)$;*
4. *if $F \subset \mathcal{H}^*(U)$ and $\widehat{\inf F}$ is lower finite, then $\widehat{\inf F} \in \mathcal{H}^*(U)$;*
5. *if $V \subset U$ open and $f \in \mathrm{NUM}(\partial U)$, then $\overline{\mathrm{H}}(U,f) \geq \overline{\mathrm{H}}(V,\overline{\mathrm{H}}(U,f))$;*
6. *if $f \in \mathrm{NUM}(\partial U)$ is semiresolutive, then $\mathrm{H}(U,f)$ is semiresolutive for every MP-set $V \subset U$ and we have $\mathrm{H}(U,f) = \mathrm{H}(V,\mathrm{H}(U,f))$;*

7. *if U is resolutive and V strong resolutive in U, then for all $f \in \mathrm{NUM}(\partial U)$ that are $\mathrm{M}(U,.)$-integrable, we have that $\mathrm{M}(U,f)$ is $\mathrm{M}(V,.)$-integrable and $\mathrm{M}(U,f) = \mathrm{M}(V,\mathrm{M}(U,f))$.*

Proof. (1) For every V, strong resolutive in U we have $\underline{\mathrm{H}}(V, \sup F) = \sup\{\mathrm{M}(V,f) | f \in F\} \leq \sup\{f | f \in F\} = \sup F$.
(2) Set $F = \{\mathrm{M}(U,g) | f \geq g \in \mathrm{C}(\partial U)\}$. Then F is upper directed and $F \subset \mathcal{H}(U) \subset \mathcal{H}^*(U)$. Hence we get $\underline{\mathrm{H}}(U,f) = \sup F \in \mathcal{H}^*(U)$.
(3) For every V, strong resolutive in U we have $\mathrm{H}(V, \sup F) = \sup\{\mathrm{H}(V,f) | f \in F\} = \sup\{f | f \in F\} = \sup F$.
(4) Suppose W resolutive in U. Then for all $f \in F$ we have $\underline{\mathrm{H}}(W, \widehat{\inf F}) \leq \underline{\mathrm{H}}(W,f) \leq f$ and hence $\underline{\mathrm{H}}(W, \widehat{\inf F}) \leq \inf F$. Furthermore since $\underline{\mathrm{H}}(W, \widehat{\inf F}) \in \mathcal{H}^*(W) \subset \mathrm{LSC}(W)$ we also have $\underline{\mathrm{H}}(W, \widehat{\inf F}) \leq \widehat{\inf F}$.
(5) We have $h \in \mathcal{H}^*(U,f) \subset \mathcal{H}^*(V, \overline{\mathrm{H}}(U,f))$.
(6) We have $\mathrm{H}(U,f) \geq \overline{\mathrm{H}}(V, \mathrm{H}(U,f)) \geq \underline{\mathrm{H}}(V, \mathrm{H}(U,f)) \geq \mathrm{H}(U,f)$.
(7) For $f \in \mathrm{C}(\partial U)$ this is evident. So, using the lower semicontinuity of $\mathrm{M}(V,.)$, we get for $f \in \mathrm{LSC}(F)$ that $\mathrm{M}^*(U,f) = \sup\{\mathrm{M}(V, \mathrm{M}(U,h)) | f \geq h \in \mathrm{C}(\partial U)\} = \mathrm{M}^*(V, \sup\{\mathrm{M}(U,h) | f \geq h \in \mathrm{C}(\partial U)\}) = \mathrm{M}^*(V, \mathrm{M}^*(U,f))$, and for $f \in \mathrm{NUM}(F)$ we get $\mathrm{M}^{**}(U,f) = \inf\{\mathrm{M}^*(V, \mathrm{M}^*(U,h)) | f \leq h \in \mathrm{LSC}(\partial U)\} \geq \mathrm{M}^{**}(V, \inf\{\mathrm{M}^*(U,h) | f \leq h \in \mathrm{LSC}(\partial U)\}) = \mathrm{M}^{**}(V, \mathrm{M}^{**}(U,f))$. □

The next theorem is crucial for almost all other results in this text. It shows the richness of the sets $\mathcal{H}^*$ and $\mathcal{H}_*$.

Theorem 3.2. *Let U, V be open, $V \subset U$, $u \in \mathcal{H}^*(U)$ and $v \in \mathcal{H}^*(V)$. Define:*

$$f = \begin{cases} u \wedge v & \text{on } V \\ u & \text{on } U \setminus V \end{cases}.$$

If f is lower semicontinuous, then $f \in \mathcal{H}^*(U)$.

Proof. Let W be strong resolutive in U. If f is lower semicontinuous, then $\hat{v} \geq u \geq \underline{\mathrm{H}}(W,u)$ on $\partial V \cap W$. Take $w \in \mathcal{H}_*(W,f)$, then we have $\hat{v} \geq \check{w}$ on $\partial(V \cap W)$ and hence $v \geq w$ on $V \cap W$. But then also $v \geq \underline{\mathrm{H}}(W,f)$ on $V \cap W$. Combining this with $\underline{\mathrm{H}}(W,f) \leq u$ on W. we get $\underline{\mathrm{H}}(W,f) \leq f$ on W. Since W was arbitrary we get $f \in \mathcal{H}^*(U)$. □

Corollary 3.3. *Let U, V be open, $\overline{V} \subset U$ and $u \in \mathcal{H}^*(U)$, then $\overline{\mathrm{H}}(V,u) = \inf\{h \in \mathcal{H}^*(U) | u \leq h \text{ on } U \setminus V\}$ on V.*

Proof. For any $g \in \mathcal{H}^*(V,u)$ we can define h by $h = g \wedge u$ on V and $h = u$ on $U \setminus V$. Then $h \in \mathcal{H}^*(U)$ and $h \geq u$ on $U \setminus V$. Hence $\inf\{h \in \mathcal{H}^*(U) | u \leq h \text{ on } U \setminus V\} \leq \overline{\mathrm{H}}(V,f)$. The other inequality follows from the presheaf property. □

Corollary 3.4. *Let U, V be open, $\overline{V} \subset U$ and $u \in \mathcal{H}^*(U)$. Define:*

$$u_V = \begin{cases} \overline{\mathrm{H}}(\widehat{V,u}) & on\ \overline{V} \\ u & on\ U \setminus \overline{V} \end{cases}.$$

If u_V is lower finite, then $u_V \in \mathcal{H}^*(U)$.

Proof. $u_V = \inf\{h \in \mathcal{H}^*(U) | u \leq h \text{ on } U \setminus V\}\hat{}$ □

Note that in the linear case u_V is the Poisson-modification if X has a countable base (and V is resolutive). Unfortunately if X has no countable base, then u doesn't need to be semiresolutive and hence u_V doesn't need to be the Poisson-modification. However, u_V has all the properties of the Poisson-modification we ever need, so the difference is not important. For more details on this see [Go].

Lemma 3.5. *Let U, V be MP-sets with $V \subset U$, $f \in \mathrm{NUM}(\partial U)$ and $u \in \mathcal{H}^*(U,f)$. Define $g \in \mathrm{NUM}(\partial V)$ to be u on U and f on ∂U. Then $\overline{\mathrm{H}}(U,f) \leq \overline{\mathrm{H}}(V,g)$ on V.*

Proof. For $v \in \mathcal{H}^*(V,g)$ define v' to be $u \wedge v$ on V and u on $U \setminus V$. Then $v' \in \mathrm{LSC}(U)$ and hence $v' \in \mathcal{H}^*(U)$ and even $v' \in \mathcal{H}^*(U,f)$. So $v \geq v' \geq \underline{\mathrm{H}}(U,f)$ on V and hence $\overline{\mathrm{H}}(V,g) \geq \overline{\mathrm{H}}(U,f)$ on V. □

Let U be relatively compact and $x \in \partial U$.

- x is called an upper regular point if $\limsup_{U \ni y \to x} \overline{\mathrm{H}}(f)(y) \leq f(x)$ for all $f \in \mathrm{C}(\partial U)$.
- x is called an upper barrierpoint if there is a filterbase τ of neighbourhoods of x such that for all $V \in \tau$, $\epsilon > 0$ and $a, b \in \mathrm{R}$ finite, there is a $g \in \mathcal{H}^*(U \cap V) \cap \mathrm{LSC}(\overline{V})$ such that $g \geq a$ on $\partial U \cap \overline{V}$, $\hat{g} \geq a$ on $\partial V \cap U$ and $\limsup_{U \cap V \ni y \to x} g(y) \leq a + \epsilon$.

Lemma 3.6. *Let $x \in \partial U$ be an upper regular point, then for all upper bounded $f \in \mathrm{NUM}(\partial U)$ we have $\limsup_{U \ni y \to x} \overline{\mathrm{H}}(f)(y) \leq \check{f}(x)$.*

Proof. Choose $\epsilon > 0$ and take $g \in \mathrm{C}(\partial U)$ such that $g \geq f$ and $g(x) \leq \check{f}(x) + \epsilon$. Then we have $\limsup_{U \ni y \to x} \overline{\mathrm{H}}(f)(y) \leq \limsup_{U \ni y \to x} \overline{\mathrm{H}}(g)(y) \leq g(x) \leq \check{f}(x) + \epsilon$. □

Lemma 3.7. *Let U be an MP-set such that all boundary points are upper regular. Then $f \in \mathrm{USC}(\partial U)$ is resolutive iff $\overline{\mathrm{H}}(U,f) \in \mathcal{H}(U)$.*

Proof. If $\overline{\mathrm{H}}(U,f) \in \mathcal{H}(U)$ then also $\overline{\mathrm{H}}(U,f) \in \mathcal{H}_*(U,h)$ and hence $\overline{\mathrm{H}}(U,h) = \underline{\mathrm{H}}(U,h) \in \mathcal{H}(U)$. The other implication is evident. □

The following two theorems show that in most practical cases, $x \in \partial U$ is an upper regular point iff it is an upper barrierpoint.

Theorem 3.8. *Suppose U is relatively compact, $h \in \mathrm{NUM}(\partial U)$ and $x \in \partial U$ an upper barrierpoint. If there is a $u \in \mathcal{H}^*(U,h)$ with $\limsup_{U \ni y \to x} u(y)$ finite, then* $\limsup_{U \ni y \to x} \overline{\mathrm{H}}(U,h)(y) \le \check{h}(x)$.

Proof. We may suppose that $\check{h}(x)$ is lower finite. Choose $\epsilon > 0$ and note that both u and h are upper bounded in a neighbourhood of x. So there is a $V \in \tau$ and $a, b \in \mathrm{R}$ finite such that $a \ge h$ on $\partial U \cap \overline{V}$, $a \le \check{h}(x) + \epsilon$ and $b \ge u$ on $\partial V \cap U$. Now take $g \in \mathcal{H}^*(U \cap V) \cap \mathrm{LSC}(\overline{V})$ such that $g \ge a$ on $\partial U \cap \overline{V}$, $\hat{g} \ge a$ on $\partial V \cap U$ and $\limsup_{U \cap V \ni y \to x} g(y) \le a + \epsilon$. Define v to be u on $U \setminus V$ and $u \wedge g$ on V. Then $v \in \mathcal{H}^*(U,h)$. But then $\limsup_{U \ni y \to x} \overline{\mathrm{H}}(U,h)(y) \le \check{h}(x) + 2\epsilon$. □

Theorem 3.9. *Suppose U and $x \in \partial U$ are such that for all W in a filterbase τ of neighbourhoods of x we have that $W \cap U$ is MP and x is upper regular in $W \cap U$. Furthermore, suppose that x has a neighbourhood V such that for all $y \in V$ and all $a, b \in \mathrm{R}$ finite, there is a $f \in \mathcal{H}_*(V) \cap \mathrm{C}(V)$ with $f(x) \le b$ and $f(y) \ge a$. Then x is an upper barrierpoint.*

Proof. Take $W \in \tau$, $\overline{W} \subset V$, $\epsilon > 0$ and $a, b \in \mathrm{R}$ finite with $a \le b$. Take $f_x \in \mathcal{H}_*(V) \cap \mathrm{C}(V)$ with $a < f_x(x) \le a + \epsilon$ and for all $y \in \partial(W \cap U) \setminus \{x\}$ take $f_y \in \mathcal{H}_*(V) \cap \mathrm{C}(V)$ with $f_y(x) \le a$ and $f_y(y) > b$. Now there is a finite set $I \subset \partial(W \cap U) \setminus \{x\}$ such that $\partial(W \cap U) \setminus \subset \cup_{y \in I}\{f_y > b\}$. Now set $f = f_x \vee \sup_{y \in I} f_y$, then $g = \overline{\mathrm{H}}(W \cap U, f)$ satisfies the required conditions. □

4. The axioms of convergence

In this section we will assume that the following properties hold:

- for all open U and all upper directed $F \subset \mathcal{H}(U)$ we have $\sup F \in \mathcal{H}(U)$ if $\sup F \in \mathrm{C}(U)$;
- for all open U and all lower directed $F \subset \mathcal{H}(U)$ we have $\inf F \in \mathcal{H}(U)$ if $\inf F \in \mathrm{C}(U)$;
- for all open U and all increasing sequences $(h_n) \subset \mathcal{H}(U)$ we have $\sup(h_n) \in \mathcal{H}(U)$ if $\sup(h_n)$ is locally bounded;
- for all open U and all decreasing sequences $(h_n) \subset \mathcal{H}(U)$ we have $\inf(h_n) \in \mathcal{H}(U)$ if $\inf(h_n)$ is locally bounded.

Note that the axioms of strong resolutivity and completeness together imply the first two properties. These properties are necessary to prove the next theorem.

Theorem 4.1. *For all open U and upper directed $F \subset \mathcal{H}(U)$ we have $\sup F \in \mathcal{H}(U)$ if $\sup F$ is locally bounded.*

Proof. Let $F \subset \mathcal{H}(U)$ be upper directed such that $\sup F$ is locally bounded. Then the supremum of every increasing sequence in F is locally bounded and hence in $\mathcal{H}(U) \subset$

C(U). But now since every increasing sequence in F has continuous supremum, also $\sup F \in \mathrm{C}(U)$. But then we can conclude $\sup F \in \mathcal{H}(U)$. □

Theorem 4.2. *Suppose $\mathcal{H}$ is a sheaf. If there are $f, g \in \mathcal{H}(\mathrm{X})$ such that $f < g$ on* X, *then* X *is locally connected.*

Proof. Let W be a component of X and $x \in W$. Define F to be the collection of all functions

$$h = \begin{cases} f & \text{on } V \\ g & \text{on } \mathrm{X} \setminus V \end{cases}$$

where V is a clopen neighbourhood of x. Then $F \subset \mathcal{H}(\mathrm{X})$ is upper directed and bounded, and thus $\sup F \in \mathrm{C}(\mathrm{X})$. Since either $\sup F(y) = f(y)$ or $\sup F(y) = g(y)$ we have that $W' = \{y \in \mathrm{X} | \sup F(y) = f(y)\} = \{y \in \mathrm{X} | \sup F(y) \neq g(y)\}$ is a minimal clopen neighbourhood of x. Since W' is minimal, it is connected and hence $W' \subset W$ and since W' is clopen we have $W \subset W'$. So $W' = W$ and hence W is open. □

In the next three lemmas U will be a resolutive set in X.

Lemma 4.3. *If $f \in \mathrm{NUM}(\partial U)$ is bounded, then $\mathrm{M}^{**}(U, f) \in \mathcal{H}(U)$.*

Proof. First suppose $f \in \mathrm{LSC}(\partial U)$ and bounded. Then $\mathrm{M}^*(U, f) = \sup\{\mathrm{M}(U, g) | f \geq g \in \mathrm{C}(\partial U)\}$, which is the supremum of a bounded upper directed set of harmonic functions, and hence harmonic itself. Now suppose $f \in \mathrm{NUM}(\partial U)$ and bounded. Then $\mathrm{M}^{**}(U, f) = \inf\{\mathrm{M}^*(U, g) | f \leq g \in \mathrm{LSC}(\partial U)\}$, which is the infimum of a bounded lower directed set of harmonic functions, and hence harmonic itself. □

Lemma 4.4. *If $f \in \mathrm{NUM}(\partial U)$ is upper bounded, then $\mathrm{M}^{**}(U, f) \in \mathrm{USC}(U)$.*

Proof. $\mathrm{M}^{**}(U, f) = \inf\{\mathrm{M}^*(U, g) | f \leq g \in \mathrm{LSC}(\partial U), g \text{ upper bounded}\}$. But if $g \in \mathrm{LSC}(\partial U)$ is upper bounded, then it is bounded and hence by the previous lemma $\mathrm{M}^*(U, g) \in \mathcal{H}(U) \subset \mathrm{C}(U)$. □

Lemma 4.5. *If $u \in \mathrm{LSC}(\overline{U})$ is locally bounded and $\underline{\mathrm{H}}(U, u) \leq u$, then $\underline{\mathrm{H}}(U, u) \in \mathcal{H}(U)$.*

Proof. Since $u \in \mathrm{LSC}(\partial U)$ we have $\underline{\mathrm{H}}(U, u) = \sup\{\mathrm{M}(U, f) | u \geq f \in \mathrm{C}(\partial U)\}$. Since we also have $\underline{\mathrm{H}}(U, u) \leq u$ we can conclude that $\underline{\mathrm{H}}(U, u)$ is the supremum of a locally bounded, upper directed set of harmonic functions, and hence harmonic itself. □

5. Harmonic spaces (the sheaf axiom)

We will say that X endowed with $\mathcal{H}^*$ and $\mathcal{H}_*$ is an harmonic space if $\mathcal{H}$ is a sheaf and all axioms of the previous sections are satisfied. That means X is an harmonic space if:

- $\mathcal{H}$ is a sheaf;
- the strong resolutive sets form a base of X;
- for all open $U \subset$ X: $\mathcal{H}^*(U) = \{u \in \mathrm{LSC}(U) | \forall V$ strong resolutive in U : $\underline{\mathrm{H}}(V, u) \leq u\}$;
- for all open $U \subset$ X: $\mathcal{H}_*(U) = \{u \in \mathrm{USC}(U) | \forall V$ strong resolutive in U : $\overline{\mathrm{H}}(V, u) \geq u\}$;
- for all open U and all increasing sequences $(h_n) \subset \mathcal{H}(U)$ we have $\sup(h_n) \in \mathcal{H}(U)$ if $\sup(h_n)$ is locally bounded;
- for all open U and all decreasing sequences $(h_n) \subset \mathcal{H}(U)$ we have $\inf(h_n) \in \mathcal{H}(U)$ if $\inf(h_n)$ is locally bounded.

In the rest of this text X will be an harmonic space.

6. Superharmonic functions

For all open U we define:

- $\mathcal{S}^*(U) = \{u \in \mathcal{H}^*(U) | \forall V$ strong resolutive in $U : \underline{\mathrm{H}}(V, u) \in \mathcal{H}(V)\}$;
- $\mathcal{S}_*(U) = \{u \in \mathcal{H}_*(U) | \forall V$ strong resolutive in $U : \overline{\mathrm{H}}(V, u) \in \mathcal{H}(V)\}$.

Functions in $\mathcal{S}^*(U)$ (resp. $\mathcal{S}_*(U)$) will be called superharmonic (resp. subharmonic) on U. As $\mathcal{H}^*$ and $\mathcal{H}_*$, also $\mathcal{S}^*$ and $\mathcal{S}_*$ are presheaves, but not necessarily sheaves.

Lemma 6.1. *Let U be open, then:*

1. *if $u \in \mathcal{H}^*(U)$ is locally bounded, then $u \in \mathcal{S}^*(U)$;*
2. *if $u \in \mathcal{S}^*(U)$ and $\in \mathcal{H}^*(U)$, then $u \wedge v \in \mathcal{S}^*(U)$;*
3. *if $f \in \mathcal{S}^*(U)$, then for all V resolutive in U : $\underline{\mathrm{H}}(V, u) \in \mathcal{H}(V)$;*
4. *if $f \in \mathcal{S}^*(U)$, then for all V resolutive in U : $\overline{\mathrm{H}}(V, u) \in \mathcal{H}(V)$.*

Proof. (1) Apply lemma 4.5.
(2) Let V be strong resolutive in U. Then $\underline{\mathrm{H}}(V, u \wedge v) = \sup\{\mathrm{M}(V, h) | u \wedge v \geq h \in \mathrm{C}(\partial U)\} \leq \underline{\mathrm{H}}(V, u) \in \mathcal{H}(V)$. So $\underline{\mathrm{H}}(V, u \wedge v)$ is the supremum of a bounded upper directed set of harmonic functions, and hence harmonic itself.
(3) For all W, strong resolutive in V we have $\underline{\mathrm{H}}(V, f) = \sup\{\mathrm{M}(V, g) | f \geq g \in \mathrm{C}(\partial V)\} \leq \underline{\mathrm{H}}(W, f) \in \mathcal{H}(W)$ on W. So by the upper convergence property $\underline{\mathrm{H}}(V, f)$ is harmonic on W and then by the sheaf property it is harmonic on V.
(4) For all W, strong resolutive in V we have $\overline{\mathrm{H}}(V, f) \geq \inf\{\underline{\mathrm{H}}(W, u) | u \in \mathcal{S}^*(V), \hat{u} \geq f$ on $\partial V\} \geq \underline{\mathrm{H}}(V, f) \in \mathcal{H}(W)$ on W. So by the lower convergence property $\overline{\mathrm{H}}(V, f)$ is harmonic on W and then by the sheaf property it is harmonic on V. □

Let U be open. $F \subset \mathcal{H}^*(U)$ is called an upper Perron set if:

- F is lower directed;
- F is minorated by a subharmonic function;
- for all $x \in U$ there is a neighbourhood V of x, strong resolutive in U, such that:
  - for all $f \in F$ we have $f_V \in F$,

– there is an $f \in F$ such that $f_V \in \mathcal{H}(V)$.

The reason for considering upper Perron sets is the following theorem.

Theorem 6.2. *The infimum of an upper Perron set is harmonic.*

Proof. Let F be an upper Perron set on U. Take $x \in U$ and take a neighbourhood V of x as in the definition. Then there is an $f \in F$ such that $f_V \in \mathcal{H}(V)$, but then for all $g \in F, g \leq f$ we have also $g_V \in \mathcal{H}(V)$. So on V we have that $\inf F = \{g_V \in F | g \leq f\}$ is the infimum of a lower directed set of harmonic functions. Furthermore, if u is a subharmonic minorant of F, so is u_V and hence $\inf F$ is locally bounded. Using the lower convergence property and the sheaf property we get $\inf F \in \mathcal{H}(U)$. □

Lemma 6.3. *Suppose that U is an MP-set, $f \in \mathrm{NUM}(\partial U)$, $\mathcal{H}^*(U,f)$ contains a superharmonic function and $\mathcal{H}_*(U,f)$ contains a subharmonic function. Then:*

1. *$\mathcal{H}^*(U,f)$ is an upper Perron set;*
2. *$\overline{\mathrm{H}}(U,f) \in \mathcal{H}(U)$;*
3. *$\overline{\mathrm{M}}^{**}(U,f) \in \mathcal{H}(U)$;*

Proof. (1) Evidently, $\mathcal{H}^*(U,f)$ is lower directed and minorated by a subharmonic function. For all V, strong resolutive in U and all $g \in \mathcal{H}^*(U,f)$ we have $g_V \in \mathcal{H}^*(U)$ and $f \leq \hat{g} = \widehat{g_V}$ on ∂U and thus $g_V \in \mathcal{H}^*(U,f)$. Furthermore, if g is the superharmonic function in $\mathcal{H}^*(U,f)$, then for all V, strong resolutive in U we have $g_V \in \mathcal{H}(V)$. So we can conclude that $\mathcal{H}^*(U,f)$ is an upper Perron set.

(2) Apply theorem 6.2.

(3) For $f \in \mathrm{C}(\partial U)$ we have $\overline{\mathrm{M}}(U,f) = \overline{\mathrm{H}}(U,f) \in \mathcal{H}(U)$. If $f \in \mathrm{LSC}(\partial U)$, then $\overline{\mathrm{M}}^*(U,f) = \sup\{\overline{\mathrm{M}}(U,g) | f \geq g \in \mathrm{C}(\partial U)\} \leq \overline{\mathrm{H}}(U,f) \in \mathcal{H}(U)$. So by the upper convergence property $\overline{\mathrm{M}}^*(U,f) \in \mathcal{H}(U)$. If $f \in \mathrm{NUM}(\partial U)$, then $\overline{\mathrm{M}}^{**}(U,f) = \inf\{\overline{\mathrm{M}}^*(U,g) | f \leq g \in \mathrm{LSC}(\partial U)\} \geq \underline{\mathrm{H}}(U,f) \in \mathcal{H}(U)$. So by the lower convergence property $\overline{\mathrm{M}}^{**}(U,f) \in \mathcal{H}(U)$. □

Let U be open and $\mathcal{C}$ be a covering of U with sets that are strong resolutive in U. Suppose $\mathcal{S}^*(U) \ni u \geq v \in \mathcal{S}_*(U)$. Define F as the collection of all functions $u_{V_1 \cdots V_n}$ with $V_1, \cdots, V_n \in \mathcal{C}$. Evidently F is an upper Perron set. It is called the upper Perron set generated by u and $\mathcal{C}$.

Theorem 6.4. *Suppose $\mathcal{S}^*(U) \ni u \geq v \in \mathcal{S}_*(U)$. Then:*

1. *u has a greatest hypoharmonic minorant $h_u \in \mathcal{H}(U)$ which is the infimum of any upper Perron set generated by u;*
2. *the set of harmonic functions on U between u and v is a complete lattice. (But not necessarily a sublattice from $\mathrm{NUM}(U)$.)*

Proof. (1) Let $\mathcal{C}$ be a covering of U with sets that are strong resolutive in U and let h_u be the infimum of the upper Perron set generated by u and $\mathcal{C}$. Then evidently h_u is an (hypo)harmonic minorant of u. Now let f be a general (hypo)harmonic minorant of

u. Then for any finite sequence $V_1, \cdots, V_n \in \mathcal{C}$ we have $f \leq f_{V_1 \cdots V_n} \leq u_{V_1 \cdots V_n}$ and thus $f \leq h_u$. So h_u is the greatest hypoharmonic minorant of u. Since $\mathcal{C}$ was arbitrary, h_u is the infimum of any upper Perron set generated by u.
(2) Let F be the harmonic functions on U between u and v. For any $h_1, h_2 \in F$ we have $\mathcal{S}^* \ni h_1 \wedge h_2 \geq v \in \mathcal{S}_*(U)$ and so $h_1 \wedge h_2$ has a greatest hypoharmonic minorant, which is evidently an element of F. So F is a lower lattice. Now let $G \subset F$. Define G' as the infima (in F) of all finite subsets of G. Then G' is a lower directed set of harmonic functions, lower bounded by h_v. So $\inf G = \inf G' \in F$ and thus F is a complete lower lattice. Dually F is a complete upper lattice. □

Not that in the linear case we usually take $v \equiv 0$. In that case h_u is the harmonic part of u and $u - h_u$ is the potential part of u.

7. Reduits

We say that $\mathcal{H}^*$ is non-degenerate in $x \in \mathrm{X}$ if for all neighbourhoods U of x and for all $f \in \mathcal{H}^*(U)$ we have that $f(x) = \sup\{f_V(x) | V$ strong resolutive neighbourhood of x in $U\}$. Furthermore, we define the upper reduit of a function $f \in \mathrm{NUM(X)}$ by $R^*(f) = \inf\{u \in \mathcal{H}^*(\mathrm{X}) | u \geq f\}$.

Lemma 7.1. *We have:*

1. *R^* is an increasing map;*
2. *If $f \in \mathrm{LSC(X)}$ then $R^*(f) \in \mathcal{H}^*(U)$;*
3. *If $F \subset \mathrm{LSC(X)}$ is upper directed, then $R^*(\sup F) = \sup R^*(F)$.*

Proof. (2) $R^*(f) \geq f \in \mathrm{LSC(X)}$ and hence $f \leq \widehat{R^*(f)} \in \mathcal{H}^*(\mathrm{X})$. So $\widehat{R^*(f)} \geq R^*(f) \geq \widehat{R^*(f)} \in \mathcal{H}^*(\mathrm{X})$.
(3) We have $R^*(F) \subset \mathcal{H}^*(\mathrm{X})$ upper directed and hence $\sup R^*(F) \in \mathcal{H}^*(\mathrm{X})$ Since also $\sup R^*(F) \geq F$ we have $\sup R^*(F) \geq R^*(\sup F)$. The other inequality is evident. □

Lemma 7.2. *Let $V \subset \mathrm{X}$ be relatively compact and $f \in \mathrm{NUM(X)}$ majorated by a superharmonic function. If $f \leq \overline{\mathrm{H}}(V, f) \in \mathcal{H}(V)$, then $R^*(f) \in \mathcal{H}(V)$.*

Proof. $\{h | f \leq h \in \mathcal{H}^*(\mathrm{X})\}$ is an upper Perron set on V. □

Lemma 7.3. *Let $f \in \mathrm{LSC(X)}$ be continuous in x and majorated by a superharmonic function. If $\mathcal{H}^*$ is non-degenerate in x and $R^*(f)(x) > f(x)$, then $R^*(f)$ is harmonic in a neighbourhood of x.*

Proof. Since $\mathcal{H}^*$ is non-degenerate in x we can find a strong resolutive neighbourhood W of x with $R^*(f)_W(x) > f(x)$. But then, since f is continuous in x and since $R^*(f) \in \mathrm{LSC(X)}$, we can find a neighbourhood V of x, strong resolutive in W such that $R^*(f)_W(V) \geq f(V)$. Now we have $f \leq R^*(f)_V \in \mathcal{H}^*(\mathrm{X})$ and thus $R^*(f) = R^*(f)_V \in \mathcal{H}(V)$. □

Theorem 7.4. *Let $f \in \mathrm{LSC}(\mathrm{X})$ be continuous and finite in x and majorated by a superharmonic function. Suppose that $\mathcal{H}^*$ is non-degenerate in x and that for all $\epsilon > 0$ there exists a function h, superharmonic in a neighbourhood U of x such that: $h \geq R^*(f)$ on U and $R^*(f)(x) + \epsilon \geq h(x) > f(x)$. Then $R^*(f)$ is continuous and finite in x.*

Proof. Take $\epsilon > 0$. Since $h(x) > f(x)$ and since $\mathcal{H}^*$ is non-degenerate in x, we can find a neighbourhood W of x, strong resolutive in U, such that $h_W(x) > f(x)$. But then, since f is continuous in x and since $h \in \mathrm{LSC}(U)$, we can find a neighbourhood V of x, strong resolutive in W, such that $h_W(V) \geq f(V)$. Now define the function h' to be $h_V \wedge R^*(f)$ on U and $R^*(f)$ on $\mathrm{X} \setminus U$. Then $h' \in \mathrm{LSC}(\mathrm{X})$ and hence $h' \in \mathcal{H}^*(\mathrm{X})$, and since also $h' \geq f$ we have $h_V(x) \geq h'(x) \geq R^*(f)(x)$. Since $h_V(x)$ is finite, $R^*(f)(x)$ is also finite. Since h_V is continuous in x we have $h(x) \geq h_V(x) = \limsup_{U \ni y \to x} h_V(y) \geq \limsup_{y \to x} R^*(f)(y)$. But then $\limsup_{y \to x} R^*(f)(y) \leq R^*(f)(x) + \epsilon$ and so $R^*(f)$ is upper semicontinuous, and hence continuous, in x. □

8. Strong harmonic spaces

In this section we will assume that $\mathcal{H}^*$ is non-degenerate in every point of X and that for all $f \in \mathcal{H}^*(\mathrm{X})$ we have $f = \inf\{g \in \mathcal{H}^*(\mathrm{X}) | g \gg f\}$.

Let $f \in \mathrm{NUM}(\mathrm{X})$. The carrier $\mathrm{C}(f)$ of f is defined as the smallest closed set K with $f \in \mathcal{H}(\mathrm{X} \setminus K)$. From the sheaf property it is clear such a set exists.

In this section $h \in \mathcal{H}(\mathrm{X})$ and $\mathcal{B}_h^* \subset \{u \in \mathcal{S}^*(\mathrm{X}) | u \geq h\}$ will be an upper directed set such that for every $x \in \mathrm{X}$ and every finite $a \in \mathrm{R}$ there is a $u \in \mathcal{B}_h^*$ with $u(x) \geq a$. Furthermore, we suppose that if $f \in \mathcal{S}^*(\mathrm{X})$ with $h \leq f \leq g \in \mathcal{B}_h^*$, then $f \in \mathcal{B}_h^*$. In the linear theory, usually $h \equiv 0$ and $\mathcal{B}_h^*$ is either the set of all positive superharmonic functions or the set of potentials.

Lemma 8.1. *For all compact $K \subset \mathrm{X}$ and all upper finite $a \in \mathrm{R}$ we can find a $u \in \mathcal{B}_h^*$ with $u > a$ on K.*

Proof. For all $x \in K$, take $h_x \in \mathcal{B}_h^*$ with $h_x > a$. Since all the h_x are lower semicontinuous, we can find for all x a neighbourhood V_x such that $h_x > a$ on V_x. Since K is compact there is a finite set $I \subset K$ such that K is covered by the V_x with $x \in I$. Now take $h \in \mathcal{B}_h^*$ with $h \geq h_x$ for all $x \in I$. □

Lemma 8.2. *Let $K \subset \mathrm{X}$ be compact and $f \in \mathrm{C}(\mathrm{X}) \cap \mathcal{S}_*(\mathrm{X} \setminus K)$ with $f \geq h$ and such that there is a $g \in \mathcal{B}_h^*$ with $g \geq f$ on $\mathrm{X} \setminus K$. Then $R^*(f) \in \mathrm{C}(\mathrm{X}) \cap \mathcal{B}_h^* \cap \mathcal{H}(\mathrm{X} \setminus K)$.*

Proof. Since f is upper bounded on K we can find $g' \in \mathcal{B}_h^*$ with $g' \geq f$ on K and $g' \geq g$. Now using lemma 7.2 we get $R^*(f) \in \mathcal{B}_h^* \cap \mathcal{H}(\mathrm{X} \setminus K)$. Now for any $x \in \mathrm{X}$ we can find $g' \in \mathcal{B}_h^*$ with $g' \geq R^*(f)$ and $g'(x) > f(x)$. Using the assumption that

$R^\star(f) = \inf\{k \in \mathcal{H}^\star(\mathrm{X}) | k \gg R^\star(f)\}$ we can make sure that $g'(x) \leq R^\star(f)(x) + \epsilon$ for any $\epsilon > 0$ we choose. So we can use theorem 7.4 to get $R^\star(f) \in \mathrm{C}(\mathrm{X})$. □

Theorem 8.3. *Let $f \in \mathcal{H}^\star(\mathrm{X})$ be a majorant of h. Then $\mathcal{V} = \{g \in \mathrm{C}(\mathrm{X}) \cap \mathcal{B}_h^\star | f \geq g$ with $\mathrm{C}(g)$ compact$\}$ is upper directed and $f = \sup \mathcal{V}$. If X has a countable base, then there is an increasing sequence in $\mathcal{V}$ converging to f.*

Proof. Take $v, w \in \mathcal{V}$. Then there is a $u \in \mathcal{B}_h^\star$ with $u \geq v \vee w$ and there is a compact set $K \subset \mathrm{X}$ with $v \vee w \in \mathcal{S}_\star(\mathrm{X} \setminus K)$. But then $R^\star(v \vee w) \in \mathcal{V}$ and so $\mathcal{V}$ is upper directed. Let F be a family of continuous functions such that $\sup F = f$ and such that for all $g \in F$ there is a compact set K with $g \equiv -\infty$ outside K. Then $R^\star(F \vee h) \subset \mathcal{V}$ and $f = \sup F \leq \sup R^\star(F \vee h) \leq \sup \mathcal{V} \leq f$. If X has a countable base, then we can take F to be an increasing sequence. □

Lemma 8.4. *Let $V \subset \mathrm{X}$ be relatively compact and $f \in \mathrm{NUM}(\partial V)$ upper bounded. Then $\mathcal{H}^\star(V, f)$ contains a continuous bounded superharmonic function.*

Proof. Take a relatively compact $W \subset \mathrm{X}$ such that $\partial V \subset W$. Then we can find a continuous function g on X such that $g = h$ on $\mathrm{X} \setminus W$ and $g \geq f$ on ∂V. But then $R^\star(g) \in \mathrm{C}(\mathrm{X}) \cap \mathcal{S}^\star(\mathrm{X})$. Since also $R^\star(g) \geq g \geq f$ on ∂V we have $R^\star(g) \in \mathcal{H}^\star(V, f)$. □

Lemma 8.5. *Let W, V be open such that $V \subset W$. Suppose $x \in \partial W \cap \partial V$ is such that $B \cap W = B \cap V$ for some neighbourhood B of x. If x is upper regular in V, then x is upper regular in W.*

Proof. Take $f \in \mathrm{C}(\partial W)$ and an upper bounded $u \in \mathcal{H}^\star(W, f)$. Define $g = f$ on ∂W and $g = u$ on W. Then $\limsup_{W \ni y \to x} \overline{\mathrm{H}}(W, f)(y) \leq \limsup_{V \ni y \to x} \overline{\mathrm{H}}(V, g)(y) \leq \check{g}(x) = f(x)$. □

References

[Be] E.M.J. Bertin, Convex potential theory. Preprint 607, University of Utrecht, 1990.

[CC] C. Constantinescu, A. Cornea, Potential theory on harmonic spaces. Springer-Verlag, Berlin 1972.

[Gi] G. Gierz, K. Hofmann, K. Keimel, J. Lawson, M. Mislove, D. Scott, A compendium of continuous lattices. Springer-Verlag, Berlin 1980.

[Go] F.A. van Gool, Non-linear potential theory. Preprint 606, University of Utrecht, 1990.

[La] I. Laine, Axiomatic non-linear potential theories. Lecture Notes in Mathematics 1344, 118–132, Springer-Verlag, Berlin 1988. I. Laine, Introduction to a quasi-linear potential theory. Ann. Acad. Scient. Fennicae 10 (1985) 339–348.

[Le] P. Lehtola, An axiomatic approach to non-linear potential theory. Thesis, Jyväskylä 1986.

[Sm] P. Smyrnelis, Axiomatique des fonctions biharmoniques. Ann. Inst. Fourier 25:1 (1975) 35–97.

The Neumann problem and Hausdorff measures

Hisako Watanabe

Dedicated to Professor M. Kishi on the occasion of his 60th birthday

Abstract

For a positive real number $\beta \leq d-1$ we introduce a countably sublinear functional γ_β on a family of functions defined on the boundary of a $C^{1,\alpha}$-domain D in $\mathbf{R}^d$ ($d \geq 3$) and a class $\mathcal{L}(\gamma_\beta, C(\partial D))$, instead of the L^p-norm and the space $L^p(\partial D)$, respectively. We will show that the normal derivative of a solution to the Neumann problem with boundary data $f \in \mathcal{L}(\gamma_\beta, C(\partial D))$ converges non-tangentially to f on ∂D except for a set of β-dimensional Hausdorff measure zero.

1980 Mathematics Subject Classification (1985 Revision): 31B20, 35J22.

1. Introduction

Let D be a bounded, smooth domain in $\mathbf{R}^d$. The classical Neumann problem is to find and describe a harmonic function u in D such that the normal derivative of u is equal to f on the boundary of D. The method of layer potentials yields the existence of a solution u to the Neumann problem with boundary data f.

In [FJR] E.B.Fabes, M.Jodeit JR. and N.M. Rivière solved the above problem with boundary data in $L^p(\partial D)$ for a C^1-domain D in the case $1 < p < +\infty$. It has been known that the same conclusion is obtained even for a Lipschitz domain D and a real number p satisfying $1 < p < 2+\epsilon$ (cf.[JK], [DK]).

In this paper we consider a bounded $C^{1,\alpha}$-domain D in $\mathbf{R}^d$ for $0 < \alpha < 1$ and find a class of boundary functions such that the normal derivative of a solution to the Neumann problem with boundary data f in the class converges non-tangentially to f on ∂D except for a set of β-dimensional Hausdorff measure zero.

For this purpose we must classify the functions in $L^1(\partial D)$ depending on the degree of continuity. To classify functions, we use a countably sublinear functional γ_β and a class $\mathcal{L}(\gamma_\beta, C(\partial D))$ of functions on ∂D instead of the L^p-norm and $L^p(\partial D)$, respectively. We denote by $B(P,r)$ the open ball in $\mathbf{R}^d$ with center P and radius r. We write $A(P,r) = B(P,r) \cap \partial D$ for $P \in \partial D$ and call it a surface ball. Let $J(\partial D)$ be the class of all extended real-valued functions on ∂D and define

$$\gamma_\beta(f) = \inf\{ \sum_{j=1}^{\infty} b_j r_j^\beta : b_j \geq 0,\ \sum_{j=1}^{\infty} b_j \chi_{A(P_j,r_j)} \geq |f| \text{ on } \partial D \}$$

for $f \in J(\partial D)$.

Potential Theory, ed. by M. Kishi

The functional γ_β is countably sublinear on $J(\partial D)$, i.e., it has the following properties:

(c_1) $\gamma_\beta(f) = \gamma_\beta(|f|)$,

(c_2) $\gamma_\beta(bf) = |b|\gamma_\beta(f)$ for every $b \in \mathbf{R}$,

(c_3) $f,\ f_n \geq 0,\ f \leq \sum_{n=1}^{\infty} f_n \Rightarrow \gamma_\beta(f) \leq \sum_{n=1}^{\infty} \gamma_\beta(f_n)$.

For a subset E of ∂D we write $\gamma_\beta(E)$ instead of $\gamma_\beta(\chi_E)$. A set E satisfying $\gamma_\beta(E) = 0$ is called γ_β-polar. We say that a property holds γ_β-q.e. if it holds on ∂D except for a γ_β-polar set.

In §4 we shall show that a Borel subset E on ∂D is γ_β-polar if and only if it is of β-dimensional Hausdorff measure zero.

Next, denote by $\mathcal{L}(\gamma_\beta, C(\partial D))$ the class of all Borel measurable functions f such that $\gamma_\beta(f - f_n) \to 0$ for some sequence $\{f_n\} \subset C(\partial D)$, where $C(\partial D)$ stands for the class of all continuous real-valued functions on ∂D.

Let $0 < \eta < 1$. For $P \in \partial D$ a non-tangential domain at P is defined by

$$\Gamma_\eta(P) = \{\, X \in D :\ \langle P - X, N_p\rangle > \eta|X - P|\,\}.$$

Denote by σ the surface measure on ∂D and by N_P the outer unit normal to the ∂D at P. In §5 we shall prove the following theorem.

Theorem 1. *Let $0 < \alpha < 1$ and D be a $C^{1,\alpha}$-domain in $\mathbf{R}^d$. Further, assume that $0 < \beta \leq d - 1$ and $0 < \eta < 1$. For every function $f \in \mathcal{L}(\gamma_\beta, C(\partial D))$ such that $\int f d\sigma = 0$ there exist a harmonic function u in D and a subset E of ∂D of β-dimensional Hausdorff measure zero such that*

$$\lim_{X \to P, X \in \Gamma_\eta(P)} \langle \nabla u(X), N_P\rangle = f(P)$$

for every $P \in \partial D \setminus E$.

In §6 it will be shown that, if $\lambda > d - 1 - \beta > 0$, then $\mathcal{L}(\gamma_\beta, C(\partial D))$ contains all functions of the form:

$$P \mapsto \int |P - Q|^{\lambda+1-d} g(Q)\, d\sigma(Q)$$

for $g \in L^1(\partial D)$.

2. Estimates of surface balls

Throught this paper let $0 < \alpha < 1$, $d \geq 3$ and D be a bounded $C^{1,\alpha}$-domain in $\mathbf{R}^d$. Points of D are generally denoted by X, Y and points on the boundary ∂D of D are represented by P, Q.

Definition. Let $0 < \alpha < 1$. A bounded domain D in $\mathbf{R}^d$ is called a $C^{1,\alpha}$-domain if to each point $Q \in \partial D$ there correspond a system of coordinates of $\mathbf{R}^d$ with origin Q and an open ball $B(Q,\delta)$ with center Q and radius δ such that with respect to this coordinates system

$$D \cap B(Q,\delta) = \{ (x,t) : \ x \in \mathbf{R}^{d-1}, \ t > \phi(x) \} \cap B(Q,\delta),$$

where $\phi \in C_o^{1,\alpha}(\mathbf{R}^{d-1})$ and $\phi(0) = D_{x_j}\phi(0) = 0$. Note that $C_o^{1,\alpha}(\mathbf{R}^{d-1})$ stands for the space of all functions g in $C^1(\mathbf{R}^{d-1})$ with compact support, satisfying

$$|D_{x_j} g(x) - D_{x_j} g(y)| \le M|x-y|^\alpha$$

for all $x, y \in \mathbf{R}^{d-1}$ and $1 \le j \le d-1$.

In the definition of γ_β in §1 we may take only surface balls with radius $r \le R$ for a positive real number R such that $\bar{D} \subset B(0, \frac{R}{2})$. Since D is a $C^{1,\alpha}$-domain, it is well-known that surface balls have the following property.

Lemma 2.1. *Let $0 < r \le R$ and $A_o = A(P,r)$. Then there is a constant, independent of P and r, such that*

$$c^{-1} r^{d-1} \le \sigma(A_o) \le c r^{d-1}.$$

Let us denote by $\mathcal{B}(\gamma_\beta)$ the class of all Borel measurable functions $f \in J(\partial D)$ such that $\gamma_\beta(f) < +\infty$. We note that if $f \in \mathcal{B}(\gamma_\beta)$, then the set $\{P \in \partial D : \ f(P) = \pm\infty\}$ is γ_β-polar.

By using Lemma 2.1 we have easily

Lemma 2.2. *Let $0 < \beta \le d-1$ and $p = \frac{d-1}{\beta}$. Then*

(i) $\gamma_\beta(E) = 0$ for a Borel set $E \Rightarrow \sigma(E) = 0$,

(ii) $f, g \in \mathcal{B}(\gamma_\beta)$, $f = g$ γ_β-q.e. $\Rightarrow f = g$ σ-a.e.,

(iii) $(\int |f|^p \, d\sigma)^{1/p} \le c\gamma_\beta(f)$ for all $f \in \mathcal{B}(\gamma_\beta)$, where c is a constant independent of f.

We have the following estimate for a surface ball.

Lemma 2.3. *Let $0 < \beta \le d-1$, $0 < r, r_1 \le R$ and set $A_1 = A(Q_1, r_1)$. Define*

$$g(P) = \sigma(A(P,r) \cap A_1).$$

Then

$$\gamma_\beta(g) \le c \min\{r^{d-1} r_1^\beta, \ r_1^{d-1} r^\beta\},$$

where c is a constant independent of r, r_1, P and Q_1.

Proof. Suppose $A(P,r) \cap A_1 \ne \emptyset$. If $r \le r_1$, then by Lemma 2.1

$$g(P) \le \sigma(A(P,r)) \le c r^{d-1} \chi_{A(Q_1, r+r_1)}(P)$$

and hence

$$\gamma_\beta(g) \le cr^{d-1}(r+r_1)^\beta \le 2^\beta cr^{d-1}r_1^\beta.$$

If $r_1 < r$, then

$$g(P) \le \sigma(A_1) \le cr_1^{d-1}\chi_{A(Q_1,r+r_1)}(P)$$

and $\gamma_\beta(g) \le 2^\beta cr_1^{d-1}r^\beta$. Therefore

$$\gamma_\beta(g) \le 2^\beta c \min(r^{d-1}r_1^\beta, r_1^{d-1}r^\beta).$$

This completes the proof. □

Furthermore we have

Lemma 2.4. *Let* $0 < \beta \le d-1$, $B_o = B(0,r)$ *with* $r \le R$, $f \in \mathcal{B}(\gamma_\beta)$ *and define*

$$g(P) = \int \chi_{B_o}(P-Q)f(Q)\,d\sigma(Q).$$

Then

$$\gamma_\beta(g) \le cr^{d-1}\,\gamma_\beta(f), \tag{2.1}$$

where c *is a constant independent of* r *and* f.

Proof. Suppose that

$$|f| \le \sum_j b_j\chi_{A_j},$$

where $b_j \in \mathbf{R}^+$ and A_j is a surface ball with radius $r_j \le R$. Setting

$$g_j(P) = \int \chi_{B_o}(P-Q)\chi_{A_j}(Q)\,d\sigma(Q),$$

we have

$$|g| \le \sum_j b_j g_j.$$

From Lemma 2.3 we deduce

$$\gamma_\beta(g) \le \sum_j \gamma_\beta(g_j) \le cr^{d-1}\sum_j b_j r_j^\beta,$$

which leads to (2.1). □

We now consider the following maximal function M_σ on ∂D with respect to the surface measure σ:

$$M_\sigma f(P) := \sup\{ r^{1-d} \int_{A(P,r)} |f|\, d\sigma : r > 0\}$$

for $f \in L^1(\partial D)$ and $P \in \partial D$.

Lemma 2.5. *Let $0 < \beta < d-1$ and A_o be the surface ball with center P_o and radius r. Then*

$$\gamma_\beta(M_\sigma \chi_{A_o}) \le c r^\beta, \tag{2.2}$$

where c is a constant, independent of P_o and r.

Proof. Putting $A_1 = A(P_o, 2r)$, we claim that

$$M_\sigma \chi_{A_o}(P) \le 2^{d-1} \sigma(A_o) |P - P_o|^{1-d} \tag{2.3}$$

for all $P \in \partial D \setminus A_1$. Indeed, since

$$d(P, A_o) = \inf\{ |P - Q| : Q \in A_o\} \ge |P - P_o| - r \ge \frac{1}{2}|P - P_o|,$$

we obtain

$$M_\sigma \chi_{A_o}(P) \le \sigma(A_o)\, d(P, A_o)^{1-d} \le 2^{d-1}\sigma(A_o)\, |P - P_o|^{1-d},$$

which shows (2.3).

Next, set

$$E_j = \{P \in \partial D \setminus A_1 : M_\sigma \chi_{A_o}(P) > \frac{c}{2^j}\}$$

with c in Lemma 2.1 for $j \ge 0$. Note that $E_o = \emptyset$. If $P \in E_j$, then by (2.3) and Lemma 2.1, we have

$$|P - P_o| \le 2^{1+j/(d-1)} r \tag{2.4}$$

for all $P \in E_j$. Denoting by r_j the right-hand side of (2.4), we obtain

$$\begin{aligned} M_\sigma \chi_{A_o} &\le \sum_{j=1}^\infty \frac{c}{2^{j-1}} \chi_{E_j \setminus E_{j-1}} + c\chi_{A_1} \\ &\le \sum_{j=1}^\infty \frac{c}{2^j} \chi_{E_j} + c\chi_{A_1} \\ &\le \sum_{j=1}^\infty \frac{c}{2^j} \chi_{A(P_o, r_j)} + c\chi_{A_1}. \end{aligned}$$

Therefore we have

$$\gamma_\beta(M_\sigma \chi_{A_o}) \leq c(\sum_{j=1}^{\infty} 2^{-j} r_j^\beta + 2^\beta r^\beta)$$
$$\leq 2^\beta c(\sum_{j=1}^{\infty} 2^{(-1+\beta/(d-1))j} + 1) r^\beta.$$

Noting that $1 - \frac{\beta}{d-1} > 0$, we obtain (2.2). □

Furthermore we have the following estimate for the maximal function.

Lemma 2.6. *(i) If $0 < \beta < d-1$ and $f \in \mathcal{L}(\gamma_\beta, C(\partial D))$, then*

$$\gamma_\beta(M_\sigma f) \leq c\gamma_\beta(f),$$

where c is a constant independent of f.

(ii) If $0 < \beta \leq d-1$, then there exists a constant c such that

$$\gamma_\beta(E_{f,b}) \leq \frac{c}{b}\gamma_\beta(f)$$

for every $f \in \mathcal{L}(\gamma_\beta, C(\partial D))$ and every positive real number b, where

$$E_{f,b} = \{P \in \partial D : \ M_\sigma f(P) > b\}.$$

Proof. From Lemma 2.5 we can easily deduce (i). If $\beta < d-1$, then it is also easy to see that (i) implies (ii). Next, suppose that $\beta = d-1$. Using a covering lemma (cf. Lemma, p.9 in [S]) and Lemma 2.1 we can show that there is a constant c such that

$$\frac{1}{c}\gamma_\beta(f) \leq \int |f| d\sigma \leq c\gamma_\beta(f)$$

for all $f \in \mathcal{L}(\gamma_\beta, C(\partial D))$. By the aid of [S, Theorem 1 in Chapter 1] we have (ii).□

Lemma 2.7. *Let $0 < \beta \leq d-1$, $0 < \lambda < d-1$ and $f \in \mathcal{B}(\gamma_\beta)$. Define*

$$h(P) = \int |P-Q|^{\lambda+1-d} |f(Q)| \, d\sigma(Q).$$

Then

$$h(P) \leq c_1 M_\sigma f(P) \tag{2.5}$$

and

$$\gamma_\beta(h) \leq c_2 \gamma_\beta(f),$$

where c_1 and c_2 are constants independent of f.

Proof. Put

$$g(X) = |X|^{\lambda+1-d}$$

and

$$E_n = \{X \in \mathbf{R}^d : \ g(X) > 2^n\}. \tag{2.6}$$

Then E_n is an open ball with center 0 and radius $r_n = 2^{n/(\lambda+1-d)}$, and

$$g \leq \sum_{n=k}^{\infty} 2^n \chi_{B(0,r_n)}$$

holds on $B(0,R)$ for some integer k depending on R. Hence we have

$$\begin{aligned} h(P) &\leq \sum_{n=k}^{\infty} 2^n \int \chi_{B(0,r_n)}(P-Q)|f(Q)|\, d\sigma(Q) = \sum_{n=k}^{\infty} 2^n \int_{A(P,r_n)} |f(Q)| d\sigma(Q) \\ &\leq (\sum_{n=k}^{\infty} 2^n r_n^{d-1}) M_\sigma f(P) = c_1 M_\sigma f(P), \end{aligned}$$

where $c_1 = \sum_{n=k}^{\infty} 2^n r_n^{d-1}$ is finite. Therefore we have the conclusion by Lemma 2.4. □

Remark. The inequality (2.5) will be used in the proof of Theorem 3.

3. The operator K^*

Let ω_d be the area of the unit sphere in $\mathbf{R}^d$. We define, for a function $f \in J(\partial D)$,

$$K^* f(P) = \frac{1}{\omega_d} \int \frac{\langle P-Q, N_P \rangle}{|P-Q|^d} f(Q)\, d\sigma(Q),$$

if it is well-defined, and set $K^* f(P) = 0$ otherwise.

Furthermore, denote by $L(\gamma_\beta, C(\partial D))$ the family of all equivalent classes relative to the relation between f and g in $\mathcal{L}(\gamma_\beta, C(\partial D))$, which is defined by $f = g$ γ_β-q.e. on ∂D. The space $L(\gamma_\beta, C(\partial D))$ is a Banach one with norm $\gamma_\beta(f)$.

The following theorem is fundamental.

Theorem 2. *Let $0 < \beta \le d-1$ and f be a Borel measurable function. Then*
(a) $\gamma_\beta(K^*f) \le c\gamma_\beta(f)$ for some constant c independent of f,
(b) K^* is a compact operator on $L(\gamma_\beta, C(\partial D))$.

Proof. (a): It suffices to assume that $f \in \mathcal{B}(\gamma_\beta)$. Moreover, noting that $f \in \mathcal{B}(\gamma_\beta)$ and $g \in C(\partial D)$ imply $fg \in \mathcal{B}(\gamma_\beta)$ and using a partition of unity, we may suppose that supp $f \subset B(Q_o, r_o)$ and $A(Q_o, r_o) = B(Q_o, r_o) \cap \partial D$ is expressed as

$$\{(x, \phi(x)) : x \in \mathbf{R}^{d-1}\} \cap B(Q_o, r_o)$$

with $\phi \in C_o^{1,\alpha}(\mathbf{R}^{d-1})$. Writing $P = (x, \phi(x))$ and $Q = (z, \phi(z))$, we have

$$\begin{aligned} |\langle P-Q, N_P\rangle| &= \frac{|\langle x-z, \nabla\phi(x)\rangle - (\phi(x)-\phi(z))|}{(1+|\nabla\phi(x)|^2)^{1/2}} \\ &\le c_1|x-z|^{\alpha+1} \le c_1|P-Q|^{\alpha+1} \end{aligned}$$

and hence

$$\frac{|\langle P-Q, N_P\rangle|}{\omega_d|P-Q|^d} \le c_2|P-Q|^{\alpha+1-d} \tag{3.1}$$

for all P, $Q \in \partial D$. Therefore, on account of Lemma 2.7 we obtain the estimate (a).

(b): Let $\{E_n\}$ be the same decreasing sequence of open balls with radius r_n as (2.6) for $\lambda = \alpha$ and choose a sequence $\{v_n\} \subset C(\mathbf{R}^d)$ such that $0 \le v_n \le 1$, $v_n = 1$ on E_n and supp $v_n \subset E_{n-1}$.

Let k be the integer obtained near the end of §2. Define, for $m \in \mathbf{N}$ and P, $Q \in \partial D$,

$$\begin{aligned} &h_m(P,Q) \\ &= \max\{\min\{\frac{\langle P-Q, N_P\rangle}{\omega_d|P-Q|^d}, c_2\sum_{n=k}^{m} 2^n v_n(P-Q)\}, -c_2\sum_{n=k}^{m} 2^n v_n(P-Q)\} \end{aligned}$$

with c_2 in (3.1) and

$$H_m f(P) = \int h_m(P,Q) f(O)\, d\sigma(Q)$$

for $f \in \mathcal{L}(\gamma_\beta, C(\partial D))$. To see that K^* maps $\mathcal{L}(\gamma_\beta, C(\partial D))$ into $\mathcal{L}(\gamma_\beta, C(\partial D))$, let f be a continuous function on ∂D. Since

$$|K^*f - H_m f| \le c_2 \sum_{n=m+1}^{\infty} 2^n \int \chi_{E_n}(P-Q)|f(Q)|d\sigma(Q),$$

we have, by Lemma 2.4,

$$\gamma_\beta(K^*f - H_m f) \le c_3 \sum_{n=m+1}^{\infty} 2^n r_n^{d-1} \gamma_\beta(f),$$

where r_n is the radius of E_n as in the proof of Lemma 2.7. This shows that $K^*f \in \mathcal{L}(\gamma_\beta, C(\partial D))$. Combining this with (a), we see that K^* maps $\mathcal{L}(\gamma_\beta, C(\partial D))$ into $\mathcal{L}(\gamma_\beta, C(\partial D))$.

Further, since $f = g$ γ_β-q.e. implies $K^*f = K^*g$, K^* is regarded as a bounded linear operator from $L(\gamma_\beta, C(\partial D))$ into $L(\gamma_\beta, C(\partial D))$.

Finally, to prove that K^* is a compact operator on $L(\gamma_\beta, C(\partial D))$, let $\{f_n\}$ be a sequence in $L(\gamma_\beta, C(\partial D))$ such that $\gamma_\beta(f_n) \le 1$. The sequence $\{H_m f_n\}_n$ is uniformly bounded. In fact if $\gamma_\beta(f) \le 1$, then we have, by Lemma 2.2,

$$|H_m f(P)| \le b_m \, \|f\|_1 \le b'_m \|f\|_p \le b''_m \, \gamma_\beta(f) \le b''_m,$$

where b_m, b'_m, b''_m are constants independent of f and $p = \frac{d-1}{\beta}$. Moreover $\{H_m f_n\}_n$ is equicontinuous on ∂D. By the diagonal method we can choose a subsequence $\{g_i\}$ of $\{f_n\}$ such that $\{H_m g_i\}_i$ is uniformly convergent for every $m \in \mathbf{N}$. Note that

$$\begin{aligned}
&|K^*g_i(P) - K^*g_j(P)| \\
&\le |K^*g_i(P) - H_m g_i(P)| + |H_m g_i(P) - H_m g_j(P)| + |H_m g_j(P) - K^*g_j(P)| \\
&\le c_2 \sum_{n=m+1}^{\infty} 2^n \int \chi_{E_n}(P-Q) g_i(Q)\, d\sigma(Q) + |H_m g_i(P) - H_m g_j(P)| \\
&\qquad + c_2 \sum_{n=m+1}^{\infty} 2^n \int \chi_{E_n}(P-Q) g_j(Q)\, d\sigma(Q) = u_1 + u_2 + u_3.
\end{aligned}$$

By the aid of Lemma 2.4 the first term u_1 satisfies

$$\gamma_\beta(u_1) \le c_3 \sum_{n=m+1}^{\infty} 2^n r_n^{d-1} \gamma_\beta(g_j) \le c_3 \sum_{n=m+1}^{\infty} 2^n r_n^{d-1}.$$

Similarly we have

$$\gamma_\beta(u_3) \le c_3 \sum_{n=m+1}^{\infty} 2^n r_n^{d-1}.$$

For u_2 we have

$$\gamma_\beta(u_2) \le \|H_m g_i - H_m g_j\|_\infty \, \gamma_\beta(1),$$

where $\| \ \|_\infty$ means the supnorm on $\partial D \times \partial D$. Noting that $\sum_{n=1}^{\infty} 2^n r_n^{d-1} < +\infty$, we see that $\{K^*g_j\}$ is a Cauchy sequence. Since $L(\gamma_\beta, C(\partial D))$ is a Banach space, $\{K^*g_j\}$ is convergent. Thus we see that K^* is compact. □

4. Single layer potentials

In this section we will study the boundary behavior of normal derivatives of single layer potentials. For this purpose we prepare a general Fatou type theorem.

Let U be an open subset of a topological space, γ be a countably sublinear functional on the class $J(\partial U)$ of all extended real-valued functions on ∂U and $\mathcal{B}(\gamma)$ be the class of all Borel measurable functions in $J(\partial U)$ such that $\gamma(f) < +\infty$. Further, let $\mathcal{H}$ be a subspace of real-valued functions in $\mathcal{B}(\gamma)$ and denote by $\mathcal{L}(\gamma, \mathcal{H})$ the family of all Borel measurable functions such that $\gamma(f - f_n) \to 0$ for some $\{f_n\} \subset \mathcal{H}$.

Suppose that to each $x \in U$, $z \in \partial U$ and $f \in \mathcal{L}(\gamma_\beta, C(\partial D))$ there corresponds a real number $\Phi_f(x, z)$ with the following properties:

(i) $f + g = h$ γ-q.e. on $\partial U \Rightarrow \Phi_f(x,z) + \Phi_g(x,z) = \Phi_h(x,z)$,

(ii) $\Phi_{bf}(x,z) = b\Phi_f(x,z)$ for $b \in \mathbf{R}$.

Furthermore, suppose to each $z \in \partial U$ there corresponds a filter $\mathcal{F}_z$ of subsets of U, converging to z.

Under these assumptions we can show the following theorem by the same method as in Theorem 1 in [W].

Theorem A. *Let T be a mapping from $\mathcal{L}(\gamma, \mathcal{H})$ to $\mathcal{B}(\gamma)$ satisfying the following properties:*

(i) $b \in \mathbf{R} \Rightarrow T(bf) = bTf$,

(ii) $f + g = h$ γ-q.e. $\Rightarrow Tf + Tg = Th$ γ-q.e.,

(iii) $\gamma(Tf) \leq c\gamma(f)$ for some $c > 0$, independent of f.

Furthermore, let $\mathcal{H}_1$ be a dense subset of $\mathcal{L}(\gamma, \mathcal{H})$ and suppose that for every $f \in \mathcal{H}_1$ there is a γ-polar set E such that

$$\mathcal{F}_z - \lim_{x \to z} \Phi_f(x, z) = Tf(z) \text{ at all } z \in \partial U \setminus E. \tag{4.1}$$

Moreover, putting

$$E_{f,b} = \{z \in \partial U : \mathcal{F}_z - \limsup_{x \to z} |\Phi_f(x,z)| > b\},$$

assume that for every $f \in \mathcal{L}(\gamma, \mathcal{H})$ and $b > 0$

$$\gamma(E_{f,b}) \leq \frac{c}{b}\gamma(f),$$

where c is a constant independent of f and b. Then for every $f \in \mathcal{L}(\gamma, \mathcal{H})$ there is a γ-polar set E such that (4.1) holds.

For a subset E of $\mathbf{R}^d$ a set function $M^{(\beta)}$ similar to the β-dimensional Hausdorff measure Λ_β is defined by

$$M^{(\beta)}(E) = \inf\{\sum_{j=1}^{\infty} r_j^\beta : \cup_{j=1}^{\infty} B(X_j, r_j) \supset E\}.$$

It is well-known that $M^{(\beta)}(E) = 0$ if and only if E is a set of β-dimensional Hausdorff measure zero.

The relation of $M^{(\beta)}$ and γ_β is as follows:

Lemma 4.1. *Let E be a Borel subset of ∂D. Then E is γ_β-polar if and only if $M^{(\beta)}(E) = 0$.*

Proof. First, we shall show that

$$\gamma_\beta(E) \leq 2^\beta M^{(\beta)}(E) \tag{4.2}$$

for every subset E of ∂D. Suppose that $E \subset \cup_j B(X_j, r_j)$. If $B(X_j, r_j) \cap E \neq \emptyset$, then we take $P_j \in B(X_j, r_j) \cap E$. Since $B(X_j, r_j) \cap \partial D \subset A(P_j, 2r_j)$, we have

$$E \subset \cup_k A(P_k, 2r_k),$$

where the index k runs over natural numbers j such that $A(P_j, r_j) \cap E \neq \emptyset$. Hence

$$\gamma_\beta(E) \leq \sum_k (2r_k)^\beta \leq 2^\beta \sum_j r_j^\beta.$$

From this and the definition of $M^{(\beta)}$ the inequality (4.2) follows.

If $M^{(\beta)}(E) = 0$, then $\gamma_\beta(E) = 0$ by (4.2). To prove the 'only if' part, suppose that $M^{(\beta)}(E) > 0$. Then there is a compact subset K of E such that $M^{(\beta)}(K) > 0$ (cf. [C, Theorem 3 in § II]). There is also a positive measure μ such that supp $\mu \subset K$, $\mu(B) \leq r^\beta$ for every ball B with radius r and

$$M^{(\beta)}(K) \leq c_1 \mu(K), \tag{4.3}$$

where c_1 is a constant, depending only on d and β (cf. [C, Theorem 1 in §II]). Furthermore we claim

$$\mu(K) \leq \gamma_\beta(K). \tag{4.4}$$

To see this, assume that $\chi_K \leq \sum_j b_j \chi_{A_j}$, where $A_j = A(P_j, r_j)$. Then

$$\mu(K) \leq \sum_j b_j \mu(A_j) \leq \sum_j b_j r_j^\beta,$$

which leads to (4.4). Combining this with (4.3), we have $\gamma_\beta(K) > 0$ and hence $\gamma_\beta(E) > 0$. □

Let us now study the properties of single layer potentials for functions in $\mathcal{L}(\gamma_\beta, C(\partial D))$. Recall that

$$\Gamma_\eta(P) = \{X \in D : \langle P - X, N_P\rangle > \eta|X - P|\}.$$

Similarly we define

$$\Gamma_\eta^e(P) = \{X \in \mathbf{R}^d \setminus D : \langle X - P, N_P\rangle > \eta|X - P|\}$$

and prove

Theorem 3. *Assume that* $0 < \beta \leq d-1$, $0 < \eta < 1$ and $f \in \mathcal{L}(\gamma_\beta, C(\partial D))$. *For* $X \in \mathbf{R}^d \setminus \partial D$, *define*

$$u_f(X) = -\frac{1}{\omega_d(d-2)} \int \frac{1}{|X - Q|^{d-2}} f(Q)\, d\sigma(Q).$$

Then u_f *has the following properties:*

(a) There exist positive real numbers c, δ, *independent of* f, *such that*

$$u_f^*(P) \leq cM_\sigma f(P),$$

where

$$u_f^*(P) = \sup\{|\langle \nabla u_f(X), N_P\rangle| : X \in \Gamma_\eta(P),\ |X - P| < \delta\}.$$

(b) There is a constant c *such that*

$$\gamma_\beta(E_{f,b}) \leq \frac{c}{b}\gamma_\beta(f) \text{ for every } b > 0,$$

where

$$E_{f,b} = \{P \in \partial D : u_f^*(P) > b\}.$$

(c) There exists a γ_β*-polar set* E *such that* $\langle \nabla u_f(X), N_P\rangle$ *converges to* $(K^* - \frac{1}{2}I)f(P)$ *for every* $P \in \partial D \setminus E$ *as* $X \to P$, $X \in \Gamma_\eta(P)$.

(d) There exists a γ_β*-polar set* E *such that* $\langle \nabla u_f(X), N_P\rangle$ *converges to* $(K^* + \frac{1}{2}I)f(P)$ *for every* $P \in \partial D \setminus E$ *as* $X \to P$, $X \in \Gamma_\eta^e(P)$.

Proof of Theorem 3. (a): By the same method as in the proof of Theorem 1.3 in [FJR] we can find c, δ such that

$$u_f^*(P) \leq c(M_\sigma f(P) + \int |P - Q|^{\alpha+1-d}|f(Q)|\, d\sigma(Q)).$$

On the other hand we see, by the aid of (2.5), that

$$\int |P - Q|^{\alpha+1-d}|f(Q)|\, d\sigma(Q) \leq c_1 M_\sigma f(P),$$

where c_1 is a constant independent of f and P. Thus we obtain the estimate (a).

(b): This is an easy consequence of (a) and Lemma 2.6.

(c): Let $f \in C^1(\partial D)$. Noting that

$$\begin{aligned}\langle \nabla u_f(X), N_P\rangle &= \frac{1}{\omega_d}\int \frac{\langle X-Q, N_P\rangle}{|X-Q|^d} f(Q)\, d\sigma(Q)\\ &= \frac{1}{\omega_d}\int \frac{\langle X-Q, N_P\rangle}{|X-Q|^d}(f(Q)-f(P))\, d\sigma(Q) - f(P)\\ &+ \frac{1}{\omega_d} f(P)\int \frac{\langle X-Q, N_P - N_Q\rangle}{|X-Q|^d}\, d\sigma(Q),\end{aligned}$$

we have

$$\begin{aligned}\lim_{X\to P, X\in \Gamma_\eta(P)} \langle \nabla u_f(X), N_P\rangle &= \frac{1}{\omega_d}\int \frac{\langle P-Q, N_P\rangle}{|P-Q|^d}(f(Q)-f(P))\, d\sigma(Q)\\ &- f(P) + \frac{f(P)}{\omega_d}\int \frac{\langle P-Q, N_P-N_Q\rangle}{|P-Q|^d}\, d\sigma(Q)\\ &= K^*(f-f(P))(P) - f(P) + K^*(1)f(P) + \frac{1}{2}f(P)\\ &= K^*f(P) - \frac{1}{2}f(P)\end{aligned}$$

for every $P \in \partial D$ and $f \in C^1(\partial D)$.

On the other hand we recall that $C^1(\partial D)$ is uniformly dense in $C(\partial D)$ and hence it is dense in $\mathcal{L}(\gamma_\beta, C(\partial D))$. On account of (b) we see that the functional $\Phi_f(X,P)$ defined by

$$\Phi_f(X,P) = \langle \nabla u_f(X), N_P\rangle$$

and the mapping T defined by

$$Tf(P) = K^*f(P) - \frac{1}{2}f(P)$$

satisfy the assumptions in Theorem A. Therefore we obtain the conclusion (c).

(d): For the approach region $\Gamma_\eta^e(P)$ we also obtain the estimates corresponding to (a) and (b). Therefore we can show (d) by the same method as in the proof of (c). □

5. Proof of Theorem 1

Let us prove Theorem 1. First, we note that the operator $K^* - \frac{1}{2}I$ is invertible on H defined by

$$H = \{f \in L(\gamma_\beta, C(\partial D)) : \int f\, d\sigma = 0\}.$$

In fact, suppose that $(K^* - \frac{1}{2}I)f = 0$ γ_β-q.e. and $\int f\, d\sigma = 0$ for $f \in \mathcal{L}(\gamma_\beta, C(\partial D))$. Then f belongs to $L^p(\sigma)$ for $p = (d-1)/\beta$ by Lemma 2.2 and $f = 0$ σ-a.e. by Theorem 2.5 in [FJR]. We note that the theorem is true for $p = 1$. Therefore we see that $K^* f(P) = 0$ for all $P \in \partial D$ and hence $f = 0$ γ_β-q.e., which shows that K^* is injective.

On the other hand the operator K^* is compact by Theorem 2 and H is closed in $L(\gamma_\beta, C(\partial D))$. Therefore the operator $K^* - \frac{1}{2}I$ on the Banach space H is invertible.

Now let f be a function in $\mathcal{L}(\gamma_\beta, C(\partial D))$ such that $\int f\, d\sigma = 0$. By the above consideration there exists $g \in \mathcal{L}(\gamma_\beta, C(\partial D))$ such that

$$(K^* - \frac{1}{2}I)g = f \ \gamma_\beta\text{-q.e. on } \partial D.$$

Define a harmonic function u_g in D by

$$u_g(X) = -\frac{1}{\omega_d(d-2)} \int \frac{1}{|X-Q|^{d-2}} g(Q)\, d\sigma(Q).$$

By the aid of Theorem 3 we see that there exists a γ_β-polar set E such that $\langle \nabla u_g(X), N_P \rangle$ converges to $(K^* - \frac{1}{2}I)g(P) = f(P)$ for every $P \in \partial D \setminus E$ as $X \to P$, $X \in \Gamma_\eta(P)$. Since we may assume that E is Borel measurable, E is a set of β-dimensional Hausdorff measure zero by Lemma 4.1. Thus we see that u_g is a desired function. □

6. The class $\mathcal{L}(\gamma_\beta, C(\partial D))$

In this section we study what functions belong to $\mathcal{L}(\gamma_\beta, C(\partial D))$. It seems that the degree of continuity of a function in $\mathcal{L}(\gamma_\beta, C(\partial D))$ is better than that of a general function in $L^p(\partial D)$. Indeed we can show that a Borel measurable function f belongs to $\mathcal{L}(\gamma_\beta, C(\partial D))$ if and only if $\gamma_\beta(f) < +\infty$ and f is γ_β-quasicontinuous. Recall that f is called γ_β-quasicontinuous if for any $\epsilon > 0$ there is a closed set F such that $\gamma_\beta(\partial D \setminus F) < \epsilon$ and the restriction of f to F is continuous.

Furthermore we have

Proposition 6.1. *Let β, λ be positive real numbers such that $\beta \le d-1$, $\lambda > d-1-\beta$. Then for every $f \in L^1(\partial D)$ the function g defined by*

$$g(P) = \int |P-Q|^{\lambda+1-d} f(Q)\, d\sigma(Q)$$

belongs to $\mathcal{L}(\gamma_\beta, C(\partial D))$.

Proof. Using a covering lemma (cf. Lemma, p.9 in [S]) we choose sequences $\{a_j\}$, $\{A_j\}$ of nonnegative real numbers and surface balls such that

$$|f| \leq \sum_j a_j \chi_{A_j} \quad \text{and} \quad \sum_j a_j \sigma(A_j) < +\infty,$$

respectively. Let $\{E_n\}$, $\{v_n\}$ and $\{H_m\}$ be the same sequences as in the proof of (*b*) of Theorem 2. Then $\{H_m f\}$ is continuous and we have

$$\begin{aligned}|g - H_m f| &\leq c_1 \sum_{n=m+1}^{\infty} 2^n \int \chi_{E_n}(P-Q)|f(Q)|\, d\sigma(Q) \\ &\leq c_1 \sum_{n=m+1}^{\infty} \sum_{j=1}^{\infty} 2^n a_j \int \chi_{E_n}(P-Q)\chi_{A_j}(Q)\, d\sigma(Q).\end{aligned}$$

This and Lemma 2.3 imply

$$\gamma_\beta(g - H_m) \leq c_2 \Big(\sum_{n=m+1}^{\infty} 2^n r_n^\beta\Big)\Big(\sum_{j=1}^{\infty} a_j \sigma(A_j)\Big)$$

for $r_n = 2^{-n/(d-1-\lambda)}$. Noting that $\sum_{n=1}^{\infty} 2^n r_n^\beta < +\infty$ by the assumption $\lambda > d-1-\beta$, we see that $\gamma_\beta(g - H_m f)$ as $m \to \infty$. Thus we have $g \in \mathcal{L}(\gamma_\beta, C(\partial D))$.□

References

[C] Carleson, L., Selected problems on exceptional sets. Van Nostrand, Princeton-New Jersey 1967

[DK] Dahlberg, B.E.J., C.E.Kenig, Hardy spaces and the Neumann problem in L^p for Laplace's equation in Lipschitz domains. Ann. Math. 125 (1987), 437–465

[FJR] Fabes, E.B., M.Jodeit JR., N.M.Rivière, Potential techniques for boundary value problems on C^1-domains. Acta Math. 141 (1978), 165–186

[JK] Jerison, D.S., C.E. Kenig, The Neumann problem on Lipschitz domains. Bull. Amer. Math. Soc. (N.S.) 4 (1981), 203–207

[S] Stein, E.M., Singular integrals and differentiability properties of functions. Princeton University Press, Princeton-New Jersey 1970

[W] Watanabe, H., Countably sublinear functionals and Hausdorff measures. Atti Sem. Mat. Fis. Univ. Modena (to appear)

On potential extension and capacity on harmonic spaces

Wu Jiong-Qi and Gao Qi-Ren

Abstract

This paper considers the potential extension property and the capacity defined by the balayage of hyperharmonic functions in a C.-C. harmonic space X without the restriction that X has a countable base and obtains some analogues of Boboc-Cornea theorem and Wallin-Ramasamy theorem.

1980 Mathematics Subject Classification (1985 Revision): 31D05.

1. Introduction and main results

Let X be a C.-C. harmonic space endowed with a hyperharmonic sheaf $\mathcal{U}$ (cf. [1]). For any numerical function f on $E \subset X$, we put $R_f^E := \inf\{u \in \mathcal{U}_+ : u|_E \geq f\}$, where $\mathcal{U}_+ := \{u \in \mathcal{U} : u \geq 0\}$. Let $\hat{g}$ denote the l.s.c. regularization of a function g on X. For a family $\mathcal{G}$ of numerical functions, we denote by $\widehat{\inf}\, \mathcal{G}$ the function $\hat{g}$ with $g = \inf \mathcal{G}$. The notation $\widehat{\lim}$ will be used in a similar manner.

Definition 1. A subset E of X is called an F-set (resp. O-set) if $R_f^E = f$ on E (resp. $R_f^E = 0$ on $X \setminus E$) for each strictly positive, continuous function f defined on E (denoted by $f \in spc(E)$). E is called an FO-set if E is both an F-set and an O-set.

Definition 2. Let Y be a locally compact Hausdorff space. A map $\Gamma : 2^Y \to \mathcal{U}_+$ is called a pre-capacity on Y if it satisfies the following
  (1) $A \subset B \subset Y \Rightarrow \Gamma(A) \leq \Gamma(B)$;
  (2) for any increasing sequence (A_n) of sets of Y we have

$$\lim_{n\to\infty} \Gamma(A_n) = \Gamma(\bigcup_{n=1}^{\infty} A_n) .$$

Our main results are the following

Theorem 1. *If X is a P-harmonic space, $K \subset X$ is a compact FO-set, then any $f \in spc(K)$ can be extended to a potential p on X such that p is continuous on K and harmonic outside an arbitrarily given neighborhood of K; moreover, if K is of type G_δ, then p can be continuous on X.*

Potential Theory, ed. by M. Kishi

Theorem 2. *If a pre-capacity $\Gamma : 2^Y \to \mathcal{U}_+$ is convex, i.e. it satisfies the following*
(3) *for any* $A, B \subset Y$, $\Gamma(A \bigcup B) + \Gamma(A \bigcap B) \leq \Gamma(A) + \Gamma(B)$,
then the map $\Gamma^* : 2^Y \to \mathcal{U}_+$

$$A \to \widehat{\inf}\{\Gamma(U) : A \subset U \subset Y, U \textit{ is open}\}$$

is a convex capacity on Y, *i.e.* Γ^* *satisfies* (1) – (3) *and*
(4) *for any decreasing sequence* (K_n) *of compact sets of* Y *we have*

$$\widehat{\lim_{n\to\infty}} \Gamma^*(K_n) = \Gamma^*(\bigcap_{n=1}^{\infty} K_n) .$$

Theorem 3. *Let* X *be a P-harmonic space. If* $u \in \mathcal{U}_+$ *is continuous, then the map* $r : E \to \hat{R}_u^E$ *is a convex capacity on* X *and* $r^*(K) = r(K)$ *for any compact set* $K \subset X$. *Consequently, for any* $\mathcal{K}$*-analytic set* $A \subset X$ *we have*

$$\hat{R}_u^A = \sup\{\hat{R}_u^K : K \textit{ compact} \subset A\} .$$

Our Theorem 3 and Theorem 1 generalize some similar results of N. Boboc, A. Cornea and Ramasamy Jesuraj by omitting the restriction that X has a countable base (cf. [1], [2], [3]).

2. Exceptional sets and potential extension

In this section, we suppose X is a P-harmonic space. Then a compact polar set is an O-set; a nowhere dense compact O-set is a polar set; for any $x \in X$, $\{x\}$ is an F-set and so an F-set is not necessarily polar. By Proposition 5.3.2 of [1], we can conclude that if for each $x \in X$, $\{x\}$ is of type G_δ, then any compact polar set is an FO-set.

Lemma 1. *If* K *is a compact F-set and* $f \in spc(K)$, *then for any neighborhood* U_0 *of* K *and any* $\varepsilon > 0$, *there is a continuous potential* p *on* X *such that* p *is harmonic on* $X \setminus U_0$ *and* $0 \leq p - f < \varepsilon$ *on* K.

Proof. Since X is a locally compact Hausdorff space, K has such a neighborhood base: $\mathcal{F} := \{U : K \subset U \subset \overline{U} \subset U_0, U \text{ is open}, \overline{U} \text{ is compact}\}$. By Tietze extension theorem, there is an $f_0 \in C(X)$ such that $f_0|_K = f$ and $f_0 \geq 0$. For any $U \in \mathcal{F}$, we set

$$U_* := \{\varphi : \varphi \in C(X),\ 0 \leq \varphi \leq f_0,\ \varphi|_K = f \text{ and } supp(\varphi) \subset U\} .$$

Obviously, $U_* \neq \emptyset$ and there exists an $U \in \mathcal{F}$ such that

$$R_f^K = w := \inf\{R\varphi : \varphi \in U_*\}, \text{ where } R\varphi := R_\varphi^X . \tag{1.1}$$

In fact, for any $u \in \mathcal{U}_+ : u|_K \geq f$, and any $\alpha \in (1, +\infty)$, by the semicontinuity of u and $-f_0$, $G := \{x \in X : \alpha u(x) > f_0(x)\}$ is open and $G \supset K$, so we have an $U \in \mathcal{F}$

such that $\overline{U} \subset G$. Now for any $\varphi \in U_*$, we have $\alpha u \geq \varphi$ and $\alpha u \geq R_\varphi$ and then $\alpha u \geq w$. As α and u are arbitrary, we obtain $R_f^K \geq w$. The opposite inequality is obviously valid and so we have (1.1).

Now for any $\varepsilon > 0$ and any $x \in K$, there is a $\varphi_x \in U_*$ such that

$$0 \leq R\varphi_x(x) - R_f^K(x) < \varepsilon ;$$

here $R\varphi_x \in C(X)$ is a potential on X. Since K is an F-set and $f \in spc(K)$, we have $R_f^K = f$ on K and so $D_x := \{y \in K : R\varphi_x(y) - f(y) < \varepsilon\}$ is a relatively open neighborhood of x on K. From the compactness of K we can get a finite subset I of K such that $\bigcup_{x\in I} D_x \supset K$. Put $g := \inf\{\varphi_x : x \in I\}$; then $g \in U_*$ and $p := Rg \in C(X)$ is a potential which is harmonic on $X \setminus U \supset X \setminus U_0$. Since $R_f^K \leq Rg \leq R\varphi_x, x \in I$, we have $0 \leq p - f < \varepsilon$ on K. □

Lemma 2. *If K is a compact FO-set, G is an open, relatively compact neighborhood of K, $f \in spc(K), f_0 \in C(X)$ is such a function that $f_0|_K = f, f_0|_{\overline{G}} > 0, f_0 \geq 0$ and $supp(f_0)$ is compact, then for any $\varepsilon > 0$ and any open set $D \supset K$, there is a potential $p \in C(X)$ satisfying the following conditions:* a) *$p \leq Rf_0$ on X;* b) *p is harmonic on $X \setminus \overline{(D \cap G)}$;* c) *$p < f_0$ on $\overline{G}$ and* d) *$p \geq f - \varepsilon$ on K.*

Proof. Assume $\varepsilon > 0$ is small enough so that $g := f - \varepsilon > 0$ on K. Set $\mathcal{F} := \{U : K \subset U \subset D \cap G,\ U \text{ is open},\ \overline{U} \text{ is compact}\}$, $U_1 := \{\varphi : \varphi \in C(X),\ \varphi|_K = g,\ 0 \leq \varphi \leq f_0,\ supp(\varphi) \subset U\}$ for $U \in \mathcal{F}$.

Since K is both an F-set and an O-set, by means of Lemma 1 and the compactness of K and $\overline{G}$, it is easy to complete the proof. □

Proof of Theorem 1. Let U be a neighborhood of K. Then there is a decreasing sequence $\{D_n\}$ of open relatively compact sets and a function $f_0 \in C(X)$ such that $U \supset \overline{D_n} \supset D_n \supset \overline{D_{n+1}} \supset D_{n+1} \supset K$ for $n = 0, 1, 2, \cdots$ and $f_0|_K = f,\ f_0 \geq 0,\ f_0|_{\overline{D_0}} > 0$ and $supp(f_0)$ is compact.

For any $\varepsilon > 0$, by Lemma 2, we can select inductively an increasing sequence $\{p_n\}$ such that for any $n = 1, 2, \cdots$, p_n is a potential $\in C(X)$ satisfying the conditions : a) $p_n \leq Rf_0$ on X ; b) p_1 is harmonic on $X \setminus D_1$, $p_{n+1} - p_n \geq 0$ is harmonic on $X \setminus D_{n+1}$; c) $p_n < f_0$ on $\overline{D_0}$ and d) $p_n > f - 2^{-n}\varepsilon$ on K. Set $p := \lim_{n\to\infty} p_n$. It is easy to verify that p is the required potential. □

Remark. Theorem 1 is considered in general P-harmonic spaces, but its main idea is essentially due to Ramasamy Jesuraj.

3. Capacity and balayage of hyperharmonic functions

Proof of Theorem 2. Let $\Gamma : 2^Y \to \mathcal{U}_+$ be a convex pre-capacity on Y and $\mathcal{B}$ be a base of X of relatively compact resolutive sets. It is easy to see $g := \inf\{\Gamma(U) : A \subset U \subset Y,\ U \text{ is open}\}$ is a $\mathcal{B}$-nearly hyperharmonic function on X and so $\Gamma^*(A) = \hat{g} \in \mathcal{U}_+$.

Obviously $\Gamma^*(A) \leq \Gamma^*(B)$ when $A \subset B \subset Y$. To verify that Γ^* satisfies the condition (2), let $(A_n)_{n\in N}$ ($N = \{1, 2, \cdots\}$) be an increasing sequence of subsets of Y, and $A_0 := \bigcup_{n\in N} A_n$. Then we have

$$\lim_{n\to\infty} (\inf \Gamma(U_n)) = \inf \Gamma(U_0) \ , \tag{2.1}$$

where U_n runs through the set of neighborhoods of A_n for each $n \in N \bigcup \{0\}$. In fact, for any $x \in X$, we have obviously

$$\lim_{n\to\infty} (\inf \Gamma(U_n))(x) \leq \inf \Gamma(U_0)(x) \ ,$$

where $\Gamma(U_n)(x) := (\Gamma(U_n))(x)$. To get opposite inequality, we may assume $\lim_{n\to\infty} (\inf \Gamma(U_n))(x) < \infty$. Since Γ is convex, for any $\varepsilon > 0$, we can select inductively a sequence $(G_n)_{n\in N}$ such that G_n is an open neighborhood of A_n and $G_n \subset G_{n+1}$ for each $n \in N$ and

$$\Gamma(G_n)(x) \leq \inf \Gamma(U_n)(x) + \sum_{i=1}^{n} 2^{-i}\varepsilon \ .$$

Put $G := \bigcup_{n\in N} G_n$. Then $G \supset A_0$ and

$$\inf \Gamma(U_0)(x) \leq \Gamma(G)(x) = \lim_{n\to\infty} \Gamma(G_n)(x) \leq \lim_{n\to\infty} (\inf \Gamma(U_n))(x) + \varepsilon \ .$$

Since x and ε are arbitrary, we obtain (2.1).

By Axiom of upper directed sets [1], we have $\lim_{n\to\infty} (\widehat{\inf} \Gamma(U_n)) \in \mathcal{U}_+$. Because $\inf \Gamma(U_n)$ is $\mathcal{B}$-nearly hyperharmonic and X endowed with the fine topology is a Baire space, (2.1) implies that

$$\lim_{n\to\infty} (\widehat{\inf} \Gamma(U_n)) = \widehat{\inf} \Gamma(U_0) \ ,$$

i.e. $\lim_{n\to\infty} \Gamma^*(A_n) = \Gamma^*(A_0)$.

Now we will show Γ^* satisfies the condition (4) and so Γ^* is a capacity on Y. Consider a decreasing sequence $(K_n)_{n\in N}$, of compact subsets of Y and put $K := \bigcap_{n\in N} K_n$. As $\lim_{n\to\infty} \Gamma^*(K_n) \geq \Gamma^*(K)$, we have $\widehat{\lim}_{n\to\infty} \Gamma^*(K_n) \geq \Gamma^*(K)$. On the other hand, for any open neighborhood U of K, we have $U \supset K_m$ for some $m \in N$, so $\lim_{n\to\infty} \Gamma^*(K_n) \leq \Gamma(U)$. Hence $\lim_{n\to\infty} \Gamma^*(K_n) \leq \inf\{\Gamma(U) : K \subset U,\ U \text{ is open}\}$, which implies $\widehat{\lim}_{n\to\infty} \Gamma^*(K_n) \leq \Gamma^*(K)$ and (4) is valid.

Finally, the convexity of Γ^* follows from the convexity of Γ and the definition of Γ^* immediately. The proof is complete. □

Proof of Theorem 3. Since X is a P-harmonic space and $u \in \mathcal{U}_+$ is continuous, for any compact subset K of X we have

$$\hat{R}_u^K = \bigwedge \hat{R}_u^U := \widehat{\inf}\{\hat{R}_u^U : K \subset U \subset X,\ U \text{ is open}\}. \tag{2.2}$$

To verify it, at first we suppose $u(x) > 0$, $x \in K$. For any $v \in \mathcal{U}_+$ satisfying $v(x) \geq u(x), x \in K$ and any $\alpha \in (1, \infty)$, it is easy to see that the set $W := \{x \in X : u(x) < \alpha v(x)\}$ is open and $K \subset W$, so we have

$$\bigwedge \hat{R}_u^U \leq \hat{R}_u^W \leq \alpha v \, .$$

Since v and α are arbitrary, we have $\bigwedge \hat{R}_u^U \leq R_u^K$. Thus, $\bigwedge \hat{R}_u^U \leq \hat{R}_u^K$. The opposite inequality is obviously valid and so we get (2.2).

Then we consider the general case that $u \geq 0$. Since X is a P-harmonic space and K is compact, there is a continuous potential p on X such that $p(x) > 0$, $x \in K$. The above result implies

$$\bigwedge \hat{R}_{u+p}^U = \hat{R}_{u+p}^K \, .$$

By a result of Brelot [1], we have $\bigwedge \hat{R}_p^U = \hat{R}_p^K$. Consequently,

$$\hat{R}_u^K + \hat{R}_p^K = \hat{R}_{u+p}^K = \bigwedge \hat{R}_{u+p}^U \geq \bigwedge \hat{R}_u^U + \bigwedge \hat{R}_p^U = \bigwedge \hat{R}_u^U + \hat{R}_p^K,$$

which implies $\hat{R}_u^K \geq \bigwedge \hat{R}_u^U$ and so (2.2) is valid.

Now by means of (2.2) and a method similar to the proof of Theorem 2, it is easy to show that the balayage $r : E \to \hat{R}_u^E$ is a capacity on X. (2.2) means $r^*(K) = r(K)$ when K is compact. As any $\mathcal{K}$-analytic set A is capacitable, we have

$$\hat{R}_u^A = \sup\{\hat{R}_u^K : K \text{ compact} \subset A\} \, . \qquad \square$$

References

[1] Constantinescu, C., Cornea, A., Potential theory on harmonic spaces, Springer-Verlag, Berlin, 1972.

[2] Boboc, N., Cornea, A., Nonnegative hyperharmonic functions. Balayage and natural order, Rev. Roumaine Math. Pures Appl. **13** (1968), 609-618.

[3] Ramasamy Jesuraj, Continuous functions on polar sets, Proc. Amer. Math. Soc. **93**:2 (1985), 262-266.

Singularities of the solution of integral equations of potential theory arising in problems of elasticity for non-homogeneous media

Stepan S. Zargaryan

Abstract

This article examines the integral equations of the theory of the logarithmic potential on a closed piecewise-smooth boundary of the cross-sectional region and the joining line of two different media. Asymptotic representations are obtained for the solutions of the integral equations in the region of the corner points formed by an intersection of external boundary and joining lines of dissimilarities, and formulas are obtained for the coefficients of these representations.

As in [1], information on the solutions of the integral equations of the potential theory is derived from the well-known results of the asymptotic form of the solutions of the internal and external Dirichlet and Neumann problems. It is shown that, irrespective of the values of the angles in the region of the corner points formed by an intersection of the external boundary and joining lines of dissimilarities, the solution of the integral equation of the problem has a singularity, whereas the solution of the boundary value problem may have no singularities for some angles. These integral equations may occur in a torsion problem of a cylindrical rod composed of different materials, in a problem of an antiplane shear of infinite non-homogeneous cylinder with stresses or displacements prescribed at the boundary, i.e., the problems where corner points occur either on the boundary line or on the joining line with no intersection were considered in previous publications [2,3].

1980 Mathematics Subject Classification (1985 Revision): 53C30, 22E15.

1. Introduction

We consider the following problem for harmonic functions

$$\Delta u^{(j)} = 0 \quad \text{in} \quad \Omega^{(j)}, \qquad j = 0, 1, 2, .., m, \tag{1.1}$$

$$u^{(0)} = u^{(j)} \quad \text{on} \quad \gamma_j \setminus K, \tag{1.2}$$

$$G_0 \frac{\partial u^{(0)}}{\partial n} - G_j \frac{\partial u^{(j)}}{\partial n} = f_j \quad \text{on} \quad \gamma_j \setminus K, \tag{1.3}$$

$$\frac{\partial u^{(0)}}{\partial n} = g_0 \quad \text{on} \quad \partial\Omega^{(0)} \setminus \cup_{j=1}^{m} \gamma_j, \tag{1.4}$$

$$\frac{\partial u^{(j)}}{\partial n} = g_j \quad \text{on} \quad \partial\Omega^{(j)} \setminus \gamma_j. \tag{1.5}$$

Potential Theory, ed. by M. Kishi

Here, we suppose $\Omega = \cup_{j=0}^{m}\Omega^{(j)}$, $\Omega^{(j)}$ being a bounded domain occupied by an elastic homogeneous material with piecewise-smooth contour $\partial\Omega^{(j)}$ which join to the main domain $\Omega^{(0)}$ occupied by another elastic homogeneous material and by a separate part γ_j of its boundary. In particular cases, γ_j may coincide with $\partial\Omega^{(j)}$ (see Fig.1); G_0 and G_j are constant values (shear modulas), specified in the domains $\Omega^{(0)}$ and $\Omega^{(j)}$ respectively; K is a union of all angular points of the boundary $\partial\Omega = \cup_{j=0}^{m}\partial\Omega^{(j)}$.

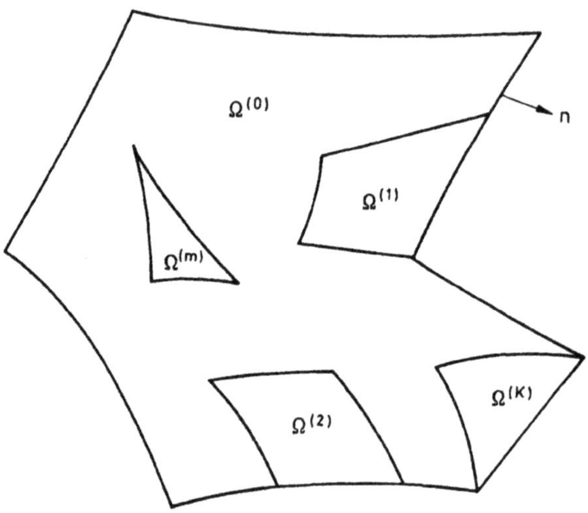

Fig. 1

The solution of the problem (1.1) – (1.5) in Ω can be expressed in the form of the potential of a single layer

$$u(x) = \int_{\partial\Omega} \nu(y)\log\frac{1}{r}d_y s = \int_{\partial\Omega^{(0)}\setminus\cup_{j=1}^{m}\gamma_j} \nu(y)\log\frac{1}{r}d_y s + \sum_{j=1}^{m}\int_{\partial\Omega^{(j)}} \nu(y)\log\frac{1}{r}d_y s \quad (1.6)$$

where $r = |x-y|, u(x) = u^{(j)}(x)$ in $\Omega^{(j)}$, and $\nu(y)$ is the layer's density distribution.

We take into account the continuity of the potential (1.6) on a joining line γ_j, and in the case when $x \in \partial\Omega$

$$\Big[\frac{\partial u}{\partial n_x}\Big]_0 = \pi\nu + \int_{\partial\Omega}\nu(y)\frac{\partial}{\partial n_x}\log\frac{1}{r}d_y s, \quad (1.7)$$

$$\Big[\frac{\partial u}{\partial n_x}\Big]_j = -\pi\nu + \int_{\partial\Omega}\nu(y)\frac{\partial}{\partial n_x}\log\frac{1}{r}d_y s, \quad (1.8)$$

substituting (1.7) and (1.8) into (1.3) – (1.5), we obtain

$$\nu(x) + \int_{\partial\Omega} K(x,y)\nu(y)d_y s = f(x), \quad x \in \partial\Omega \setminus K. \quad (1.9)$$

Here, n is the normal vector exterior to $\Omega^{(0)}$ and interior to $\Omega^{(j)}$ on γ_j, also it is exterior to $\Omega^{(0)}$ on $\partial\Omega^{(0)} \setminus \cup_{j=1}^{m}\gamma_j$ and $\partial\Omega^{(j)} \setminus \gamma_j$. Moreover,

$$K(x,y) = \frac{G_0 - G_j}{\pi(G_0 + G_j)} \frac{\partial}{\partial n_x} \log\frac{1}{r}, \quad y \in \gamma_j,$$

$$K(x,y) = \pi^{-1}\frac{\partial}{\partial n_x}\log\frac{1}{r}, \quad y \in \partial\Omega \setminus \cup_{j=1}^{m}\gamma_j,$$

$$f(x) = \begin{cases} [\pi(G_0 + G_j)]^{-1} f_j(x) & \text{on} \quad x \in \gamma_j, \\ \pi^{-1} g_0 & \text{on} \quad \partial\Omega^{(0)} \setminus \cup_{j=1}^{m}\gamma_j, \\ \pi^{-1} g_j & \text{on} \quad \partial\Omega^{(j)} \setminus \gamma_j. \end{cases}$$

The integral equation (1.9) is not Fredholm's equation because of the singularities of the kernel at the corner points of the contour.

The aim of this article is to determine the asymptotic forms of the solutions of the integral equation (1.9) in the neighborhood of the corner points, formed by the intersection of the external boundary and joining lines of two different materials. These asymptotic forms are useful in obtaining the numerical solutions of integral equations.

Traditionally, singularities of the solution of integral equations were determined by means of direct study of the properties of integral operators [4]. However, relying upon the results of earlier research [1], solutions of the auxiliary boundary value problems became effective in the representation of the solution of integral equations. Meantime, the solvability of integral equations follows from the solvability of the corresponding boundary value problem for interior and exterior regions.

The asymptotic form of the solution of the integral equations (AFSIE) (1.9) can be determined by the formula

$$\nu = \begin{cases} \frac{1}{2\pi}\Big(\frac{\partial u^{(0)}}{\partial n} - \frac{\partial u^{(j)}}{\partial n}\Big) & \text{on} \quad \gamma_j \setminus K, \\ \frac{1}{2\pi}\Big(\frac{\partial u^{(j)}}{\partial n} - \frac{\partial u_{m+1}}{\partial n}\Big) & \text{on} \quad \partial\Omega^{(j)} \setminus \gamma_j, \\ \frac{1}{2\pi}\Big(\frac{\partial u^{(0)}}{\partial n} - \frac{\partial u_{m+1}}{\partial n}\Big) & \text{on} \quad \partial\Omega^{(0)} \setminus \cup_{j=1}^{m}\gamma_j, \end{cases} \tag{1.10}$$

where u_{m+1} is a harmonic function, determined in free space outside of Ω.

2. Asymptotic form of the solution of the integral equation

We consider the cases of non-homogeneous bodies when the joining line of two different elastic homogeneous materials intersect at the boundary of the cross-sectional region of the rod in the angular point p_{2j-1}. For simplicity, suppose that cross-section of the composite beam coincides with the sector $\{x = x_1 + ix_2 = \rho e^{i\omega};\ 0 < \rho < \delta,\ -\beta < \omega < \alpha\}$. In fact, one component, with an elastic shear modulus G_0, occupies the sector $\{0 < \rho < \delta;\ -\beta < \omega < 0\}$ and another component, with the shear modulus G_j, occupies the sector $\{0 < \rho < \delta;\ 0 < \omega < \alpha\}$.

According to [5], the asymptotic form of a solution of the problem (1.1) – (1.5) may be introduced by

$$\begin{aligned} u^{(0)} &\sim \rho(A\cos\lambda\omega + B\sin\lambda\omega), \quad -\beta \le \omega \le 0 \\ u^{(j)} &\sim \rho(C\cos\lambda\omega + D\sin\lambda\omega), \quad 0 \le \omega \le \alpha, \end{aligned} \tag{2.1}$$

where $\alpha + \beta < 2\pi$. The boundary conditions on $\partial\Omega^{(0)} \setminus \cup_{j=1}^{m}\gamma_j$ $(\omega = -\beta)$, on γ_j $(\omega = 0)$, and on $\partial\Omega^{(j)} \setminus \gamma_j$ $(\omega = \alpha)$ yield

$$\begin{aligned} A\sin\lambda\beta + B\cos\lambda\beta &= 0 \\ \mu_j A\sin\lambda\alpha - B\cos\lambda\alpha &= 0 \\ A = C, \quad B &= \mu_j D \end{aligned} \tag{2.2}$$

where $\mu_j = G_j/G_0$ is a ratio of the shear modulas of different materials. The existence of a nontrivial solution of the system (2.2) is possible if λ is a root of the equation

$$\mu_j \tan\lambda\alpha + \tan\lambda\beta = 0. \tag{2.3}$$

In [6] it was shown that roots of the above equation are real numbers, and the smallest positive root depends on μ_j, α and β in the ranges as shown in Fig.2.

Bearing in mind the asymptotic forms (2.1), according to (1.10) we find that on γ_j $(\omega = 0)$ as $\rho \to 0$

$$\nu \sim \frac{\lambda}{2\pi}(\mu_j - 1)D\rho^{\lambda-1},$$

when $\lambda \in \mathrm{SC}$ $(\lambda < 1)$. In the case of stresses in a dissimilar body one can notice that the domain with stress concentration (SC), when $0 < \lambda < 1$, demarcates from a domain of bounded stresses (BS), when $\lambda \ge 1$, by the line $\mu_j = -\tan\beta/\tan\alpha$. This line is obtained from (2.3) when $\lambda = 1$.

If $\lambda \in \mathrm{SC}$

$$\begin{aligned} u^{(0)}(x) &= u^{(0)}(0) - \mu_j D\rho^{\lambda}\frac{\cos\lambda(\beta+\omega)}{\sin\lambda\beta} + O(\rho), \quad -\beta \le \omega \le 0, \\ u^{(j)}(x) &= u^{(0)}(0) + D\rho^{\lambda}\frac{\cos\lambda(\alpha-\omega)}{\sin\lambda\alpha} + O(\rho), \quad 0 \le \omega \le \alpha. \end{aligned} \tag{2.4}$$

In case of $\lambda \in \mathrm{BS}$

$$u^{(0)}(x) = u^{(0)}(0) + O(\rho), \qquad u^{(j)}(x) = u^{(0)}(0) + O(\rho). \tag{2.5}$$

The state of BS does not mean the absence of singularities at the corner points of the solution of integral equation (1.9).

To determine the asymptotic form of the solution of the integral equation (AFSIE) in a neighbourhood of the point p_{2j-1}, first we construct a solution of the Dirichlet problem in free space Ω_{m+1} outside of Ω

$$\Delta u^{m+1} = 0 \text{ in } \Omega_{m+1}, \quad u^{(m+1)} = u^{(0)} \text{ on } \partial\Omega^{(0)} \setminus \cup_{j=1}^{m}\gamma_j, \\ u^{(m+1)} = u^{(j)} \text{ on } \partial\Omega^{(j)} \setminus \gamma_j, \tag{2.6}$$

where $u^{(0)}$ and $u^{(j)}$ have an asymptotic form (2.4).

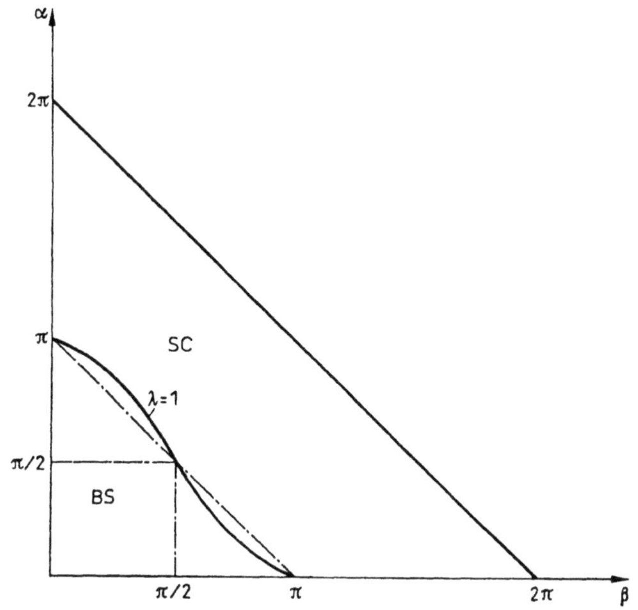

Fig. 2

The asymptotic form of the solution of the problem (2.6) as $\rho \to 0$ is

$$u^{(m+1)} = u^{(0)} + B\rho^{\lambda_0} \frac{\sin \rho_0(\omega - \alpha)}{\cos \lambda_0 \alpha} + \\ + D\rho^{\lambda}(A_1 \cos \lambda\omega + A_2 \sin \lambda\omega) + O(\rho), \ \alpha \le \omega \le 2\pi - \beta$$

where

$$A_1 = \frac{1}{\sin \lambda(2\pi - \alpha - \beta)}\left(\frac{\sin \lambda(2\pi - \beta)}{\sin \lambda\alpha} + \frac{\sin \lambda\alpha}{\sin \lambda\beta}\right),$$

$$A_2 = -\frac{1}{\sin \lambda(2\pi - \alpha - \beta)}\left(\mu_j \frac{\cos \lambda\alpha}{\sin \lambda\beta} + \frac{\cos \lambda(2\pi - \beta)}{\sin \lambda\alpha}\right),$$

$\lambda_0 = \pi/(2\pi - \alpha - \beta)$, and $\lambda \in$ SC (Fig.2) is the root of equation (2.3).

If $\lambda_0 < \lambda < 1$ and $\rho \to 0$, according to (1.10) AFSIE in this case is given by

$$\nu \sim \pm \frac{\lambda_0}{2\pi \cos \lambda_0 \alpha} B \rho^{\lambda_0 - 1} \tag{2.7}$$

where the plus sign corresponds to the ray $\omega = \alpha$, and the minus sign corresponds to the ray $\omega = 2\pi - \beta$. This case is possible only if $\alpha + \beta < \pi$. For instance, if $\alpha = 0.2\pi$ and $\beta = 0.6\pi$, then $\lambda = 0.9$ if $\mu_j = 12.57$ and $\lambda_0 = 0.8(3)$.

In the case $\lambda < \lambda_0 < 1$ and $\lambda \in$ SC, as $\rho \to 0$ we have on $\partial\Omega^{(0)} \setminus \cup_{j=1}^m \gamma_j$

$$\nu \sim \frac{\lambda}{2\pi} D \rho^{\lambda - 1}(-A_1 \sin \lambda(2\pi - \beta) + A_2 \cos \lambda(2\pi - \beta))$$

and on $\partial\Omega^{(j)} \setminus \gamma_j$

$$\nu \sim \frac{\lambda}{2\pi}(-A_1 \sin \lambda\alpha + A_2 \cos \lambda\alpha). \tag{2.8}$$

This case is tied with the stress concentration even when $\alpha + \beta < \pi$ (which is not possible in a homogeneous media). The asymptotic form (2.8) holds in cases $\pi < \alpha + \beta < \frac{3}{2}\pi$ if $\lambda \in$ SC and when $\frac{3}{2}\pi \le \alpha + \beta < 2\pi$ with $\lambda < 1$ and arbitrary μ_j.

Next we consider the case when $0 < \alpha \le \pi/2$ and $0 < \beta < \pi/2$. It is easy to see from Fig.2 that $\lambda > 1$ for arbitrary μ_j. The low-stressed state in a neighbourhood of an angular point p_j described by asymptotic form (2.5) does not mean the absence of singularities of the integral equation (1.9). From the solution of Dirichlet problem (2.6) we have

$$u^{(m+1)} = u^{(0)}(0) + B\rho^{\lambda_0} \frac{\sin \lambda_0(\omega - \alpha)}{\cos \lambda_0 \alpha} + O(\rho). \tag{2.9}$$

Due to this equality and (1.10) AFSIE may be written in the form (2.7).

The smallest positive root of the equation (2.3) is larger than 1 depending on μ_j even though $\pi < \alpha + \beta < \frac{3}{2}\pi$. In this case $\lambda \in$ BS and the system of the integral equations (1.9) has no singularities because the asymptotics of the BVP (1.1)–(1.5) and the Dirichlet problem (2.6) is given by (2.5) and (2.9) with $\lambda_0 > 1$. That is why according to (1.10) when $\rho \to 0$ on γ_j, $\partial\Omega^{(j)} \setminus \gamma_j$ and $\partial\Omega^{(0)} \setminus \cup_{j=1}^m \gamma_j$

$$\nu = O(\rho).$$

The coefficient D in the asymptotic form (2.4) may be determined by the formula

$$D = \frac{G_0}{G_0 K_0 - G_j K_j} \int_{\partial\Omega^{(0)} \setminus \cup_{j=1}^m \gamma_j} \zeta^{(0)} g_0 d_y s -$$

$$- \frac{1}{G_0 K_0 - G_j K_j} \sum_{k=1}^m \{G_k \int_{\partial\Omega^{(k)} \setminus \gamma_k} \zeta^{(k)} g_k d_y s - \int_{\gamma_k} \zeta^{(k)} f_k d_y s\}, \tag{2.10}$$

where

$$K_0 = \frac{\lambda\beta\mu_j^2}{\sin^2 \lambda\beta} + \frac{\mu_j^2 \sin 4\lambda\beta}{4 \sin^2 \lambda\beta} + \mu_j^2 \sin 2\lambda\beta,$$

$$K_j = -\alpha\lambda(1 + \mu_j^2 \cot^2 \lambda\beta) - \frac{1}{2}\mu_j^2 \cot^2 \lambda\beta \sin 2\lambda\alpha + 2\mu_j \cot \lambda\beta \sin^2 \lambda\alpha + \frac{1}{2} \sin 2\lambda\alpha.$$

This formula is obtained by the method described in [7]. The contour integration in (2.10) is through the path $\partial\Omega^{(0)}$ which leaves the domain $\partial\Omega^{(j)}$ to the left and the path $\cup_{j=1}^{m}\partial\Omega^{(0)} \setminus \gamma_j$ which leaves the domain $\partial\Omega^{(j)}$ to the right. The functions $\zeta^{(j)}$ are solutions of the conjugate to (1.1) – (1.5) boundary value problem (for $\alpha + \beta < 2\pi$)

$$\Delta\zeta^{(j)} = 0 \text{ in } \Omega^{(j)}, \quad j = 0, 1, 2, .., m,$$

$$\zeta^{(0)} = \zeta^{(j)} \text{ and } G_0 \frac{\partial\zeta^{(0)}}{\partial n} - G_j \frac{\partial\zeta^{(j)}}{\partial n} = 0 \text{ on } \gamma_j \setminus K, \tag{2.11}$$

$$\frac{\partial\zeta^{(0)}}{\partial n} = 0 \text{ on } \partial\Omega^{(0)} \setminus \cup_{j=1}^{m}\gamma_j, \quad \frac{\partial\zeta^{(j)}}{\partial n} = 0 \text{ on } \partial\Omega^{(j)} \setminus \gamma_j.$$

This has an asymptotic form at a neighbourhood of the point p_{2j-1} presented by

$$\zeta^{(0)} \sim \rho^{-\lambda}\mu_j \frac{\cos\lambda(\beta + \omega)}{\sin\lambda\beta}, \quad -\beta \leq \omega \leq 0,$$

$$\zeta^{(j)} \sim \rho^{-\lambda}(\mu_j \cot\lambda\beta \cos\lambda\omega - \sin\lambda\omega), \quad 0 \leq \omega \leq \alpha.$$

However, it is not possible, with this above procedure, to extend the solutions in Ω to Ω_{m+1}. Indeed, according to [1] the coefficient B in the asymptotics of the Dirichlet problem is given by the formula

$$B = -\frac{1}{\pi} \int_{\partial\Omega^{(m+1)}} [f(x) - f(0)] \frac{\partial\zeta^{(m+1)}}{\partial n} ds \tag{2.12}$$

where

$$f = \begin{cases} u^{(0)} & \text{on } \partial\Omega^{(0)} \setminus \cup_{j=1}^{m}\gamma_j, \\ u^{(j)} & \text{on } \partial\Omega^{(j)} \setminus \gamma_j, \quad j = 1, 2, .., m \end{cases}$$

and $\zeta^{(m+1)}$ is a solution of the problem

$$\Delta\zeta^{(m+1)} = 0 \text{ in } \Omega^{(m+1)}, \quad \zeta^{(m+1)} = 0 \text{ on } \partial\Omega^{(m+1)} \setminus K,$$

with the asymptotics

$$\zeta^{(m+1)} \sim -\rho^{-\lambda_0} \frac{\sin\lambda_0(\omega - \alpha)}{\cos\lambda_0\alpha}$$

The boundary values of the problem (1.1)–(1.5) will be definite only after the solution of the above problem. Thus, in (2.12) the functions under the integral are not expressed through boundary values of the problem (1.1)–(1.5). In order to be consistent let us construct solution of the problem

$$\Delta z^{(j)} = 0 \text{ in } \Omega^{(j)}, \quad j = 0, 1, 2, .., m,$$

$$z^{(0)} = z^{(j)} \quad \text{and} \quad G_0 \frac{\partial z^{(0)}}{\partial n} - G_j \frac{\partial z^{(j)}}{\partial n} = f_j \quad \text{on} \quad \gamma_j \setminus K, \tag{2.13}$$
$$\frac{\partial z^{(0)}}{\partial n} = \frac{\zeta^{(m+1)}}{\partial n} \quad \text{on} \quad \Omega^{(0)} \setminus \cup_{j=1}^{m} \gamma_j, \qquad \frac{\partial z^{(j)}}{\partial n} = \frac{\zeta^{(m+1)}}{\partial n} \quad \text{on} \quad \Omega^{(j)} \setminus \gamma_j,$$

with the asymptotics at the point p_{2j-1}

$$z^{(0)} \sim \rho^{-\lambda_0}(a \cos \lambda_0 \omega + b \sin \lambda_0 \omega),$$
$$z^{(j)} \sim \rho^{-\lambda_0}(c \cos \lambda_0 \omega + d \sin \lambda_0 \omega),$$

where

$$\lambda_0 = \pi/(2 - \alpha - \beta), \qquad a = c = \Delta^{-1}(\cos \lambda_0 \alpha + \mu_j \cos \lambda_0 \beta), \qquad b = \mu_j d,$$

$$d = \Delta^{-1}(\sin \lambda_0 \alpha + \sin \lambda_0 \beta), \qquad \Delta = \cos^2 \lambda_0 \alpha \cos^2 \lambda_0 \beta (\tan \lambda_0 \beta + \mu_j \tan \lambda_0 \alpha).$$

The functions $z^{(0)}, z^{(j)}$ and $\zeta^{(m+1)}$ have non zero flow through the circle $|x - p_j| = \varepsilon$, $\varepsilon = \text{const}$. It is easy to verify that

$$\int_{-\beta}^{0} \frac{\partial z^{(0)}}{\partial n} ds + \int_{0}^{\alpha} \frac{\partial z^{(j)}}{\partial n} ds + \int_{\alpha}^{2\pi-\beta} \frac{\partial \zeta^{(m+1)}}{\partial n} ds = -\varepsilon^{-\lambda_0} d(\mu_j - 1),$$

which is equal to zero only if $\mu_j = 1$. This is in the case of a homogeneous body. Consequently, it is not possible to realize the harmonic extension mentioned above.

3. Determination of AFSIE on external boundary

In order to construct AFSIE (1.9) on $\partial\Omega^{(0)} \setminus \cup_{j=1}^{m} \gamma_j$ and $\partial\Omega^{(j)} \setminus \gamma_j$ in case of $0 < \alpha \le \pi/2$ and $0 < \beta \le \pi/2$ when $\rho \to 0$ we try a different approach.

To use the equation (1.10) in determining AFSIE we need to extend definitions of the functions $u^{(j)}, j = 0, 1, .., m$, outside of $\Omega = \cup_{j=1}^{m} \Omega_j$ into domains $\Omega_{m+1}^{(j,j+1)}$ and $\Omega_{m+1}^{(j)}$ in order to construct harmonic continuation. Instead of the problem (2.6) let us consider a new BVP

$$\Delta w^{(j)} = 0 \quad \text{in} \quad \Omega_{m+1}^{(j)}, \qquad j = 1, 2, .., m,$$
$$\Delta w^{(j-1,j)} = 0 \quad \text{in} \quad \Omega_{m+1}^{(j-1,j)}, \qquad j = 1, 2, .., m+1,$$
$$w^{(j-1,j)} = u^{(0)} \quad \text{on} \quad \ell_{2j-2,2j-1}, \qquad j = 1, 2, .., m+1,$$
$$w^{(j)} = u^{(j)} \quad \text{on} \quad \partial\Omega^{(j)} \setminus \gamma_j, \qquad j = 1, 2, .., m, \tag{3.1}$$
$$w^{(j-1,j)} = w^{(j)} \quad \text{on} \quad L_{2j-1}, \qquad w^{(j-1,j)} = w^{(j-1)} \quad \text{on} \quad L_{2j-2},$$

$$G_0 \frac{\partial w^{(j-1,j)}}{\partial n} - G_j \frac{\partial w^{(j)}}{\partial n} = 0 \text{ on } L_{2j-1} \setminus p_{2j-1},$$

$$G_0 \frac{\partial w^{(j-1,j)}}{\partial n} - G_{j-1} \frac{\partial w^{(j-1)}}{\partial n} = 0 \text{ on } L_{2j-2} \setminus p_{2j-2}, \quad j = 1, 2, .., m+1$$

Here $\Omega_{m+1}^{(j-1,j)}$ and $\Omega_{m+1}^{(j)}$ are infinite domains that border with $\Omega^{(0)}$ and$\Omega^{(j)}$ through $\partial\Omega^{(0)} \setminus \cup_{j=1}^{m} \gamma_j$ and $\partial\Omega^{(j)} \setminus \gamma_j$, respectively (Fig.3). These regions have common boundaries L_{2j-1} and L_{2j}, which are smooth continuations of boundaries γ_j outside of Ω in such a way that tangents in their points of intersection p_{2j-1} and p_{2j} with $\partial\Omega^{(0)}$ are continuous functions; $\ell_{2j-2,2j-1}$ are segments of boundary $\partial\Omega^{(0)} \setminus \cup_{j=1}^{m} \gamma_j$ with the ends p_{2j-2} and p_{2j-1}.

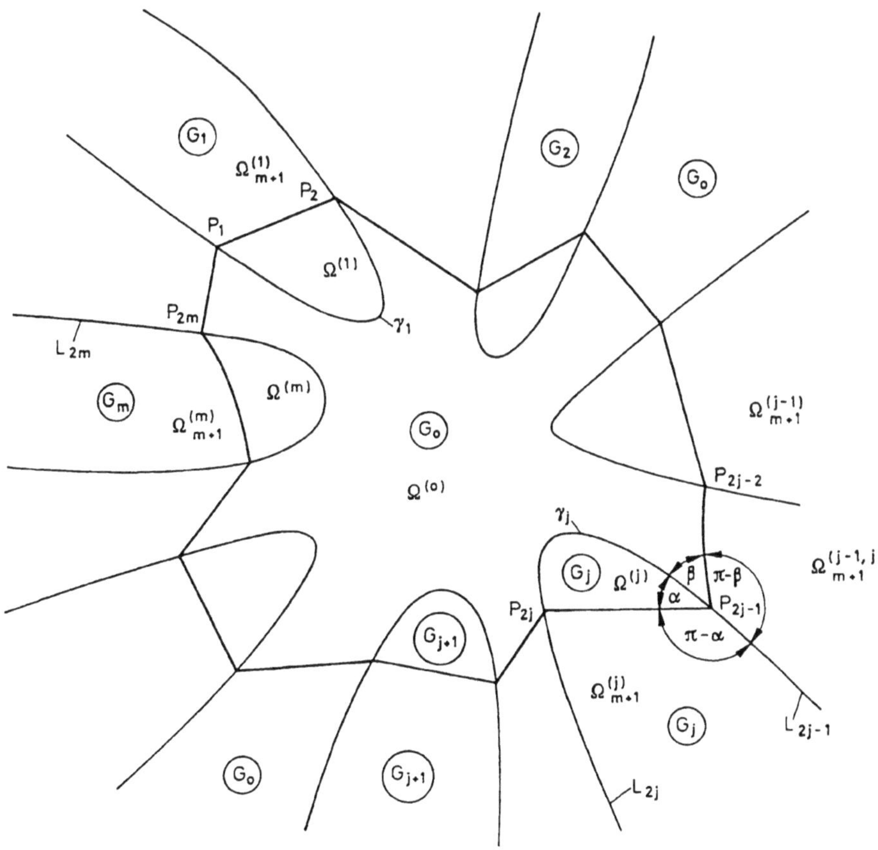

Fig. 3

Obviously, $\Omega_{m+1}^{(m,m+1)} = \Omega_{m+1}^{(m,1)}$, $w^{(0,1)} = w^{(m,m+1)} = w^{(m,1)}$, and $\ell_{2m,2m+1} = \ell_{2m,1} = \ell_{0,1}$; we also assume that the shear modulus G_j corresponds to $\Omega_{m+1}^{(j)}$ and the shear modulus G_0 to $\Omega_{m+1}^{(j-1,j)}$. In (3.1) n is the external normal vector to $\partial\Omega^{(0)}$ and $\partial\Omega_{m+1}^{(j)}$ and the internal normal vector to $\Omega^{(j)}$ and $\Omega_{m+1}^{(j-1,j)}$.

In the vicinity of the intersection point p_{2j-1} the four regions $\Omega^{(0)}, \Omega^{(j)}, \Omega_{m+1}^{(j)}$ and $\Omega_{m+1}^{(j-1,j)}$ converge, which presumably coincide with the sectors $\{(x_1,x_2) = \rho e^{i\theta};\ 0 < \rho < \delta,\ -\beta < \theta < 0\}$, $\{(x_1,x_2) = \rho e^{i\theta};\ 0 < \rho < \delta,\ 0 < \theta < \alpha\}$, $\{(x_1,x_2) = \rho e^{i\theta};\ 0 < \rho < \delta,\ -(\pi-\alpha) < \theta < 0\}$, and $\{(x_1,x_2) = \rho e^{i\theta};\ 0 < \rho < \delta,\ 0 < \theta < \pi-\beta\}$, respectively.

As $\rho \to 0$ the asymptotic form of BVP (3.1) will be represented in the form

$$w^{(j)} \sim \rho^\lambda (B_1 \cos\lambda\omega + B_2 \sin\lambda\omega), \quad -(\pi-\alpha) \le \omega \le 0,$$

$$w^{(j-1,j)} \sim \rho^\lambda (B_3 \cos\lambda\omega + B_4 \sin\lambda\omega), \quad 0 \le \omega \le \pi-\beta.$$

Because $0 < \alpha \le \pi/2$ and $0 < \beta \le \pi/2$ the solution of the BVP (1.1) – (1.5) has no singularities and is described by (2.5). Therefore, the asymptotic form of BVP (3.1) will determine from the homogeneous conditions, which give the following system of equations with respect to B_k, $k = 1,2,3,4$,

$$B_1 \cos\lambda(\pi-\alpha) - B_2 \sin\lambda(\pi-\alpha) = 0,$$
$$B_3 \cos\lambda(\pi-\beta) + B_4 \sin\lambda(\pi-\beta) = 0,$$
$$B_1 = B_3, \quad \mu_j B_2 = B_4.$$

The solvability of this equation requires that $\lambda = \lambda^e$ is a root of the equation

$$\tan\lambda^e(\pi-\alpha) + \mu_j \tan\lambda^e(\pi-\beta) = 0. \tag{3.2}$$

Hence, if $0 < \lambda^e < 1$ as $\rho \to 0$

$$w^{(j)} = u^{(0)}(0) + B_4 \rho^{\lambda^e} \mu_j^{-1} [\tan\lambda^e(\pi-\alpha)\cos\lambda\omega + \sin\lambda\omega] + O(\rho), \quad -(\pi-\alpha) \le \omega \le 0,$$
$$w^{(j-1,j)} = u^{(0)}(0) + B_4 \rho^{\lambda^e} [-\tan\lambda^e(\pi-\beta)\cos\lambda\omega + \sin\lambda\omega] + O(\rho), \quad 0 \le \omega \le \pi-\beta, \tag{3.3}$$

where λ^e is the smallest positive root of the equation (3.2).

At the same time, it is necessary to consider the conjugate to (3.1) BVP

$$\Delta\eta^{(j)} = 0 \ \text{ in } \Omega_{m+1}^{(j)}, \quad j = 1,2,..,m,$$

$$\Delta\eta^{(j-1,j)} = 0 \ \text{ in } \Omega_{m+1}^{(j-1,j)}, \quad j = 1,2,..,m+1,$$
$$\eta^{(j-1,j)} = 0 \ \text{ on } \ell_{2j-2,2j-1}, \quad j = 1,2,..,m+1,$$
$$\eta^{(j)} = 0 \ \text{ on } \partial\Omega \setminus \gamma_j, \quad j = 1,2,..,m,$$
$$\eta^{(j-1,j)} = \eta^{(j)} \ \text{ on } L_{2j-1}, \quad \eta^{(j-1,j)} = \eta^{(j-1)} \ \text{ on } L_{2j-2},$$

$$G_0 \frac{\partial \eta^{(j-1,j)}}{\partial n} - G_j \frac{\partial \eta^{(j)}}{\partial n} = 0 \text{ on } L_{2j-1} \setminus p_{2j-1},$$
$$G_0 \frac{\partial \eta^{(j-1,j)}}{\partial n} - G_{j-1} \frac{\partial \eta^{(j-1)}}{\partial n} = 0 \text{ on } L_{2j-2} \setminus p_{2j-2}, \quad j = 1,2,..,m+1,$$
$$\eta^{(m,m+1)} = \eta^{(m,1)}, \quad \ell_{2m,2m+1} = \ell_{2m,1}, \quad G_{m+1} = G_1$$

with the asymptotic form, as $\rho \to 0$

$$\begin{aligned} \eta^{(j)}_{m+1} &\sim -\mu_j^{-1} \rho^{-\lambda^e} (\tan \lambda^e(\pi - \alpha) \cos \lambda^e \omega + \sin \lambda^e \omega), \\ \eta^{(j-1,j)}_{m+1} &\sim \rho^{\lambda^e} (\tan \lambda^e(\pi - \beta) \cos \lambda^e \omega - \sin \lambda^e \omega). \end{aligned} \tag{3.4}$$

The concept of the flows of functions $z^{(0)}, z^{(j)}, \eta^{(j)}$ and $\eta^{(j-1,j)}$ through the circle $|x - p_{2j-1}| = \varepsilon$ should be checked again. Let $S_\varepsilon^{(j-1,j)} = S^{(j-1,j)} \cap \{|x - p_{2j-1}| > \varepsilon\}$, $\varepsilon > 0$, where $S^{(j-1,j)} = \Omega_{m+1}^{(j-1,j)} \cap \{|x - p_{2j-1}| < R\}$, R is a radius of the circle with the center p_{2j-1}. The domain Ω completely lies in γ_R. Applying Green's formula to the functions $w^{(j-1,j)} - u^{(0)}(0)$ and $\eta^{(j-1,j)}$ over the area $S_\varepsilon^{(j-1,j)}$ and assuming that $R \to \infty$ and $\varepsilon \to 0$, we obtain

$$K_0 B_4 = -\int_{\partial \Omega_{m+1}^{(j-1,j)}} \Big\{ [w^{(j-1,j)} - u^{(0)}(0)] \frac{\partial \eta^{(j-1,j)}}{\partial n} - \eta^{(j-1,j)} \frac{\partial w^{(j-1,j)}}{\partial n} \Big\} d_y s. \tag{3.5}$$

Applying the same procedure to the functions $w^{(j)}_{(j)} - u^{(0)}_{(j)}(0)$ and $\eta^{(j)}$ over the area $S_\varepsilon^{(j)} = S_{m+1}^{(j)} \cap \{|x - p_{2j-1}| > \varepsilon\}$, $\varepsilon > 0$, where $S_{m+1}^{(j)} = \Omega_{m+1}^{(j)} \cap \{|x - p_{2j-1}| < R\}$, we have

$$K_j B_4 = -\int_{\partial \Omega_{m+1}^{(j)}} \Big\{ [w^{(j)} - u^{(0)}(0)] \frac{\partial \eta^{(j)}}{\partial n} - \eta^{(j)} \frac{\partial w^{(j)}}{\partial n} \Big\} d_y s. \tag{3.6}$$

Passing $\partial \Omega^{(0)} \setminus \cup_{j=1}^m \gamma_j$ and $\partial \Omega^{(j)} \setminus \gamma_j$ in (3.5) and (3.6) takes place in a way that leaves Ω to the left.

Because the point $x = p_{2j-1}$ is out of $\Omega_{m+1}^{(k-1,k)}$ and $\Omega_{m+1}^{(j)}$ when $j \neq k$, Green's formula gives for $k = 1,2,..,m+1$, $k \neq j$

$$\int_{\partial \Omega_{m+1}^{(k-1,k)}} \Big\{ w^{(k-1,k)} \frac{\partial \eta^{(k-1,k)}}{\partial n} - \eta^{(k-1,k)} \frac{\partial w^{(k-1,k)}}{\partial n} \Big\} d_y s = 0, \tag{3.7}$$

and for $k = 1,2,..,m$, $k \neq j$

$$\int_{\partial \Omega_{m+1}^{(k)}} \Big\{ w^{(k)} \frac{\partial \eta^{(k)}}{\partial n} - \eta^{(k)} \frac{\partial w^{(k)}}{\partial n} \Big\} d_y s = 0. \tag{3.8}$$

Multiplying (3.5) by G_0, (3.6) by G_j, (3.7) by G_0 for all $k = 1,2,..,m,m+1$, and (3.8) by G_k for $k = 1,2,..,m$, and taking into account of the boundary conditions of the problem (3.4) we construct the following expression

$$(G_0 K_0 + G_j K_j) B_4 = -G_0 \sum_{k=1}^{m+1}{}' \int_{\ell_{2k-2,2k-1}} w^{(k-1,k)} \frac{\partial \eta^{(k-1,k)}}{\partial n} d_y s -$$

$$- \sum_{k=1}^{m}{}' G_k \int_{\partial\Omega^{(k)}\setminus\gamma_k} w^{(k)} \frac{\partial\eta^{(k)}}{\partial n} d_y s - G_j \int_{\partial\Omega^{(j)}\setminus\gamma_j} [w^{(k)} - u^{(0)}(0)] \frac{\partial\eta^{(j)}}{\partial n} d_y s - \tag{3.9}$$
$$- G_0 \int_{\ell_{2j-2,2j-1}} [w^{(j-1,j)} - u^{(0)}(0)] \frac{\partial\eta^{(j-1,j)}}{\partial n} d_y s.$$

Integrals over the lines L_{2j-1} and L_{2j-2}, $j = 1, 2, .., m, m+1$, cancel each other. In (3.7) and (3.8) minus signs are preserved because the passage $\partial\Omega^{(0)} \setminus \cup_{j=1}^{m} \gamma_j$ and $\partial\Omega^{(j)} \setminus \gamma_j$ takes place in a way which leaves Ω to the left.

The accent over the summations in (3.9) means the absence of the term with the number $k = j$. In order to express the constant in terms of the boundary values of the problem (1.1) – (1.5) let construct a harmonic extension of the functions $\eta^{(j)}$ in the region Ω

$$\begin{gathered} \Delta z^{(j)} = 0 \text{ in } \Omega^{(j)}, \quad j = 0, 1, .., m, \\ \frac{\partial z^{(j)}}{\partial n} = \frac{\partial \eta^{(j)}}{\partial n} \text{ on } \partial\Omega^{(j)} \setminus \gamma_j, \\ \frac{\partial z^{(0)}}{\partial n} = \frac{\partial\eta^{(j-1,j)}}{\partial n} \text{ on } \ell_{2j-2,2j-1}, \quad j = 1, 2, .., m, \\ \ell_{0,1} = \ell_{2m,1} = \ell_{2m,2m+1} \\ z^{(j)} = z^{(0)} \text{ and } G_0 \frac{\partial z^{(0)}}{\partial n} - G_j \frac{\partial z^{(j)}}{\partial n} = 0 \text{ on } \gamma_j \setminus K. \end{gathered} \tag{3.10}$$

The asymptotics of the solutions of the BVP (3.10) in a neighbourhood of the point p_{2j-1} when $\rho \to 0$ may be expessed as

$$\begin{aligned} z^{(j)} &\sim \rho^{-\lambda^e}(D_1 \cos\lambda^e\omega + D_2 \sin\lambda^e\omega), \quad 0 \le \omega \le \alpha, \\ z^{(0)} &\sim \rho^{-\lambda^e}(D_3 \cos\lambda^e\omega + D_4 \sin\lambda^e\omega), \quad -\beta \le \omega \le 0. \end{aligned} \tag{3.11}$$

Taking into account the asymptotics (3.4), after satisfaction of the boundary conditions of the problem (3.1) we obtain

$$\begin{gathered} -D_1 \sin\lambda^e\alpha + D_2 \cos\lambda^e\alpha = \tan\lambda^e(\pi-\beta)\sin\lambda^e(\pi-\alpha) - \mu_j^{-1}\cos\lambda^e(\pi-\alpha), \\ D_1 \sin\lambda^e\beta + \mu_j D_2 \cos\lambda^e\beta = -\tan\lambda^e(\pi-\beta)\sin\lambda^e(\pi-\beta) - \cos\lambda^e(\pi-\beta), \\ D_1 = D_3, \quad D_4 = \mu_j D_2, \end{gathered}$$

or in view of (3.2) we have

$$\begin{aligned} -D_1 \sin\lambda^e\alpha + D_2 \cos\lambda^e\alpha &= (-\mu_j \cos\lambda^e(\pi-\alpha))^{-1}, \\ D_1 \sin\lambda^e\beta + \mu_j D_2 \cos\lambda^e\beta &= (-\cos\lambda^e(\pi-\beta))^{-1}. \end{aligned} \tag{3.12}$$

We will denote by $\gamma_\varepsilon^\alpha$, $\gamma_\varepsilon^{(\pi-\alpha)}$, $\gamma_\varepsilon^{(\pi-\beta)}$, and γ_ε^β parts of the circle γ_ε of radius ε with center at the point p_{2j-1} situated in $\Omega^{(j)}$, $\Omega_{m+1}^{(j)}$, $\Omega_{m+1}^{(j-1,j)}$ and $\Omega^{(0)}$ respectively, where

ε is a small positive number. The flows of the solutions $z^{(j)}$, $\eta^{(j)}$, $\eta^{(j-1,j)}$ and $z^{(0)}$ through γ_ε are

$$\Pi_1(\eta^{(j)}) = G_j \int_{-(\pi-\alpha)}^{0} \frac{\partial\eta^{(j)}}{\partial n} d_y s = \varepsilon^{-\lambda^e} \mu_j^{-1} G_j (1 - \cos^{-1}\lambda^e(\pi-\alpha)) + O(\varepsilon^{\lambda^e}),$$

$$\Pi_2(\eta^{(j-1,j)}) = G_0 \int_0^{\pi-\beta} \frac{\partial\eta^{(j-1,j)}}{\partial n} d_y s = -\varepsilon^{-\lambda^e} G_0 (1 - \cos^{-1}\lambda^e(\pi-\beta)) + O(\varepsilon^{\lambda^e}),$$

$$\Pi_3(z^{(0)}) = G_0 \int_{-\beta}^{0} \frac{\partial z^{(0)}}{\partial n} d_y s = \varepsilon^{-\lambda^e} G_0 (-\mu_j D_2 + D_1 \sin\lambda^e\beta + \mu_j D_2 \cos\lambda^e\beta) + O(\varepsilon^{\lambda^e}),$$

$$\mu_j = G_j / G_0,$$

$$\Pi_4(z^{(j)}) = G_j \int_0^{\alpha} \frac{\partial z^{(j)}}{\partial n} d_y s = \varepsilon^{-\lambda^e} G_j (D_1 \sin\lambda^e\alpha - D_2 \cos\lambda^e\alpha + D_2) + O(\varepsilon^{\lambda^e}).$$

In view of (3.12)

$$\Pi_1(\eta^{(j)}) + \Pi_2(\eta^{(j-1,j)}) + \Pi_3(z^{(0)}) + \Pi_4(z^{(j)}) = O(\varepsilon^{\lambda^e}).$$

Hence, the flows of the solutions $z^{(j)}$, $\eta^{(j)}$, $\eta^{(j-1,j)}$ and $z^{(0)}$ through coutour γ_ε are equal to zero. In comparison with the extension offered by BVP (2.6), the continuation of the solution of BVP given in Ω into $\Omega_{m+1}^{(j)}$ and $\Omega_{m+1}^{(j-1,j)}$ realized by (3.1) is possible.

According to the notations introduced by (3.1) and (3.10), from (3.9) we have

$$\begin{aligned}(G_0 K_0 + G_j K_j) B_4 = {} & -G_0 \sum_{k=1}^{m+1}{}' \int_{\ell_{2k-2,2k-1}} u^{(0)} \frac{\partial z^{(0)}}{\partial n} d_y s - \\ & - G_0 \int_{\ell_{2j-2,2j-1}} [u^{(0)} - u^{(0)}(0)] \frac{\partial z^{(0)}}{\partial n} d_y s - \\ & - \sum_{k=1}^{m}{}' \int_{\partial\Omega^{(k)}\setminus\gamma_k} u^{(k)} \frac{\partial z^{(k)}}{\partial n} d_y s - G_j \int_{\partial\Omega^{(j)}\setminus\gamma_j} [u^{(j)} - u^{(0)}(0)] \frac{\partial z^{(j)}}{\partial n} d_y s,\end{aligned} \tag{3.13}$$

where $\ell_{2m,2m+1} = \ell_{2m,1}$.

Using Green's formula we have

$$\begin{aligned}\int_{\partial\Omega^{(k)}\setminus\gamma_k} u^{(k)} \frac{\partial z^{(k)}}{\partial n} d_y s + \int_{\gamma_k} u^{(k)} \frac{\partial z^{(k)}}{\partial n} d_y s = {} & \int_{\partial\Omega^{(k)}\setminus\gamma_k} z^{(k)} \frac{\partial u^{(k)}}{\partial n} d_y s + \\ & + \int_{\gamma_k} z^{(k)} \frac{\partial u^{(k)}}{\partial n} d_y s, \quad k = 1, 2, .., m,\end{aligned} \tag{3.14}$$

$$\begin{aligned}\int_{\partial\Omega^{(0)}\setminus\cup_{k=1}^m\gamma_k} u^{(0)} \frac{\partial z^{(0)}}{\partial n} d_y s + \int_{\cup_{k=1}^m\gamma_k} u^{(0)} \frac{\partial z^{(0)}}{\partial n} d_y s = {} & \int_{\partial\Omega^{(0)}\setminus\cup_{k=1}^m\gamma_k} z^{(0)} \frac{\partial u^{(0)}}{\partial n} d_y s + \\ & + \int_{\cup_{k=1}^m\gamma_k} z^{(0)} \frac{\partial u^{(0)}}{\partial n} d_y s,\end{aligned} \tag{3.15}$$

where

$$u^{(k)} = \begin{cases} u^{(k)} - u^{(0)}(0) & \text{on } \gamma_k \text{ if } k = j \\ u^{(k)} & \text{on } \gamma_k \text{ if } k \neq j \end{cases}$$

$$u^{(0)} = \begin{cases} u^{(0)} - u^{(0)}(0) & \text{on } \ell_{2j-2,2j-1} \\ u^{(0)} & \text{on } \Omega^{(0)} \setminus (\ell_{2j-2,2j-1} \cup \gamma_j) \end{cases} \tag{3.16}$$

and $u^{(0)} = u^{(k)}$ on γ_k, $k = 1, 2, .., m$, according to (1.2).

Multiplying (3.14) by G_k and summarizing over all k, and adding (3.15) multiplied by G_0 we obtain

$$\sum_{k=1}^{m} G_k \int_{\partial\Omega^{(k)}\setminus\gamma_k} u^{(k)} \frac{\partial z^{(k)}}{\partial n} d_y s + \sum_{k=1}^{m} G_k \int_{\gamma_k} u^{(k)} \frac{\partial z^{(k)}}{\partial n} d_y s +$$
$$+ G_0 \int_{\partial\Omega^{(0)}\setminus\cup_{k=1}^{m}\gamma_k} u^{(0)} \frac{\partial z^{(0)}}{\partial n} d_y s + \sum_{k=1}^{m} G_0 \int_{\gamma_k} u^{(0)} \frac{\partial z^{(0)}}{\partial n} d_y s =$$
$$= \sum_{k=1}^{m} G_k \int_{\partial\Omega^{(k)}\setminus\gamma_k} z^{(k)} \frac{\partial u^{(k)}}{\partial n} d_y s + \sum_{k=1}^{m} G_k \int_{\gamma_k} z^{(k)} \frac{\partial u^{(k)}}{\partial n} d_y s +$$
$$+ G_0 \int_{\partial\Omega^{(0)}\setminus\cup_{k=1}^{m}\gamma_k} z^{(0)} \frac{\partial u^{(0)}}{\partial n} d_y s + \sum_{k=1}^{m} G_0 \int_{\gamma_k} z^{(0)} \frac{\partial u^{(0)}}{\partial n} d_y s.$$

In view of (3.10) and (1.2), the second and fourth items of the left part must be canceled because the integration over γ_k takes place in opposite directions according to Green's formulas (3.14) and (3.15). It is easy to see that the left part of this equality is $-B_4(G_0K_0 + G_jG_j)$. Therefore, taking into account (3.10) and (1.1) – (1.5), (3.13) may be rewritten in the form

$$B_4 = -\frac{G_0}{(G_0K_0 + G_jK_j)} \int_{\partial\Omega^{(0)}\setminus\cup_{j=1}^{m}\gamma_j} z^{(0)} g_0 d_y s - \frac{1}{(G_0K_0 + G_jK_j)} \sum_{k=1}^{m} \int_{\gamma_k} z^{(0)} f_k d_y s -$$
$$- \frac{1}{(G_0K_0 + G_jK_j)} \sum_{k=1}^{m} G_k \int_{\partial\Omega^{(k)}\setminus\gamma_k} z^{(k)} g_k d_y s, \tag{3.17}$$

where

$$K_0 = -\lambda^e \tan^2 \lambda^e(\pi - \beta)[\lambda^e(\pi - \beta) + \frac{1}{2}\sin 2\lambda^e(\pi - \beta)] +$$
$$+ \tan\lambda^e(\pi - \beta)[1 - \cos 2\lambda^e(\pi - \beta)] - \lambda^e(\pi - \beta) + \frac{1}{2}\sin 2\lambda^e(\pi - \beta),$$

$$K_j = -[\frac{\lambda^e}{\mu_j^2}(\pi - \alpha) + \frac{\sin 2\lambda^e(\pi - \beta)}{2\mu_j^2}]\tan^2\lambda^e(\pi - \beta) +$$
$$+ \frac{2}{\mu_j^2}\tan\lambda^e(\pi - \beta)\sin^2\lambda^e(\pi - \beta) - \frac{\lambda^e}{\mu_j^2}(\pi - \alpha) + \frac{\sin 2\lambda^e(\pi - \beta)}{2\mu_j^2}.$$

Equations (3.3) and (3.17) define the principal term of the asymptotic form of the functions $w^{(j)}$ and $w^{(j-1,j)}$ when $0 < \alpha \le \pi/2$ and $0 < \beta \le \pi/2$.

Since $w^{(j-1,j)}$ and $w^{(j)}$ are identical with a potential of single layer (1.6) outside of Ω so that on $\partial\Omega^{(0)} \setminus \cup_{j=1}^{m}\gamma_j$

$$\nu = \frac{1}{2\pi}\Big(\frac{\partial u^{(0)}}{\partial n} - \frac{\partial w^{(j-1,j)}}{\partial n}\Big),$$

and on $\partial\Omega^{(j)} \setminus \gamma_j$

$$\nu = \frac{1}{2\pi}\Big(\frac{\partial u^{(j)}}{\partial n} - \frac{\partial w^{(j)}}{\partial n}\Big).$$

Taking into account (3.3), we have on $\partial\Omega^{(0)} \setminus \cup_{j=1}^{m}\gamma_j$ close to the point p_{2j-1} as $\rho \to 0$

$$\nu \sim \frac{\lambda^e}{2\pi} B_4 \rho^{\lambda^e - 1} \cos^{-1} \lambda^e(\pi - \beta), \tag{3.18}$$

and on $\partial\Omega^{(j)} \setminus \gamma_j$

$$\nu \sim \frac{\lambda^e}{2\pi} B_4 \rho^{\lambda^e - 1} \mu_j^{-1} \cos^{-1} \lambda^e(\pi - \alpha). \tag{3.19}$$

Expressions (3.18), (3.19) and (3.17) define the principal terms of the AFSIE (1.9) in a neighbourhood of the angular point on $\partial\Omega^{(0)} \setminus \cup_{j=1}^{m}\gamma_j$ and $\partial\Omega^{(j)} \setminus \gamma_j$ when $0 < \alpha \leq \pi/2$ and $0 < \beta \leq \pi/2$.

References

[1] Zargaryan S.S. and Maz'ya V.G., The asymptotic form of the solutions of the integral equations of potential theory in the neighbourhood of the corner points of a contour, Prikl. Matem. Mekhan., 48, pp.169-174, 1984 (in Russian). Translation in PMM U.S.S.R., vol.48, No.1, pp.120-124, 1984. Printed in Great Britan, Pergamon Press Ltd, 1985.

[2] Zargaryan S.S., Torsion of rods consisting of composites having ribs on the joining surfaces of the different media, Mekhanika Kompozitnykh Materialov, Riga, no.2, pp.320-327, 1984.

[3] Gevorgyan G.G. and Zargaryan S.S., Torsion of a cylindrical rod reinforced with a tetrahedral prism made of another material, Izvestiya Akademii Nauk Armyanskoi SSR. Mekhanika, vol.39, No.4, pp.12-14, 1986 (in Russian). Translation in Izvest. AN Arm.SSR, Mech., vol.39, no.4, pp.12-26, 1986. Allerton Press. 1987.

[4] Muskelishvili N.I., Singular Integral Equations, P.Noordholff Ltd., Groningen, Holland, 1969.

[5] Kondrat'ev V.A., Boundary value problems for elliptic equations in the region of conical or corner points, Trudy Mosk. Mat. O-va, 16, 1967.

[6] Chobanyan K.S., Phenomena of the low stress of the periphery of a contact surface boundary of a loaded compound body, Diploma for Discovery No.102. Discoveries, Inventions, Industrial Models, Trademarks, no.36, p.3, 1971.

[7] Maz'ya V.G. and Plamenevskii B.A., The coefficients in asymptotic form of the solutions of elliptic boundary value problems in regions with conical points, Math. Nachr. 76, 1977.

List of participants

David R. ADAMS
Department of Mathematics, University of Kentucky, Lexington, KY 40506-0027, U.S.A.

Hiroaki AIKAWA
Department of Mathematics, Gunma University, Kiryu, 376, Japan

Ömer AKIN
Department of Mathematics, Faculty of Sciences, Ankara University, 06100-Tandogan, Ankara, Turkey

Tateo BABA
Sanhowaito M507-23, 5-17-14 Minami-Naruse, Machida, 194, Japan

Martha BĂNULESCU
Sos. Iancului, NR 33, Bloc 105A, Sc. Ḃ, AP. 71, Bucharest, 73376, Romania

Heinz BAUER
Mathematisches Institut, Universität Erlangen-Nürnberg, Bismarckstrasse 1 1/2, D-8520 Erlangen, F.R.G

Alexander D. BENDIKOV
Flat 85, Komarov Str. 24, Rostov-on-Don, 344092, USSR

Christian BERG
Department of Mathematics, University of Copenhagen, Universitetsparken 5, DK-2100 Copenhagen ϕ, Denmark

Emile M. J. BERTIN
Mathematisch Instituut, R. U. U., P.O.Box 80.010, NL-3508 TA, Utrecht, Netherlands

Marco BIROLI
Department of Mathematics, Politecnico di Milano, Piazza Leonardo da Vinci, 32-20133 Milano, Italy

Andrej V. BITSADZE
V. A. Steklov Institute of Mathematics, Academy of Sciences of USSR, 117966 Moscow, B-33, USSR

Göran BJÖRCK
Kvarnvägen 13, S-191 51 Sollentuna, Sweden

Jürgen BLIEDTNER
Fachbereich Mathematik, Johann Wolfgang Goethe-Universität, Robert-Mayer-Str. 6-10, D–6000 Frankfurt am Main, F.R.G.

Nicu BOBOC
Department of Mathematics, University of Bucharest, Str. Academiei 14, Bucharest, Romania

Andre BOIVIN
Department of Mathematics, University of Western Ontario, London, Ontario, Canada, N6A 5B7

Abderrahman BOUKRICHA
Département de Mathématiques, Faculté des Sciences de Tunis, Campus Universitaire, 1060 Tunis, Tunisie

Mikhail BRODSKY
Department of Statistics, University of California, Berkeley, CA 94706, U.S.A.

Gheorghe BUCUR
Department of Mathematics, INCREST, Bd. Pàcii 220, Bucharest, Romania

Huy-Qui BUI
Department of Mathematics, University of Canterbury, Christchurch 1, New Zealand

Huaihui CHEN
Department of Mathematics, Nanjing Normal University, Nanjing, 210024, P.R. of China

Allen T. CHWANG
Institute of Hydraulic Research, University of Iowa, Iowa City, Iowa 52242, U.S.A.

Aurel CORNEA
Richard-Strauss-Str. 50, D-8078 Eichstaett, F.R.G.

Arnaud de la PRADELLE
71, rue Saint Antoine, 75004 Paris, France

Claude DELLACHERIE
169, rue Danguy, 76230 Bois Guillaume, France

Vladimir Ya. EIDERMAN
Apt. 165, ul. Volodarskogo, 26/32, 109172 Moscow, USSR

Sirkka-Liisa A. ERIKSSON-BIQUE
Department of Mathematics, University of Joensuu, P.O.Box 111, SF-80101 Joensuu, Finland

Matts ESSÉN
Department of Mathematics, University of Uppsala, Thunbergsvägen 3, S-75238 Uppsala, Sweden

Bent FUGLEDE
Department of Mathematics, University of Mathematics, Universitetsparken 5, DK-2100 Copenhagen ϕ, Denmark

Tatsuo FUJI'I'E
10-234, Shakadani, Ohmiya, Kita-ku, Kyoto, 603, Japan

Masatoshi FUKUSHIMA
Department of Mathematical Science, Faculty of Engineering Science, Osaka University, Toyonaka, 560, Japan

Jerzy A. GAWINECKI
ul. Pirenejska 14/50, 01-493 Warsaw, Poland

Myron GOLDSTEIN
5914, South Newberry, Tempe, Arizona, 85283, U.S.A.

Yasuhiro GOTOH
Department of Mathematics, Faculty of Sciences, Kyoto University, Kyoto, 606, Japan

Kohur GOWRISANKARAN
Department of Mathematics & Statistics, McGill University, Montreal, Canada, H3A 2K6

Nikolai V. GRACHEV
Chair of Mathematics, Polytechnical Institute, Tolstoi str. 4-100, Pskov, 180680, USSR

Alexander A. GRIGOR'YAN
Institute of Control Sciences, Profsouznaja 65, 117806 Moscow, USSR

Archil GULISASHVILI
Institute of Mathematics, Georgian Academy of Sciences, Rukhadze str. 1, 380093 Tbilisi, USSR

Ashot S. HAKOBYAN
Institute of Mechanics, Armenian Academy of Sciences, Bagramyan Ave. 28, Yerevan, USSR

A. Kersti HALISTE
Department of Mathematics, Umeå University, S-90187 Umeå, Sweden

Wolfhard HANSEN
Fakultät für Mathematik, Universität Bielefeld, Postfach 8640, D-4800 Bielefeld 1, F.R.G.

Heihachiro HARA
213-4, Sugata-cho, Matsue, 690, Japan

Masaru HARA
Faculty of Sciences and Technology, Meijo University, Tenpaku-ku, Nagoya, 468, Japan

Yuji HASHIMOTO
Department of Mathematics, Aichi Institute of Technology, Yagusa, Toyota, 470-03, Japan

Morisuke HASUMI
Department of Mathematics, Faculty of Sciences, Ibaragi University, Bunkyo, Mito, 310, Japan

Kaoru HATANO
362-79, Nishimochida-cho, Matsue, 690, Japan

John HAWKES
Flat 3, 8, Compayne Gardens, London, NW6 3DH, U.K.

Kazumichi HAYASHI
4-6-10, Minami-Ogikubo, Suginami-ku, Tokyo, 167, Japan

Walter K. HAYMAN
Department of Mathematics, University of York Heslington, York, YO1 5DD, U.K.

Lars I. HEDBERG
Department of Mathematics, University of Linköping, S-58183 Linköping, Sweden

Juha HEINONEN
Department of Mathematics, University of Michigan, Ann Arbor, MI 48109, U.S.A.

Isao HIGUCHI
Department of Mathematics, Aichi Institute of Technology, Yagusa, Toyota, 470-03, Japan

Imsik HONG
4-7-12-505, Takanawa, Minato-ku, Tokyo, 108, Japan

Toshio HORIUCHI
Department of Mathematics, Ibaraki University, Mito, 310, Japan

Alfred O. HUBER
Freudenbergstrasse 101, CH-8044 Zuerich, Switzerland

Jun Shung HWANG
Institute of Mathematics, Academia Sinica Taipei, 11529, Taiwan, R. O. C.

Kanji ICHIHARA
Department of Mathematics, College of General Education, Nagoya University, Nagoya, 464-01, Japan

Teruo IKEGAMI
Department of Mathematics, Osaka City University, 3-3-138, Sugimoto, Sumiyoshi-ku, Osaka, 558, Japan

Hideo IMAI
Department of Mathematics, Daido Institute of Technology, 10-3, Takiharu, Minami-ku, Nagoya, 457, Japan

Katsuya ISHIZAKI
Department of Mathematics, Tokyo College of Technology, 1220-2, Kunugida-machi, Hachioji, 192, Japan

Masayuki ITÔ
Department of Mathematics, College of General Education, Nagoya University, Nagoya, 464-01, Japan

Yoshifusa ITO
Toyohashi University of Science & Technology, Tenpaku-cho, Toyohashi, 440, Japan

Niels JACOB
Mathematisches Institut, Universität Erlangen-Nürnberg, Bismackstr. 1 1/2, D-8520 Erlangen, F.R.G.

Klaus JANSSEN
Institut für Statistik und Dokumentation der Universität Düsseldorf, Universitätsstr. 1, D–4000 Düsseldorf 1, F.R.G.

Naondo JIN
Department of Mathematics, Gakushuin University, 1-5-1, Mejiro, Toshima-ku, Tokyo, 171, Japan

Vadim A. KAIMANOVICH
Department of Applied Mathematics, Leningrad Shipbuilding Institute, Lotsmanskaya 3, Leningrad, 190008, USSR

Shigeyasu KAMIYA
Department of Mechanical Science, Okayama University of Science, 1-1, Ridai-cho, Okayama, 700, Japan

Mamoru KANDA
Department of Mathematics, University of Tsukuba, Tsukuba, 305, Japan

Hiroshi KANEKO
Department of Mathematics, Tokyo University of Science, 26, Wakamiya, Shinjuku-ku, Tokyo,162, Japan

Michihiko KAWAMURA
Department of Mathematics, Faculty of Education, Gifu University, Gifu, 501-11, Japan

Carlos E. KENIG
Department of Mathematics, University of Chicago, Chicago, IL 60632, U.S.A.

Ronald A. KERMAN
Department of Mathematics, Brock University, St.Catharines, Ontario, Canada, L2S 3A1

Tero KILPELÄINEN
Department of Mathematics, University of Jyväskylä, Seminaarink 15, 40100 Jyväskylä, Finland

Masanori KISHI
Department of Mathematics, College of General Education, Nagoya University, Nagoya, 464-01, Japan

Vachtang KOKILASHVILI
Institute of Mathematics, Georgian Academy of Sciences, Rukhadze str. 1, 380093 Tbilisi, USSR

Boris KORENBLUM
Department of Mathematics, State University of New York at Albany, Albany, NY 12222, U.S.A.

Jacob KOREVAAR
Faculty of Mathematics and Computer Science, University of Amsterdam, Plantage Muidergracht 24, 1018 TV Amsterdam, Netherlands

Toshiaki KORI
Department of Mathematics, Waseda University, 3-4-1, Okubo, Shinjuku-ku, Tokyo, 160, Japan

Victor I. KRUGLIKOV
Fl. 3, Artom St. 93, Donetsk, 340000, USSR

Takahide KUROKAWA
Department of Mathematics, College of General Education, Kagoshima University, Kagoshima, 890, Japan

Toshiko KUROKAWA
Department of Mathematics, Faculty of Education, Mie University, Kamihama-cho, Tsu, 514, Japan

Yukio KUSUNOKI
7-2, Osagi-cho, Iwakura, Sakyo-ku, Kyoto, 606, Japan

Kazuhiro KUWAE
Department of Mathematics, Faculty of Sciences, Osaka University, Toyonaka, 560, Japan

Ilpo E. LAINE
Department of Mathematics, University of Joensuu, P.O.Box 111, SF-80101 Joensuu, Finland

Julian ŁAWRYNOWICZ
Institute of Mathematics of the Polish Academy of Sciences, Łódź Branch, ul. Narutowicza 56, PL-90-136 Łódź, Poland

Gottlieb LEHA
Faculty of Mathematics & Information, University of Passau, Innstr. 33, D-8390 Passau, F.R.G.

Lars P. LINDQVIST
Institute of Mathematics, Helsinki University of Technology, SF-02150 Espoo, Finland

Eric R. LOVE
Department of Mathematics, University of Melbourne, Parkville, Victoria, 3052, Australia

Jaroslav LUKEŠ
MFF UK, Department of Mathematical Analysis, Sokolovská 83, 18600 Praha 8, Czechoslovakia

Zhi Ming MA
BiBoS, Fakultät für Physik, Universität Bielefeld, D-4800 Bielefeld 1, F.R.G.

Fumi-Yuki MAEDA
Department of Mathematics, Faculty of Sciences, Hiroshima University, Hiroshima, 730, Japan

Fumio MAITANI
Department of Mechanical and System Engineering, Kyoto Institute of Technology, Sakyo-ku, Kyoto, 606, Japan

Nikolai G. MAKAROV
LOMI, Fontanka 27, Leningrad, 191011, USSR

Juan J. MANFREDI
Department of Mathematics, University of Pittsburgh, Pittsburgh, PA 15260, U.S.A.

Donald E. MARSHALL
Department of Mathematics, University of Washington, Seattle, WA 98195, U.S.A.

Olli T. MARTIO
Department of Mathematics, University of Jyväskylä, P.O.Box 35, SF-40351 Jyväskylä, Finland

Hiroaki MASAOKA
Department of Mathematics, Osaka City University, Sugimoto, Sumiyoshi-ku, Osaka, 558, Japan

Makoto MASUMOTO
Department of Mathematics, Faculty of Sciences, Hiroshima University, Hiroshima, 730, Japan

Shigeki MATSUMOTO
Department of Mathematics, Faculty of Sciences, Hiroshima University, Hiroshima, 730, Japan

Pertti MATTILA
Department of Mathematics, University of Jyväskylä, Seminaarinkstr. 15, 40100 Jyväskylä, Finland

Yoichi MIYAZAKI
Toyohashi University of Science & Technology, Tenpaku-cho, Toyohashi, 440, Japan

Hisao MIZUMOTO
3-39-22, Koi-Osako, Nishi-ku, Hiroshima, 733, Japan

Yoshihiro MIZUTA
Department of Mathematics, Faculty of Integrated Arts and Sciences, Hiroshima University, Hiroshima, 730, Japan

Gabriel MOKOBODZKI
10, rue Raynouard, 75016 - Paris, France

Charles N. MOORE
Department of Mathematics, Washington University, St. Louis, MO 63130, U.S.A.

Hisashi MORINAKA
Department of General Education, Osaka Institute of Technology, 5-16-1, Omiya, Asahi-ku, Osaka, 535, Japan

Takafumi MURAI
Department of Mathematics, School of Sciences, Nagoya University, Nagoya, 464-01, Japan

Minoru MURATA
Department of Mathematics, Faculty of Science, Kumamoto University, Kumamoto, 860, Japan

Tadashi MURAZAWA
Department of Applied Mathematics, Kyoto Prefectural University, Shimogamo, Sakyo-ku, Kyoto, 606, Japan

Hideo NAGAI
Department of Mathematics, College of General Education, Nagoya University, Nagoya, 464-01, Japan

Linda NAÏM
6, rue Jules Lejeune, B-1060, Bruxelles, Belgium

Mitsuru NAKAI
Department of Mathematics, Nagoya Institute of Technology, Gokiso, Showa-ku, Nagoya, 466, Japan

Raimo NÄKKI
Department of Mathematics, University of Jyväskylä, 40100 Jyväskylä, Finland

Junichiro NARITA
201, Miyuki, 7 Yamagoe, Nagakute, Aichi, 480-11, Japan

Ivan NETUKA
Mathematical Institute, Charles University, Sokolovská 83, 186 00 Praha 8, Czechoslovakia

Duc Tuan NGUYEN
Institute of Mathematics, P.O.Box 631, Bo Ho, 10000 Hanoi, Vietnam

Xuan-Loc NGUYEN
Institute of Computer Science and Cybernetics, Lieu Giai. Ba dinh. Hanoi, Vietnam

Nobuyuki NINOMIYA
808-321, Higashi-ga-oka, Kishiwada, 596, Japan

Nobutaka NISHIKAWA
2-215, 4-10-2, Shioyaki, Ichikawa, 272-01, Japan

Masaharu NISHIO
Department of Mathematics, Osaka City University, 3-3-138, Sugimoto, Sumiyoshi-ku, Osaka, 558, Japan

Ryohei NOZAWA
5-7, W-25, 7-16-402, Chuo-ku, Sapporo, 064, Japan

Anthony G. O'FARRELL
Maynooth College, Co. Kildare, Ireland

Hiromi OHTAKE
Department of Mathematics, Faculty of Sciences, Kyoto University, Kyoto, 606, Japan

Makoto OHTSUKA
Department of Mathematics, Faculty of Sciences, Gakushuin University, Toshima-ku, Tokyo, 171, Japan

Masami OKADA
Department of Mathematics, College of General Education, Tohoku University, Sendai, 980, Japan

Masayuki OSADA
Department of Mathematics, Hokkaido University of Education, 5-3-1, Ainosato, Kita-ku, Sapporo, 060, Japan

Yoichi OSHIMA
94-5, Hosui, Shisui-cho, Kikuchi, Kumamoto, 861-12, Japan

Wellington H. OW
Department of Mathematics, Michigan State University, East Lansing, Michigan, 48824, U.S.A.

Ram S. PATHAK
Department of Mathematics, Banaras Hindu University, Varanasi 221 005, India

Igor PAVLOV
Rostov Civil Engineering Institute, Sotsialisticheskaya str. 162,
344022 Rostov-on-Don, USSR

Igor M. PETRUSHKO
Department of Mathematics, MEI, Krasnokazarmennaja 14, Moscow, 105835, USSR

Massimo A. PICARDELLO
Department of Mathematics, University of Roma "Tor Vergata" od33 Roma, Italy

Aleksey I. PRILEPKO
Moscow Engneering Physics Institute, Kashirskoe Shosse 31, 115409 Moscow, USSR

Shuxi QIU
Department of Mathematics, Xiamen University, Xiamen, Fujian, P.R. of China

Sundararaja RAMASWAMY
Statistics & Mathematics Unit, Indian Statistical Institute, 8th Mile, Mysore Road,
Bangalore-560059, India

Alexandr Yu. RASHKOVSKIĬ
Institute for Low Temperature Physics & Engineering, Ukr. SSR Acad. Sci.,
Lenin Av. 47, 310164 Kharkov,USSR

Philip J. RIPPON
Faculty of Mathematics, Open University, Milton Keynes, MK7 6AA, U.K.

Michael RÖCKNER
Mathematisches Institut der Universität, Wegelerstr. 10, D-5300 Bonn, F.R.G.

Makoto SAKAI
Department of Mathematics, 1-1 Minami-Ōsawa, Hachiōji, 192-03,
Tokyo Metropolitan University, Japan

Ken-ichi SAKAN
Department of Mathematics, Osaka City University, Sugimoto, Sumiyoshi-ku, Osaka,
558, Japan

Evgueni G. SALTYKOV
Department of Computational Mathematics and Cybernetics,
Moscow State University, Leninskie Gory V-234, Moscow GSP 119899, USSR

Stephan G. SAMKO
Department of Mathematics, Rostov State University, ul. Engelsa, 105,
Rostov-on-Don 344006, USSR

Hiroki SATO
Department of Mathematics, Faculty of Sciences, Shizuoka University, Shizuoka, 422,
Japan

Ken-iti SATO
Department of Mathematics, College of General Education, Nagoya University, Nagoya, 464-01, Japan

Sadao SATO
Department of Mathematics, Tokyo Denki University, 2-2, Kanda-Nishiki, Chiyoda-ku, Tokyo, 101, Japan

Hitoshi SAWAI
3-12-22, Kita-Ohtsuka, Toshima-ku, Tokyo, 170, Japan

Shigeo SEGAWA
Department of Mathematics, Daido Institute of Technology, 10-3 Takiharu, Minami-ku, Nagoya, 457, Japan

Valeri S. SEROV
Faculty of Computational Mathematics and Cybernetics, Moscow State University, 119899 Moscow, USSR

Yuri V. SHESTOPALOV
Faculty of Computational Mathematics and Cybernetics, Moscow State University, 119899 Moscow, USSR

Masakazu SHIBA
Department of Mathematics, Hiroshima University, Hiroshima, 730, Japan

Keiichi SHIBATA
Department of Mathematics, Okayama University of Science, 1-1 Ridai-cho, Okayama, 700, Japan

Abdullah SHIDFAR
Department of Mathematics, Iran University of Science and Technology, Narmak, Tehran-16, Iran

Katsunori SHIMOMURA
Department of Mathematics, Ibaraki University, Mito, Ibaraki, 310, Japan

Peter SJÖGREN
Department of Mathematics, Chalmers University of Technology, S-412 96 Göteborg, Sweden

Wayne S. SMITH
Department of Mathematics, University of Hawaii, Honolulu, HI 96822, U.S.A.

Su Hung SONG
4-3-5, Kita, Katayama, Ikuno-ku, Osaka, 544, Japan

David A. STEGENGA
Department of Mathematics, University of Hawaii, Honolulu, HI 96822, U.S.A.

Kenneth R. STEPHENSON
Department of Mathematics, University of Tennessee, Knoxville, TN 37996-1300, U.S.A.

Karl T. STURM
Institut für Angewandte Mathematik des Universität Zürich, Rämistrasse 74, CH-8001 Zürich, Switzerland

Toshiyuki SUGAWA
Department of Mathematics, Faculty of Sciences, Kyoto University, Sakyo-ku, Kyoto, 606, Japan

Kunihiko SUGIMURA
Department of Mathematics, School of Sciences, Nagoya University, Chikusa-ku, Nagoya, 464-01, Japan

Nobuyuki SUITA
Department of Mathematics, Tokyo Institute of Technology, Ohokayama, Meguro-ku, Tokyo, 152, Japan

Noriaki SUZUKI
Department of Mathematics, College of General Education, Nagoya University, Nagoya, 464-01, Japan

Toshimasa TADA
Department of Mathematics, Daido Institute of Technology, Daido, Minami-ku, Nagoya, 457, Japan

Masayoshi TAKEDA
Department of Mathematics, Himeji Institute of Technology, Shosya 2167, Himeji, 671-22, Japan

Promarz M. TAMRAZOV
Institute of Mathematics, Ukrainian Academy of Sciences, ul. Repina, 3, 252601, Kiev-4, GSP, USSR

Hidematu TANAKA
Department of Mathematics, Okayama University of Science, 1-1 Ridai-cho, Okayama, 700, Japan

Hiroshi TANAKA
Department of Mathematics, Joetsu University of Education, Joetsu, 943, Japan

Masahiko TANIGUCHI
Department of Mathematics, Faculty of Sciences, Kyoto University, Kyoto, 606, Japan

Setsuo TANIGUCHI
Department of Applied Science, Faculty of Engineering, Kyushu University, Fukuoka, 812, Japan

John C. TAYLOR
Department of Mathematics & Statistics, McGill University, Montreal, QC, Canada, H3A 2K6

Nobushige TODA
Department of Mathematics, Nagoya Institute of Technology, Gokiso, Showa-ku, Nagoya, 466, Japan

Hongji TU
Department of Mathematics, Fuzhou University, Fuzhou, Fujian 350002, P.R. of China

Hironobu URABE
Department of Mathematics, Kyoto University of Education, Fushimi-ku, Kyoto, 612, Japan

Frans van GOOL
Mathematical Institute, State University of Utrecht, P.O.Box 80.010, 3508 TA Utrecht, Netherlands

Jiří VESELÝ
Faculty of Mathematics and Physics, MÚUK Sokolovská 83, 186 00 Praha 8, Czechoslovakia

Sergey VODOP'JANOV
Novosibirsk State University, Pirogova 2, 630090 Novosibirsk, USSR

Hisako WATANABE
Department of Mathematics, Ochanomizu University, 2-1-1, Ohtsuka, Bunkyo-ku, Tokyo, 112, Japan

Günther WILDENHAIN
Universität Rostock, Sektion Mathematik, Universitätsplatz 1, 2500 Rostock, F.R.G.

Jiongqi WU
Department of Mathematics, Zhangzhou Teachers College, Zhangzhou, Fujian 363000, P.R. of China

Jang-Mei WU
Department of Mathematics, University of Illinois, 1409 West Green St., Urbana, IL 61801, U.S.A.

Hiroshi YAMAGUCHI
Department of Mathematics, Faculty of Education, Shiga University, Otsu, 520, Japan

Hiromichi YAMAMOTO
Department of Mathematics, Faculty of Sciences, Kochi University, Kochi, 780, Japan

Takahisa YAMANO
Mezon Fuji 103, 3-24-21, Minami-Ikebukuro, Toshima-ku, Tokyo, 171, Japan

Maretsugu YAMASAKI
Department of Mathematics, Shimane University, Matsue, 690, Japan

Niro YANAGIHARA
Department of Mathematics, Faculty of Sciences, Chiba University, 1-33, Yayoi-cho, Chiba, 260, Japan

Hidenobu YOSHIDA
Department of Mathematics, Faculty of Sciences, Chiba University, 1-33, Yayoi-cho, Chiba, 260, Japan

Lawrence ZALCMAN
Department of Mathematics, Bar-Ilan University, 52100 Ramat-Gan, Israel

Stepan ZARGARYAN
Polytechnic Institute of Yerevan, Teryan str. 105, Yerevan, 375009, USSR

Shunyan ZHANG
Department of Mathematics, Peking University, 100871, Beijing, P.R. of China

Natalia V. ZORIĬ
Institute of Mathematics, Ukrainian Academy of Sciences, ul. Repina, 3, 252601, Kiev-4, GSP, USSR

List of non-participating contributors

Qi-Ren GAO
Department of mathematics, Xiamen University, Xiamen, Fujian, P.R. of China

M. A. HOSSEINI
University of Science and Technology, Narmak, Tehran-16, Iran

Wolfgang WOESS
Dipartimento di Matematica, Università di Milano, 20133, Milano, Italy

List of lectures

Invited lectures

D. R. ADAMS
L^p potential theory techniques and nonlinear PDE

J. BLIEDTNER
Applications of Choquet theory to potential theory

C. DELLACHERIE
A nonlinear version of Hunt's theorem

W. K. HAYMAN
Strict isoperimetric inequalities and asymmetry

L. I. HEDBERG
Nonlinear potential theory

C. E. KENIG
The Dirichlet and Neumann problems for elliptic equations with non-smooth coefficients

N. G. MAKAROV
Density properties of harmonic measure

O. T. MARTIO
Potential theory and quasiconformal mappings

G. MOKOBODZKI
Potential theory and ergodic pointwise convergence theorems

I. NETUKA
The boundary behaviour of solutions of the Dirichlet problem

A. G. O'FARRELL
Capacities in function theory

M. RÖCKNER
Potential theory on non-locally compact spaces via Dirichlet forms

J. C. TAYLOR
The Martin compactification of the polydisc at the bottom of the positive spectrum

J.-M. WU
Boundary density and the Green function

Short communications

H. AIKAWA
Oscillating Poisson integral

Ö. AKIN
On some properties of the positive solutions of the generalized Weinstein equation

M. BĂNULESCU
Potential theory on locally compact non-abelian groups

A. D. BENDIKOV
Potential theory on the group

C. BERG
Liouville operators

E. M. J. BERTIN
Nonlinear potential theory

M. BIROLI
Energy decay for local solutions of equations relative to Hörmander's square type operators

A. V. BITSADZE
The generalized Neumann problem

G. BJÖRCK
The energy integral resulting from a kernel which is a positive power of the distance

N. BOBOC
Compressions in excessive structures II

A. BOIVIN
Weighted L^p-capacities and holomorphic approximation in the complex plane

A. BOUKRICHA
Characterizations of the Picard principle for rotation free densities and applications

M. BRODSKY
Generalizations of the Schwarz function and their application to the inverse potential problem

G. BUCUR
Compressions in excessive structures I

H.-Q. BUI
On weighted Hardy spaces of harmonic functions

H. CHEN
An integral bound for the spherical derivative

A. T. CHWANG
Application of potential theory in interaction hydrodynamics

A. CORNEA
Functions having the mean value property

A. de la PRADELLE
Slim sets and quasi-norm principle for Gaussian measures

V. Ya. EIDERMAN
Relative Carleson's measures and capacity of exceptional sets arising in estimations of a δ-subharmonic functions

S.-L. A. ERIKSSON-BIQUE
Balayage spaces and standard H-cones

M. ESSÉN
On minimal thinness, reduced functions and Green potentials

J. A. GAWINECKI
On thermoelastic potential in three-dimensional hyperbolic thermoelasticity theory

M. GOLDSTEIN
Quadrature formulae

K. GOWRISANKARAN
Polydiscs and non-tangential limits

N. GRACHEV
Representation and estimates for inverse operators of the potential theory integral equations on polyhedron

A. A. GRIGOR'YAN
Harmonic functions with finite Dirichlet integral on Riemannian manifolds

A. GULISASHVILI
On the existence of singular integrals almost everywhere with respect to Hausdorff measures

A. S. HAKOBYAN
Use of single layer potential in mixed problems of elasticity

W. HANSEN
Continuous solutions of the Schrödinger equation

H. HARA
Determination of the modulus of quadrilaterals by finite element methods

J. HEINONEN
Harmonic morphisms in non-linear potential theory

J. S. HWANG
On the falsity of Riemann's hypothesis

N. JACOB
Some remarks on the potential theory of the Kolmogorov equation

K. JANSSEN
Potential theory for polyexcessive measures

V. KAIMANOVICH
Bounded harmonic functions and entropy

M. KANDA
Two applications of Dellacherie, Feyel and Mokobodzki's measure

R. A. KERMAN
Positive multilinear operators acting on weighted L^p spaces

T. KILPELÄINEN
Nonlinear fine potential theory

V. KOKILASHVILI
Weighted norm inequalities for some integral operators

B. KORENBLUM
Smoothness and factorization theorems for the Bergman space

J. KOREVAAR
Fields of electrons on the sphere and quadrature problems

V. I. KRUGLIKOV
Capacities and mappings quasiconformal in the mean

T. KUROKAWA
Beppo Levi functions and Riesz potentials

I. E. LAINE
Harmonic morphisms in the quasi-linear potential theory

J. ŁAWRYNOWICZ
Potential-theoretical questions connected with foliations by complex manifolds involving the complex Hessian

G. LEHA
On stationary distributions of diffusion processes in Hilbert spaces

P. LINDQVIST
Conjugate p-harmonic and q-harmonic functions in the complex plane:

$$f = u_p + iv_q, \quad 1/p + 1/q = 1$$

E. R. LOVE
Expansions in series of Legendre functions

J. LUKEŠ
Weakened forms of the fine Dirichlet problem

Z. M. MA
Potential theory for non-regular Dirichlet forms

F. MAITANI
Accola's theorem for reproducing forms on Riemannian manifolds

J. J. MANFREDI
Estimates for p-harmonic mappings

H. MASAOKA
On the behavior of non-negative finely superharmonic functions at the Martin boundary

M. MASUMOTO
Integrability of superharmonic functions on plane domains

P. MATTILA
Singular Cauchy integrals and rectifiability of measures in the plane

Y. MIZUTA
Boundary limits of harmonic functions

C. N. MOORE
Subgaussian estimates for harmonic functions

M. MURATA
Martin boundaries for Schrödinger equations

D. T. NGUYEN
Spectral representations of some infinitely divisible process

X.-L. NGUYEN
The differentiability properties of finely harmonic functions and applications

W. H. OW
Uniform harmonic approximation on closed sets with continuous extension to the boundary

R. S. PATHAK
Distributional boundary values of analytic functions

I. PAVLOV
Spaces of martingales: Applications to the potential probability theory

I. M. PETRUSHKO
On behavior near the boundary for solutions of parabolic and degenerating parabolic equations

M. A. PICARDELLO
Martin boundaries of products of Markov chains

A. I. PRILEPKO
Inverse problems in mathematical physics and potential theory

S. QIU
Survey of the studies on potential theory in China

S. RAMASWAMY
Comparison of fine limsup and ordinary limsup

A. Y. RASHKOVSKIĬ
On radial projections of harmonic measures of a star-shaped domain

P. J. RIPPON
On surface integrals of harmonic functions in the unit ball

M. SAKAI
Unbounded quadrature domains

E. G. SALTYKOV
Green's function for inhomogeneous medium

S. G. SAMKO
Inversion theorems for potentials in R^n and on S^{n-1}

V. S. SEROV
Some problems of the potential theory for Schrödinger operators with singular potential

Y. V. SHESTOPALOV
Potential theory for Helmholtz operator and nonlinear eigenvalue problems

K. SHIBATA
On stationary points of the energy integral

A. SHIDFAR
Applications of potential theory in biharmonic analysis

D. A. STEGENGA
The quasihyperbolic metric is exponentially integrable on Hölder domains in R^n

K. R. STEPHENSON
The use of discrete potential theory in circle packing

K. T. STURM
Fine properties of Schrödinger operators

N. SUZUKI
Nonintegrability of positive subharmonic functions on Lipschitz domains

P. M. TAMRAZOV
Extension of functions to subharmonic and plurisubharmonic ones in Euclidean and topological vector spaces

H. TANAKA
Kuramochi boundaries of Riemannian manifolds

S. TANIGUCHI
On the closable parts of pre-Dirichlet forms and the fine supports of underlying measures

H. TU
Some new applications of potential theory to function theory

F. van GOOL
On non-linear potential theory

S. VODOP'JANOV
Weighted L^p-potential theory with respect to generalized kernels and its applications

H. WATANABE
The Neumann problem and Hausdorff measures

G. WILDENHAIN
Aspects of approximation in potential theory

J. WU
On potential extension and capacity on harmonic spaces

H. YAMAGUCHI
Variations of harmonic modules of domains in R^3

M. YAMASAKI
Nonlinear potentials on an infinite network

S. S. ZARGARYAN
Singularities of solution of integral equations of potential theory arising in theory of elasticity for non-homogeneous media

S. ZHANG
Explicit bounds in Schottky's theorems

N. V. ZORIĬ
The minimum Riesz energy problem for space condensers

Linguistische
Arbeiten 83

Herausgegeben von Herbert E. Brekle, Hans Jürgen Heringer, Christian Rohrer, Heinz Vater und Otmar Werner